The Australian vegetation is the end result of a remarkable history of climate change, latitudinal change, continental isolation, soil evolution, interaction with an evolving fauna, fire and most recently human impact. This book presents a detailed synopsis of the critical events that led to the evolution of the unique Australian flora and the wide variety of vegetational types contained within it. The first part of the book details the past continental relationships of Australia, its palaeoclimate, fauna and the evolution of its landforms since the rise to dominance of the angiosperms at the beginning of the Cretaceous period. A detailed summary of the palaeobotanical record is then presented. The palynological record gives an overview of the vegetation and the distribution of important taxa within it, while the complementary macrofossil record is used to trace the evolution of critical taxa.

This book will interest graduate students and researchers interested in the evolution of the flora of this fascinating continent.

History of the Australian vegetation: Cretaceous to Recent

History of the Australian vegetation:
Cretaceous to Recent

EDITED BY

Robert S. Hill

Professor in Plant Science
University of Tasmania

CAMBRIDGE
UNIVERSITY PRESS

Published by the Press Syndicate of the University of Cambridge
The Pitt Building, Trumpington Street, Cambridge CB2 1RP
40 West 20th Street, New York, NY 10011-4211, USA
10 Stamford Road, Oakleigh, Melbourne 3166, Australia

First published 1994

Printed in Great Britain at
the University Press, Cambridge

A catalogue record for this book is available from the British Library

Library of Congress cataloguing in publication data

History of the Australian vegetation : Cretaceous to Recent / edited
by Robert S. Hill.
p.   cm.
Includes index.
ISBN 0-521-40197-6
1. Paleobotany – Australia.   2. Plants – Evolution.   I. Hill,
Robert S.
QE948.A1H57   1994
561'.1994 – dc20   93-32747 CIP

ISBN 0 521 40197 6 hardback

RO

Contents

[vii]

Contributors

DR N. F. ALLEY
Department of Mines and Energy, PO Box 151, Eastwood, South
Australia 5063, Australia

PROFESSOR M. ARCHER
Vertebrate Palaeontology Laboratory, School of Biological Sciences,
University of New South Wales, PO Box 1, Kensington, New South Wales
2033, Australia

DR D. T. BLACKBURN
Kinhill Engineers, 186 Greenhill Road, Parkside, Adelaide, South
Australia 5063, Australia

MR P. J. BROWN
11 Hooper Place, Flynn, ACT 2615, Australia

DR R. J. CARPENTER
Department of Plant Science, University of Tasmania, PO Box 252C,
Hobart, Tasmania 7001, Australia

DR D. C. CHRISTOPHEL
Department of Botany, University of Adelaide, PO Box 498, Adelaide,
South Australia 5005, Australia

DR M. E. DETTMANN
Department of Botany, University of Queensland, St Lucia, Queensland
4072, Australia

DR J. G. DOUGLAS
42 Sunhill Road, Mt Waverley, Victoria 3149, Australia

MR H. GODTHELP
Vertebrate Palaeontology Laboratory, School of Biological Sciences,
University of New South Wales, PO Box 1, Kensington, New South Wales
2033, Australia

DR D. R. GREENWOOD
Paleobiology Department, NHB MRC 121, National Museum of Natural
History, Smithsonian Institution, Washington DC 2056, USA

DR S. J. HAND
Verbetrate Palaeontology Laboratory, School of Biological Sciences, University of New South Wales, PO Box 1, Kensington, New South Wales 2033, Australia

PROFESSOR R. S. HILL
Department of Plant Science, University of Tasmania, PO Box 252C, Hobart, Tasmania 7001, Australia

DR G. S. HOPE
Department of Biogeography and Geomorphology, Research School of Pacific Studies, Australian National University, Canberra, ACT 0200, Australia

DR G. J. JORDAN
Department of Plant Science, University of Tasmania, PO Box 252C, Hobart, Tasmania 7001, Australia

DR A. P. KERSHAW
Department of Geography and Environmental Science, Monash University, Clayton, Melbourne, Victoria 3168, Australia

DR H. A. MARTIN
School of Biological Science, University of New South Wales, PO Box 1, Kensington, New South Wales 2033, Australia

DR J. R. C. McEWEN MASON
125-1, Yasuda, Inamori B-2, Aomori-shi, Aomori-ken, 038 Japan

DR M. K. MACPHAIL
20 Abbey Street, Gladesville, New South Wales 2111, Australia

DR P. G. QUILTY
Australian Antarctic Division, Channel Highway, Kingston, Tasmania 7050 Australia

DR I. R. K. SLUITER
Department of Conservation and Natural Resources, State Government Offices, 253 Eleventh Street, Mildura, Victoria 3500, Australia

PROFESSOR G. TAYLOR
School of Resource and Environmental Science, University of Canberra, PO Box 1, Belconnen, ACT 2617, Australia

DR E. M. TRUSWELL
Division of Continental Geology, Bureau of Mineral Resources, Geology and Geophysics, PO Box 378, Canberra, ACT 2601, Australia

DR G. E. WILFORD
88 Gouger Street, Torrens, ACT 2607, Australia

1 The Australian fossil plant record: an introduction

R. S. HILL

The living Australian flora is a complex mixture of species with widely varying distributions and interactions, covering the range from arid zone grassland to rainforest, alpine heath to mangrove swamp. The latitudinal range of Australia spans tropical to cool temperate climatic zones, and this is reflected in the extant vegetation, which is enormously complex (see Groves, 1981). Attempts to explain the distribution of Australian vegetation based solely on prevailing variables have been less than satisfactory. Australia, in all its aspects, is a product of its past. That is especially true of its flora and, unless the fossil record is properly considered, all attempts to explain vegetation patterns will be incomplete. Despite the complexity of the living vegetation, there has been a tendency to consider the past vegetation, especially that of the pre-Quaternary, as consisting in widespread, monotonously uniform communities. The recent explosion in information on Australian fossil plants shows that this perception was largely the result of a highly incomplete database, where the unknown areas were assumed to be the same as the small areas that were relatively well understood. It is now clear that past vegetation complexity, at least during the Cenozoic, was as high as, or possibly higher than, that seen at present. The past complexity is abundantly illustrated in this book, which may well represent the last occasion on which a thorough review of such a large period of time can be accomplished for the whole of Australia. Data are accumulating at a rapid rate, and almost every new site produces much that is novel, causing a reassessment of the prevailing hypotheses.

There is a long history of attempts to explain the origin and evolution of the Australian flora; some of these are well known, others more obscure. Hooker's (1860) discussion of the Australian flora provided a base of the highest quality, which slowly evolved into the invasion theory, perhaps best argued by Burbidge (1960). However, this theory crumbled soon after Burbidge's work, due to the broad acceptance of plate tectonics and the massive new collection of fossil data. Unfortunately, the fossil record has been given scant treatment by most Australian botanists, but the angiosperm fossil record has a long history, beginning in earnest with the work of von Ettingshausen (1888). The cosmopolitan theory espoused by von Ettingshausen led to a bitter debate, particularly involving Deane (e.g. 1900), which was well summarised by Maiden (1922). This seems to have gone largely unnoticed outside the palaeobotanical world, and in Australia had little impact on other areas of research. Palaeobotanical research that impinged on the living flora went through the doldrums following the early activity of Deane and of Chapman (e.g. 1921) in particular, but was almost single-handedly resurrected by Isabel Cookson, who set palaeobotanical studies on their modern course (for a list of her publications, see

Baker, 1973). There are currently many researchers working on plant remains in post-Jurassic Australian sediments. While their results, which are summarised here, suggest an increasingly complex picture, they also show an exciting prospect, where the enormous changes in Australia's past position and climate will make it a fertile ground for syntheses on factors involved in plant evolution at all levels in the years to come.

There are a number of features of this book that require a brief introduction, which will be attempted here. Some areas are incomplete or absent, but that is more the nature of current knowledge than a deliberate omission. The breadth of the topic of this book is such that some aspects are bound to be covered in less detail than may have been desired by some readers.

GEOLOGICAL TIME

There is not a standard time scale in use throughout this work. While that is desirable in theory, it proved impossible in practice because different research groups are tied to different interpretations of geological time. The inconsistencies are, however, relatively small, and in my view do not substantially affect the vegetation history presented here.

More critical is the choice of starting time for this work. The beginning of the Cretaceous was selected for two main reasons. Firstly, few pre-Cretaceous plants had a major, direct impact on the extant flora, and this work is meant primarily to explain the underlying causes of the make-up of the living vegetation. While some pre-Cretaceous genera, e.g. *Glossopteris* and *Dicroidium*, are very well known, they belong to another era of earth's history and cannot really be considered as 'Australian' plants in any meaningful way. Secondly, the history of Australia as an individual entity is a post-Jurassic phenomenon and therefore it seems legitimate to consider the Australian flora as having roughly that starting time.

ENVIRONMENTAL VARIABLES

Climatic change

Since the Jurassic, Australia has separated from Gondwana and moved over thousands of kilometres to its present location. Movement of Gondwanic land masses has, coincidentally, caused major climatic shifts, which have been vitally important for the development of the extant flora. Those changes are detailed in this book, although there is still much to be learned. There is a strong impression that climatic change has been the major factor shaping the extant vegetation, but there have been other factors that are more difficult to document, and the effects of which are more difficult to predict.

Photoperiod

One of the critical features of the Australian environment that has changed dramatically since the Cretaceous is photoperiod. For a long time the extreme southern part of Australia was at more than latitude 60° S (Wilford & Brown, Chapter 2, this volume). This means that the vegetation was growing under conditions of winter darkness and continuously light summers. The effect of this on the vegetation is difficult to predict, since there is no comparable situation in the modern landscape where the climate is suitable for large-tree growth. However, there is a good fossil record from both high southern (e.g. Jefferson, 1982) and northern (e.g. Francis & MacMillan, 1987) latitudes that suggests large trees were able to thrive as long as there was a suitable climate. Since the sun remains quite close to the horizon during much of the growing season, the trees must have been widely spaced and probably cone shaped to intercept the maximum possible incoming radiation (Francis & McMillan, 1987). The canopy layer is therefore effectively vertical or subvertical instead of horizontal as in modern tropical forests. While there is good evidence for the spacing of trees in these forests, the canopy shape can only be inferred. However, this suggests a forest structure very different from that which currently prevails in Australia, and this may have had many subtle but far-reaching effects. There are as yet no direct data on the extent of these effects.

Carbon dioxide levels

Evidence is accumulating for widely fluctuating CO_2 levels during the Cretaceous–Recent, from extreme highs of about 1000 (Walker *et al.*, 1981) or even more than 1500 p.p.m. (Berner, 1990) during the Cretaceous to lows of just over 180 p.p.m. between 40 000 and 160 000 years ago (Barrett, 1991). Such levels of CO_2 must have had extreme effects on plant growth. Many of the broad-leaved conifers from Cretaceous and early Tertiary sediments in southern Australia have extraordinarily thick cuticles (e.g. Cantrill, 1991), which are well beyond the range of that exhibited by any extant conifers. The leaves are often quite intact and well preserved, suggesting the presence of clean abscission, and sometimes leaf bases demonstrate a clean abscission zone. The robust nature of the leaves and their extremely thick cuticles, in the context of extant CO_2 levels, suggest they were evergreen, since leaves of extant deciduous plants tend to be very thin, with an almost insignificant cuticle. However, with CO_2 levels above 1000 p.p.m., growth rates may have been so rapid that other features (e.g. resistance to herbivory) may have been more critical than seasonal loss of carbohydrate. Certainly the winter darkness in a mild climate would seem to suit the deciduous habit, and more effort should be directed toward determining whether these leaves were in fact winter deciduous.

There is evidence, largely unpublished, that winter deciduousness was more prevalent in Australia during the early Tertiary than at present. Only one winter deciduous species survives in Australia today (the Tasmanian endemic *Nothofagus gunnii*).

NOMENCLATURE

There are few major nomenclatural problems in Australian palaeobotany, since most researchers work to a common system. Occasional conflicts in this book do not increase the difficulty of understanding the data presented. However, there is one exception which must be explained in some detail. One of the dominant genera in the Australian fossil record, particularly that of pollen, is *Nothofagus*. Pollen of this type was initially split into two morphological types by Cranwell (1939), with a third added later by Cookson (1952) and Cookson & Pike (1955). These types were informally categorised as the *Nothofagus brassii*, *N. fusca* and *N. menziesii* types. One of the major problems with these pollen types was that the species assigned to the types did not closely match the formal infrageneric divisions of the time (van Steenis, 1953). Dettmann *et al.* (1990) revised the pollen of fossil and living *Nothofagus* and recognised eight types, four of which were produced by extant species. At about the same time Hill & Read (1991) revised the infrageneric classification of the extant species and proposed four subgenera, which, in species make-up, matched the new pollen groupings exactly. This paved the way for the use of formal subgeneric names in place of the informal pollen names, and that procedure has been adopted in this book. Unfortunately, the history of the infrageneric nomenclature of *Nothofagus* is very complex, and there are errors in Hill & Read's names. Hill & Jordan (1993) have corrected two of the subgeneric names, so that the names used here and their equivalents in Hill & Read's nomenclature are as follows (Hill & Read's names in parentheses where they differ): *Nothofagus* subgenus *Nothofagus*, *N.* subgenus *Fuscospora* (*Fuscaspora*), *N.* subgenus *Lophozonia* (*Menziesospora*) and *N.* subgenus *Brassospora*. These four subgeneric names are used throughout the text for fossil pollen and macrofossils.

THE CURRENT STATE OF RESEARCH

Research on various aspects of Australian Cretaceous and Cenozoic palaeobotany is proceeding at an all time high. However, not all areas are at the same level of understanding. This is very strongly reflected in this book. Some topics, such as Paleogene palynology, may be reviewed on the basis of an enormous data set. However, at the other end of the spectrum, Neogene macrofossil palaeobotany is very poorly known and is represented here by a combined study of the micro- and macrofossil record of the Latrobe Valley brown coal. While this flora is very well understood, there are only a few isolated and poorly studied macrofloras in this age range outside of southeastern Australia. The range between these extremes is clear in the various chapters. Our understanding of the evolution of the flora is expanding, but such is the complexity of the problem that it will be many years before an overall sense of order prevails.

REFERENCES

BAKER, G. (1973). Dr Isabel Clifton Cookson. In *Mesozoic and Cainozoic Palynology: Essays in Honour of Isabel Cookson*, ed. J. E. Glover & G. Playford. *Geological Society of Australia Special Publication*, 4, iii–x.

BARRETT, P. J. (1991). Antarctica and global climate change: a geological perspective. In *Antarctica and Global Climate Change*, ed. C. M. Harris & B. Stonehouse, pp. 35–50. London: Bellhaven Press.

BERNER, R. A. (1990). Atmospheric carbon dioxide levels of Phanerozoic time. *Science*, 249, 1382–6.

BURBIDGE, N. T. (1960). The phytogeography of the Australian region. *Australian Journal of Botany*, 8, 75–211.

CANTRILL, D. J. (1991). Broad leafed coniferous foliage from the Lower Cretaceous Otway Group, southeastern Australia. *Alcheringa*, 15, 177–90.

CHAPMAN, F. (1921). A sketch of the geological history of Australian plants: the Cainozoic flora. *Victorian Naturalist*, 37, 115–32.

COOKSON, I. C. (1952). Identification of Tertiary pollen grains with those of New Guinea and New Caledonian beeches. *Nature*, 170, 127.

COOKSON, I. C. & PIKE, K. M. (1955). The pollen morphology of *Nothofagus* Bl. sub-section *Bipartitae* Steen. *Australian Journal of Botany*, 3, 197–206.

CRANWELL, L. M. (1939). Southern beech pollens. *Auckland Institute Museum, Records*, 2, 175–96.

DEANE, H. (1900). Observations on the Tertiary flora of Australia, with special reference to Ettingshausen's theory of the Tertiary cosmopolitan flora. *Proceedings of the Linnean Society of New South Wales*, pp. 463–75.

DETTMANN, M. E., POCKNALL, D. T., ROMERO, E. J. & ZAMALOA, M. de C. (1990). *Nothofagidites* Erdtman ex Potonié, 1960: a catalogue of species with notes on the paleogeographic distribution of *Nothofagus* Bl. (southern Beech). *New Zealand Geological Survey Paleontological Bulletin*, 60.

ETTINGSHAUSEN, C. von (1888). Contributions to the Tertiary flora of Australia. *Memoirs of the Geological Survey of New South Wales, Palaeontology*, 2, 1–189.

FRANCIS, J. E. & McMILLAN, N. J. (1987). Fossil forests in the far north. *GEOS*, 16, 6–9.

GROVES, R. H. (1981). *Australian Vegetation*. Cambridge: Cambridge University Press.

HILL, R. S. & JORDAN, G. J. (1993). The evolutionary history of *Nothofagus* (Nothofagaceae). *Australian Systematic Botany*, 6, 111–26.

HILL, R. S. & READ, J. (1991). A revised infrageneric classification of *Nothofagus* (Fagaceae). *Botanical Journal of the Linnean Society*, 105, 37–72.

HOOKER, J. D. (1860). Introductory essay. In *Botany of the Antarctic Voyage of H. M. Discovery Ships 'Erebus' and 'Terror', in the Years 1839–1843*. III. *Flora Tasmaniae*, vol. I *Dicotyledones*. London: Lovell Reeve.

JEFFERSON, T. H. (1982). Fossil forests from the Lower Cretaceous of Alexander Island, Antarctica. *Palaeontology*, 25, 681–708.

MAIDEN, J. H. (1922). *A Critical Revision of the Genus Eucalyptus*. vol. IV. Sydney: Government Printer.

STEENIS, C. G. G. J. van (1953). Results of the Archbold expeditions. Papuan *Nothofagus*. *Journal of the Arnold Arboretum*, 34, 301–74.

WALKER, J. C. G., HAYS, P. B. & KASTING, J. F. (1981). A negative feedback mechanism for the long-term stabilization of the Earth's surface temperature. *Journal of Geophysical Research*, 86, 976–82.

2 Maps of late Mesozoic–Cenozoic Gondwana break-up: some palaeogeographical implications

G. E. WILFORD & P. J. BROWN

The nature and positions of neighbouring land areas have been significant factors in the evolution of the Australian flora, both directly in determining migration routes and indirectly in influencing ocean currents and climate. The maps (Figures 2.3 to 2.10) show the approximate positions of the continents at 10 million years (Ma) intervals from 150 Ma onwards, based on the data of Scotese & Denham (1988), with modifications referred to in the notes below. Separation of continental fragments by sea-floor spreading was commonly preceded by rifting. Where the relative motion of the fragments was oblique, some fault blocks were uplifted and eroded and others were deeply buried by sediment, resulting in zones with a varied and changing mosaic of complex environments by comparison, for instance, with adjacent interior areas. These zones, peripheral to the Australian land mass, are shown together with some of the larger areas of sedimentation and volcanicity (from BMR Palaeogeography Group, 1990) which would have influenced soil type and vegetation. The time scale shown in Figure 2.1 has been used for the reconstructions. The key for Figures 2.3–2.10 is shown in Figure 2.2.

150 Ma (Figure 2.3)

At about 150 Ma the 'Antarctic' coastline of Gondwana was positioned close to the South Pole

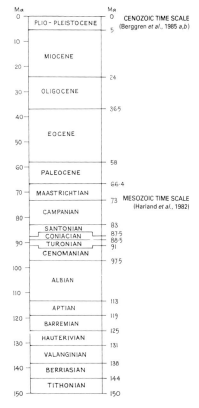

Figure 2.1 Time scale on which the following palaeogeographical maps are based.

[5]

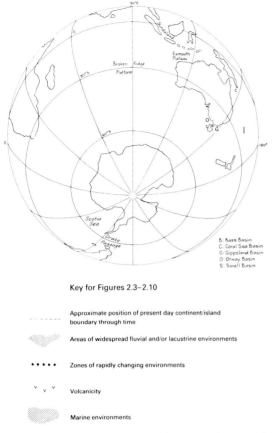

Key for Figures 2.3–2.10

- - - - -   Approximate position of present day continent/island
            boundary through time

░░░░   Areas of widespread fluvial and/or lacustrine environments

• • • • •   Zones of rapidly changing environments

ᵛ ᵛ ᵛ   Volcanicity

▨▨   Marine environments

Figure 2.2 Reference to localities referred to in the text, and the key to symbols used in Figures 2.3–2.10.

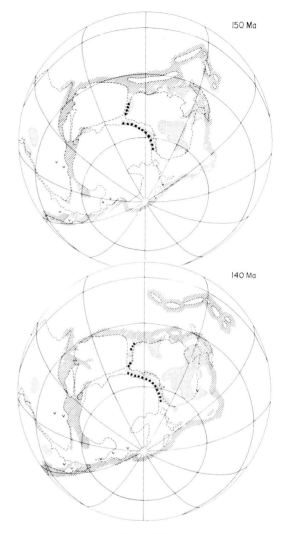

Figure 2.3

and western Tethys was connected to the proto-Pacific Ocean. Continental slivers from the northeast margin (Northwest Shelf region of Australia) and eastern tip (New Guinea region) of Gondwana may have rifted off periodically and been carried northwestwards from Permian time onwards. Such fragments now lie in South Tibet and various parts of Southeast Asia, but little definitive information exists on their time of movement and even less on their palaeogeographical configuration during transit across Tethys. Following Audley-Charles *et al.* (1988), who favoured a later, rather than earlier migration of fragments, the maps show them as a string of possible islands, assuming most departed after the Callovian (*c.* 167

Ma) sea-floor spreading event along that part of the edge of Gondwana. The fragments accreted to the southeastern part of the then Asian continent to form a promontory (Sundaland), which has remained in equatorial latitudes to the present day. It is shown on the maps by a group of 'hypothetical' islands to indicate possible 'stepping stone' links between Australian Gondwana and the Asian mainland, although the palaeogeography of the area will have changed quite dramatically with time.

At about 150 Ma, the present-day eastern Australian coastline was separated from the proto-Pacific Ocean by a strip of terrain now marked by the present-day Lord Howe Rise, Queensland Plateau and New Caledonia.

Gondwana break-up was preceded, in early to mid-Jurassic time, by widespread basaltic magmatic activity (dolerites of Tasmania, southeastern Australia; Karoo dolerites of South Africa; Ferrar Super Group, Dufek Intrusion in Antarctica), reflecting a major thermal event. By 150 Ma, movement between Africa–South America (West Gondwana) and Antarctica–Madagascar–Greater India–Australia (East Gondwana) had been initiated. However, the break-up was dominated by strike–slip movement, narrow basins were formed and it is likely that land connections between the two major fragments existed from time to time. Rifting along the southern margin of Australia was initiated at about this time (Willcox & Stagg, 1990).

140 Ma (Figure 2.3)

By 140 Ma narrow seaways existed along much of the split between Africa and Antarctica–Madagascar–Greater India–Australia and the fragments from the northeastern margin of Gondwana had moved towards the Equator. The incipient separation of Australia from Antarctica and Greater India was marked by the continued developments of rifts along the former's western and southern boundaries, although evidence for sedimentation in East Antarctica is virtually restricted to recycled palynomorphs (Truswell, 1983). Rifting along Australia's southern margin was accompanied by initial extension in a northwest to southeast direction of about 300 km and this was accomplished by about 120 Ma (Willcox & Stagg, 1990). New Zealand at this time was part of a considerable land mass flanking East Antarctica and southeast Australia. This land mass probably reached its greatest extent in the Early Cretaceous as a result of the Rangitata Orogeny (Stevens, 1989).

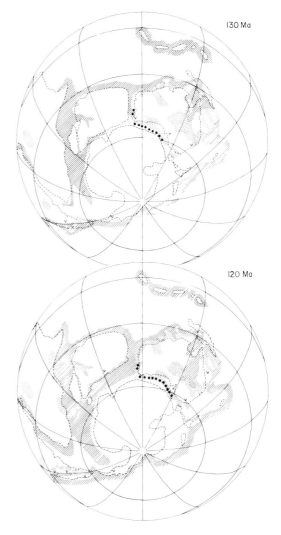

Figure 2.4

130 Ma (Figure 2.4)

By 130 Ma, Greater India, Australia and Antarctica had begun to move apart, although no significant seaways had developed between them (Powell *et al.*, 1988), the rate of separation between the last two being of the order of only a few millimetres per year. Break-up began at the southwest margin of the Exmouth Plateau at about 132 Ma and progressed southwards, then eastwards along the southern margin of Australia. The South Atlantic started to open about this time, propagating northwards (Nurnberg & Muller, 1991).

120 Ma (Figure 2.4)

By 120 Ma, substantial seaways existed along the east coast of Africa and between Greater India and Antarctica–western Australia. The slow continental extension between Australia and Antarctica continued but the earlier NW–SE direction changed to NNE–SSW with movement in the east of 120 km leading to the formation of the Gippsland and Bass Basins and modifying the development of the Otway and Sorell Basins (Willcox & Stagg, 1990).

110 Ma (Figure 2.5)

The map shows the Australian continent at the height of the marine transgression that culminated in the Aptian (*c.* 116–113 Ma), when extensive areas were covered by shallow seas. At about this time rifting commenced along the southeastern ('Australian') margin of Gondwana, and along the western margin of New Zealand, the former eventually leading to the formation of the Tasman Sea (Stevens, 1989). Rifting, together with the rotation of crustal blocks locally, affected the Antarctic Peninsula and adjacent areas of West Antarctica until about 100 Ma but had no major effect on the overall geography (Storey *et al.*, 1988). Spreading between Australia and Antarctica allowed the proto-Indian Ocean to enter from the west, initiating the formation of the Southern Ocean.

100 Ma (Figure 2.5)

By 100 Ma, erosion and subsidence had reduced the land masses around New Zealand, allowing the sea to flood a number of rift zones, although a land connection to Antarctica persisted in the south (Stevens, 1989). Uplift of the Australian Eastern Highlands may have started about this time, associated with the subsequent opening of the Tasman Sea (Wellman, 1987). At about 95

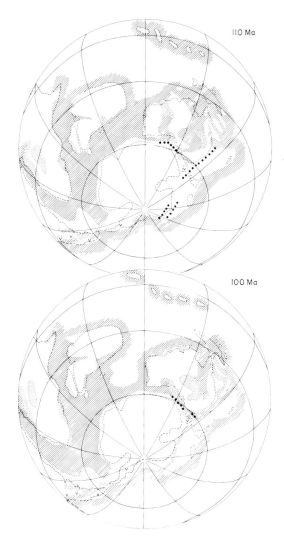

110 Ma

100 Ma

Figure 2.5

Key for Figures 2.3–2.10

- - - - - Approximate position of present day continent/island boundary through time

Areas of widespread fluvial and/or lacustrine environments

• • • • • Zones of rapidly changing environments

ᵛ ᵛ ᵛ Volcanicity

Marine environments

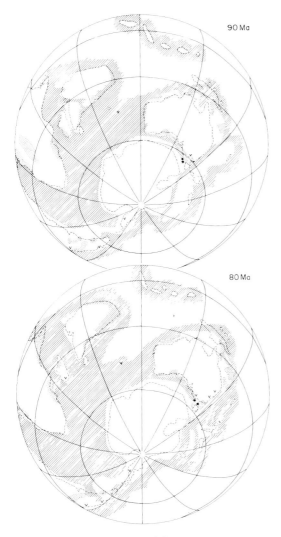

90 Ma

80 Ma

Figure 2.6

open ocean in the Tasman Sea and to the south of New Zealand isolating 'New Zealand' and 'New Caledonia' from the remainder of Gondwana (Stevens, 1989). Following Stevens (1989, figs. 5, 6), we show a land link between 'New Zealand' and 'New Caledonia' until about 70 Ma, although direct evidence for this is lacking. Subaerial volcanism existed in the open ocean west of Australia from 90 to 60 Ma, sited at the northern edge of the Broken Ridge platform (Rea *et al.*, 1990), and might have provided a 'stepping stone' for floral migration.

At about this time sea-floor spreading was under way between northeastern Australia and eastern New Guinea, leading to the formation of the Coral Sea Basin by the end of the Paleocene. Coeval sea-floor spreading in the vicinity of eastern New Guinea may have given rise to the separation of continental fragments that, in post-Paleocene times, were carried westwards by Pacific plate movements and are now located in the eastern islands of Indonesia (Pigram & Symonds, 1991).

80 Ma (Figure 2.6)

Madagascar separated from greater India at about this time, both being isolated in a surrounding ocean. Tasmania was still close to Antarctica and connected to Australia, the present-day Bass Strait area being occupied by mainly lowland, depositional environments until flooded by the sea in the Oligo-Miocene.

70 Ma (Figure 2.7)

The paucity of clastic sediments of this and earlier Late Cretaceous ages on and around the Australian continent (except for parts of the tectonically active southern margin), together with the widespread presence of thick, weathered profiles, indicate that relatively uniform humid climates were typical of much of Late Cretaceous time (G. E. Wilford, R. P. Langford & E. M. Truswell, unpublished data).

Ma the rate of Greater India's northward movement rapidly increased.

90 Ma (Figure 2.6)

Slow sea-floor spreading in a NNW direction between Antarctica and Australia continued until, by about 90 Ma, a tongue of the proto-Southern Ocean extended almost to Tasmania. Sea-floor spreading also resulted in the establishment of

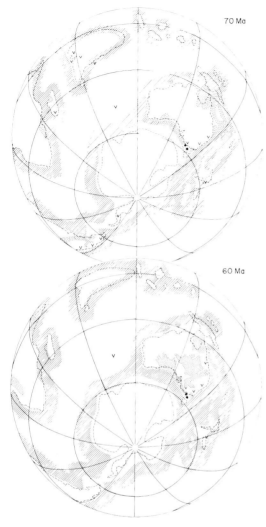

Figure 2.7

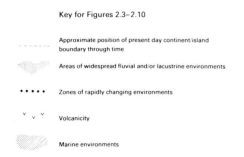

Key for Figures 2.3–2.10

‒ ‒ ‒ ‒   Approximate position of present day continent/island boundary through time

   Areas of widespread fluvial and/or lacustrine environments

• • • • •   Zones of rapidly changing environments

ˇ   ˇ   ˇ   Volcanicity

   Marine environments

60 Ma (Figure 2.7)

By about 60 Ma, sea-floor spreading ceased in the Tasman Sea, although Australia's slow northward drift away from Antarctica continued. In the early Cenozoic, one or more cooling events, probably associated with ice build-up on Antarctica, triggered widespread erosion and deposition across much of Australia. Broad, swampy, alluvial channels were formed in the west, and extensive sandy fans accumulated in the east (G. E. Wilford, R. P. Langford & E. M. Truswell, unpublished data). These conditions continued intermittently until about Early Oligocene.

50 Ma (Figure 2.8)

Uplift of the Transantarctic Mountains at about this time (Fitzgerald & Gleadow, 1988) would have encouraged accumulation of snow and ice which was to become a major feature of Antarctica for much of the remainder of the Cenozoic.

40 Ma (Figure 2.8)

At about this time movement of the Pacific Plate changed from a northerly to a more west-northwesterly direction, resulting eventually in collision with the Australian Plate in New Guinea and mountain building there from the mid-Oligocene onwards (Pigram & Symonds, 1991) and subduction of the Australian Plate beneath Sundaland. At about 44 Ma, the spreading rate between Australia and Antarctica increased as the former moved northwards, and by about 38 Ma a deep marine strait had formed between Tasmania and Antarctica (Kennett, 1980), allowing circum Antarctic oceanic flow for the first time and resulting in the increased cooling of Antarctica and the adjacent oceans.

Until about this time microcontinental blocks formed a link between the Antarctic Peninsula and South America. They have subsequently become dispersed by plate movements and sea-floor spreading, associated with the opening of the Scotia Sea. However, considerable uncertainty

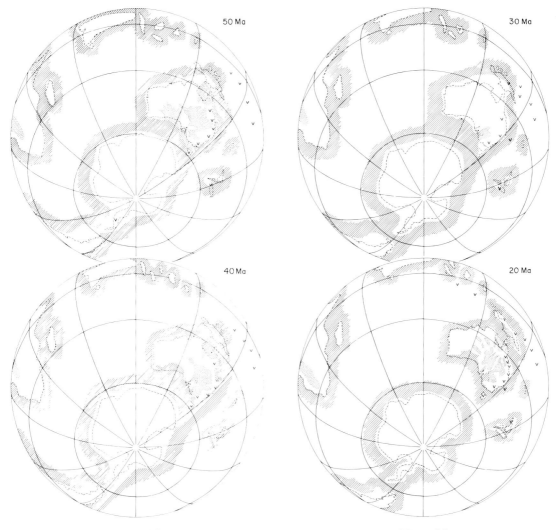

Figure 2.8

Figure 2.9

surrounds the exact timing of the opening of Drake Passage, the oceanic seaway between the Antarctic Peninsula and southern South America. Lawver *et al.* (1985) suggested the Passage opened between 64 and 34 Ma, whilst Barker & Burrell (1977) favoured a younger age of 30–25 Ma.

A possible 'stepping stone' island on which Cretaceous limestone and chert were exposed existed in the vicinity of Broken Ridge, a submarine feature 2000 km west of Perth, about 42 Ma but was probably short lived (Rea *et al.*, 1990).

30 Ma (Figure 2.9)

Large-scale ice sheet development became established in East Antarctica by Early Oligocene (38–36 Ma) times (Ocean Drilling Program, 1988), causing a change to a much drier climate over much of Australia and the formation of widespread duricrusts (G. E. Wilford, R. P. Langford & E. M. Truswell, unpublished data).

Although there is no evidence for a land connection between New Caledonia and Australia during the Tertiary, volcanic islands between the

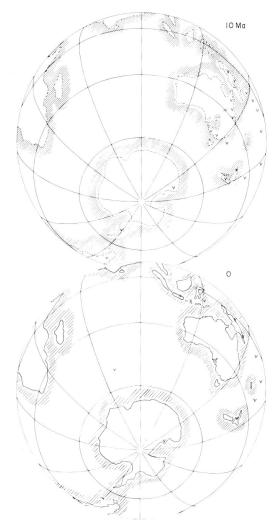

Figure 2.10

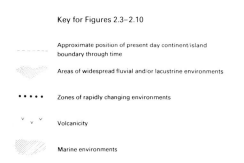

Key for Figures 2.3–2.10

- - - - - Approximate position of present day continent/island
          boundary through time

          Areas of widespread fluvial and/or lacustrine environments

• • • • • Zones of rapidly changing environments

ᵛ ᵛ ᵛ   Volcanicity

          Marine environments

two from about 30 Ma onwards could have influenced plant migration during the Neogene. Except for Lord Howe Island, just over 6 million years old, these islands are now submerged and form seamounts in the Tasman Sea (McDougall & Duncan, 1988).

Major mountain building in New Guinea dates from about 30 Ma (Pigram & Symonds, 1991).

20 Ma (Figure 2.9)

By 20 Ma, Australia's drift towards the equator had brought its northern fringes within the reaches of a tropical monsoonal climate, resulting in bauxite formation at well-drained coastal sites. Elsewhere, rivers flowed in channels and basins that were active in the early Tertiary, but less rainfall and a partly duricrusted surface reduced erosion. Lakes were common.

As the Australian plate continued to collide with that of Southeast Asia, sporadic volcanism along the junction provided island connections between the Southeast Asian mainland and the Australian continent that have persisted, albeit in a changing form, to the present day.

10 Ma (Figure 2.10)

By 10 Ma and despite the continued equatorwards movement of the Australian continent, the expansion of Antarctic ice sheets began to cause increasing aridity, particularly in the interior of Australia. This resulted in less widespread erosion and deposition compared to that in earlier periods.

The Australian plate's northward movement has continued to the present day, with accompanying volcanic activity along the junction with the Pacific plate in the rugged mountainous terrain of Papua New Guinea.

During the Quaternary the waxing and waning of polar ice caps produced sea level fluctuations that resulted in large areas of the continental shelf periodically becoming dry and both Tasmania and Papua New Guinea being continuous with the continent during these times.

The authors thank the following colleagues within the Bureau of Mineral Resources, Canberra for helpful comments: C. J. Pigram, H. I. M. Struckmeyer, R. J. Tingey, A. M. Walley and M. Yeung.

REFERENCES

AUDLEY-CHARLES, M. G., BALLANTYNE, P. D. & HALL, R. (1988). Mesozoic–Cenozoic rift-drift sequence of Asian fragments from Gondwanaland. *Tectonophysics*, **155**, 317–30.

BARKER, P. F. & BURRELL, J. (1977). The opening of the Drake Passage. *Marine Geology*, **25**, 15–34.

BERGGREN, W. A., KENT, D. V. & FLYNN, J. J. (1985*a*). Neogene geochronology and chronostratigraphy. In *Geochronology and the Geological Time Scale*, ed. N. J. Snelling. *Geological Society of London Memoir*, **10**, 211–50.

BERGGREN, W. A., KENT, D. V. & FLYNN, J. J. (1985*b*). Paleogene geochronology and chronostratigraphy. In *Geochronology and the Geological Time Scale*, ed. N. J. Snelling. *Geological Society of London Memoir*, **10**, 141–95.

BMR Palaeogeography Group (1990). *Australia: Evolution of a Continent*. Canberra: Bureau of Mineral Resources.

FITZGERALD, R. G. & GLEADOW, A. J. W. (1988). Fission-track geochronology, tectonics and structure of the Transantarctic Mountains in northern Victoria Land, Antarctica. *Chemical Geology (Isotope Geoscience sections)*, **73**, 169–98.

HARLAND, W. B., COX, A. V., LLEWELLYN, P. G., PICKTON, C. A. G., SMITH, A. G. & WALTERS, R. (1982). *A Geologic Time Scale*. Cambridge: Cambridge University Press.

KENNETT, J. P. (1980). Palaeoceanographic and biogeographic evolution of the Southern Ocean during the Cenozoic, and Cenozoic microfossil datums. *Palaeogeography, Palaeoclimatology, Palaeoecology*, **31**, 123–52.

LAWVER, L. A., SCLATER, J. G. & MEINKE, L. (1985). Mesozoic and Cenozoic reconstruction of the South Atlantic. *Tectonophysics*, **114**, 233–54.

McDOUGALL, I. & DUNCAN, R. A. (1988). Age progressive volcanism in the Tasmantid Seamounts. *Earth and Planetary Science Letters*, **89**, 207–20.

NURNBERG, D. & MULLER, R. D. (1991). The tectonic evolution of the South Atlantic from Late Jurassic to present. *Tectonophysics*, **191**, 27–53.

OCEAN DRILLING PROGRAM (1988). Ocean drilling program: early glaciation of Antarctica. *Nature*, **333**, 303–4.

PIGRAM, C. J. & SYMONDS, P. A. (1991). A review of the timing of the major tectonic events in the New Guinea Orogen. *South East Asian Journal of Earth Sciences*, **6**, 307–18.

POWELL, C. McA., ROOTS, S. R. & VEEVERS, J. J. (1988). Pre-breakup continental extension in East Gondwanaland and the early opening of the eastern Indian Ocean. *Tectonophysics*, **155**, 261–83.

REA, D. K., PEHN, J., DRISCOLL, N. W. *et al.* (1990). Palaeoceanography of the eastern Indian Ocean from ODP Leg 121 drilling on Broken Ridge. *Geological Society of America Bulletin*, **102**, 679–90.

SCOTESE, C. R. & DENHAM, C. R. (1988). *User's Manual for Terra Mobilis: Plate Tectonics for the Macintosh*.

STEVENS, G. R. (1989). The nature and timing of biotic links between New Zealand and Antarctica in Mesozoic and early Cenozoic times. In *Origins and Evolution of the Antarctic Biota*, ed. J. A. Crame, *Geological Society Special Publication*, **47**, 141–66.

STOREY, B. C., DALZIEL, I. W. D., GARRETT, S. W., GRUNOW, A. M., PANKHURST, R. J. & VENNUM, W. R. (1988). West Antarctica in Gondwanaland: crustal blocks, reconstruction and breakup processes. *Tectonophysics*, **155**, 381–90.

TRUSWELL, E. M. (1983). Geological implications of recycled palynmorphs in continental shelf sediments around Antarctica. In *Antarctic Earth Science*, ed. R. L. Oliver, R. R. James & J. B. Jags, pp. 394–9. Canberra: Australian Academy of Science.

WELLMAN, P. (1987). Eastern Highlands of Australia: their uplift and erosion. *BMR Journal of Australian Geology and Geophysics*, **10**, 277–86.

WILLCOX, J. B. & STAGG, H. M. J. (1990). Australia's southern margin: a product of oblique extension. *Tectonophysics*, **173**, 269–81.

3 The background: 144 million years of Australian palaeoclimate and palaeogeography

P. G. QUILTY

The preface to Pittock *et al.* (1978) discussed the question 'What is climate?' (without saying so) and refers to its main feature – its variability. And this was in relation to modern climate! How difficult then to address the question of palaeoclimate. We must recognise and accept that any reconstruction of palaeoclimate is at best an approximation, and, even then, is averaged over time scales of hundreds of thousands to millions of years, a length in which modern 'short term' variability is not decipherable. Many of the elements measured as part of modern climate are not recognisable in the geological record or must be implied from various features of the record. Although this may seem a somewhat negative way to introduce the subject, there are several firm conclusions that can be made about past climates and the way in which they have changed.

The purposes of this chapter are:

1. To review the climate of Australia over the last 144 million years (Ma) (Cretaceous to Recent).
2. To provide a background for other studies of Australian fauna and flora over the same interval.
3. To stimulate discussion on the directions for further research into the Cretaceous–Cenozoic of Australia.

The figure of 144 million years is taken because

that is what is understood to be the age of the boundary between the Jurassic and Cretaceous periods. It is based on our understanding of that boundary dated by isotopic methods and using constants that are accepted at present. It is not an absolute date and can be expected to vary as our understanding of the isotopic dating method improves. This range of time is taken because it is the one during which Australia underwent its major change from part of a supercontinent (Gondwana), to an isolated interval in which the modern fauna and flora evolved, to another when it is colliding with Asia and also undergoing changes associated with the influence of humanity.

To achieve its aims this chapter is divided into two parts, describing background and principles, and reconstructions, respectively.

BACKGROUND AND PRINCIPLES

Past climate is inferred almost entirely from evidence incorporated into the sedimentary rock record and particularly from the contained fossils. Australia is a continent on which continuous onshore sedimentary records, particularly of those deposits that accumulated in a nonmarine environment, are rare, and students must depend on the offshore marine sequences, themselves incomplete, around the continent margins augmented by other diverse sporadic records. For much of the time, Australia has had a relatively low humidity regime with low runoff. Even in the offshore, much of the area has been characterised since the mid-Cretaceous by carbonate deposition with low input, by global standards, of terrigenous (land-derived) detritus. Figure 3.1 shows the basins that have been active during the Cretaceous–Cenozoic and contain sections that are providing the data for reconstructions of palaeogeography and climate.

While these records provide excellent, and globally exceptional, sections for use in interpreting offshore history (tectonic and climatic), they are frustratingly incomplete insofar as records of onshore conditions. In this summary, offshore terrigenous material is taken as indicative of onshore high humidity and runoff. While this approach can be criticised for reconstruction of short-term conditions, it is probably valid over longer intervals as close inshore sediment reservoirs (such

as estuarine systems common today) fill, and are bypassed, on a time scale longer than a few thousand years.

The time scale (Figure 3.2) used is that assumed in the compilation of the Australian palaeogeographical series by the Bureau of Mineral Resources, Geology and Geophysics (Bradshaw *et al.*, 1993) and by Wilford & Brown (Chapter 2, this volume). It is based on that of Harland *et al.* (1982) and the Australian application is discussed in more detail by Burger (1989) for the Cretaceous and by Truswell *et al.* (1989) for the Cenozoic. Different versions of the time scale were given by Haq *et al.* (1987) and Veevers (1984).

The topic of climate history is of critical importance at present as humanity endeavours to come to grips with the question of global change, usually in the context of the next few decades. The story documented below shows that the Australian continent, which is tectonically very stable over long periods in global terms, has changed climate

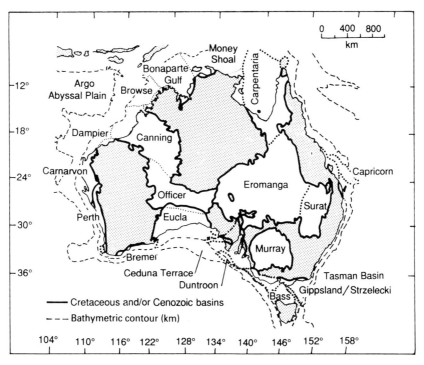

Figure 3.1 Sedimentary basins that have been active during the Cretaceous and Cenozoic in Australia (and in Antarctica during the Early Cretaceous).

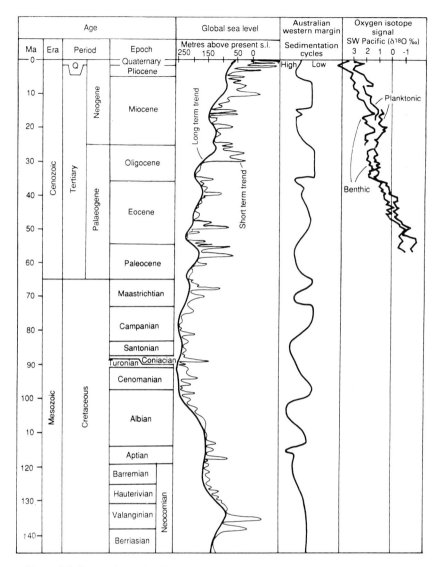

Figure 3.2 Time scale, sea level (s.l.) (global and Australian) and oxygen isotope data from southeast of Australia. Q, Quaternary. Scale for global sea level is metres above present sea level (redrawn from Haq *et al.*, 1987). Australian sea level curve is the western margin curve redrawn from Veevers (1984). Oxygen isotope curves are based on separate analysis of benthic and planktonic foraminifera to differentiate temperature changes on the sea-floor and surface, respectively. Breaks in individual curves reflect study on samples from different Deep Sea Drilling Project holes (redrawn from Shackleton & Kennett, 1975). For a fascinating review that includes the origin of the time terms, see Rudwick (1972).

greatly, naturally, over that time and that much global change is natural. The critical quest is to understand natural variation and to differentiate it from human-induced change.

At the time of commencement of this story, Australia was part of an almost complete Gondwana, and the early part of the interval was a time of dismemberment of this long-standing supercontinent. The details of the break-up of Gond-wana have been the subject of many studies, including those by Audley-Charles (1987), Barron *et al.* (1978), Veevers (1984; Veevers *et al.*, 1991) and Wilford & Brown (Chapter 2, this volume). The details of the Australia–Antarctica–New Zealand part of the break-up story also have received a great deal of attention such as that from Griffiths (1971), Hayes & Ringis (1973), Weissel *et al.* (1977), Cande & Mutter (1982) and Mutter *et al.* (1985). Thus, the evolution of the climate of Australia until some 90 Ma ago is inextricably linked with that of Antarctica and New Zealand, and even since then there have been important influences from the south during the development of the Southern Ocean circulation system. It was not until about 30 Ma ago that Antarctica and Australia became totally independent as the South Tasman Rise cleared North Victoria Land, the Circumpolar Current system became established and both Antarctica and Australia became isolated from each other and from the rest of the world. Since then, the influence of Antarctica has been distant, as a source of weather, and thus ultimately climate, for southern Australia. The role of the dispersion of continents and continental fragments to the north and northwest of Australia has been the subject of reviews by Audley-Charles (1987) and Truswell *et al.* (1987), the latter examining the migration paths of vegetation through this pathway.

The topic of Australian climate has been reviewed on many occasions, usually as part of compilations of longer-term history, such as those by Kemp (1981), Quilty (1982), Frakes *et al.* (1987*a, b*), and Frakes & Vickers-Rich (1991), and the Bureau of Mineral Resources, Geology and Geophysics has produced a general synthesis of which climate is part (BMR Palaeogeographic Group, 1990). Bradshaw *et al.* (1988) have produced a compilation of the evolution of the North West Shelf and this can be viewed as a subset of the Palaeogeographic Atlas series. Veevers (1984) also referred to the topic, although not as his major thesis. There have been few major changes in interpretation since the last of these was published, but there has been a great deal of refine-ment, and a significant growth in the database due to continuing routine studies and particularly the drilling of Ocean Drilling Program (ODP) legs around Australia. Unfortunately, the results of these legs are not yet available for integration into the story. Perhaps the main recent change over the time under consideration is the recognition that there may have been a cool interval in the Early Cretaceous with sea ice as a transporting mechanism for pebbles in the Bulldog Shale (Frakes & Francis, 1988). Another major development has been the concentrated emphasis on the Murray Basin, because of concern over the effects of humanity's activities there (Brown, 1989*a*).

In recent years, there has been a great growth in the study of vertebrate faunas and the results have been summarised by Vickers-Rich *et al.* (1991). To date, studies have concentrated on the taxonomy and evolutionary significance of the fossils. In future they can be expected to contribute to the environmental synthesis.

Cretaceous and Cenozoic sections are separated everywhere in Australia by an unconformity, and thus this continent seems not to contribute to the debate concerning events at this boundary (Alvarez *et al.*, 1980; Emiliani *et al.*, 1981).

Controls on climate

The approach taken here is to accept that the main elements that decide climate are long-term temperature and hydrological characteristics. Long-term temperature is determined by a series of factors such as latitude (= distance from the South Pole) and oceanic circulation patterns; hydrological characteristics depend on ocean circulation, sea level, degree of inundation, relief, and wind direction and intensity. There are other factors that are outside the scope of this chapter. They include variation in solar constant, cloudiness, etc. Some of the elements influencing climate are identifiable from the geological record, either directly or indirectly, and others, while not normally included specifically as elements of climate, are reviewed here because they are important in influencing, or being influenced by, climate. Many determinants of climate depend on the tectonic history of the region and this has been reviewed elsewhere (Wilford & Brown, Chapter 2, this volume).

Many of the elements listed in combination constitute the palaeogeography of a region, and in this work both palaeoclimate and palaeogeography are discussed.

Palaeolatitude

Figure 3.3 shows the Apparent Polar Wander Path (APWP) of the South Pole relative to Australia, although in reality it is the continent that

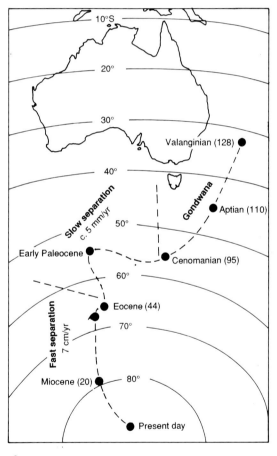

Figure 3.3 Apparent Polar Wander Path (APWP) of the South Pole relative to Australia over the interval Cretaceous–Cenozoic. APWP is shown by thick dashed line. Intervals indicated include those during which Australia was part of Gondwana, or was separating slowly or quickly from Antarctica. (From Embleton & McElhinny, 1982; Veevers, 1984).

moves in relation to the Pole. The assumption is that the measured palaeomagnetic pole is statistically the same as the geographical Pole.

The position relative to the Pole controls latitude, and thus general temperatures (oceanic and atmospheric), major meteorological features (such as continent position relative to lows and fronts and their frequency), and the proportion of the continent under tropical influences. Also relevant, but nonmeasurable at this stage in the geological record, are the effects of albedo including questions of constancy (day-to-day and annual fluctuation especially during intervals when there may have been winter snow). In contrast, only nonquantifiable albedo variations on a long time scale can be guessed at, due to changes in vegetation/aridity factors such as the contrast between a vegetated Middle Miocene (low albedo), due to a humid environment, and the Late Cretaceous when the continent was thought to be arid and thus paler (higher albedo). Also nonquantifiable geologically are the impacts of changes of sun angle due to differing latitudes and the summer/winter day duration factors.

During this interval, the relationship between Australia and the south geographical Pole (or axis of rotation of the earth) as measured by the APWP changed radically. Throughout the Cretaceous, the Pole lay to the southeast or south of Australia but relative movement was slow and did not cause palaeolatitude to change greatly. The most significant change followed the more rapid separation of the continents commencing in the Paleocene, leading to relatively fast movement of Australia north through various climate zones and ever decreasing latitude. The movement of Antarctica during this time has been minor. The situation has been reviewed in detail by Embleton & McElhinny (1982) and formed the basis for interpretations by Veevers (1984).

Palaeoceanography

Nearshore temperature and precipitation are very strongly influenced by ocean temperature, wind direction and strength, and terrestrial relief. The interaction is too complex and poorly known to explore here in detail. It is probably true also that atmospheric composition and consequently some elements of vegetation distribution and evolution are controlled by events occurring in the oceans. It is in the area of oceanography particularly that the global milieu impacts on local or regional evolution.

This interval is seen by many authors (e.g. Frakes *et al.*, 1987*a*) to have been a key event in the evolution of the oceans to their modern mode. Initially they were regions of depressed high to low latitude temperature gradient, very low shallow- to deep-water temperature gradient, and with much less differentiation and mobility than today. Cretaceous oceans were characterised by a series of 'anoxic events' but the number, duration and distribution of the resultant sediments are subjects of some debate. The oceans were generally warmer, even at depth, than they are today and thus may have had a lower O_2 solubility. The change to the modern style seems to have begun in the Late Cretaceous.

The palaeoceanography of the Australasian region is very poorly known except in the most general way, and the reconstructions shown on Figure 3.4 are very tentative. Control on the understanding of past water temperature is mainly biological (presence or absence of large foraminifera, features of planktonic foraminifera such as diversity, presence or absence of keels, etc.), and not on directly measured parameters such as oxygen and carbon isotopes, except for southeast of the continent, and then only for part of the Cenozoic. The understanding of the situation around Australia is worse than that of the area around New Zealand and the only significant results to date stem from Deep Sea Drilling Project (DSDP) material (e.g. Shackleton & Kennett, 1975). There has been a considerable amount of oxygen isotope study performed earlier (Dorman & Gill, 1959; Bowen, 1961; Dorman, 1966, 1968) but this suffers the disadvantage of having been conducted on spot samples rather than from continuous sections.

The palaeoceanographic reconstructions employed are based on those by Burns (1977), Edwards (1975), Kennett (1980) and Savin

(1977), which are generally consistent throughout the Eocene and younger and have been the source for many more recent compilations. They differ in the Tasman Sea at 21 Ma, where Kennett proposed a clockwise current (shown on Figure 3.4). At ages greater than 21 Ma, the current direction in the Tasman Sea is taken to be from south to north, in contrast to the situation since. This is believed to result from the formation of the Tasman Sea and Lord Howe Rise and the resultant channelling effect. Earlier than 63 Ma, the reconstructions are largely speculative and my own and are based on the assumption that each major southern hemisphere ocean basin was characterised by a large counterclockwise current system. Smaller current systems are not yet differentiated and the general evolution in the Indian Ocean is unknown. Earlier than 95 Ma, the position of the continents relative to the South Pole was significantly different than later and there is little basis even for conjecture.

Temperature

As noted above, palaeotemperature estimates in Australia do not rest on a good basis of many oxygen isotope profiles. Australian reconstructions depend on data from nearby (e.g. New Zealand; Clayton & Stevens, 1967; Stevens & Clayton, 1971), from more scattered oxygen isotope data sets, from sediment features (Frakes & Francis, 1988), and on other information from fossils, such as features that are temperature dependent.

The most important oxygen isotope data set available is that from Leg 29 of the DSDP from south and southeast of Tasmania (Shackleton & Kennett, 1975). The curves for both planktonic and benthic foraminifera are shown on Figure 3.2.

Sea level/inundation

Sea level varies significantly from time to time and Figure 3.2 shows the variation throughout the Cretaceous and Cenozoic as understood at present (Quilty, 1977, 1980; Vail *et al.*, 1977; Haq *et al.*, 1987). The subject is one that is evolving

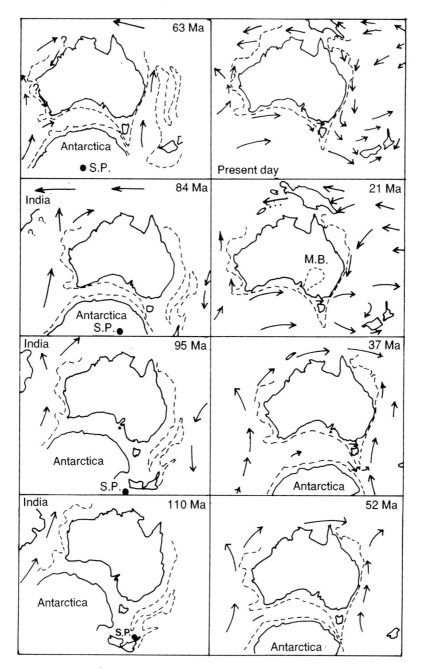

Figure 3.4 Variation in oceanic current patterns through the Cretaceous and Cenozoic of Australia, from many sources. S.P., South Pole; M.B., Murray basin. Arrows indicate positions of surface currents. Continent outlines in continuous lines are as present and bear little relation to positions of past coastlines. Dashed lines represent the continent–ocean boundaries.

rapidly. The global sea level change curve shown in Figure 3.2 illustrates two orders of variation, a long-term component that ignores short-term change, and a short-term component that is made up of more rapid variations. For most of the time, there is not a great disparity between the two curves but there is a marked change at about 30 Ma (mid-Oligocene) when the two diverge. This probably marks the time at which the short-term component became controlled by variation in the volume of sea water locked up in polar icecaps, allowing variation to be within ± 100 m (modern sea level is about 65–70 m below maximum, and 18 000 years ago, during the last glacial maximum, was some 130 m below present). This ice-volume-controlled eustacy does not seem to be a valid explanation for much of the variation prior to the mid-Oligocene, and an explanation is still required for much sea level variation. Tectonics seem to control the long-term component and may also be responsible for some of the short-term fluctuations. The equivalent curve for the western margin of Australia (Veevers, 1984), although constructed on a totally different basis, is grossly similar to a smoothed version of the short-term global curve, but differs markedly in the Oligocene, when the Australian data suggest a major sea level fall at the end of the Eocene, whereas the global curve would have the same change occurring in the mid-Oligocene. The marked increase in $\delta^{18}O$ at the Eocene/Oligocene boundary (Shackleton & Kennett, 1975) seems to suggest that the Australian curve (e.g. Quilty, 1977, 1980) reflects the time of onset of some Antarctic glaciation and resultant sea level fall better than does the global curve described by Haq et al. (1987).

Inundation is caused by two factors: sea level rise and tectonics. Sea level change on a coast of low relief has a marked effect on the area to be covered by inundation, on the biota there and the climate it lives in. Significant sea level rise causes depression of the crust due to the effects of hydro-isostasy and this leads to enhanced inundation. This effect can augment the sea level rise effect by as much as 30%. Tectonics can cause depression of continental crust, allowing inun-

dation even at times of globally depressed sea level. Although Australia is generally of low relief and tectonically stable over the interval discussed, the continent did undergo major inundations in the Aptian and Albian (Frakes et al., 1987b) during the Potoroo Regime described by Veevers (1984). The cause was probably largely tectonic, as too much negative relief was generated to be ascribed to a simple sea level rise and global sea level at the time was generally depressed except at the end of the interval. Following this interval, the Australian history seems almost to be the antithesis of the global scenario in the later Cretaceous.

Relief

Relief is the opposite of tectonic depression referred to above and has a major controlling influence on precipitation and on the direction of runoff. It is controlled by tectonics.

The feature that is very important for the Cretaceous–Cenozoic history of Australia, and even more so for Antarctica, is that compressional tectonics has been almost irrelevant, except to the extent that it played a part in the development of the southeastern margin in New Zealand, and the northern margin through Papua New Guinea. The virtual absence of compressional features has meant the absence of very high mountains and their associated andesitic volcanism except in southeastern Australia during the Jurassic/Cretaceous transition, when the basins along southern Victoria did accumulate very significant amounts of andesitic volcanics. The Australian relief through the interval under discussion has therefore been important but not extreme due to its origin in divergent plate tectonics.

Australia is now a large land area of generally low relief, with a few marked exceptions such as Tasmania, and areas marginal to the continent in eastern Australia and northwestern Australia. This has also generally been true throughout the time under consideration, but with marked exceptions, particularly in the early part. There is considerable evidence from the Perth Basin (Johnstone et al., 1973; Veevers, 1984) to indicate that, during the latest Jurassic and earliest Cretaceous, the Darling Range stood much higher than it does now, as a result of the evolution of the divergent margin during the separation of India from southwestern Australia. Veevers (1984) stated that there were predecessors to the Flinders Ranges (among others) in the Cretaceous. The details of the evolution of the Eastern Highlands are poorly known but they seem to have had forebears during most of the Cretaceous (especially those in Queensland in the Late Cretaceous) and may owe their origin in part to the Australia–New Zealand separation. There is little known of the timing of commencement of the highlands (Stirling Ranges, etc.) along the southern margin of Western Australia.

Drainage

Drainage depends on the presence of relief and enough precipitation to provide excess surface water. In the interval under consideration there is no evidence to suggest that there was any significant contribution from snowfall except perhaps in the earlier Cretaceous in southeastern Australia. Patterns of drainage have varied greatly, and included times when there is no evidence of any drainage at all, to times when drainage was dominantly to the central south coast, or to inland depocentres (past equivalents of Lake Eyre), or to a vast inland ocean via short streams radial to a series of smaller land masses. The best summary to date is that edited by Veevers (1984). There is not enough information available so far to decipher the history of most of the river systems flowing to the east of the Eastern Highlands, but, for most of the rest of the Australian margin, oil exploration drilling has yielded data that provide some clues.

Volcanism

Although continental Australia is devoid of active volcanoes at present, throughout most of the time covered by this chapter, volcanism has been a major feature of the landscape over an arc from northern Queensland, in a zone a few hundred kilometres wide around the eastern margin of the continent to Mt Gambier in South Australia, where dates as young as 4600 years before present (4.6 ka) have been recorded, suggesting that volcanism in some areas should be regarded as dormant rather than extinct. Tasmania has a major history of volcanism during a large part of the Cenozoic.

The volcanic activity has been responsible for a great deal of the geomorphology of eastern Australia, including lava plain, volcanic cones with fresh appearance and plugs that are positive geomorphic features left by the erosion of the softer surrounding material. Much of the Australian gemstone industry is based on the products of this phase of volcanism.

The controls on this volcanism are not fully understood, but a major element can be ascribed to movement of the Australian plate over a series of 'hotspots' that periodically cause lava to make its way to the surface. The product of this type of control is a linear series of volcanic vents and rocks. This feature is well documented but is not enough to explain all occurrences of volcanics. The volcanics are spatially associated with the highlands of eastern Australia and to the Australian plate margin. Johnson (1989) discussed the volcanism over this time in much greater detail. The occurrence of volcanoes is recorded wherever relevant in the second part of this chapter.

EVOLUTION OF AUSTRALIAN PALAEOCLIMATE AND PALAEOGEOGRAPHY

This section consists of a series of reconstructions (Figures 3.5–3.13) that illustrate the palaeogeography of Australia at discrete times throughout the Cretaceous and Cenozoic. The palaeoclimate is discussed in relation to these reconstructions.

The use of regimes as a basis for discussion

Veevers (1984) introduced the concept of regimes as a means of discussing the geological history of Australia. It is an adaptation of an earlier concept applied to the historical evolution of North

America (Sloss, 1963), and was also employed by Haq *et al.* (1987). Two of the three Phanerozoic regimes identified by Veevers are relevant to discussion of the Cretaceous–Cenozoic. They are the Innamincka (320–90 Ma, terminating in the Cenomanian, using the time scale of Veevers, 1984) and Potoroo (90 Ma to Recent) Regimes. The boundary between the regimes is identified by two features: a major retreat of marine conditions from central Australia during the Ceno-

manian, and a widespread change in depositional regime in the offshore from terrigenous (before) to dominantly biogenic carbonate (after) sediment.

Cretaceous part of Innamincka Regime
(Figures 3.5–3.7)

When the Cretaceous began, terrigenous sedimentation was generally a continuation of that in

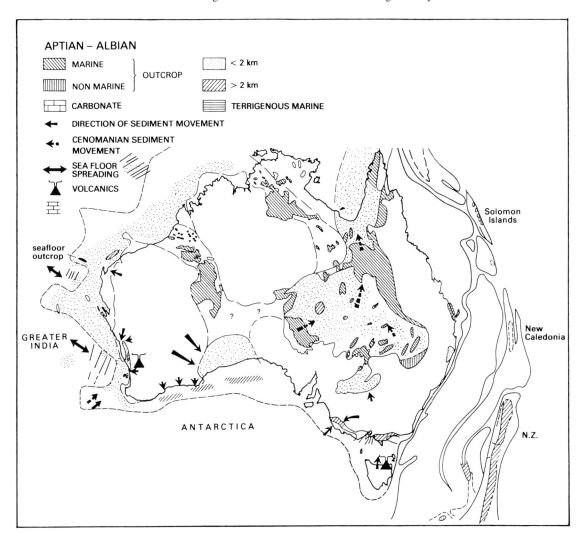

Figure 3.5 Reconstruction of the palaeogeography of Australia during the Aptian–Albian. Identified are areas of marine sedimentation, volcanism, direction of sediment movement, locus of sedimentation, and an indication of sediment thickness. NZ, New Zealand. (Modified from Quilty, 1982.)

Figure 3.6 Coastline variation during the earlier Cretaceous part of the Innamincka Regime. 1, Latest Jurassic–early Neocomian (Berriasian); 2, mid-Neocomian (Valanginian–Hauterivian). (After Frakes *et al.*, 1987*a*.)

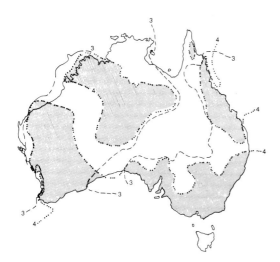

Figure 3.7 Coastline variation during the Innamincka Regime, towards its end. 3, Aptian; 4, late Albian. Hatched areas were not inundated during this interval. In addition to the marine sedimentation within the coastlines indicated, there was a great deal of nonmarine accumulation outside these coastlines. (After Frakes *et al.*, 1987*a*.)

the Jurassic, dominantly in active graben structures around the Australian margin and in a large gentle depression in 'central' Australia that formed the Great Australian Basin. This was the last time that great thicknesses of detrital sedi-

ment accumulated because this interval marked (1) the final act in the tensional development of depocentres related to the dismemberment of Gondwana and (2) the last long-term interval of detrital sedimentation in Australia, these conditions giving way in the mid-Cretaceous to a dominantly biogenic carbonate regime in the marine environment. Most deposition after this time was on subsiding continental margins without an obvious basin style geometry.

Volcanism was a major source of sediments around southern and southeastern Australia and the Bunbury Basalt of southwestern Australia was extruded onto an erosional surface during this time. Other evidence of volcanism in the form of plugs, flows and volcaniclastic sediment of this age is not well documented.

In northwestern Australia, the southern part of the west coast, south and southeast coasts and the Great Australian Basin, deposition of great volumes of sediment indicates that river systems were active and well developed and thus that there was significant precipitation continentwide. In the Great Australian Basin and southeast basins (Otway and Gippsland/Strzelecki), the land-derived material was accompanied, and sometimes dominated, by andesitic volcanic input related to the tension associated with the break-up of Gondwana. Drainage was concentrated to two major deposition sites, one a series of small basins along the south coast, and the other in central Australia. The system to the central south coast (then a depocentre for drainage from both Antarctica and Australia) was fed in part by the salt lake system of Western Australia and South Australia, which is taken as a relict of this interval and perhaps earlier. At the same time there was active tectonism in the Perth Basin and onshore, with the present Darling Ranges markedly uplifted causing an influx of coarse detrital sediments with fresh feldspars to the Perth Basin, all suggesting rapid erosion and deposition with transport by active short path streams. Sediment thickness in the offshore Perth Basin is very great but the depocentres are relatively small. The other side of the ranges was serviced by much longer, slower streams carrying finer material to the Eucla Basin.

The major drainage during the final stages of the Innamincka Regime was into the central Australian depression, which accumulated the bulk of the sediment deposited during this time.

Marine sediments were initially restricted to the western margin as the Indian Ocean evolved, but marine conditions spread progressively eastward along the south coast, and from an initially north–south incursion (Valanginian–Hauterivian) into the Great Australian Basin from the north, as shown on Figure 3.6. Marginally marine microfossils are sporadic in sediments older than Aptian but the first major evidence for advancing marine conditions is in sediments of the Eucla Basin in the Aptian as the earliest transgression of the sea began between Australia and Antarctica. In the Aptian–Albian, conditions continentwide became marine and Australia was divided into three large islands (Figures 3.5, 3.7) by an inundation, the like of which had not occurred before in the Phanerozoic, and has not occurred since. Drainage was radial from these islands but also generally to the depocentre to the south of the central south coast. Deposition was in shallow-water intracratonic basins that were very widespread. The general Aptian–Albian transgression may have been two transgressions separated by a small regression, and this is indicated on the Australian sea level curve on Figure 3.2. Reducing conditions were widespread, radiolaria an important constituent in the sediments, and planktonic foraminifera rare, or if present of low diversity and small size (Scheibnerova, 1976). Molluscs are abundant and widespread indicators of this marine phase (see e.g. Ludbrook, 1966). Marine conditions became more restricted by the end of the Albian to small areas of the Great Australian Basin and northwestern Australia. At the end of the Early Cretaceous, marine conditions retreated from central Australia, and have not returned. The position of the coastline throughout the Cretaceous has been documented in considerable detail by Frakes *et al.* (1987*a*). Figures 3.6 and 3.7 summarise the variation in position of the coastline for this part of the Cretaceous.

The microfossil evidence cited above suggests that marine conditions may have been cool throughout much of the Aptian–Albian, consistent with the location of the southeastern quarter of Australia inside the Antarctic Circle. Firmer conclusions come from other sources. Douglas (1969) identified a cool temperate climate, with high rainfall in the range 760–1140 mm, with a seasonal dry period as normal in southeastern Victoria during most of the Early Cretaceous. Waldman (1971) suggested cold winters with ice cover in restricted environments during part of this interval, and Frakes & Francis (1988) also proposed that winter sea ice was present more generally and was a sediment-depositing agent. Sheard (1990) detailed the occurrence of glendonites in the Bulldog Shale and called for very cold to frigid conditions at the time. Jell & Duncan (1986) questioned the need for winter ice but agreed with the other elements of Waldman's hypothesis. While conditions were cool generally across Australia, perhaps with a gradient towards increased coolness in the southeast, there is no evidence of nearby widespread glaciation. Rich *et al.* (1989) have reviewed in detail the results of studies of fossils from the Victorian Early Cretaceous and included data from other sources, to make a summary that Early Cretaceous polar atmosphere may have had a temperature range of -5 to $+8\,°C$. They also quote evidence to suggest cool, humid conditions with seasonal climate, consistent with earlier hypotheses. These cool climate scenarios are in conflict with some other hypotheses (Barron *et al.*, 1981; Hallam, 1984, 1985) that the Cretaceous was an ice-free period, but have not yet developed to the stage of suggesting the presence of any polar icecaps. There is a need to search the Antarctic for any evidence of cold conditions, even of the presence of an icecap and sediment transport by icebergs.

Cretaceous part of the Potoroo Regime (including the Cenomanian Interregnum) (Figures 3.8–3.9)

This is an interval for which data from central Australia are very few and much reliance for climate must be on well-studied marine sections

Figure 3.8 Variation of coastline through the Late Cretaceous. Over most of the western margin, variation in position throughout was very minor. Areas of difference are in the Gulf of Carpenteria and Eucla Basin. 1, Cenomanian coastline where it deviates from average; 2, average coastline throughout most of the Late Cretaceous; 3, Campanian–Maastrichtian coastline along the south coast; 4, Cenomanian, Turonian, Santonian, Maastrichtian coastline along southwest Victoria. (After Frakes *et al.*, 1987*a*.)

along the western margin and less well-known sections along the south coast. The interpretations are thus weaker for central regions than for other intervals during the Cretaceous–Cenozoic.

Veevers (1984) separated the Innamincka and Potoroo Regimes by a 5 Ma Cenomanian Interregnum, which here is taken as part of the later Cretaceous history. At the end of the Albian, the sea withdrew from the Great Australian Basin (except for the northernmost area around the margin of the Gulf of Carpenteria) and nonmarine deposition replaced marine in the central parts of the Basin for a time. This transition set the pattern for Australian climate and palaeogeography until the present day. The sea entered the narrow rift opening between Antarctica and Australia, and terrigenous sedimentation along the west coast gave way to biogenic carbonate sediment, reflecting reduced river activity and perhaps precipitation continentwide. In contrast with the global sea level curve (Figure 3.2), Australia was not subject to a transgression in the Cenomanian,

but was characterised by regression of marine conditions. The coastline during each Late Cretaceous sedimentation cycle along the western margin was very uniform in position (Figure 3.8) and the only areas where there was significant variation in position were the Gulf of Carpenteria and the Eucla Basin. The Cenomanian coastline in these two areas is a relic of the Albian transgression and along the south coast depended on elements of the Australia–Antarctica break-up history for its variation. During this interval Australia and New Zealand separated with the opening of the Tasman Sea, and slow spreading began between Australia and Antarctica.

Volcanism occurred through several areas of eastern Australia but is neither well dated nor well understood. The Cenozoic volcanism, so well developed and understood, had its origin in the Late Cretaceous. Dates of around 70 Ma are known for volcanic rocks from Queensland and New South Wales but are not widespread. Group 1 rocks (Johnson, 1989) of this age are known from central Victoria. Lavas at Cape Portland in Tasmania may have been extruded at the earliest part of this interval or perhaps a little earlier (Sutherland & Corbett, 1974).

Most of continental Australia was bypassed by sediment except for the Cenomanian Winton Formation and possibly the Mt Howie Sandstone (Figure 3.9). Sediment sources included uplifted Eastern Highlands in Queensland and southeastern Australia. The former seem to have been the source of a large amount of volcanic debris, some of which made its way into the Winton Formation and possibly the central south coast depocentre, which by this time had ceased to be a site of deposition for Antarctic-sourced sediment. Otherwise, the terrigenous component of sediments changed from the earlier volcanogenic type to a more quartzose, continental source.

The central Australian depocentre declined dramatically in importance and was overridden by drainage to the central south coast, which now became the site of deposition of sediments from the drainage of approximately two thirds of Australia, including that by ancestors of the Murray and Darling Rivers, which seem to have been

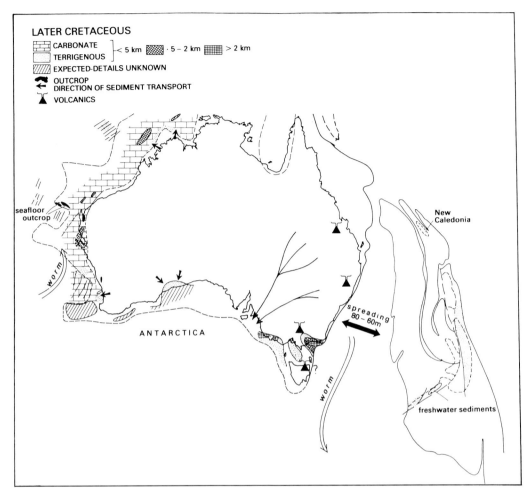

Figure 3.9 Reconstruction of the palaeogeography of Australia during the Late Cretaceous. (Modified from Quilty, 1982.)

initiated during this time. The change in drainage character was effected by the movement north, almost to the Gulf of Carpenteria, of the drainage divide that had been in place earlier. Instead of deposition in the region, much of it seems to have been subject to intense chemical weathering, perhaps reflecting humid conditions and temperatures that were warmer than they had been earlier. This is consistent with global reconstructions for the time (Hallam, 1984).

Douglas *et al.* (1976) have suggested that, in southeastern Australia, cool conditions characteristic of the earlier Cretaceous continued until the Santonian, when temperature and humidity increased through to the end of the Cretaceous. The evidence for considerable humidity exists in the form of coals, although, in total, the coal swamp environment was less widespread than in earlier parts of the Mesozoic. Dettmann (1981) suggested cool to mild temperatures and humid conditions for this interval, whereas Truswell (1990) noted that it was during the Late Cretaceous that 'temperate forest elements like *Notho-fagus*' evolved, the evidence coming from the Otway and Gippsland Basins.

Along the western margin, the pre-existing

basin geometry gave way to deposition of a sediment wedge thickening generally seawards with a prograding geometry. The best sections known to date are from the Carnarvon Basin and offshore Northwest Shelf (Apthorpe, 1979), and sporadic records exist for the Perth Basin (e.g. Quilty, 1978; McNamara *et al.*, 1988). Sediments are dominated by biogenic carbonate, of which the foraminifera are the only part that has been the subject of detailed published studies.

Cold water indices are absent from the region, as are tropical fossils such as rudistid bivalves, large foraminifera or coral reefs, although Glaessner (1960) recorded large foraminifera from nearby New Guinea, indicating that the region was very close to areas where tropical conditions prevailed. Most of Late Cretaceous time is represented by sediments containing abundant, diverse planktonic foraminifera, with double keels and other complexities indicating warm to tropical conditions. Faunas belong to Sliter's (1976) Transitional or Tethyan Faunal Provinces. Much needs to be done to document allocation to Faunal Provinces in detail but marine temperatures were considerably higher than during the earlier Cretaceous. During the Santonian–Campanian, the faunas appear to have been Transitional, but at other times they were Tethyan and thus represent subtropical–tropical conditions. While sediments are dominantly biogenic carbonate, virtually all contain a fine terrigenous component and there is sandstone sporadically in the Perth Basin and more widely around the Kimberley Block, attesting to some activity of drainage systems in those areas, particularly in the north.

Cenozoic

The Cenozoic encompasses the last 66.4 Ma of geological history and includes the Quaternary in which humanity lives. Cenozoic rocks are very widespread in Australia and include not only sedimentary rocks and volcanic accumulations but other features not so evident from earlier times, such as soils and other weathering effects including the almost ubiquitous laterite (taken here as a generic term to include silcrete). Climatically the Cenozoic was a continuation of the regime established in the Late Cretaceous, but developed further with a general trend towards more biogenic carbonate in the marine realm, and reduced internal drainage, although with some important exceptions.

There is no evidence for the presence in Australia of sediment sections representing continuity from Cretaceous to Cenozoic, and thus no evidence can be provided to contribute to the debate on the immediate causes of the major extinction that took place at the end of the Cretaceous. The Australian oceanic biota changed in the same way as did others elsewhere in the world. A great deal needs to be done before onshore changes to fauna and flora are documented and understood. Generally the Cenozoic climate change is better understood than that of other parts of geological time because the outcrop is better and more accessible, closer to the centres of population, and organisms are more similar to modern ones, and thus provide better analogues than do older examples. In addition, this is the time for which reliable oxygen isotope records (e.g. those of Shackleton & Kennett, 1975) became available and formed an important determinant in reconstructions of global as well as Australian climate evolution.

During this interval, Australia began to migrate northward from Antarctica at a greater rate than it did earlier (Cande & Mutter, 1982) and the nexus between the two continents was finally broken.

During the Cenozoic, the biogenic carbonate regime expanded progressively to encompass the southwestern margin and eventually the entire south and east coasts. Uplift along the south coast caused cessation of major sedimentation in the central south coast and the Murray and Birdsville Basins came into being.

Several authors such as Quilty (1977, 1980) and McGowran (1979) have recognised a convenient subdivision of the Australian Cenozoic record into four intervals based on the concept of 'unconformity-bounded stratigraphic units' (Ki Hong Chang, 1975) or sedimentation cycles. Consequently, this has been found to be a useful discussion framework and was employed by Quilty (1982), Veevers (1984), Frakes *et al.* (1987*b*), and Apthorpe (1989). It is employed here again! The time units are

Paleocene–Early Eocene

Middle–Late Eocene

Oligocene–Middle Miocene

Late Miocene and younger

Paleocene–Early Eocene (Figure 3.10)

Savin *et al.* (1975; Savin, 1977) proposed that this interval globally was of warm marine conditions with bottom temperatures generally of the order of 14–15 °C. At the Paleocene/Eocene boundary, water temperatures were higher than at any time since, and had been rising slowly throughout the Paleocene after a brief cool period near the Cretaceous/Tertiary boundary. Marine faunal evidence shows that there was less tropical–polar

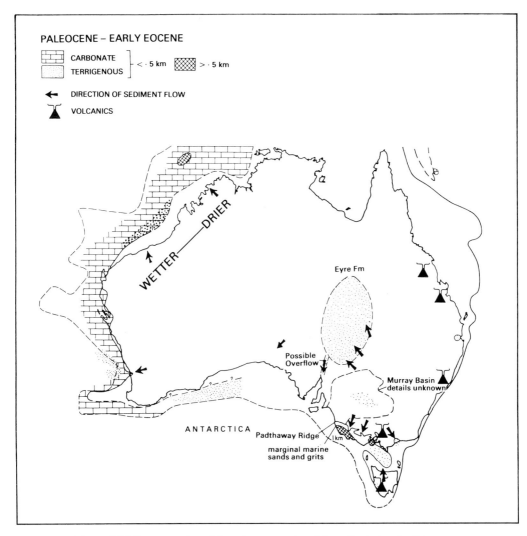

Figure 3.10 Reconstruction of the palaeogeography of Australia during the Paleocene–Early Eocene. (Modified from Quilty, 1982.)

differentiation of climate zones than now. Kemp (1981), in a comprehensive review of the palynological evidence for reconstruction of Tertiary climates of Australia, indicated that control of precipitation in southern Australia would have been by westerly winds, and farther north the circulation pattern would have been less regular and somewhat chaotic inland. The evidence for warm marine conditions is consistent with the interval being one of considerable humidity.

There was considerable volcanism with diverse products in southeastern Queensland and northeastern and southeastern New South Wales. In Victoria, Group 1 and Group 2 volcanics were emplaced in the central and southwest parts of the state. This interval seems to have been a minor source of volcanics in Tasmania (southeast only) and the geometry of the volcanics has been used to indicate that the present topography of Tasmania had been generated prior to this time (Johnson, 1989).

Quilty (1982) suggested that generally this interval was one of low significant drainage but a great deal of information is now available to suggest that this view should be changed. Apthorpe (1989) divided this cycle into two (her subcycles 1A and B), and presented evidence that a gradient existed in the northern part of northwestern Australia from a 'relatively wet climate in the south' to drier in the northeast, and that water temperatures were 'at least' warm temperate. The later subcycle coincides, at its earliest, with the most widespread part of transgression in this cycle towards the youngest part of the Paleocene. Apthorpe suggested that conditions initially were warmer than in subcycle 1A with hot arid conditions in the northeast, perhaps with seasonal precipitation. In the Carnarvon Basin area, waters may have been warm temperate–subtropical and there may have been a coastal countercurrent analogous to the Leeuwin Current of today. At the end of the subcycle 1B, within the Early Eocene, Apthorpe recorded evidence of significant cooling in oceanic temperatures and an increase in runoff from the Carnarvon Basin hinterland. This cooling follows immediately the time of warmest marine con-

ditions worldwide (Savin et al., 1975; Savin, 1977; Haq, 1981; Kennett & Barker, 1990). The northern parts, in the Bonaparte Basin, continued tropical and arid. As noted by Apthorpe, this cooling seems to pre-date by some time the global cooling described by Savin (1977), but the evidence for the timing of the global cooling is not very precise and it may be that the evidence from northwestern Australia is the best available, except for oxygen isotope-based studies such as that by Kennett & Barker (1990). Davies et al. (1991) indicated that this scenario was also applicable to northeastern Australia, where the Early Eocene was an interval of the warmest conditions of the Paleogene (warm enough for colonial coral growth) but not as warm as at present. Warm temperate conditions probably applied also during sedimentation in the Perth Basin (Quilty, 1974a,b), with significant rainfall and runoff (Kemp, 1978).

In southeastern Australia, there is evidence of considerable precipitation and runoff (Deighton et al., 1976), and Kemp (1978) referred to temperate conditions with high rainfall; Harris (1965) mentioned cool, moist conditions for the Paleocene in the Otway Basin. The Murray River and many of its tributaries (including the Darling River) may have been in existence since the Late Paleocene (Stephenson & Brown, 1989) but its mouth moved throughout history, depending on the position of the coast, itself determined by both structural movements and sea level variation, which controlled the extent of marine transgression into the Murray Basin.

The Eyre Formation is thin and widespread in the Birdsville Basin of central Australia and has been taken to indicate a climate that was 'warm temperate with seasonally high precipitation' (Wopfner et al., 1974) and represents the depocentre of a significant drainage system. Sluiter (1991) has provided a much more detailed analysis of material from this region and differentiated a Late Paleocene–earliest Early Eocene climate with mean annual temperatures of 18–19 °C and rainfall of about 1400 mm/year, giving way later in the Early Eocene to warmer conditions (20 °C) and increased precipitation. By the Middle

Eocene, the temperatures had declined to 17–18 °C and rainfall to 1500 mm/year. Conditions were consistently warmer than in the Gippsland Basin, but the difference varied with time.

Middle–Late Eocene (Figure 3.11)

All major reviews (Savin *et al.*, 1975; Savin, 1977; Grant-Mackie, 1979; McGowran, 1979, 1989) of this interval have suggested that it was a time of

marked general decrease in oceanic water temperature, with surface and bottom water temperature decreasing by more than 5 deg.C, although the timing of the change, as measured by oxygen isotope studies, occurred at slightly different times or with different gradients. This general decrease followed a major stepwise change towards heavier ocean water near the close of the prior cycle (see e.g. Kennett & Barker, 1990). Oberhänsli (1986) commented that this change at the Early/Middle

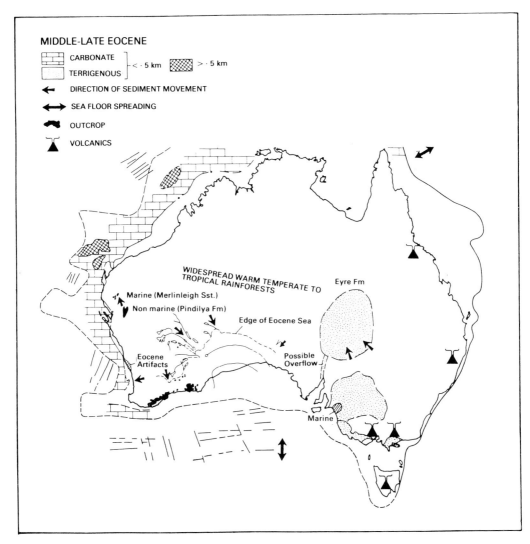

Figure 3.11 Reconstruction of the palaeogeography of Australia during the Middle–Late Eocene. (Modified from Quilty, 1982.) Fm, formation; sst, sandstone.

Eocene transition is the 'most striking feature in the oxygen isotope record' in the Indian Ocean.

Volcanism seems to have been relatively minor in northern Queensland and throughout New South Wales, except around Monaro. The so-called Group 2 volcanics of southern Victoria were produced largely during this time and were voluminous in the totality of Australian volcanics. Volcanic rocks of this interval occur in Tasmania but this pre-dates the main episode of Tasmanian Cenozoic volcanism (Johnson, 1989).

The Australian sediment record for this interval is better than for most times in the Late Cretaceous–Cenozoic and the sections are very widely distributed. The sea entered between Australia and Antarctica as the continents moved far enough apart for the Southern Ocean to become established, even if not with unrestricted access south of Tasmania to the Pacific Ocean. Australia moved farther away from the South Pole (Figure 3.3) and thus into a different atmospheric circulation regime (Kemp, 1981), with the influence of the westerlies decreasing even along the southern reaches of the continent.

The marine carbonate regime established along the western margin expanded into the Eucla Basin and to the basins of South Australia. Generally there is a major hiatus along both western and southern margins between the sediments of the Paleocene–Early Eocene and those of the Middle–Late Eocene, although Apthorpe (1989) suggested that there may be continuous sequences in parts of the Northwest Shelf. The simple two subcycle concept of cycle 2 introduced by Quilty (1977) has been subject to revision, particularly by McGowran (1989) for southern Australia. He recognised four transgressions within the cycle and related these to intervals of high sea level (due to a variety of causes) and of warmer water, but these cannot be differentiated in northwestern Australia (Apthorpe, 1989). The interval appears to have been one of decreasing oceanic temperatures worldwide, but there was a short reversal of this trend during the later Eocene around Australia. Tropical conditions were widespread over much of the western margin during times of sedimentation and even penetrated to the Bremer

Basin, to judge by the occurrence there of characteristic larger foraminifera (Cockbain, 1967). Even though generally warm, there is evidence of periodic cool water incursions along the south coast (McGowran & Beecroft, 1985). Conditions warm enough for the growth of larger foraminiferids existed along the northeastern margin (Davies et al., 1991) during the early part of the Middle Eocene but then deteriorated to the extent that no tropical carbonates accumulated until the Miocene.

For the northern half of the continent, vegetation data are very sparse. Vegetation was subtropical to tropical rainforest across the south (Blackburn, 1981) and western coasts, even to the northwest, and into central Australia (Lange, 1982; Truswell & Harris, 1982). Truswell (1990) presented distribution maps of the various vegetation groups she had differentiated for this time, including coastal mangroves. The pattern is consistent with the hypothesis of deep weathering prior to formation of the laterite and silcrete surface material so widespread throughout southern and western Australia. Apthorpe (1989) urged caution in the too ready application of the tropical concept to the northwest as some of the planktonic foraminiferal faunas are more consistent with warm temperate conditions. She suggested that, in that region, rainfall decreased throughout the interval and she differentiated more terrigenous Middle and more carbonate-rich Late Eocene subcycles, consistent with the views of Quilty (1982). Central Australia continued to accumulate sandstone as a result of internal drainage to the Birdsville Basin, and all is consistent with the warm–hot humid climate hypothesis.

In southeastern Australia, there may have been less tropical conditions, with changes in vegetation reflecting the changes in marine water temperature. Kemp (1981) referred to a lack of tropical forms there by the Late Eocene. Martin (1989) indicated that this interval was one of rainforest with high year-round humidity, perhaps with rainfall of the order of 1500–2000 mm/year, and Macphail & Truswell (1989) referred to the flora as being 'complex evergreen forest dominated by *Nothofagus*' with high year-round precipitation.

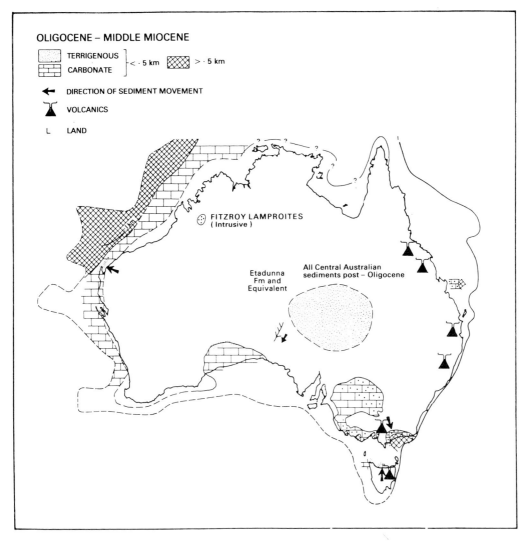

Figure 3.12 Reconstruction of the palaeogeography of Australia during the Oligocene–Middle Miocene. (Modified from Quilty, 1982.)

Oligocene–Middle Miocene (Figure 3.12)

This is a key interval and marks the transition from the essentially continuous Cretaceous–Eocene regime described above to the Oligocene–Neogene, which is strongly influenced by a markedly different marine and atmospheric circulation regime around Australia and Antarctica and led directly to the modern climate of Australia. The oceanographic choke between Australia and Antarctica, and the Drake Passage between South America and Antarctica opened in the mid-Oligocene, allowing, for the first time, waters to circulate freely around Antarctica. Australia became genuinely isolated from influences from both Antarctica and South America.

Volcanism was concentrated in central coastal Queensland, the highlands of New South Wales and central and eastern Victoria. The principal Cenozoic volcanics of Tasmania were formed in this time, considerably after the uplift of the eastern highlands. The diversity of volcanic rocks and forms of volcanism were very great during this interval (Johnson, 1989).

There was significant cooling initially and return of very warm water conditions for a brief time at the end of this interval. Sea level varied widely, and transgressions came and went. It was a time of great variation in marine conditions around Australia, perhaps more variable than had been the case until that time. Consequently, because of variation in sea level and sea water temperature, it would be expected to be a time of great variation in precipitation and runoff.

Oxygen isotope data (see e.g. Shackleton & Kennett, 1975) suggest that after a marked change at the Eocene/Oligocene boundary, conditions were approximately stable in the marine environment south of Australia throughout the rest of this interval.

There is a major hiatus in sedimentation around Australia at the beginning of the Oligocene (Quilty, 1977, 1982) described by Apthorpe (1989), and Quilty (1975). This has been taken to coincide with the main interval of formation of the extensive laterite surface in large areas of Western Australia (Johnstone et al., 1973), perhaps indicating seasonal humidity. The sedimentation history around Australia in the Oligocene suggests that sea level fell at the end of the Eocene and rose in the mid-Oligocene. As pointed out by Apthorpe (1989), this is difficult to reconcile with the sea level curves for the time proposed by Vail et al. (1977) and Haq et al. (1987), which indicate exactly the opposite to the Australian pattern. The available data from the DSDP (Oberhänsli, 1986) and the ODP results around Antarctica (Kennett & Barker, 1990) indicate that the Indian Ocean generally and the oceans around Antarctica experienced major cooling at the same time as the Australian sections suggest sea level fall. There is a consistent pattern here. The apparent conflict between Australian and global sea level curves needs to be studied in detail and resolved.

There is ample evidence to indicate that the Oligocene was the time when Antarctica began its glacial phase, with sea ice development. The degree of development of an icecap is still debated. Forthcoming ODP data will add greatly to the database for interpretation. Whatever the case, the cooling of Antarctica clearly led to increased gradients in temperature between tropics and pole, and these increased gradients led to greater wind strength. The Oligocene seems to be a globally cool interval to judge from the virtual absence of warm water morphological features on such abundant marine fossils as the planktonic foraminiferids.

Marine sedimentation during this interval was widespread around the Australian margin but did not extend far inland anywhere except in the Murray Basin (Ludbrook, 1961). Two subcycles can be recognised and they are separated by a brief regression at the end of the Early Miocene (N6–N7, according to Apthorpe (1989) and Quilty (1982)). The products of the latter of the two subcycles are the more widespread.

In the Late Oligocene, large warm water foraminiferids reappeared around north-western Australia (Chaproniere, 1984) as a major new transgression occurred there. From this time on, the faunas indicate tropical conditions across northern Australia during the time of initial collision of Australia with Southeast Asia and structural development, perhaps of the Kimberley Block and some offshore structures, as a result. The transgression is recorded elsewhere (Quilty, 1972, 1974b; Brown, 1986; Shafik, 1991) around the Australian margin but information for much of this interval is lacking over much of the continent. At the end of the Early Miocene, subtropical–tropical waters surrounded Australia and *Lepidocyclina* faunas became established briefly as a result, in virtually all sedimentary basins where marine sediments of this age are recorded.

Off northeastern Australia, conditions were temperate to cool temperate early in this interval and no tropical carbonates were formed (Davies et al., 1991), although during the Early Miocene warmer conditions reappeared and coral reefs grew once again.

Palynological data from southern and south-eastern Australia are consistent with humid, cool temperate conditions (Harris, 1971; Stover & Partridge, 1973; Kemp, 1978, 1981; Truswell, 1990) during the Oligocene, and macro- and micro-fossil analysis (Macphail *et al.*, 1991) suggests humid, cool conditions at about 700 m elevation in the Late Oligocene–Early Miocene of central Tasmania. Vegetation records show that temperature fell from that of the Eocene (Truswell, 1990), perhaps associated with some decrease in humidity. *Nothofagus* forests expanded at the expense of others and there is also the first evidence of Australian alpine flora at this time. Towards the end of the Oligocene, there seems to have been a decrease in rainfall in the Murray Basin region (Martin, 1989), although rainforest continued as the dominant vegetation. Macphail & Truswell (1989) found that the area had the greatest diversity of habitats in southeastern Australia at the time.

The Middle Miocene was a very different interval, the last time when there is evidence of drainage and significant vegetation in Central Australia. Vertebrate fossil localities there (Kemp, 1978; Callen, 1977; Stirton *et al.*, 1967) contain evidence for the existence of large lakes and watercourses lined by rainforest trees and otherwise more open grassland. This indicates humid conditions, the source of the precipitation being the northern reaches of the westerlies. Conditions were probably more equable there than at any time since. Martin (1989) recorded a change during the Middle Miocene to seasonal conditions with a marked dry period, and Macphail & Truswell (1989) agreed with this judgement. This change led to decreased importance of the *Nothofagus*-dominated forest and development of 'drier and more open forest types' (Macphail & Truswell, 1989). Fire seems to have become an important element in the environment (Martin, 1989).

Late Miocene–Recent (Figure 3.13)

Galloway & Kemp (1981) commented that the database from Australia for interpreting the history of Australian climate through the Pliocene and younger part of this interval is surprisingly poor, and this is still so, although some effort is now being concentrated here (Williams *et al.*, 1991). Some inferences can be made from global patterns, from deep-sea carbonate sections, and from Antarctic records, which are improving. Hydrocarbon exploration wells generally are not good sources of data as often this part of the column is thin, usually drilled without sample taking and confined behind surface casing. For the Quaternary, some intervals are better represented, but there needs to be more effort to improve the database, which is derived mainly in southeastern Australia from the Murray Basin (Brown, 1989*a*), and such sources as the Lake George drillholes (Singh *et al.*, 1981; McEwen Mason, 1991).

During this time, Australia continued its steady northward progression at about 6 cm/year with relation to fixed 'hotspots' and some 6.5–7 cm/year with relation to Antarctica (McDougall & Duncan, 1988; Quilty, 1993). Collision with Southeast Asia continued and provided an access to Australia, not previously possible, of faunal and floral influences. The modern circulation patterns for both ocean and atmosphere became established as the seaway between Australia and Antarctica widened, and that between Australia and Asia became more constricted. In the Quaternary, these patterns varied rapidly. Open vegetation became the norm for Australia, grasslands became established, and in places there were rapid alternations between rainforest and sclerophyllous vegetation (Truswell, 1990). As Australia dried out, the area inhabited by rainforest steadily decreased but seems to have been roughly constant throughout the Quaternary.

Volcanism was concentrated in northern Queensland and eastern Victoria, with some activity in the highlands of New South Wales. Activity in Tasmania was slight and marked the end of volcanism in that state. A considerable amount of volcanism in western Victoria and southeastern South Australia formed the Western Plains of Victoria and such volcanic edifices as Mt Schank and Mt Gambier. A significant amount of volcanic activity has occurred in the last 25 ka, even up to 4.5 ka.

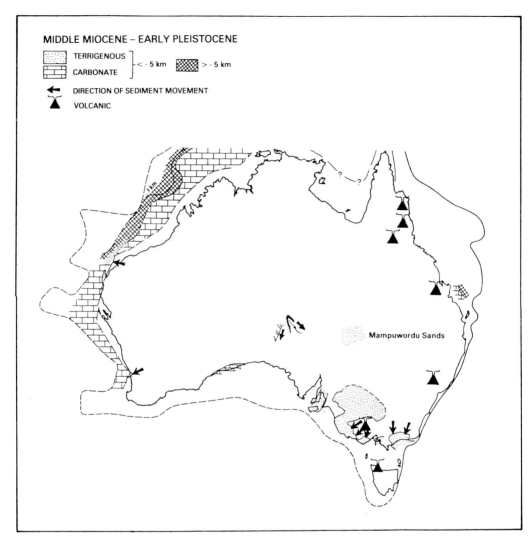

Figure 3.13 Reconstruction of the palaeogeography of Australia during the Middle Miocene–Early Pleistocene. (Modified from Quilty, 1982.)

Marine water temperatures declined dramatically south of Australia (Shackleton & Kennett, 1975), even well north of the Antarctic Convergence as Antarctica entered its present glacial phase, probably at about 2.4–2.6 Ma. Sediments around Australia generally became more carbonate rich as the influence of outflow to the oceans decreased, reflecting the spread of arid conditions and perhaps the soil-binding role of grasses. Around northern Australia, tropical conditions

continued, with a northerly migration of reef growth (Davies *et al.*, 1991) during a cooler Late Miocene. In the Pliocene, such reef growth once more moved south and the modern pattern of high rainfall restricted to southeastern Australia became established.

In northwest Australia, marine data give a clue to some changes that occurred in the marine environment (Quilty, 1974*b*). There, changes in planktonic percentage and coiling ratio patterns in planktonic foraminiferids suggest that the Early Pliocene was warmer than the Late Miocene and Late Pliocene and younger, and that there was a marked shallowing of the continental shelf area, either as a global sea level fall or as a migration of the continental shelf edge over the area. Apthorpe (1989) documented in significant detail what is known of rocks of this age farther north from the sections recorded by Quilty (1974b) and showed that the influence of tectonics is strong in controlling the diachronous nature of the base of the section, and that sea level fluctuations throughout the interval were important in influencing sedimentation, as was the strong tidal pattern that is now so well known in the region. It is tempting to link some of these changes to those globally, but time differentiation is inadequate at present at some key times.

In the Early Pliocene in the Murray Basin (Martin, 1989), there was a brief resurgence of the *Nothofagus*-dominated rainforest with high, nonseasonal precipitation (the 'warm, wet phase' of Truswell (1990) in the Lake George region), and this gave way to wet sclerophyllous forest during the mid and Late Pliocene as the area began to take on its present dry phase. This change to drier conditions in the Murray Basin may have coincided with the change recorded by McEwen Mason (1991) in the Lake George region early in the Late Pliocene, when the modern regime became established. Lake Bungunnia developed (Brown, 1989b) between 2.5 and 0.7 Ma as the southern reaches of the Murray River became dammed because of structural movements (Stephenson & Brown, 1989). This dam was breached at the end of this interval and the Murray River assumed its present form.

Recent discoveries in the Pliocene of Antarctica (R. S. Hill and E. M. Truswell, personal communication) of angiosperm fossils suggest that Antarctica had a warm interval at the same time.

During the Late Pliocene, probably at about 2.4–2.6 Ma, the present polar glaciation pattern became established and this seems to be controlled by features of the earth's orbit around the sun – the Milankovitch hypothesis (Imbrie, 1984). This pattern consists of an approximately 100 ka cycle of increasing icecap volume, with associated sea level fall (about 130 m lower than present in the last two glacial maxima), followed by a collapse of the icecap (dominantly in the northern hemisphere) over some 10 ka. After an interval of stable icecap volume and sea level (perhaps about 5–10 ka; we seem to be in one now), the cycle begins again. Fink & Kukla (1977) identified 17 such cycles over the last 1.8 Ma.

Williams (in Veevers, 1984) summarised this history as far as it applies to Australia and referred to a near surface air temperature change of 5–10 °C between glacial and interglacial phases of each cycle, commonly over 10 ka. This rate of change is great compared with any other change throughout the Tertiary (and incidentally with any suggested change due to human-induced 'greenhouse' change). The pattern consists of alternation between the modern style environment in interglacial phases (e.g. over the last 10 ka), and a glacial phase climate (e.g. about 18 ka ago) when the continental arid zone expanded to its maximum, sea surface temperatures were lower in the south, west and northwest, windblown dust level increased in marine sediment, there was a much reduced incidence of tropical cyclones in northern Australia, and Australia was generally colder, windier and more arid. There was a higher snowfall in Tasmania and the Mt Kosciusko area and some significant glaciation there. Precipitation in the lee of the Tasmanian icefield was lower and the area shows evidence of being more arid. With the lower sea level, migration between presently low altitude areas via exposed continental shelves was enhanced and, during the last glacial maximum, Aboriginal people could traverse across a dry eastern Bass Strait between Victoria and Tasmania.

It is clear that the changes within Australia between glacial and interglacial times are very great but the detailed chronology and history is difficult to establish except in the most recent cycles. At present, geological studies have trouble differentiating time scale events this short, except for the last few hundred thousand years, and also correlating the events within each cycle consistently.

CONCLUSIONS

A gross understanding of the variation of Australian climate throughout the Cretaceous and Cenozoic has emerged over the last few decades but there are still very few data to document short-term variations in general, particularly from sections in central and northern Australia where such sections are very sporadic, poorly dated and otherwise little studied at present. Data exist from the offshore regions, particularly from oil exploration activities, but the sections have not yet been studied in sufficient depth to make use of all the information they contain.

It is clear that, during a significant part of the Cretaceous, global glaciation in its modern style is rare and occupies only a small proportion of geological time, even as an Antarctic phenomenon. The earth is today in one of the few glacial periods in its history, but within the modern glacial regime the globe is interglacial.

Early Cretaceous conditions in southeastern Australia appear to have been cool and humid with some seasonality. Winter sea ice was a feature but no major glaciation is obvious. Volcanism was widespread throughout eastern and southeastern Australia and minor in south-western Australia. There was a gradient of increasing but still cool temperature towards the northwest. The continent was characterised by considerable runoff and sediments were very dominantly of terrigenous material. The continent was subject to the greatest marine transgression of the Phanerozoic, with generally increasing sea coverage throughout this interval.

The Late Cretaceous saw a general retreat of marine conditions from central Australia and data for an interpretation of conditions there become very rare. The drainage divide moved north and runoff of most of Australia was generally to the south central coast with a significant volcanogenic content from the highlands of eastern Queensland feeding through the ancestral Darling River system. In southeastern Australia, cool conditions continued until the Santonian, when the area became much warmer and humid until the end of the Cretaceous. Along the west coast, sedi-

ments are dominantly of biogenic carbonates with fossils, indicating warm, and at times subtropical, waters. Drainage was local around the cratonic blocks in the region, but was relatively minor except in the north.

In the Paleocene–Early Eocene, southern Australia was a region of high rainfall, probably seasonal, and temperate climate, while along the northwest coast, there was a gradient from wetter warm temperate to subtropical conditions in the Carnarvon Basin to hotter, drier and even arid conditions with seasonal rainfall to the northeast. Central Australia was an area of internal drainage to the newly formed Birdsville Basin.

There is a need for a great deal of oxygen isotope analysis of onshore and offshore Cretaceous and Cenozoic carbonate sections from many areas around the Australian margin before Australian marine temperature variation can be understood to the extent it is in other parts of the world. The same can be said for palaeoceanography, the study of which is in its infancy in this country. The offshore seismic stratigraphy around the margin needs to be analysed using the same approach as that employed by Haq et al. (1987) and Vail et al. (1977) so that a better basis for comparison of Australian and global sea level variation can be assembled.

Antarctic sediments need to be searched for evidence of glacial conditions, particularly during the Early Cretaceous when seasonal sea ice appears to have been present in parts of Australia.

It is essential that every effort be made to increase the pool of data through scientifically led and controlled data-gathering programmes such as the ODP. Onshore equivalents are needed in areas where data are rarest. Reviews akin to that by Rich et al. (1989) need to be conducted for other areas and ages throughout Australia, particularly from areas poorly known at present.

A particularly pressing need is for more data in the Miocene, Pliocene and Quaternary, where the database is very poor. The sections that exist through this interval in hydrocarbon exploration wells and in foundation drilling for many engineering purposes should be studied in detail to document changes in vegetation, precipitation and temperature, and there should be a comprehensive, dedicated coring programme from geoscience research vessels in the marine environment. Onshore sections should be drilled and studied in detail for the same reasons. The last five million years of this interval is of interest to the International Geosphere Biosphere Program, the Intergovernmental Panel on Climate Change and the ODP.

Not only are samples of continuous sections needed, but dating systems need to be established in some cases so that ages can be more refined and correlation effected between samples. Some of these studies will need development of technologies that are not readily available yet in this country and also funding to implement them.

I thank Prof. Hill for the invitation to contribute to this compilation and Mr H. Hutchinson, Regional Director, Bureau of Meteorology, Tasmania, for discussion on the elements that contribute to climate. Dr D. Griffin, Director, The Australian Museum, Sydney kindly gave approval to use the palaeogeographical reconstructions originally published by Quilty (1982) as a basis for most of Figures 3.5–3.13. Dr H. G. Cogger of the same institution also helped in this regard. Drs M. Bradshaw and E. M. Truswell, Bureau of Mineral Resources, Geology and Geophysics, Canberra, helped with general information and that on the state of forthcoming palaeogeographical publications from that organisation.

REFERENCES

ALVAREZ, L. W., ALVAREZ, W., ASARO, F. & MICHEL, H. V. (1980). Extraterrestrial cause for the Cretaceous–Tertiary extinction. *Science*, **208**, 1095–108.

APTHORPE, M. C. (1979). Depositional history of the Upper Cretaceous of the northwest shelf, based upon foraminifera. *Journal of the Australian Petroleum Exploration Association*, **19**, 74–89.

APTHORPE, M. (1989). Cainozoic depositional history of the North West Shelf. In *The North West Shelf Australia: Proceedings of the Petroleum Exploration Society of Australia Symposium*, ed. P. G. Purcell & R. R. Purcell, pp. 55–84. Perth: Petroleum Exploration Society of Australia.

AUDLEY-CHARLES, M. G. (1987). Dispersal of Gondwanaland: relevance to evolution of the angiosperms. In *Biogeographical Evolution of the Malay Archipelago*, ed. T. C. Whitmore, pp. 5–25. Oxford: Clarendon Press.

BARRON, E. J., HARRISON, C. G. A., & HAY, W. W. (1978). A revised reconstruction of the southern continents. *EOS*, **59**, 436–49.

BARRON, E. J., THOMPSON, S. L. & SCHNEIDER, S. H. (1981). An ice-free Cretaceous? Results from climate model simulations. *Science*, **212**, 501–8.

BLACKBURN, D. T. (1981). Tertiary megafossil flora of Maslin Bay, South Australia: numerical taxonomic study of selected leaves. *Alcheringa*, **5**, 9–28.

BMR Palaeogeographic Group (1990). *Australia: Evolution of a Continent*. Canberra: Australian Government Publishing Service.

BOWEN, R. (1961). Paleotemperature analyses of Mesozoic Belemnoidea from Australia and New Guinea. *Bulletin of the Geological Society of America*, **72**, 769–74.

BRADSHAW, M. T., BURGER, D., YEUNG, M. & BEYNON, R. M., (1993). *Palaeogeographic Atlas of Australia*. Canberra: Bureau of Mineral Resources, Geology and Geophysics.

BRADSHAW, M. T., YEATES, A. N., BEYNON, R. M., BRAKEL, A. T., LANGFORD, R. P., TOTTERDELL, J. M. & YEUNG, M. (1988). Palaeogeographic evolution of the North West Shelf region. In *The North West Shelf, Australia: Proceedings of the Petroleum Exploration Society of Australia Symposium, Perth 1988*, ed. P. G. Purcell & R. R. Purcell, pp. 29–54. Perth: Petroleum Exploration Society of Australia.

BROWN, B. R. (1986). Offshore Gippsland Silver Jubilee. In *Second South-eastern Australia Oil Exploration Symposium*, ed. R. C. Glenie, pp. 29–56. Melbourne: Petroleum Exploration Society of Australia.

BROWN, C. M. (compiler) (1989a). Papers from Murray Basin 88. *BMR Journal of Australian Geology and Geophysics*, **11**, 127–395.

BROWN, C. M. (1989b). Structural and stratigraphic framework of groundwater occurrence and surface discharge in the Murray Basin, southeastern Australia. *BMR Journal of Australian Geology and Geophysics*, **11**, 127–46.

BURGER, D. (1989). Australian Phanerozoic timescales. 9. Cretaceous. *Record, Bureau of Mineral Resources, Geology and Geophysics, Canberra*, 1989/39, 1–35.

BURNS, D. A. (1977). Major features of oceanographic development of the southeast Indian and southwest Pacific Oceans interpreted from microfossil evidence. *Marine Geology*, **25**, 35–59.

CALLEN, R. A. (1977). Late Cainozoic environments of part of northeastern South Australia. *Journal of the Geological Society of Australia*, **24**, 151–69.

CANDE, S. C. & MUTTER, J. C. (1982). A revised identification of the oldest sea floor spreading anomalies between Australia and Antarctica. *Earth and Planetary Science Letters*, **58**, 151–60.

CHAPRONIERE, G. C. H. (1984). Oligocene and Miocene larger Foraminiferida from Australia and New Zealand. *Bulletin of the Bureau of Mineral Resources, Geology and Geophysics*, **188**, 1–98.

CLAYTON, R. N. & STEVENS, G.R. (1967). Palaeo-temperatures of New Zealand Jurassic and Cretaceous. *Tuatara*, **16**, 3–7.

COCKBAIN, A. E. (1967). *Asterocyclina* from the Plantagenet Beds near Esperance, W. A. *Australian Journal of Science*, **30**, 68.

DAVIES, P. J., SYMONDS, P. A., FEARY, D. A. & PIGRAM, C. J. (1991). The evolution of the carbonate platforms of northeast Australia. In *The Cainozoic in Australia: A Re-appraisal of the Evidence*, ed. M. A. J. Williams, P. De Deckker & A. P. Kershaw, *Geological Society of Australia Special Publication*, **18**, 44–78

DEIGHTON, I., FALVEY, D. A. & TAYLOR, D. J. (1976). Depositional environments and geotectonic framework: southern Australia continental margin. *Journal of the Australian Petroleum Exploration Association*, **16**, 25–35.

DETTMANN, M. E. (1981). The Cretaceous flora. In *Ecological Biogeography of Australia*, ed. A. Keast, pp. 355–75. The Hague: Junk.

DORMAN, F. H. (1966). Australian Tertiary paleotemperatures. *Journal of Geology*, **74**, 49–61.

DORMAN, F. H. (1968). Some Australian oxygen isotope temperatures and a theory for a 30 million year world-temperature cycle. *Journal of Geology*, **76**, 297–313.

DORMAN, F. H. & GILL, E. D. (1959). Oxygen isotope palaeotemperature measurements on Australian fossils. *Proceedings of the Royal Society of Victoria*, **71**, 73–98.

DOUGLAS, J. G. (1969). The Mesozoic floras of Victoria, Parts 1 and 2. *Memoirs of the Geological Survey of Victoria*, **28**.

DOUGLAS, J. G., ABELE, C., BENEDEK, S., DETTMANN, M. E., KENLEY, P. R. & LAWRENCE, C. R. (1976). Mesozoic. In *Geology of Victoria*, ed. J. G. Douglas & J. A. Ferguson, pp. 143–76. Geological Society of Australia.

EDWARDS, A. R. (1975). Southwest Pacific Cenozoic paleoceanography and an integrated Neogene paleocirculation model. In *Initial Reports of the Deep Sea Drilling Project*, ed. J. E. Andrews, G. Packham & J. Herring vol. 30, pp. 667–84. Washington: US Government Printing Office.

EMBLETON, B. J. J. & McELHINNY, M. W. (1982). Marine magnetic anomalies, palaeomagnetism and the drift history of Gondwanaland. *Earth and Planetary Science Letters*, **58**, 141–50.

EMILIANI, C., KRAUS, E. B. & SHOEMAKER, E. M. (1981). Sudden death at the end of the Mesozoic. *Earth and Planetary Science Letters*, **55**, 317–34.

FINK, J. & KUKLA, G. J. (1977). Pleistocene climates in Central Europe: at least 17 interglacials after the Olduvai Event. *Quaternary Research*, **7**, 363–71.

FRAKES, L. A., BURGER, D., APTHORPE, M., WISEMAN, J., DETTMANN, M., ALLEY, N., FLINT, R., GRAVESTOCK, D., LUDBROOK, N., BACKHOUSE J., SKWARKO, S., SCHEIBNEROVA, V.,

McMINN, A., MOORE, P. S., BOLTON, B. R., DOUGLAS, J. G., CHRIST, R., WADE, M., MOLNAR, R. E., McGOWRAN, B., BALME, B. E. & DAY, R. A. (1987*a*). Australian Cretaceous shorelines, stage by stage. *Palaeogeography, Palaeoclimatology, Palaeoecology*, **59**, 31–48.

FRAKES, L. A. & FRANCIS, J. E. (1988). A guide to Phanerozoic cold polar climates from high-latitude ice-rafting in the Cretaceous. *Nature*, **333**, 547–9.

FRAKES, L. A., McGOWRAN, B. & BOWLER, J. M. (1987*b*). Evolution of Australian environments. In *Fauna of Australia. General Articles*, ed. G. R. Dyne & D. W. Walton, pp. 1–16. Canberra: Australian Government Printer.

FRAKES, L. A. & VICKERS-RICH, P. V. (1991). Palaeoclimatic setting and palaeogeographic links of Australia in the Phanerozoic. In *Vertebrate Palaeontology of Australasia*, ed. P. Vickers-Rich, J. M. Monaghan, R. F. Baird & T. H. Rich, pp. 111–45. Melbourne: Pioneer Design Studios in Cooperation with Monash University Publications Committee.

GALLOWAY, R. W. & KEMP, E. M. (1981). Late Cainozoic environments of Australia. In *Ecological Biogeography of Australia*, ed. A. Keast, pp. 53–80. The Hague: Junk.

GLAESSNER, M. F. (1960). Upper Cretaceous larger foraminifera from New Guinea. *Special Volume, Tohoku University Scientific Reports, Second Series (Geology)*, **4**, 37–44.

GRANT-MACKIE, J. A. (1979). Cretaceous–Recent plate tectonic history and paleoceanographic development of the southern hemisphere. *Proceedings of the International Symposium on Marine Biogeography and Evolution in the Southern Hemisphere*, **1**, 27–42.

GRIFFITHS, J. R. (1971). Reconstruction of the southwest Pacific margin of Gondwanaland. *Nature*, **234**, 203–7.

HALLAM, A. (1984). Continental humid and arid zones during the Jurassic and Cretaceous. *Palaeogeography, Palaeoclimatology, Palaeoecology*, **47**, 195–223.

HALLAM, A. (1985). A review of Mesozoic climates. *Journal of the Geological Society*, **142**, 433–45.

HAQ, B. U. (1981). Paleogene oceanography: Early Cenozoic oceans revisited. *Oceanologia Acta*, No. SP, pp. 71–82.

HAQ, B. U., HARDENBOL, J. & VAIL, P. R. (1987). Chronology of fluctuating sea levels since the Triassic. *Science*, **235**, 1156–67.

HARLAND, W. B., COX, A. V., LLEWELLYN, P. G., PICKTON, C. A. G., SMITH, A. G. & WALTERS, R. (1982). *A Geologic Time Scale*. Cambridge: Cambridge University Press.

HARRIS, W. K. (1965). Tertiary microfloras from Brisbane, Queensland. *Report of the Geological Survey of Queensland*, **10**.

HARRIS, W. K. (1971). Tertiary stratigraphic palynology, Otway Basin. *Special Bulletin, Geological Surveys of South Australia and Victoria*, pp. 67–87.

HAYES, D. E. & RINGIS, J. (1973). Seafloor spreading in the Tasman Sea. *Nature*, 243, 454–8.

IMBRIE, J. (ed.) (1984). *Milankovitch and Climate*. NATO ASI series, 2 vols.

JELL, P. A. & DUNCAN, P. M. (1986). Invertebrates, mainly insects, from the freshwater, Lower Cretaceous, Koonwarra Fossil Bed (Korumburra Group), South Gippsland, Victoria. *Memoirs, Association of Australasian Palaeontologists*, 3, 111–205.

JOHNSON, R. W. (1989). *Intraplate Volcanism in Eastern Australia and New Zealand*. Cambridge: Cambridge University Press.

JOHNSTONE, M. H., LOWRY, D. C. & QUILTY, P. G. (1973). The geology of southwestern Australia: a review. *Journal of the Royal Society of Western Australia*, 56, 5–15.

KEMP, E. M. (1978). Tertiary climatic evolution and vegetation history in the Southeast Indian Ocean region. *Palaeogeography, Palaeoclimatology, Palaeoecology*, 24, 169–208.

KEMP, E. M. (1981). Tertiary palaeogeography and the evolution of Australian climate. In *Ecological Biogeography of Australia*, ed. A. Keast, pp. 33–49. The Hague: Junk.

KENNETT, J. P. (1980). Paleoceanographic and biogeographic evolution of the Southern Ocean during the Cenozoic, and Cenozoic microfossil datums. *Palaeogeography, Palaeoclimatology, Palaeoecology*, 31, 123–52.

KENNETT, J. P. & BARKER, P. F. (1990). Latest Cretaceous to Cenozoic climate and oceanographic developments in the Weddell Sea, Antarctica: an ocean-drilling perspective. *Proceedings of the Ocean Drilling Program, Scientific Results*, 113, 937–60.

KI HONG CHANG (1975). Unconformity-bounded stratigraphic units. *Bulletin of the Geological Society of America*, 86, 1544–52.

LANGE, R. T. (1982). Australian Tertiary vegetation: evidence and interpretation. In *A History of Australasian Vegetation*, ed. J. M. B. Smith, pp. 44–89. New York: McGraw-Hill.

LUDBROOK, N. H. (1961). Stratigraphy of the Murray Basin in South Australia. *Bulletin, Geological Survey of South Australia*, 36, 1–95.

LUDBROOK, N. H. (1966). Cretaceous biostratigraphy of the Great Artesian Basin in South Australia. *Bulletin of the Geological Survey of South Australia*, 40.

MACPHAIL, M. K., HILL, R. S., FORSYTH, S. M. & WELLS, P. M. (1991). A Late Oligocene–Early Miocene cool climate flora in Tasmania. *Alcheringa*, 15, 87–106.

MACPHAIL, M. K. & TRUSWELL, E. M. (1989). Palynostratigraphy of the central west Murray Basin. *BMR Journal of Australian Geology and Geophysics*, 11, 301–31.

McDOUGALL, I., & DUNCAN, R. A. (1988). Age progressive volcanism in the Tasmantid Seamounts. *Earth and Planetary Science Letters*, 89, 207–20.

McEWEN MASON, J. R. C. (1991). The late Cainozoic magnetostratigraphy and preliminary palynology of Lake George, New South Wales. In *The Cainozoic in Australia: A Re-appraisal of the Evidence*, ed. M. A. J. Wiliams, P. De Deckker & A. P. Kershaw. *Geological Society of Australia Special Publication*, 18, 195–209.

McGOWRAN, B. (1979). The Tertiary of Australia: foraminiferal overview. *Marine Micropaleontology*, 4, 235–64.

McGOWRAN, B. (1989). The later Eocene transgressions in southern Australia. *Alcheringa*, 13, 45–68.

McGOWRAN, B. & BEECROFT, A. (1985). *Guembelitria* in the Early Tertiary of southern Australia and its palaeoceanographic significance. In *Stratigraphy, Palaeontology, Malacology Papers in Honour of Dr Nell Ludbrook*, ed. J. M. Lindsay. *Special Publication South Australia Department of Mines and Energy*, 5, 247–61.

McNAMARA, K. J., REXILIUS, J. P., MARSHALL, N. G. & HENDERSON, R. (1988). The first record of a Maastrichtian ammonite from the Perth Basin, Western Australia. *Alcheringa*, 12, 163–8.

MARTIN, H. A. (1989). Vegetation and climate of the late Cainozoic in the Murray Basin and their bearing on the salinity problem. *BMR Journal of Australian Geology and Geophysics*, 11, 291–9.

MUTTER, J. C., HEGARTY, K. A., CANDE, S. C. & WEISSEL, J. K. (1985). Breakup between Australia and Antarctica: a brief review in the light of new data. *Tectonophysics*, 114, 255–79.

OBERHÄNSLI, H. (1986). Latest Cretaceous–Early Neogene oxygen and carbon isotopic record at DSDP sites in the Indian Ocean. *Marine Micropaleontology*, 10, 91–115.

PITTOCK, A. B., FRAKES, L. A., JENSSEN, D., PETERSON, J. A. & ZILLMAN, J. W. (1978). *Climatic Change and Variability. A Southern Perspective*. Cambridge: Cambridge University Press.

QUILTY, P. G. (1972). The biostratigraphy of the Tasmanian marine Tertiary. *Papers and Proceedings of the Royal Society of Tasmania*, 106, 25–44.

QUILTY, P. G. (1974a). Cainozoic stratigraphy in the Perth area. *Journal of the Royal Society of Western Australia*, 57, 16–31.

QUILTY, P. G. (1974b). Tertiary stratigraphy of Western Australia. *Journal of the Geological Society of Australia*, 21, 301–18.

QUILTY, P. G. (1975). Late Jurassic to Recent geology of the western margin of Australia. In *Deep Sea Drilling in Australasian Waters*, ed. J. J. Veevers, pp. 15–17, 20–23. Sydney: Challenger Symposium.

QUILTY, P. G. (1977). Cenozoic sedimentation cycles in Western Australia. *Geology*, 5, 336–40.

QUILTY, P. G. (1978). The Late Cretaceous–Tertiary section in Challenger No.1 (Perth Basin): details and implications. *Bulletin of the Bureau of Mineral Resources, Geology and Geophysics, Canberra*, 192, 109–34.

QUILTY, P. G. (1980). Sedimentation cycles in the Cretaceous and Cenozoic of Western Australia. *Tectonophysics*, **63**, 349–66.

QUILTY, P. G. (1982). Mesozoic and Cenozoic history of Australia as it affects the Australian biota. In *Arid Australia*, ed. H. G. Cogger & E. E. Cameron, pp. 7–56. Sydney: Australian Museum.

QUILTY, P. G. (1993). Tasmantid and Lord Howe Seamounts: biostratigraphy and palaeoceanographic significance. *Alcheringa*, **17**, 27–53.

RICH, T. H., RICH, P. V., WAGSTAFF, B., McEWEN MASON, J., DOUTHITT, C. B. & GREGORY, R. T. (1989). Early Cretaceous biota from the northern side of the Australo-Antarctic rift Valley. In *Origins and Evolution of the Antarctic Biota*, ed. J. A. Crame, pp. 121–30. London: Geological Society.

RUDWICK, M. J. S. (1972). *The Meaning of Fossils*. London: Macdonald.

SAVIN, S. M. (1977). The history of the Earth's surface temperature during the past 100 million years. *Annual Review of Earth and Planetary Science*, **5**, 319–55.

SAVIN, S. M., DOUGLAS, R. G. & STEHLI, F. G. (1975). Tertiary marine paleotemperatures. *Bulletin of the Geological Society of America*, **86**, 1499–1510.

SCHEIBNEROVA, V. (1976). Cretaceous foraminifera of the Great Artesian Basin. *Memoir of the Geological Survey of New South Wales*, **17**, 1–265.

SHACKLETON, N. J. & KENNETT, J. P. (1975). Paleotemperature history of the Cenozoic and the initiation of Antarctic glaciation: oxygen and carbon isotope analyses in DSDP sites 277, 279, and 281. *Initial Reports of the Deep Sea Drilling Project*, **29**, 743–55.

SHAFIK, S. (1991). Upper Cretaceous and Tertiary stratigraphy of the Fremantle Canyon, South Perth Basin: a nannofossil assessment. *BMR Journal of Australian Geology and Geophysics*, **12**, 65–91.

SHEARD, M. J. (1990). Glendonites from the southern Eromanga Basin in South Australia: palaeoclimatic indicators for Cretaceous ice. *Quarterly Geological Notes, Geological Survey of South Australia*, **114**, 17–23.

SINGH, G., OPDYKE, N. D. & BOWLER, J. M. (1981). Late Cainozoic stratigraphy, palaeomagnetic chronology and vegetational history from Lake George, NSW. *Journal of the Geological Society of Australia*, **28**, 435–52.

SLITER, W. V. (1976). Cretaceous foraminifers from the southwest Atlantic Ocean, Leg 36, Deep Sea Drilling Project. *Initial Reports of the Deep Sea Drilling Project*, **36**, 519–73.

SLOSS, L. L. (1963). Sequences in the cratonic interior of North America. *Bulletin of the Geological Society of America*, **74**, 93–113.

SLUITER, I. R. K. (1991). Early Tertiary vegetation and climates, Lake Eyre region, northeastern South Australia. In *The Cainozoic in Australia: A Re-appraisal of the Evidence*, ed. M. A. J. Williams, P. De Deckker & A. P. Kershaw. *Geological Society of Australia Special Publication*, **18**, 99–118.

STEPHENSON, A. E. & BROWN, C. M. (1989). The ancient Murray River system. *BMR Journal of Australian Geology and Geophysics*, **11**, 387–95.

STEVENS, G. R. & CLAYTON, R. N. (1971). Oxygen isotope studies on Jurassic and Cretaceous belemnites from New Zealand and their biogeographic significance. *New Zealand Journal of Geology and Geophysics*, **14**, 829–97.

STIRTON, R. A., TEDFORD, R. H. & WOODBURNE, M. O. (1967). A new Tertiary formation and fauna from the Tirari Desert, South Australia. *Records of the South Australian Museum*, **15**, 427–62.

STOVER, L. E. & PARTRIDGE, A. D. (1973). Tertiary and Late Cretaceous spores and pollen from the Gippsland Basin, southeastern Australia. *Proceedings of the Royal Society of Victoria*, **85**, 237–86.

SUTHERLAND, F. L. & CORBETT, K. D. (1974). The extent of Upper Mesozoic igneous activity in relation to lamprophyric intrusions in Tasmania. *Papers and Proceedings of the Royal Society of Tasmania*, **107**, 175–90.

TRUSWELL, E. M. (1990). Australian rainforests: the 100 million year record. In *Australian Tropical Rainforests: Science–Value–Meaning*, ed. L. J. Webb & J. Kikkawa, pp. 7–22. Melbourne: CSIRO.

TRUSWELL, E. M., CHAPRONIERE, G. C. H. & SHAFIK, S. (1989). Australian Phanerozoic timescales. 10. Cenozoic. *Record, Bureau of Mineral Resources, Geology and Geophysics*, **1989/40**, 1–16.

TRUSWELL, E. M. & HARRIS, W. K. (1982). The Cainozoic palaeobotanical record in arid Australia: fossil evidence for the origins of an arid-adapted flora. In *Evolution of the Flora and Fauna of Arid Australia*, ed. W. R. Barker & P. J. M. Greenslade, pp. 67–76. Frewville: Peacock Press.

TRUSWELL, E. M., KERSHAW, A. P. & SLUITER, I. R. (1987). Flowering plant origin and dispersal: evidence from the palaeobotanical record. In *Biogeographical Evolution of the Malay Archipelago*, ed. T. C. Whitmore, pp. 32–49. Oxford: Clarendon Press.

VAIL, P. R., MITCHUM, R. M. Jr & THOMPSON, S. III (1977). Seismic stratigraphy and global changes of sea level. *Memoir of the American Association of Petroleum Geologists*, **26**, 63–81.

VEEVERS, J. J. (ed.) (1984). *Phanerozoic Earth History of Australia*. Oxford: Clarendon Press.

VEEVERS, J. J., POWELL, C. McA. & ROOTS, S. R. (1991). Review of seafloor spreading around Australia. 1. Synthesis of the patterns of spreading. *Australian Journal of Earth Sciences*, **38**, 373–89.

VICKERS-RICH, P., MONAGHAN, J. M., BAIRD, R. F. & RICH, T. H. (1991). *Vertebrate Palaeontology of Australasia*. Lilydale: Pioneer Design Studios.

WALDMAN, M. (1971). Fish from the Lower Cretaceous of Victoria, Australia, with comments on the palaeoenvironment. *Special Papers in Palaeontology*, **9**, 1–124.

WEISSEL, J. K., HAYES, D. E. & HERRON, E. M. (1977). Plate tectonics synthesis: the displacements between Australia, New Zealand, and Antarctica since the Late Cretaceous. *Marine Geology*, **25**, 231–77.

WILLIAMS, M. A. J. (1984). Quaternary environments. In *Phanerozoic Earth History of Australia*, ed. J. J. Veevers, pp. 42–7. Oxford: Clarendon Press.

WILLIAMS, M. A. J., DE DECKKER, P. & KERSHAW, A. P. (1991). *The Cainozoic in Australia: a Re-*appraisal of the Evidence. *Geological Society of Australia Special Publication*, **18**.

WOPFNER, H., CALLEN, R. & HARRIS, W. K. (1974). The Lower Tertiary Eyre Formation of the southwestern Great Artesian Basin. *Journal of the Geological Society of Australia*, **21**, 17–51.

4 Palaeobotanical evidence for Tertiary climates

D. R. GREENWOOD

Climate is the primary framework within which plant populations grow, reproduce and, in the longer time frame, evolve. However, reconstructions of palaeoclimate are often based on the marine record (e.g. Quilty, 1984 and Chapter 3, this volume) or on computer modelling (Sloan & Barron, 1990). The latter type, in particular, suffer from their dependence on simplified scenarios and must be tested against the terrestrial palaeontological record. Individual species and whole plant communities are morphologically and physiologically adapted to their physical environment, most strongly to climate, and so plant macrofossils are a proxy record of past climates (Wolfe, 1971, 1985; Upchurch & Wolfe, 1987). Several studies have indicated that plant macrofossils provide a potentially accurate record of Tertiary terrestrial palaeotemperatures (Wolfe, 1979, 1990; Read *et al.*, 1990). Interpretation of the climatic signal preserved in plant fossil assemblages is dependent on understanding (1) how plants (and vegetation) interact with climate, (2) how plant fossil assemblages were formed, and (3) how these assemblages relate to the original vegetation.

The use of terms such as tropical, subtropical and temperate in palaeoclimatic discussions is potentially confusing, as these terms are rarely defined climatically and have geographical (latitudinal) connotations that are inapplicable for much of the Tertiary. Following Wolfe (1979,

1985; Upchurch & Wolfe, 1987), the terms megathermal, mesothermal and microthermal are used here to describe the temperature characteristics of the vegetation of the main climatic zones. The definitions of Nix (1982; Kershaw & Nix, 1988) are used here,[*] although Wolfe's (1979, 1985) vegetation classification uses slightly different definitions.

This discussion is restricted to the Tertiary, since pre-Tertiary floras are dominated by plant groups with no or few modern analogues and so palaeobotanical indices of climate are (as yet) unreliable. Numerous accounts of the Tertiary climate history of Australia based on palaeobotanical evidence have been published (Kemp, 1978, 1981; Christophel, 1981, 1988; Nix, 1982; Truswell & Harris, 1982; Christophel & Greenwood, 1989), in some cases as part of global accounts (Axelrod, 1984; Wolfe, 1985). Many of these interpretations have been based solely on nearest-living-relative analogy (NLR) and provide qualitative reconstructions of climate.

Many North American studies have used a strong correlation between the proportion of woody plant species with entire leaf margins and mean annual temperature (MAT) in modern mesic forests (leaf margin analysis, Wolfe, 1979, 1990), and other aspects of foliar morphology, to

---

[*] Modified from Nix (1982): megathermal, > 24 °C (MAT); meso-megathermal interzone, 20–24 °C; mesothermal, 14–20 °C (14–24 °C if the interzone is included); microthermal, < 14 °C.

reconstruct both the regional distribution of physiognomically defined vegetation types and palaeoclimates for the Late Cretaceous and Tertiary (e.g. Wolfe, 1985; Upchurch & Wolfe, 1987; Parrish & Spicer, 1988). Wolfe (1971, 1979, 1985) has argued that foliar physiognomic approaches are preferred to NLR methodologies, primarily due to inaccurate systematic treatments for many Tertiary floras. However, foliar physiognomic analysis is used here according to the methodology proposed by Greenwood (1991a; Christophel & Greenwood, 1988, 1989), in combination with recent quantitative analyses of key NLRs, of Tertiary taxa (Nix, 1982; Kershaw & Nix, 1988; Read & Hope, 1990; Read et al., 1990) to interpret early Tertiary macrofloras from Australia (Figure 4.1). Wood anatomy (Frakes &

Francis, 1990) and epiphyllous fungi found on leaf cuticle (Lange, 1978, 1981, 1982) are other sources of climatic information from plant fossil assemblages. These approaches provide quantitative estimates for critical components of climate that can be tested against estimates from the marine record and climatic models.

THE NATURE OF THE PLANT MACROFOSSIL RECORD

Leaf accumulations in streams or lakes constitute a biassed, but nevertheless detailed, record of the parent vegetation. This bias results from a number of processes acting on the plant parts (primarily leaves) from the time of abscission or traumatic loss to their eventual entombment in sediment and lithification (Ferguson, 1985; Spicer, 1989; Greenwood, 1991a):

1. Deciduous and evergreen trees may have different representation in an assemblage if leaf-fall is asynchronous with periods of maximum sedimentation.
2. Different plant species may have leaves that are dehiscent or nondehiscent (e.g. many palms, herbs and forbs). Nondehiscent leaves are less likely to be fossilised than are dehiscent leaves.
3. Plant organs may have varying capacities for transport from the plant body. For example, leaves with a low weight per unit area tend to travel further in air than do denser leaves.
4. Leaves of individual species (e.g. sun versus shade leaves) and different species decay at different rates.
5. The varying productivity of the plants is significant, with canopy trees producing copious amounts of leaves, swamping the litter-fall.

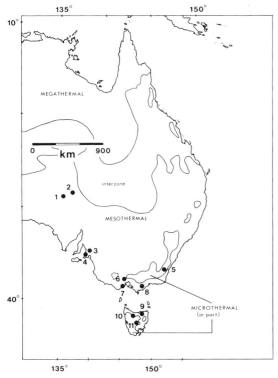

Figure 4.1 Location map of Australian Tertiary macrofloras discussed in the text, and modern thermal regimes (adapted from Nix, 1982; Christophel & Greenwood, 1989): 1, Stuart Creek (silcrete floras); 2, Nelly Creek; 3, Golden Grove; 4, Maslin Bay; 5, Nerriga; 6, Bacchus Marsh; 7, Anglesea; 8, Latrobe Valley (Yallourn & Morwell); 9, Pioneer; 10, Cethana; 11, Monpeelyata.

Fossil leaf accumulations therefore reflect only a subset of the original surrounding vegetation, and in most situations represent only the local plant communities (Greenwood, 1991a). Furthermore, smaller plant macrofossil accumulations may represent geologically instantaneous records of past vegetation. Individual sedimentary facies often reflect different components of the local vegetational mosaic (see e.g. Christophel et al.,

Table 4.1. *Foliar physiognomic characteristics of leaf-litter from modern Australian rainforest*

A. Mean % leaves

Forest type[a]	Leaf size			% entire leaf margins
	Micro-	Noto-	Mesophylls	
CMVF	6–25	50–70	11–39	90.0–98.3
CNVF	16–55	35–77	2–22	34.3–57.0
SNVF	59–85	14–36	0–11	87.0–95.3 (Qld)
				7.0–37.0 (NSW)
MV-FF[b]	74–80	19–24	1–2	56.7–96.4
MFF[b]	90–96	4–10	0	1.0–13.0
NMF[c]	96–99	1–4	0	not available

B. % taxa

Forest type[a]	Leaf size			LSI	% entire leaf margins
	Micro-	Noto-	Mesophylls		
CMVF	11–18	58–78	11–30	47–59	82.4–95.0
CNVF	30–46	14–63	0–36	33–43	50.0–81.8
SNVF	40–75	25–50	0–20	13–38	84.2–90.0 (Qld)
					14.3–25.0 (NSW)
MFF	100	0	0	0	33.3–40.0

[a]CMVF, complex mesophyll vine forest; CNVF, complex notophyll vine forest; SNVF, simple notophyll vine forest; MV–FF, microphyll vine–fern forest; MFF, microphyll fern forest; LSI, leaf size index (from Webb, 1959; Tracey, 1982).
[b]Based on a single site (four samples).
[c]Based on Macphail *et al.*, 1991, Table 3 (Lake Dobson and Precipitous Bluff).
Data from Greenwood, 1991*a* (Table 1, in part) and 1992.

1987). However, stratigraphically correlated macrofloras may be separated in time (as a record of the original vegetation, and hence climate) by tens of thousands of years.

Palaeobotanical evidence of climate

Terrestrial plant fossil assemblages record local climate usually over small time intervals. Each plant assemblage represents a single datum of climate, with separate assemblages from a single flora providing information about local variation. However, continuous deposition of terrigenous sediments within a single basin (e.g. the Latrobe Valley) preserve climate records spanning millions of years.

The present analysis uses a dataset derived from modern Australian rainforest leaf-litter and fluvially deposited beds (Christophel & Greenwood, 1987, 1988, 1989; Greenwood, 1992; Table 4.1). A multivariate foliar physiognomic analysis for the Australian Middle Eocene macrofloras is in preparation (D. R. Greenwood, unpublished data), however, here temperature characteristics are derived primarily from univariate comparisons (e.g. Figure 4.2). Other Australian Tertiary macrofloras are discussed qualitatively using leaf size and margin data only, as full data are not available. These analyses are supplemented by quantitative and qualitative estimates based on NLR analyses from Read *et al.* (1990; Read, 1990) and Nix (1982; Kershaw &

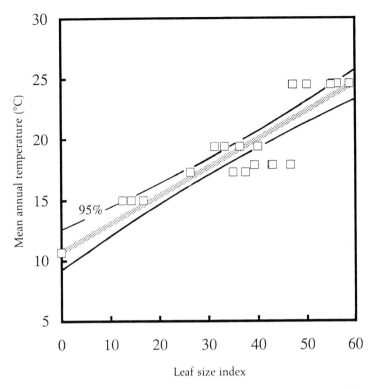

Figure 4.2 Relationship (least-squares regression) between leaf size index of forest-floor leaf-litter from Australian rainforests and mean annual temperature (MAT), $r = 0.94$. (Adapted from Greenwood, 1992.)

Nix, 1988). Forest nomenclature and typology in this discussion follows those of Webb (1959, 1968), Tracey (1982; Table 4.1) and Wolfe (1979).

Foliar physiognomy and climate

Greenwood (1992; Greenwood & Christophel, 1990, unpublished data) suggests that the relationship between leaf margin type and MAT for Australian forests is different from, and more complex than, that derived by Wolfe (1979) for east Asian forests. Leaf size is strongly correlated with MAT in Australian rainforest canopies (Webb, 1968), and, since the leaf size bias between the canopy and forest floor litter is consistent (Greenwood, 1991a), modern leaf-litter from humid vegetation shows a strong relationship between leaf size, calculated as the leaf size index

(LSI = (% microphylls + 2% notophylls + 3% mesophylls − 100)/2; Wolfe, 1978; Burnham, 1989), and MAT (Greenwood, 1991a, 1992; Figure 4.2). Mean leaf length (specimens) is also strongly correlated with MAT (Greenwood, 1992).

The LSI of the Tertiary macrofloras (where applicable) has been used to calculate the prevailing MAT using a linear regression equation from modern litter (Figure 4.2). These MAT values must be considered to be minimum estimates; the actual values are likely to be 1–3 °C higher, as the taphonomic size bias is always towards smaller leaves than are found in the parent vegetation. The spread of MAT values in Figure 4.2 indicates the uncertainty of this magnitude. New data from the Australian rainforest litter are also used here to calculate MAT using leaf margin analysis. The use of separate foliar physiognomic indices to calculate MAT provides a check on the fidelity of the estimates.

AUSTRALIAN TERTIARY MACROFLORAS AND PALAEOCLIMATES

This discussion is restricted to a small group of macrofloras and is not a general review of the Australian Tertiary climate. The macrofossil sites (Figure 4.1) represent a few key points in time: Middle–Late Eocene (Anglesea, Maslin Bay, southern Lake Eyre Basin silcrete and Nelly Creek, Nerriga), Oligocene (Cethana), Late Oligocene–Early Miocene (Pioneer, Monpeely-ata) and the Early–Middle Miocene (Bacchus Marsh, Latrobe Valley), and can be considered to be individual 'snapshots', where climate can be inferred for local areas. Late Tertiary macrofloras are scarce in Australia, and, apart from Bacchus Marsh and the Latrobe Valley, are not considered here.

Middle Eocene

Axelrod (1984) concluded that thermal regimes in southern Australia in the Late Paleocene–Eocene were comparable to the present regime. Nix (1982) depicted a latitudinal zonation of thermal regimes across Australia in the Middle Eocene similar to that of the present, although with MAT values in southern Australia '2°–4 °C warmer than at present'. Wolfe (1985), however, suggested a more southerly latitudinal extension of tropical rainforest (megathermal), with para-tropical rainforest (mesothermal–megathermal interzone *sensu* Nix (1982)), in the southern half of Australia, and notophyllous forest (mesothermal) on the adjacent part of Antarctica, in contrast to Nix's (1982) suggestion of microthermal climates in southernmost Australia and Antarctica. Early–Late Eocene macrofloras from the Antarctic Peninsula suggest that either cool temperate *Nothofagus* and podocarp rainforest (nanophyll mossy forest, NMF) or deciduous forests were dominant over much of Antarctica at that time (Tokarski *et al.*, 1987; Case, 1988).

A common element in Nix's (1982), Axelrod's (1984) and Wolfe's (1985) interpretations is for uniformly low seasonality (equable) climates in the Eocene. More recently, Sloan & Barron (1990) have suggested that the Eocene mean annual range of temperature (MART) in continental interiors should have been high. The new synthesis presented here differs from Axelrod's (1984) and Wolfe's (1985) interpretations in some important details, and provides a quantitative comparison with the estimations of Nix (1982) and of Sloan & Barron (1990).

The Middle Eocene Maslin Bay (North Maslin Sands; McGowran *et al.*, 1970), Anglesea (Eastern View Formation; Abeles *et al.*, 1976) and Golden Grove (North Maslin Sands: Alley, 1987) macrofloras (Figure 4.1) occur in mudstone lenses within fluvial sands (Christophel & Blackburn, 1978; Christophel & Greenwood, 1987, 1989; Christophel *et al.*, 1987). Each clay lens is interpreted as an infilled abandoned channel (oxbow) in a meandering stream sequence. A diverse Middle Eocene macroflora also occurs in well-laminated lake sediments (Titringo Siltstone) at Nerriga (Hill, 1982). These macrofloras provide a record of vegetation at near sea level (Maslin Bay and Anglesea) and at moderate elevations (Golden Grove and Nerriga) for the continental margin from which climate can be inferred (see Table 4.4, below). Eocene macrofloras from the Lake Eyre Basin provide information on vegetation and climate in the continental interior (Lange, 1982; Greenwood *et al.*, 1990; Christophel *et al.*, 1992).

The continental margin Middle Eocene macrofloras have many taxa in common at the generic level (Christophel & Greenwood, 1989) and are typified by abundant leaf remains of Lauraceae. A distinctive feature of the Nerriga macroflora compared to other Australian Eocene macrofloras is the common presence of hastate vine leaves of Menispermaceae (Hill, 1989). This may reflect disturbance in the local vegetation as vines are commonly enriched in modern tropical rainforest canopies after canopy disruption (Tracey, 1982), or it may reflect vine-rich lake-edge vegetation. Many of the identified common plants from these Eocene macrofloras have NLRs with

Table 4.2. *The foliar physiognomic characteristics of the macrofloras (specimen-based observations)*

Site[a]	No. of leaves[b]	% entire leaf margins	Leaf size (%)			Mean length (mm)	No. of taxa[c]
			Micro-	Noto-	Mesophylls		
Middle Eocene							
Anglesea							
(1)	81	73.0	32.0	46.0	22.0	—	41
(2)	1548	79.7	82.9	16.1	0.9	49	*ca* 35
Maslin Bay							
(1)	1046	84.0	25.0	54.0	21.0	—	57
(2)	90	76.7	43.2	41.1	15.7	88	—
(3)	210	70.0	24.0	53.0	23.0	?100	80
Golden Grove							
(6)	338	47.4	52.6	40.9	6.5	77	*ca* 30
Nerriga							
(1)	129	66.0	24.0	55.0	21.0	—	38
(4)	112	44.1	22.7	59.2	18.1	?102	25
	576	39.6	(including fragmented leaves)				
Deane's Marsh	—	—	100.0	0.0	0.0	—	—
Nelly Creek (5)	160	88.0	80.0	20.0	0.0	—	16
Oligocene							
Cethana	267	13.0	95.0	5.0	0.0	34	—
Pioneer	308	9.7	100.0	0.0	0.0	25	5
Early Miocene							
Bacchus Marsh	—	—	100.0	0.0	0.0	—	6
Monpeelyata	630	57.5	100.0	0.0	0.0	7	>7

[a]Cethana, Pioneer and Monpeelyata data; R. S. Hill & R. J. Carpenter, personal communication.
(1), Christophel, 1981; Christophel & Blackburn, 1978.
(2), Christophel & Greenwood, 1987, 1989.
(3), L. J. Scriven, personal communication.
(4), Hill, 1982, and personal communication.
(5), Christophel *et al.*, 1992.
(6), D. Barrett, personal communication; Greenwood, 1991*a*.
All other data: Christophel & Greenwood, 1987, 1988, 1989 and unpublished results.
[b]Number of whole leaves used to calculate the values.
[c]Number of leaf taxa encountered in sample used; whole macroflora may be higher.

mesothermal–megathermal modern ranges (Kershaw & Nix, 1988; Christophel & Greenwood, 1988, 1989; Hill, 1990*a*).

Christophel (1981; Christophel & Blackburn, 1978) originally considered the Anglesea, Nerriga and Maslin Bay macrofloras analogous to modern complex notophyll vine forest (CNVF) or simple mesophyll vine forest (SMVF). The initial analysis of Maslin Bay was based on a large sample (1046 leaves); however, only the tallies presented by Christophel & Blackburn (1978) were available

here. The Anglesea analysis was based on a small sample (81 leaves) and Christophel's interpretations of all of the macrofloras were based on direct comparisons with canopy values (from Webb, 1959). Burnham (1989; Burnham *et al.*, 1992) and Greenwood (1991*a*, 1992) suggested that sample sizes in excess of 200–350 leaves are required to sample leaf assemblages adequately. Subsequent analysis of the Anglesea and Golden Grove macrofloras used much larger samples (267–1548 leaves; Tables 4.2, 4.3) and a more

Table 4.3. *The foliar physiognomic characteristics of the macrofloras (taxon-based observations)*

Site[a]	No. of leaves	No. of taxa	% entire leaf margins	Leaf size (%)			LSI[b]
				Micro-	Noto-	Mesophylls	
Middle Eocene							
Anglesea							
(1)	793	30	70.0	60.0	40.0	0	20.0
(2)	81	41	78.0	23.0	50.0	27.0	49.0
Maslin Bay (2)	1046	57	79.0	25.0	47.0	28.0	51.5
Golden Grove (5)	338	24	—	53.3	46.7	0	25.0
Nerriga							
(1)	129	38	66.0	25.0	45.0	30.0	52.5
(4)	116	25	80.0	56.5	39.1	4.3	23.8
Nelly Creek (3)	160	16	75.0	60.0	40.0	0	20.0
Oligocene							
Cethana	267	—	—	?100	0	0	—
Late Oligocene–Early Miocene							
Pioneer	308	5	60.0	100	0	0	0
Monpeelyata	630	>7	>60.0	100	0	0	0
Early Miocene							
Bacchus Marsh	—	6	66.7	90.0	10.0	0	10.0

[a]Cethana, Pioneer and Monpeelyata data; R. S. Hill & R. J. Carpenter, personal communication.
(1), Christophel & Greenwood, 1987, 1988, 1989.
(2), Christophel, 1981.
(3), Christophel *et al.*, 1992.
(4), Hill, 1982, and personal communication.
(5), D. Barrett, personal communication; Greenwood, 1991*a*.
All other data: Christophel & Greenwood, 1987, 1988, 1989 and unpublished data.
[b]LSI, leaf size index.

detailed comparison with modern leaf-litter (Christophel & Greenwood, 1987, 1988, 1989). In these later analyses, the Anglesea Site II macroflora (Christophel *et al.*, 1987) matched simple notophyll vine forest (SNVF) litter, and the Nerriga, Golden Grove and Maslin Bay macrofloras compared most closely to CNVF litter; however, in this instance the Maslin Bay analysis was based on a sample of only 90 leaves (Tables 4.3, 4.4).

The Middle Eocene continental margin macrofloras are dominated by simple, narrowly elliptic (length: width > 2.0), microphyllous to notophyllous leaves (Anglesea) or notophylls (Christophel & Greenwood, 1988), with entire margins more common than nonentire (Tables 4.2, 4.3). The leaves in litter from SNVF, CNVF and complex

mesophyll vine forest (CMVF) from Queensland are typically wider and there are fewer leaf specimens and leaf species with nonentire margins present than the leaves from the fossil macrofloras (Christophel & Greenwood, 1988; Tables 4.1–4.3). Leaves from New South Wales SNVF and CNVF are as narrow as the Anglesea and Golden Grove leaves, but the proportion of leaves and species with nonentire margins is much higher in the New South Wales SNVF than in the Anglesea macroflora. A significant number of hastate-base and broad leaves (mostly Menispermaceae) gives the Nerriga macroflora a unique foliar physiognomic signature compared to the other Eocene macrofloras; however, narrowly elliptic leaves also dominate this flora. The tend-

Table 4.4. *Summaries of the sedimentological and geographic setting of each macroflora and the inferred palaeoenvironment*

Site	Sedimentary environment	Modern elevn (m)	% entire leaf margins	LSI	MAT (°C)	Vegetation type[a]
Middle Eocene						
Anglesea	Oxbow pond	50	70	20.0	15–18	SNVF
Maslin Bay	Oxbow pond	65	79	51.5	23–26	CMVF
Golden Grove	Oxbow pond	120	—	25.0	17–20	CNVF
Nerriga	Lake	535–570	66	23.8	16–21	CNVF
Nelly Creek	River channel	8	75	20.0	>20	ScF/MVF
Oligocene						
Cethana	Lake	300	—	?0	10–12	MVFF
Late Oligocene–Early Miocene						
Pioneer	Alluvial fan	Sea-level	60	0	10–12	MFF/SNVF
Monpeelyata	Lake	920	>60	0	4–8	NMF/T
Early Miocene						
Bacchus Marsh	Oxbow pond	400	67	10	12–14	MFF

elevn, elevation; LSI, leaf size index; MAT, mean annual temperature.
[a]CMVF, complex mesophyll vine forest (mesothermal–megathermal rainforest)
MVF, (?semideciduous) mesophyll vine forest (monsoon forest)
ScF, sclerophyllous forest (?marked dry season)
CNVF, complex notophyll vine forest (mesothermal rainforest)
SNVF, simple notophyll vine forest (mesothermal rainforest)
MVFF, microphyll vine–fern forest (microthermal rainforest)
MFF, microphyll fern forest (microthermal *Nothofagus* rainforest)
NMF/T, nanophyll mossy forest (subalpine rainforest or wet thicket).

ency towards narrower leaves in the Eocene macrofloras versus analogous modern Queensland mesic vegetation (SNVF, CNVF and CMVF) may reflect latitudinally controlled climatic differences, as indicated by the narrower leaves found in New South Wales CNVF and SNVF; however, the exact relationship is unclear.

The leaf size index for Anglesea (Table 4.3; Figure 4.2) suggests a MAT of 15–18 °C. Using leaf margin analysis (Wolfe, 1979, 1985; data from Greenwood, 1992) the MAT is estimated to be 17 °C. The LSI values for Maslin Bay and Golden Grove suggest MAT values of 23–26 °C and 17–20 °C respectively (Figure 4.2). Modern SNVF has a range of MAT of 12–18 °C (mesothermal), CNVF MAT of 17–22 °C (mesothermal) and CMVF MAT of 20–27 °C (mesothermal–megathermal interzone to megathermal). Fungal epiphyllous germlings recorded

on Anglesea leaf cuticles are of type V, and type 6 manginuloid hyphae are also present, which indicate a humid climate of the type restricted to modern montane tropical rainforest (Lange, 1981, 1982). The floristic character of the Anglesea macroflora is consistent with this interpretation (Christophel & Greenwood, 1989; Christophel, Chapter 11, this volume). Leaf cuticles from Golden Grove and Maslin Bay have high grade fungal epiphyllous germlings (A. I. Rowett, personal communication) indicative of warm humid climates (Lange, 1978, 1982).

The leaf size index (Table 4.3) of the Nerriga macroflora using Hill's (1982, and personal communication) data suggests a MAT of 16–21 °C (mesothermal). If Christophel's (1981) data are used, the LSI is much higher, giving a MAT of 25–28 °C. Christophel's (1981) original figures were preliminary and probably overestimate species number (Hill, 1982; 38 versus *ca* 20 species) and LSI is sensitive to both species number and the proportion of mesophylls (Wolfe & Upchurch, 1987; Greenwood, 1992). Leaf margin analysis of the Nerriga macroflora based on Hill's data indicates a MAT of 21 °C. The Nerriga macroflora sediments outcrop today at moderate elevation (Hill, 1982; Table 4.4). If this modern elevation is indicative of the Eocene elevation the Nerriga macroflora can be expected to indicate a lower MAT than do lowland Middle Eocene sites at similar palaeolatitude (e.g. Maslin Bay). The MAT estimate for Nerriga (16–21 °C) compared to Golden Grove and Maslin Bay (Table 4.4) is slightly higher than would be expected on the basis of their present elevations.

Lake sediments may be enriched with the smaller sun-leaves of the upper canopy (Roth & Dilcher, 1978; Spicer, 1981) and so the MAT for Nerriga based on LSI may be an underestimate. However, the significant number of entire-margined vine leaves may also bias the leaf margin analysis, producing a MAT estimate that is too high. Greenwood (1992; Greenwood & Christophel, 1990) has demonstrated that taphonomic and ecological factors introduce a significant uncertainty in leaf margin analysis estimates of MAT. The actual MAT value for Nerriga probably lies between the estimates derived from the LSI (Hill, 1982) and from leaf margins. In either case, a mesothermal humid climate is indicated.

There are several explanations for the relatively high estimated MAT for Nerriga, and the apparent discrepancy between Maslin Bay and Golden Grove values. The Nerriga macroflora is older than the other Middle Eocene macrofloras, so climatic change over this interval may be the cause. However, there is debate over the timing and degree of uplift of the Great Dividing Range (Ollier, 1986), including the upland area of the Nerriga macroflora. It is possible that the Nerriga macroflora was deposited at a lower elevation than at present, as significant uplift in the area may have occurred during and since the Oligocene (Ollier, 1986). If this is so, the apparent similarity between the Nerriga and Golden Grove MAT

estimates relative to Maslin Bay (Table 4.4) can be explained.

The MAT values and elevation of the Maslin Bay and Golden Grove macrofloras suggests that either two distinct and climatically controlled vegetation types occurred within only a short distance, or that taphonomic bias is causing a significant reduction in mean leaf size, resulting in an underestimate of MAT for Golden Grove (an overestimate for Maslin Bay is highly unlikely; Christophel & Greenwood, 1988). Christophel & Greenwood (1987) have suggested that there may be a significant collection bias towards smaller curated leaves for the Golden Grove macroflora. The Maslin Bay macroflora is based solely on compressions and impressions *in situ* on the mudstone matrix, whereas all of the other macrofloras are based on mummified leaves that have been isolated from the matrix. It is possible that this preservational difference may enhance leaf size taphonomic biases. Casual inspection of compression–impression facies of the Golden Grove macroflora (see Christophel & Greenwood, 1987) was equivocal on this point. These factors act to underscore the degree to which taphonomic biases influence resolution of quantitative variables.

The Lake Eyre Basin contains extensive fluvial and fluvio-lacustrine sequences of Tertiary sediments (Wopfner *et al.*, 1974; Ambrose *et al.*, 1979). An outcrop of the Paleocene–Eocene Eyre Formation (Sluiter, 1991) within the southern margin of present Lake Eyre (e.g. Nelly Creek site, Figure 4.1) contains leaf and fruit macrofloras (Wopfner *et al.*, 1974). These macrofloras are usually silicified and abundant capsular myrtaceous fruits, including putative *Eucalyptus* (Lange, 1978, 1982), have been recorded from correlated sediments in the same region (Ambrose *et al.*, 1979; Greenwood *et al.*, 1990). Middle Eocene sediments of the Eyre Formation at Nelly Creek contain mummified leaves and fruits (Christophel *et al.*, 1992).

A significant number of taxa in the Lake Eyre silcretes have leaves with broad lamina, large size and entire margins, suggesting that they grew in mesic environments; other taxa have very narrow, small leaves, with coarse serrations (to spinose), or narrow-lobed leaves, which suggest dry or seasonally dry climates (Givnish, 1984; Wolfe, 1985). The silcrete floras suggest a mosaic of sclerophyllous vegetation and broad-leaved gallery forests (Greenwood *et al.*, 1990; Greenwood, 1991*b*; Christophel *et al.*, 1992). Beadle (1967) suggested that adaption to low soil fertility during the Tertiary may have produced sclerophyllous plants, pre-adapting many lineages of Australian plants to seasonally dry and arid climates. It is plausible therefore, that the sclerophyllous elements in the Lake Eyre Basin floras do not represent dry climate vegetation, but low fertility soils.

The foliar physiognomy of the Nelly Creek macroflora partly matches the silcrete macrofloras; however, the presence of sunken stomata and thick cuticles (Christophel *et al.*, 1992) on the leaves of this mummified macroflora emphasise the sclerophyllous and/or xerophyllous character. The floristic composition of both the silcrete floras and the Nelly Creek macroflora also support this interpretation, with some taxa commonly associated with seasonally dry environments (e.g. *Brachychiton*) and others associated with mesic–humid environments (Christophel, Chapter 11, this volume). Only scarce, low grade (types I and II), epiphyllous fungal germlings occur on Nelly Creek leaf cuticles (Christophel *et al.*, 1992), indicating that rainfall was low (Lange, 1978, 1982) or markedly seasonal. The common presence of both *Gymnostoma* and *Brachychiton* in some exposures of Eyre Formation (Greenwood *et al.*, 1990; Greenwood, 1991*b*) indicate low seasonal and diurnal temperature variation in central Australia in the Eocene as these genera today are restricted to frost-free mesotherm–megatherm areas (data from Webb & Tracey, 1982). On the basis of MAT estimates for more southerly Middle Eocene macrofloras (e.g. Maslin Bay, Golden Grove) and the common presence of mesotherm–megatherm taxa (e.g. *Gymnostoma*), the Lake Eyre Basin macrofloras probably reflect a MAT of > 20 °C.

The foliar physiognomy of the Nelly Creek macroflora is distinct from the Middle Eocene macrofloras at continental margin sites (Christophel *et al.*, 1992). Nelly Creek leaves are significantly smaller overall (Tables 4.2, 4.3), with a different assortment of leaf types. Important floristic differences are discussed by Christophel (Chapter 11, this volume). Lobed leaves and species with toothed leaf margins are common in both disturbed environments (early succession), and in strongly seasonal environments (Givnish, 1984; Upchurch & Wolfe, 1987). The foliar physiognomy of the Nelly Creek and silcrete macrofloras compared to the other Eocene macrofloras suggests the presence in central Australia of seasonally dry (?monsoonal) palaeoclimates where mesic broad-leaved vegetation occurred in riparian corridors (Greenwood *et al.*, 1990; Greenwood, 1991*b*; Christophel *et al.*, 1992). This interpretation is supported by sedimentological evidence that suggests periods of aridity in central Australia in the early Tertiary (Quilty, 1984).

Oligocene

The best-documented Early Oligocene macroflora occurs at Cethana in Tasmania (Hill & Carpenter, 1991; Figure 4.1) in ?slumped lake sediments at a present altitude of 300 m (R. J. Carpenter, personal communication). The Late Oligocene–Early Miocene Pioneer macroflora occurs in lenses of organic-rich clays in coarse braided fan (alluvial) deposits close to sea level in north-eastern Tasmania (Hill & Macphail, 1983). This macroflora represents the lowland vegetation of the area at that time (Table 4.4). The Late Oligocene–Early Miocene Monpeelyata macroflora occurs at 920 m above sea level and it is probable that the original palaeovegetation occurred at this or a higher elevation (Hill & Gibson, 1986; Macphail *et al.*, 1991).

The Cethana macroflora is species rich and contains a diverse association of microthermal lineages including three of the four modern subgenera of *Nothofagus* (Hill, 1984, 1990*a*, 1991). Many of the taxa found in the Cethana assemblage are today restricted to the montane tropical rainforests of New Guinea and New Caledonia (Hill,

1990a). Physiognomically, these modern forests are analogous to Australian (microthermal) microphyll vine–fern forest (MV-FF) of upland areas (>1000 m) in the northeast Queensland Humid Tropical Region, and New Guinea montane forests.

The Pioneer macroflora is dominated by leaves of *Nothofagus*, with abundant conifer material and several species of angiosperm. The single site used for this analysis is less diverse than the Cethana macroflora.

Initial analysis of the foliar physiognomy of the Cethana and Pioneer macrofloras suggested a match with Australian microphyll fern forest (MFF) (Hill & Macphail, 1983; Christophel & Greenwood, 1989). MFF in south eastern Australia and Tasmania is species poor relative to the tropical microthermal rainforests (MV-FF) and appears physiognomically distinct (Webb, 1959, 1968; Tracey, 1982). Hill & Macphail (1983) recorded grade V germlings (Lange, 1976) from Pioneer and thus suggested that rainfall was in excess of 1500 mm annually. The relative frequency of the three main leaf size categories (Tables 4.1–4.3) in both macrofloras matches MFF. These Tasmanian macrofloras have a higher frequency of species with entire leaf margins than do modern Australian MFF, and SNVF from New South Wales, and reflect more the situation seen in modern MV-FF. Important floristic differences contribute to these discrepancies which in part reflect extinction events (Hill & Read, 1987; Carpenter *et al.*, Chapter 12, this volume), and in part may reflect climatic differences (such as a change in MART).

Leaf sizes are predominantly microphyllous in both macrofloras, although leptophyllous conifers are also prominent. The LSIs of the Cethana and Pioneer macrofloras are not meaningful as both are zero. On the basis of modern leaf-litter studies (Greenwood, 1992), mesic vegetation above 12 °C MAT will have a LSI value > 0, and a LSI of zero below 10–12 °C MAT (Figure 4.2). Modern MFF in Australia ranges from a MAT of 4–15 °C, with *Nothofagus moorei* occurring over the range 8.9–16.1 °C (Read, 1990). The presence of leaves similar to those of modern *N. moorei*,

and the predominance of microphylls in both macrofloras, suggest a MAT of 10–12 °C. On the basis of analogous modern forests (MFF and MV-FF) and particularly key taxa such as *Nothofagus* (Read, 1990), the MART of the Oligocene palaeovegetation (particularly at Cethana) was probably low, < 10 °C. The presence of climatically sensitive taxa (e.g. *Nothofagus* subgenus *Brassospora*) in the Cethana macroflora suggests that freezing temperatures were unlikely (Hill, 1990a; Read *et al.*, 1990).

The Monpeelyata macroflora is characterised by entire-margined leaves (Table 4.2) and imbricate foliage conifers, including *Araucaria* (Hill & Gibson, 1986; Hill, 1990b). Significant numbers of small-sized toothed *Nothofagus* leaves (*N. microphylla*) related to *N. cunninghamii* are present (Hill, 1991), in addition to leaves of the modern deciduous species, *N. gunnii*. Overall leaf size is very small, with nanophylls dominant (< 25 mm length). This macroflora may represent alpine shrub vegetation or low alpine woodland (Hill & Gibson, 1986). Macphail *et al.* (1991) have suggested that the macroflora represents local cool temperate rainforest (nanophyll mossy forest, NMF) or rainforest scrub (nanophyll mossy thicket, NMT), with a lakeside coniferous microphyllous–angiosperm shrubby vegetation. Significantly also, Macphail *et al.* (1991) suggested that the Monpeelyata flora was deposited below the alpine treeline. On the basis of the small size of the *N. microphylla* leaves and the modern tolerances of *N. gunnii* (Read, 1990), a MAT of 4–8 °C is indicated, and a MART of *ca* 15 °C. The presence of subalpine vegetation at Monpeelyata reflects the elevation at the time (Hill & Gibson, 1986; Macphail *et al.*, 1991).

Miocene

An Early Miocene macroflora occurs at Bacchus Marsh in Victoria (Figure 4.1) in a single clay lens in fluvial sands (Werribee Formation; Abele *et al.*, 1976). It is likely this site was at an elevation similar to the present situation (400 m) in the Miocene (Christophel, 1985). The macroflora is poorly known but is dominated by leaves from only a few species (Christophel, 1985). Leaf sizes range from microphyllous to nanophyllous, with toothed leaves (*Nothofagus*) dominating, although the second most common leaf type (Myrtaceae) has an entire margin. Some nanophyllous conifers are present, including *Araucaria* and *Dacrycarpus* (Christophel, 1985, and personal communication). Today, *Araucaria* may be found in association with *N. moorei* in the SNVF–MFF ecotone at Mt Lamington in southeast Queensland under microthermal regimes (Webb & Tracey, 1982), and in Chile and Argentina with a winter deciduous species of *Nothofagus* under winter snow climates (Veblen & Schlegel, 1982). Comparison of the macroflora with modern leaf assemblages indicates a match with *N. moorei* – MFF (Tables 4.2–4.3). This vegetation type (although with *N. cunninghamii*, not *N. moorei*) is found today near to the Bacchus Marsh site, implying a similar climate in the Miocene. The Bacchus Marsh macroflora LSI value indicates a MAT of 12–14 °C.

In the Latrobe Valley (Figure 4.1), a more-or-less continuous sequence of coal formation from the Eocene to the Pliocene has preserved a record of swamp and terrestrial vegetation (Blackburn & Sluiter, Chapter 14, this volume). In contrast to the Latrobe Valley coal floras, most species from the Middle Miocene Yallourn Clay macroflora show no sclerophyllous features; cuticles are thin with few or no trichomes and possess surficial stomata (D. R. Greenwood, unpublished data). However, there are floristic elements such as *Dacrycarpus latrobensis* (Hill & Carpenter, 1991) and *Agathis* common to the Late Miocene coals and the clays. The Clay flora typically contains mainly microphyllous entire-margined leaves. In some samples microphyllous toothed leaves of *N. moorei* type (aff. *N. tasmanica*) were common. The cuticle flora is diverse (20–30 cuticle morphotypes) and is dominated by several species in each of the Lauraceae, Myrtaceae (aff. *Syzygium*), Cunoniaceae and Proteaceae.

The predominantly microphyllous nature of the macroflora of the Yallourn Clays, together with the occasional presence of *N. moorei* type leaves suggests that the Middle Miocene palaeoclimate was similar to the present climate of the area,

although perhaps less seasonal. SNVF dominated by *Syzygium* and MFF with *N. cunninghamii* are found in the nearby Strzelecki Ranges. The relatively higher diversity of the Yallourn Clays cuticle and macroflora compared to Victorian MFF and SNVF, and the presence of *N. moorei* type leaves, however, suggests a better parallel with New South Wales MFF and SNVF, and indicates a MAT of 9–16 °C (see e.g. Read, 1990).

CONCLUSIONS

From the review and new analyses presented here, it is apparent that climates were not uniform across Australia during the Middle Eocene, and that local topography in the Oligocene and the Early Miocene influenced the climate and plant communities in southeastern Australia (Table 4.4). Nix's (1982) conclusion that increasing seasonality in both temperature extremes and rainfall and, less so, marked changes in the thermal regime, was the primary driving force of biotic change within Australia during the early to mid-Tertiary is supported by the palaeobotanical evidence of climate and vegetation patterns (e.g. Christophel & Greenwood, 1989; Read *et al.*, 1990).

The Middle Eocene macrofloras at Maslin Bay, Golden Grove, Anglesea and Nerriga (Figures 4.1 and 4.3) reveal predominantly mesothermal–megathermal humid climates ranging from MAT values of about 16 °C (Anglesea, mesothermal) to perhaps as high as 25 °C (Maslin Bay, megathermal) in the coastal lowlands and at moderate elevations. Climates in the Middle Eocene were markedly warmer in southern Australia than at present. Rainforests analogous to SNVF, CNVF and CMVF grew under these climates (Christophel & Greenwood, 1989). There is evidence that a transitional zone from mesothermal to microthermal climates to the south, and mesothermal to megathermal climates to the north, existed in an irregular band across southeastern Australia (e.g. Nix, 1982; Christophel & Greenwood, 1989); however, this interpretation and the latitudinal position of such a zone remains speculative until more macrofloras are described in

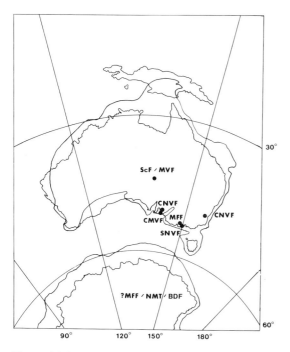

Figure 4.3 Reconstruction of Australia and region for the Middle Eocene showing palaeocoastline superimposed on modern coastline. Based in part on Christophel & Greenwood (1989). NMT, nanophyll mossy thicket or subalpine woodland (microthermal); BDF, broad-leaved deciduous forest (microthermal); MFF, microphyll fern forest (microthermal); SNVF/CNVF, simple/complex notophyll vine forest (mesothermal); CMVF, complex mesophyll vine forest (mesothermal–megathermal); ScF, sclerophyllous forest (?megathermal); MVF, ?semideciduous mesophyll vine forest (monsoonal megathermal).

detail from a wider geographical area. The Lake Eyre Basin macrofloras indicate a development by the Middle Eocene of seasonally wet/dry ?megathermal climates in the continental interior, and that vegetation physiognomically pre-adapted to aridity was in place in the interior at that time (Greenwood *et al.*, 1990; Christophel *et al.*, 1992).

Wolfe (1985) speculated on the nature of Australian Early Miocene vegetation and climate. However, the limited number of well-documented Oligocene and Miocene macrofloras from southeastern Australia precludes any extraregional synthesis. What is clear is that this region experienced cooler temperatures in the Oligocene than in the Eocene (see e.g. Hill & Gibson, 1986; Hill,

1990*a*). Microthermal lineages (*sensu* Nix, 1982), particularly *Nothofagus*, were of increasing importance, compared to the situation in the Middle Eocene (Hill, 1990*a*; Read *et al.*, 1990). The foliar physiognomic and floristic character of these macrofloras suggests humid microthermal climates, although modern Australasian tropical montane communities are better analogies for the Cethana Oligocene vegetation than are modern southern Australian humid microthermal forests such as the *Nothofagus*-dominated MFF (Hill, 1990*a*; Read *et al.*, 1990). In contrast, the Pioneer and Miocene Bacchus Marsh macrofloras suggest microthermal to modern *Nothofagus*-dominated MFF, particularly the *N. moorei*-dominated forests of montane New South Wales. The absence from the Early Miocene macrofloras of some key Oligocene taxa suggests increasing seasonality with perhaps greater temperature extremes during the Early Miocene than in the Oligocene (Hill & Read, 1987; Hill, 1990*a*). Subalpine woodlands occurred at this time at moderate altitudes in Tasmania (Hill & Gibson, 1986; Macphail *et al.*, 1991).

The Middle Miocene Yallourn Clay macroflora indicates mainly microthermal climates in southeastern Australia. Mesothermal humid vegetation (SNVF) probably persisted in parts of the Latrobe Valley during the Middle Miocene, as it does today. There is anecdotal evidence to suggest that Middle Miocene climates in the Latrobe Valley area were more equable than today, but that seasonal dryness became established by the Late Miocene, and perhaps much earlier in the continental interior (Lange, 1982; Quilty, 1984; Greenwood *et al.*, 1990).

The poor knowledge of macrofloras of suitable age and preservation over much of the continent prevents a continental synthesis at this time. There are further areas of research on palaeobotanical climatic evidence, such as wood anatomy and dispersed cuticle (e.g. Upchurch & Wolfe, 1987), which remain largely untapped. Important information on Australian Tertiary climates can be gained from analyses of macrofloras. Work continues on the well-preserved early to mid-Tertiary macrofloras of southeastern Australia and Tasmania; however, there is an urgent need to find and analyse macrofloras from outside of these areas and to apply a wider range of approaches.

This research was completed partly at the Botany Department, University of Adelaide, South Australia, and partly at the Biology Department, University of the South Pacific, Fiji Islands. The manuscript was completed during a NSERC International Fellowship at the University of Saskatchewan, Canada. The Anglesea data were collected with D. C. Christophel by Earthwatch volunteers in 1986. Data from Golden Grove (in part) were supplied by D. J. Barrett from research funded by the South Australian Department of Mines & Energy, and Monier Pty Ltd. Additional funding was provided by the Australia Council (to Christophel). I thank D. C. Christophel, A. I. Rowett, L. J. Scriven, R. S. Hill and R. J. Carpenter for supplying unpublished data, and J. F. Basinger, B. Pratt, M. E. Collinson and B. A. LePage for critically reading the manuscript.

REFERENCES

ABELE, C., GLOE, C. S., HOCKING, J. B., HOLDGATE, G., KENLEY, P. R., LAWRENCE, C. R., RIPPER, D. & THRELFAL, W. F. (1976). Tertiary. In *Geology of Victoria*, ed. J. G. Douglas & J. A. Ferguson. *Geological Society of Australia Special Publication*, **5**, 177–274.

ALLEY, N. F. (1987). Middle Eocene age of the megafossil flora at Golden Grove, South Australia: preliminary report, and comparison with the Maslin Bay flora. *Transactions of the Royal Society of South Australia*, **111**, 211–12.

AMBROSE, G. J., CALLEN, R. A., FLINT, R. B. & LANGE, R. T. (1979). *Eucalyptus* fruits in stratigraphic context in Australia. *Nature*, **280**, 387–9.

AXELROD, D. I. (1984). An interpretation of Cretaceous and Tertiary biota in polar regions. *Palaeogeography, Palaeoclimatology, Palaeoecology*, **45**, 105–47.

BEADLE, N. C. W. (1967). Soil phosphate and its role in molding segments of the Australian flora and vegetation, with special reference to xeromorphy and sclerophylly. *Ecology*, **47**, 992–1007.

BURNHAM, R. J. (1989). Relationships between standing vegetation and leaf litter in a paratropical forest: implications for paleobotany. *Review of Palaeobotany and Palynology*, **58**, 5–32.

BURNHAM, R. J., WING, S. L. & PARKER, G. G. (1992). The reflection of deciduous forest communities in leaf litter: implications for autochthonous litter assemblages from the fossil record. *Paleobiology*, **18**, 30–49.

CASE, J. A. (1988). Paleogene floras from Seymour Island, Antarctic Peninsula. *Geological Society of America Memoir*, **169**, 523–30.

CHRISTOPHEL, D. C. (1981). Tertiary megafossil floras as indicators of floristic associations and palaeoclimate. In *Ecological Biogeography of Australia*, ed. A. Keast, pp. 379–90. The Hague: W. Junk.

CHRISTOPHEL, D. C. (1985). First record of well preserved megafossils of *Nothofagus* from mainland Australia. *Proceedings of the Royal Society of Victoria*, **97**, 175–8.

CHRISTOPHEL, D. C. (1988). Evolution of the Australian flora through the Tertiary. *Plant Systematics and Evolution*, **162**, 63–78.

CHRISTOPHEL, D. C. & BLACKBURN, D. T. (1978). The Tertiary megafossil flora of Maslin Bay, South Australia: a preliminary report. *Alcheringa*, **2**, 311–19.

CHRISTOPHEL, D. C. & GREENWOOD, D. R. (1987). A megafossil flora from the Eocene of Golden Grove, South Australia. *Transactions of the Royal Society of South Australia*, **111**, 155–62.

CHRISTOPHEL, D. C. & GREENWOOD, D. R. (1988). A comparison of Australian tropical rainforest and Tertiary fossil leaf-beds. In *The Ecology of Australia's Wet Tropics*, ed. R. L. Kitching. *Proceedings of the Ecological Society of Australia*, **15**, 139–48.

CHRISTOPHEL, D. C. & GREENWOOD, D. R. (1989). Changes in climate and vegetation in Australia during the Tertiary. *Review of Palaeobotany and Palynology*, **58**, 95–109.

CHRISTOPHEL, D. C., HARRIS, W. K. & SYBER, A. K. (1987). The Eocene flora of the Anglesea locality, Victoria. *Alcheringa*, **11**, 303–23.

CHRISTOPHEL, D. C., SCRIVEN, L. J. & GREENWOOD, D. R. (1992). An Eocene megaflora from Nelly Creek, South Australia. *Transactions of the Royal Society of South Australia*, **116**, 65–76.

FERGUSON, D. K. (1985). The origin of leaf-assemblages – new light on an old problem. *Review of Palaeobotany and Palynology*, **46**, 117–88.

FRAKES, L. A. & FRANCIS, J. E. (1990). Cretaceous palaeoclimates. In *Cretaceous Resources. Events and Rhythms*, ed. R. N. Ginsberg & B. Beaudoin, pp. 273–87. Dordrecht: Kluwer Academic publishers.

GIVNISH, T. J. (1984). Leaf and canopy adaptations in tropical forests. In *Physiological Ecology of Plants of the Wet Tropics*, ed. E. Medina & C. Vazquez-Yanes. *Tasks for Vegetation Science*, **12**, 5–84.

GREENWOOD, D. R. (1991a). The taphonomy of plant macrofossils. In *The Processes of Fossilization*, ed. S. K. Donovan, pp. 141–69. London: Belhaven Press.

GREENWOOD, D. R. (1991b). Middle Eocene megafloras from central Australia: earliest evidence for Australian sclerophyllous vegetation. *American Journal of Botany Supplement*, **78**, 114–15.

GREENWOOD, D. R. (1992). Taphonomic constraints on foliar physiognomic interpretations of Late Cretaceous

58   D. R. GREENWOOD

and Tertiary palaeoclimates. *Review of Palaeobotany and Palynology*, **71**, 142–96.

GREENWOOD, D. R., CALLEN, R. A. & ALLEY, N. F. (1990). *The correlation and depositional environment of Tertiary strata based on macrofloras in the Southern Lake Eyre Basin.* South Australian Department of Mines & Energy Report No. 90/15.

GREENWOOD, D. R. & CHRISTOPHEL, D. C. (1990). *Leaf Margin Analysis of Cretaceous and Tertiary climates: modern Southern Hemisphere rainforests as analogues.* Abstracts. Geological Society of America Annual Meeting, Dallas, Texas, p. 350.

HILL, R. S. (1982). The Eocene megafossil flora of Nerriga, New South Wales, Australia. *Palaeontographica* Abt. B, **181**, 44–77.

HILL, R. S. (1984). Tertiary *Nothofagus* macrofossils from Cethana, Tasmania. *Alcheringa*, **8**, 81–6.

HILL, R. S. (1989). Early Tertiary leaves of Menispermaceae from Nerriga, New South Wales. *Alcheringa*, **13**, 37–44.

HILL, R. S. (1990a). Evolution of the modern high latitude southern hemisphere flora: evidence from the Australian macrofossil record. In *Proceedings of the Third IOP Conference*, ed. J. G. Douglas & D. C. Christophel, pp. 31–42. Melbourne: A-Z Printers.

HILL, R. S. (1990b). *Araucaria* (Araucariaceae) species from Australian Tertiary sediments – a micromorphological study. *Australian Systematic Botany*, **3**, 203–20.

HILL, R. S. (1991). Tertiary *Nothofagus* (Fagaceae) macrofossils from Tasmania and Antarctica and their bearing on the evolution of the genus. *Botanical Journal of the Linnean Society*, **105**, 73–112.

HILL, R. S. & CARPENTER, R. J. (1991). Evolution of *Acmopyle* and *Dacrycarpus* (Podocarpaceae) foliage as inferred from macrofossils in south-eastern Australia. *Australian Systematic Botany*, **4**, 449–79.

HILL, R. S. & GIBSON, N. (1986). Macrofossil evidence for the evolution of the alpine and subalpine vegetation of Tasmania. In *Flora and Fauna of Alpine Australasia: Ages and Origins*, ed. B. A. Barlow, pp. 205–17. Melbourne: CSIRO.

HILL, R. S. & MACPHAIL, M. K. (1983). Reconstruction of the Oligocene vegetation at Pioneer, northeast Tasmania. *Alcheringa*, **7**, 281–99.

HILL, R. S. & READ, J. (1987). Endemism in Tasmanian cool temperate rainforest: alternative hypotheses. *Botanical Journal of the Linnean Society*, **95**, 113–24.

KEMP, E. M. (1978). Tertiary climatic evolution and vegetation history in the southeast Indian Ocean region. *Palaeogeography, Palaeoclimatology, Palaeoecology*, **24**, 169–208.

KEMP, E. M. (1981). Tertiary palaeogeography and the evolution of Australian climate. In *Ecological Biogeography of Australia*, ed. A. Keast, pp. 31–50. The Hague: W. Junk.

KERSHAW, A. P. & NIX. H. A. (1988). Quantitative palaeoclimatic estimates from pollen data using bioclimatic profiles of extant taxa. *Journal of Biogeography*, **15**, 589–602.

LANGE, R. T. (1976). Fossil epiphyllous 'germlings', their living equivalents and their palaeohabitat value. *Neues Jahrbuch für Geologie und Paläontologie*, **151**, 142–65.

LANGE, R. T. (1978). Southern Australian Tertiary epiphyllous fungi, modern equivalents in the Australasian region, and habitat indicator value. *Canadian Journal of Botany*, **56**, 532–41.

LANGE, R. T. (1981). Arnaud's 'enigmatic little marks': an extension (type 6) to the manginuloid hyphae series of epiphyllous fungi. *Experientia*, **37**, 720.

LANGE, R. T. (1982). Australian Tertiary vegetation, evidence and interpretation. In *A History of Australasian Vegetation*, ed. J. M. B. Smith, pp. 44–89. Sydney: McGraw-Hill.

MACPHAIL, M. K., HILL, R. S., FORSYTH, S. M. & WELLS, P. M. (1991). A Late Oligocene–Early Miocene cool climate flora in Tasmania. *Alcheringa*, **15**, 87–106.

McGOWRAN, B, HARRIS, W. K. & LINDSAY, J. M. (1970). The Maslin Bay flora, South Australia, 1. Evidence for an Early Middle Eocene age. *Neues Jahrbuch für Geologie und Paläontologie*, **8**, 481–5.

NIX, H. (1982). Environmental determinants of biogeography and evolution in Terra Australis. In *Evolution of the Flora and Fauna of Arid Australia*, eds W. R. Barker & P. J. M. Greenslade, pp. 47–66. Frewville: Peacock Publications.

OLLIER, C. D. (1986). The origin of alpine landforms in Australasia. In *Flora and Fauna of Alpine Australasia*, ed. B. A. Barlow, pp. 3–26. Melbourne: CSIRO.

PARRISH, J. T. & SPICER, R. A. (1988). Late Cretaceous terrestrial vegetation: a near-polar temperature curve. *Geology*, **16**, 22–5.

QUILTY, P. G. (1984). Mesozoic and Cenozoic history of Australia as it affects the Australian biota. In *Arid Australia*, ed. H. G. Cogger & E. E. Cameron, pp. 7–56. Sydney: Australian Museum.

READ, J. (1990). Some effects of acclimation temperature on net photosynthesis in some tropical and extratropical Australasian *Nothofagus* species. *Journal of Ecology*, **78**, 100–12.

READ, J. & HOPE, G. S. (1990). Foliar frost resistance of some evergreen tropical and extratropical Australasian *Nothofagus* species. *Australian Journal of Botany*, **37**, 361–73.

READ, J., HOPE, G. S. & HILL, R. S. (1990). Integrating historical and ecophysiological studies in *Nothofagus* to examine the factors shaping the development of cool rainforest in southeastern Australia. In *Proceedings of the Third IOP Conference*, ed. J. G. Douglas & D. C. Christophel, pp. 97–106. Melbourne: A-Z Printers.

ROTH, J. L. & DILCHER, D. L. (1978). Some considerations in leaf size and margin analysis of fossil leaves. *Courier Forschung-Institut Senckenberg*, **30**, 165–71.

SLOAN, L. C. & BARRON, E. J. (1990). 'Equable' climates during Earth history? *Geology*, **18**, 489–92.

SLUITER, I. R. K. (1991). Early Tertiary vegetation and climates, Lake Eyre region, northeastern South Australia. In *The Cainozoic of Australia: A Re-appraisal of the Evidence*, ed. M. A. J. Williams, P. De Dekker & A. P. Kershaw. *Geological Society of Australia Special Publication*, **18**, 90–118.

SPICER, R. A. (1981). The sorting and deposition of allochthonous plant material in a modern environment at Silwood Lake, Silwood Park, Berkshire, England. *United States Geological Survey Professional Paper*, **1143**.

SPICER, R. A. (1989). The formation and interpretation of plant fossil assemblages. *Advances in Botanical Research*, **16**, 96–191.

TOKARSKI, A. K., DANOWSKI, W. & ZASTAWN-IAK, E. (1987). On the age of fossil flora from Barton Peninsula, King George Island, West Antarctica. *British Polar Research*, **8**, 293–302.

TRACEY, J. G. (1982). *The Vegetation of the Humid Tropics of North Queensland*. Melbourne: CSIRO.

TRUSWELL, E. M. & HARRIS, W. K. (1982). The Cainozoic palaeobotanical record in arid Australia: fossil evidence for the origin of an arid-adapted flora. In *Evolution of the Flora and Fauna of Arid Australia*, ed. W. R. Barker & P. J. M. Greenslade, pp. 67–76. Frewville: Peacock Publications.

UPCHURCH, G. R. & WOLFE, J. A. (1987). Mid-Cretaceous to Early Tertiary vegetation and climate: evidence from fossil leaves and woods. In *The Origin of Angiosperms and their Biological Consequences*, ed. E. M. Friis, W. G. Chaloner & P. R. Crane, pp. 75–105. Cambridge: Cambridge University Press.

VEBLEN, T. T. & SCHLEGEL, F. M. (1982). Reseña ecologica de los bosques del sur de Chile. *Bosque*, **4**, 73–115.

WEBB, L. J. (1959). A physiognomic classification of Australian rainforests. *Journal of Ecology*, **47**, 551–70.

WEBB, L. J. (1968). Environmental determinants of the structural types of Australian Rain Forest Vegetation. *Ecology*, **49**, 296–311.

WEBB, L. J. & TRACEY, J. G. (1982). *Rainforests: Data on Floristics and Site Characteristics*. Canberra: Biogeographic Information System, Bureau of Flora and Fauna.

WOLFE, J. A. (1971). Tertiary climatic fluctuations and methods of analysis of Tertiary floras. *Palaeogeography, Palaeoclimatology, Palaeoecology*, **9**, 27–57.

WOLFE, J. A. (1978). A paleobotanical interpretation of Tertiary climates in the Northern Hemisphere. *American Scientist*, **66**, 694–703.

WOLFE, J. A. (1979). Temperature parameters of humid to mesic forests of eastern Asia and relation to forests of other regions of the Northern Hemisphere and Australasia. *United States Geological Survey Professional Paper*, **1106**.

WOLFE, J. A. (1985). The distribution of major vegetational types during the Tertiary. In *The Carbon Cycle and Atmospheric CO₂. Natural Variation Archaen to Present*, ed. E. T. Sundquist & W. S. Broeker. *American Geophysical Union Monograph*, **32**, 357–75.

WOLFE, J. A. (1990). Palaeobotanical evidence for a marked temperature increase following the Cretaceous/Tertiary boundary. *Nature*, **343**, 153–6.

WOLFE, J. A. & UPCHURCH, G. R. (1987). North American non-marine climates and vegetation during the Late Cretaceous. *Palaeogeography, Palaeoclimatology, Palaeoecology*, **61**, 33–77.

WOPFNER, H., CALLEN, R. A. & HARRIS, W. K. (1974). The Lower Tertiary Eyre Formation of the southwestern Great Artesian basin. *Journal of the Geological Society of Australia*, **21**, 17–52.

5   Landscapes of Australia: their nature and evolution

G. TAYLOR

It is oft said that 'Australia is an old continent'. Certainly Australia has many ancient landscapes which have been exposed to the processes of weathering and soil formation for very long periods under very stable tectonic conditions. The results of this are that many of our landscapes have well-leached and infertile soils, a major factor in the evolution of the Australian flora. Having said that, it is not true that all Australian landscapes are ancient. Many are very young, and of course all our landscapes are still undergoing modification, albeit some very slowly. Many landscapes that were thought to be comparatively young, such as the Southeastern Highlands (Andrews, 1911; Browne, 1969; Hill, 1975), have recently been shown to be comparatively ancient (Wellman, 1987; Bishop, 1988; Taylor *et al.*, 1990*a*).

The Australian continent attained its present outline between 150 and 50 million years (Ma) ago (Wilford & Brown, Chapter 2, this volume), but many of our landscapes are even older. Comparison of the major landform regions (Figure 5.1) and the major geological structure of the crust or tectonic provinces (Figure 5.2) demonstrates this well.

The tectonic provinces of Australia can be divided into two along the Tasman Line (Figure 5.2; Veevers, 1984). West of this line the continent is dominated by Precambrian blocks and fold belts overlain by thin Phanerozoic basins, while to the

east of it there are mainly Phanerozoic fold belts overlain by younger basins. Broadly these fundamental geological divisions correspond to landscape regions. The Precambrian blocks correspond to plateaux at elevations of up to about 500 m, the fold belts to upland areas up to 2000 m and the basins to lowland plains with elevations of generally less than 200–300 m (Figure 5.3). The distinction between these major landform regions is reflected in the present drainage networks. Most integrated drainage occurs along the coastal margins, particularly along the eastern Phanerozoic fold belt. The drainage systems in the western parts of the continent are, however, generally uncoordinated. The Eastern Highlands are wetter now than the western two thirds of the continent and Veevers (1984) pointed out that this has been a feature of the continent throughout the Cenozoic, although the abundance of lignitic sediments of early Tertiary age in the west, south and centre of the continent suggest that the Eastern Highlands were no wetter. Veevers also suggests that in general the present drainage systems reflect those of the Tertiary.

While it is difficult to speculate about the evolution of soils in Australia much before the Cenozoic, it is possible to reconstruct some soil history throughout the Cenozoic, as there are sufficient remnants of older soils to give some insight. The soils naturally reflect the interaction between

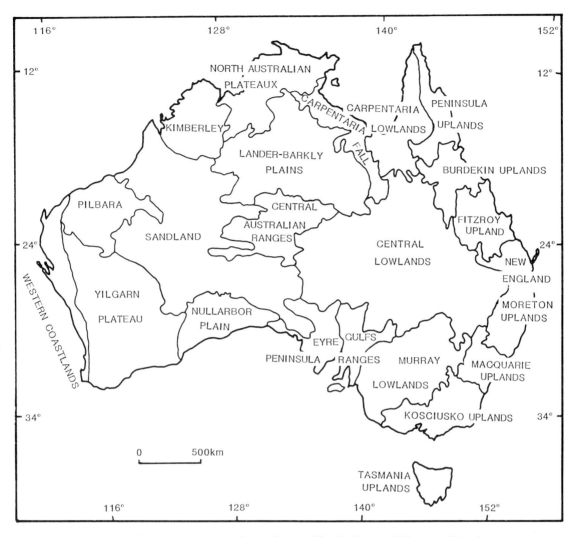

Figure 5.1 Australia's major landscape elements. The Kosciusko and Macquarie Uplands comprise what is referred to in the text as the southern sector of the Eastern Highlands, and the Moreton, New England, Fitzroy, Burdekin and Peninsula Highlands the northern sector. (Modified from Jennings & Marbutt, 1977.)

parent material, climate and rates of erosion or deposition, the last of these factors being generally tectonically controlled.

This chapter begins with the relationship between tectonic provinces and landscapes but because data from many regions are sparse the coverage is somewhat selective, concentrating on areas for which data are available. Towards the end of the chapter the evolution of regoliths and soils is discussed. Because the evolution of climate in Australia is discussed in detail by Quilty, (Chapter 3, this volume), it is not discussed here, except where climatic events are critical in landscape development.

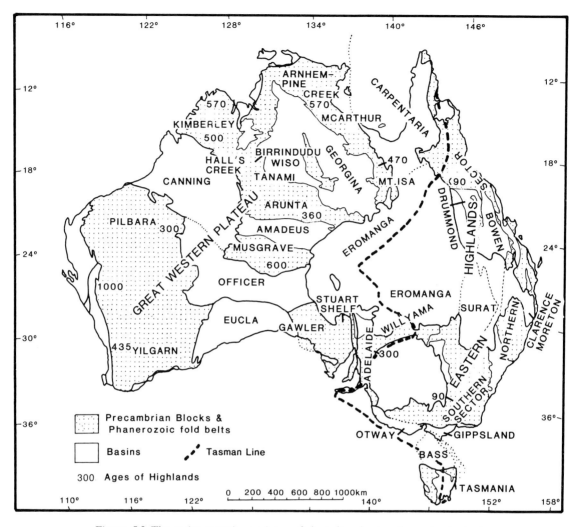

Figure 5.2 The major tectonic provinces of Australia, showing the major lowland depositional basins, Palaeozoic accretionary terrains (separated from the cratonic blocks to the west by the Tasman Line), and the cratonic blocks which make up much of the Great Western Plateau. The ages of the highland and cratonic blocks are from Veevers (1984) and Taylor. (Modified from Veevers, 1984.)

GEOLOGICAL EVOLUTION AND LANDSCAPE REGIONS

The geological (or tectonic) evolution of Australia is closely related to the nature and age of the major landscapes. The division of the continent into two along the Tasman Line is a convenient division between the older landscapes of the western two thirds of the continent and the younger higher-relief eastern third. In both the eastern and western regions the older rocks are partly covered by comparatively thinner and younger sedimentary basins, which form the major lowland landscapes. It is convenient to maintain these major subdivisions of the continent in the discussion that follows.

The maximum possible age for landscapes is determined by the 'age of exposure' of the rocks

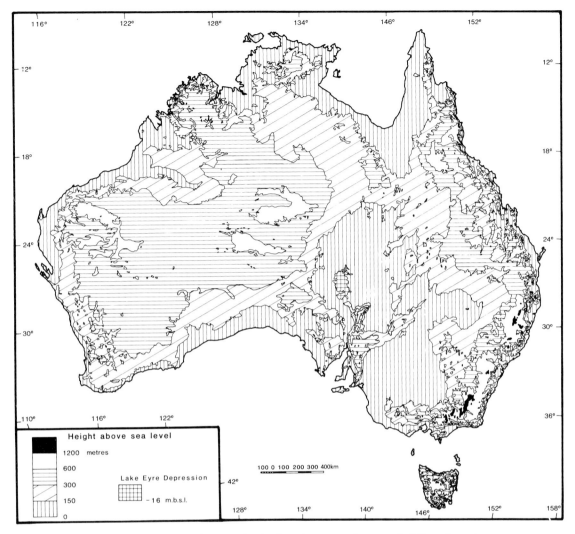

Figure 5.3 A topographic map showing how the topography generally follows the major tectonic provinces. m.b.s.l., metres below sea level.

to terrestrial processes or when major episodes of terrestrial sedimentation ceased. Figure 5.4 is a map of the 'age of exposure' compiled by Beckmann (1983). This map does not give the age of the landscape because after exposure the original surface may be partially or wholly stripped by erosion and there may be multiple erosional or depositional surfaces of varying ages in the landscape. None the less the map does reflect the gross geology and landscape ages of Australia.

Geological evolution and landscape ages of the western plateaux

The cratonic blocks that so dominate the western region of the continent are all made up predominantly of Precambrian granitic, metamorphic and sedimentary rocks that comprise the Yilgarn–Pilbara, Kimberley, Arnhem–Pine Creek–McArthur–Mt Isa and the Musgrave–Amadeus–Arunta Blocks (Figure 5.2) and

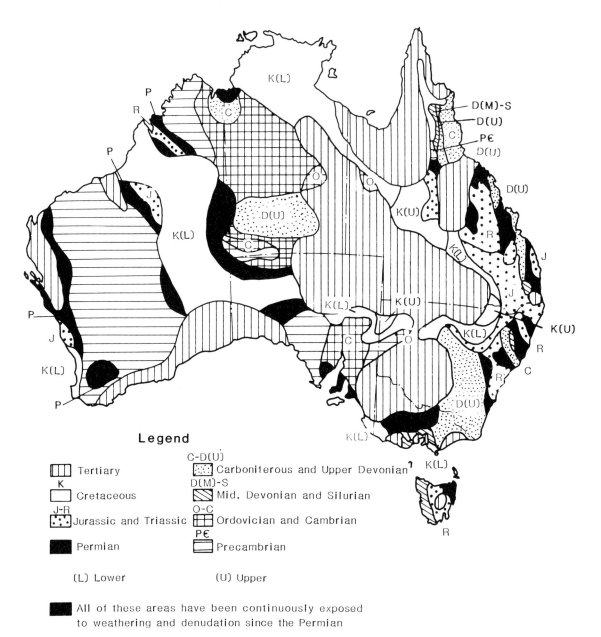

Figure 5.4 A map portraying the 'age of exposure' to weathering of the continent. While this does not include Tertiary and Quaternary volcanics it gives a good general idea of the maximum age of the landscapes of Australia. The solid tone includes areas of Permian rocks and areas known to have been exposed since the Permian but not on Permian rocks. (Modified from Beckmann, 1983.)

which correspond to major landscape provinces (Figure 5.1). These regions have been prominent landscape elements for most of the Phanerozoic, forming important sediment sources for the Phanerozoic basins which surround them.

Since their exposure the landscapes of the western region blocks have been substantially modified. The Yilgarn–Pilbara Plateau has been degraded by 600–700 m or more throughout the Phanerozoic (Veevers, 1984), probably a substantial portion of this during the Late Carboniferous and Permian, when glaciers swept across the plateau from south to north. Veevers & Wells (1962) reported glaciated pavements from this time exposed in the Pilbara region, where relics of glacially carved valleys can still be seen, and Butt (1989) reported Permian glacial deposits in valleys on the northern Pilbara. Substantial sediment was also shed west into the Perth Basin between the Permian and Cretaceous, although much of the Mesozoic sediment may be derived from uplifted parts of the basin. Van de Graaf (1981) calculated rates of denudation of between 4.5 and 5 m per Ma, which suggests that considerable uplift must have occurred during the Phanerozoic. It is unlikely that little if any significant uplift occurred after the Permian glaciation, as tills of that age remain on the surface. Jackson & van de Graaf (1981) argued that the relict drainage of the Yilgarn, now represented by chains of lakes, developed between the late Mesozoic and Eocene and involved some 100 m of erosion, but R. P. Langford, G. E. Wilford & E. M. Truswell (unpublished data) date the drainage as Permian, which is in accord with other data. Van de Graaf (1981) suggested that, during the late Cenozoic, erosion has been only a few centimetres per Ma. The Yilgarn–Pilbara Plateau has therefore been uplifted and eroded significantly during the Palaeozoic – early Cenozoic but since then it has been extremely stable.

The Kimberley and Northern Australian Plateaux and the Carpentaria Fall are underlain by the Kimberley and Arnhem Blocks, the Pine Creek Inlier, McArthur Basin and Mt Isa Block, which collectively consist largely of geosynclinal sediments and metamorphic and granitic rocks of

Precambrian age, and have formed landscapes throughout much of the Phanerozoic (Veevers, 1984). Again these ancient high landscapes have been uplifted through much of the Phanerozoic, providing sediments for their marginal basins (Figure 5.2). The present coastline of this area approximates closely to that which has existed through much of the Phanerozoic, except during the Early Cretaceous, when the sea swept briefly across the Pine Creek Inlier and Daly and McArthur Basins, leaving a quartzose sheet sand that has since been substantially removed (Skwarko, 1966). This indicates the long-term lateral stability of these cratonic areas over the last 500 Ma or so. Permian glacial debris in the Fitzroy Trough includes material derived from the Kimberley Plateau to the north, showing that, during the late Carboniferous and Permian the plateau, like the Yilgarn, underwent some glacial erosion. Young (1986) demonstrated that, in addition to physical removal of detritus, the effects of solution weathering can be substantial in Devonian quartz sandstones of the Ord Basin, and it would be surprising if similar chemical effects were not just as prominent in landscape denudation on the adjacent cratons.

The Central Australian Ranges (Figure 5.1) have the highest elevation of any of the Precambrian cratonic areas of the western region, with some peaks reaching 1500 m. The ranges are made up of three major geological entities: the Precambrian Musgrave Block, the Arunta Block and the late Proterozoic–Palaeozoic Amadeus Basin (Figure 5.2). To the northwest these regions are connected to the Kimberley Block by the Tanami Block, Birrindudu Basin and Hall's Creek Province. All these are composed of sedimentary, metamorphic and granitic rocks. The Central Australian Ranges are overlain around the margins by Phanerozoic sediments of younger sedimentary basins and partially divided by Palaeozoic sediments in the Amadeus Basin (Figure 5.2). Cambrian and Devonian sediments of the marginal basins lap on to the Arunta Block (Veevers, 1984), indicating that it was a feature of some relief during the Cambrian and that its uplift was complete by the Devonian. Both blocks acted as sediment sources for their surrounding basins. Glacial pavements in the highlands and glaciogene sediments in the surrounding basins indicate that glaciers eroded the Central Australian Ranges. Erosion during the late Mesozoic and early Cenozoic etched a southeasterly drainage network that is now partially filled with sediments (Twidale & Harris, 1977). It is this drainage that established features such as Uluru and the Olgas, proving the antiquity of such landscape elements. Minor fault-bounded Cenozoic basins occur within the Central Australian Ranges, illustrating the continued role of tectonism in these landscapes.

The Precambrian Gawler and Willyama Blocks, with the Adelaide Fold Belt and Stuart Shelf (Figure 5.2) make up the core of the Eyre Peninsula and Gulf Ranges landscape region (Figure 5.1). These regions are made up of volcanics, metasediments, flat-lying and folded sediments, and granitic rocks surrounded by basins of Phanerozoic sediments that lap on to the older cratonic platforms. This region has formed high ground on and off since the late Carboniferous (Veevers, 1984), but parts of the area formed highlands well before this. This high ground provided sediments to the marginal Phanerozoic basins (Figure 5.2) since their inception. Glacial activity during the Permian significantly eroded the highlands. Renewed late Cenozoic faulting in the Gulf Ranges region produced their present form (Callen & Telford, 1976).

Geological evolution and landscape ages of the Eastern Highlands

The Eastern Highlands of Australia, a complex landscape region (Figure 5.1) about 400–500 km wide, extends from Tasmania to North Queensland. It is a relatively high region with average elevations in excess of 500 m and large areas in excess of 1000 m. An obvious feature of the highlands is the asymmetric drainage, with short, steep, east-flowing networks and very long, low-gradient, westerly drainage with long sedimentary records in the basins flanking the inland margin of the highlands.

The Eastern Highlands are dominated by Early to mid-Palaeozoic geosynclinal deep and shallow marine sediments and granitoids with abundant Cenozoic basalts. Some Mesozoic and Cenozoic basins also occur within the highlands. They are flanked by sedimentary basins that lap on to the Palaeozoic rocks of the highlands. The highlands have been providing sediment to these basins throughout the Cenozoic.

The date of formation of the highlands is a matter of considerable debate. Early workers (Andrews, 1911; Browne, 1969; Hill, 1975) believed most of the uplift was comparatively recent (since the Late Miocene). These workers based their conclusions primarily on the southern highlands; however, more detailed work in the highlands and flanking basins has shown that considerable differences may exist between the histories of the northern (Queensland and northern New South Wales) and southern (southern New South Wales, Victoria and Tasmania) highlands. Wellman (1987) reviewed the erosional and uplift history of the whole highlands and Bishop (1988) provided a similar review, predominantly of the southern portion.

Ollier (1977) showed that, in the southern sector, the highlands had a relief of at least 600 m in the Eocene and Taylor et al. (1985, 1990a) demonstrated a relief of at least 500 m and probably 800 m by the Paleocene. Further studies of the marginal basins (Macumber, 1978; Wooley, 1978; Veevers, 1984; Brown, 1989) show that sediments derived from the southern highlands began depositing during the late Mesozoic in the Gippsland and Otway Basins and during the Paleocene in the Murray Basin. Nott et al. (1991) and Young & McDougall (1982) showed that the southern New South Wales coastal strip was established by the mid-Tertiary and probably earlier. Williams (1989) recorded Late Cretaceous and Paleocene sediments in basins along the northern coast of Tasmania. These data clearly show that the southern sector of the highlands was uplifted sufficiently by the earliest Tertiary or Late Cretaceous to provide a major sediment source for the flanking basins. There is now a general consensus that the uplift of the southern sector is at least late Mesozoic, but there is little evidence to suggest how much older it may be. The absence of sediment in the flanking basins older than late Mesozoic suggests the major uplift dates from about 95 Ma, as suggested by Jones & Veevers (1982) and favoured by Wellman (1987). Lambeck & Stephenson (1986) however, argued on the basis of a passive isostatic rebound model that the highlands date from the late Palaeozoic and have risen more or less continuously since, in response to erosional unloading. In contrast Jones & Veevers (1982) argued for periodic uplift through the Cenozoic, associated with volcanism and accompanying sedimentation/sea level fluctuations in the flanking basins. The Tasmanian segment of the highlands possibly began to develop its present form in the Triassic and the present configuration was well established by the Late Cretaceous (Williams, 1989). Whichever is the correct model, the southern sector of the highlands was certainly well established by the Cenozoic.

The northern sector of the Eastern Highlands has Mesozoic sediments preserved on the highland summit, which suggests uplift was after the Early Cretaceous in Queensland and after the Triassic near Sydney (Wellman, 1987). The presence of abundant andesitic debris in sediments as young as Cenomanian (90 Ma) in the Eromanga Basin along the western highland flanks but derived from volcanoes east of the present coast (Veevers, 1984) shows that the highlands did not exist until after this time. The presence of quartzose sediments overlying the deeply weathered Late Cretaceous–Paleocene surface (Doutch, 1976) suggests that by the Eocene the highlands were shedding sediment into the flanking basins. Grimes (1980) identified three periods of uplift and basaltic volcanism, during the Paleocene–Eocene, Late Oligocene to mid-Miocene and Pliocene–Quaternary, similar to those in the southern sector identified by Jones & Veevers (1982). He also notes that Quaternary uplift with basaltic volcanism occurred as recently as 13 ka ago in northern Queensland. In the Ebor area of New England, retreat of the Great Escarpment must postdate the formation of the Ebor Volcano

18.5 Ma ago (Gleadow & Ollier, 1987). Elsewhere Ollier (1982) argued that uplift in this region post-dates 18 Ma. Schmidt & Embleton (1976) suggested that the Sydney Basin was uplifted and eroded between 100 and 180 Ma.

The southern sector of the Eastern Highlands is probably older in general than the northern sector but both contain landscapes of considerable antiquity with superimposed younger landscapes in many areas. There is the unifying feature of widespread occurrence of basaltic landscapes, which, although highly dissected, frequently preserve the topography and soils over which the lavas flooded (Bishop, 1988). These basalts range in age from Holocene in Victoria to Late Cretaceous in central Queensland and northern New South Wales. The overall rate of eruption was more or less constant through the Cenozoic. Lava field volcanism dominated until about 35 Ma, while after that time most activity was associated with central volcanic eruptions (Wellman & McDougall, 1974).

History of landscapes in the basins flanking highland regions

While many of the sedimentary basins which flank the highland regions (Figure 5.2) have histories which extend well back into the Palaeozoic, most were inundated by the sea during the Early to mid-Cretaceous (Figure 5.5). During this period sedimentation alternated between marine and terrestrial due to global sea level changes (Morgan, 1980) or to uplift of the newly formed continent (Veevers, 1984; Wilford & Brown, Chapter 2, this volume). During the marine phases the highland areas were separated by shallow seaways, which Wasson (1982) suggested would have increased spatial and habitat diversity on the continent. Certainly the shallow seas in which sea levels oscillated would have led to a diversity of marginal marine habitats through this period. On the retreat of the Cretaceous seas during the late Aptian, the basins generally had a low relief surface, which, because of limited post-Cretaceous tectonism, was retained through most of the Cenozoic. There are remnants of this mid-

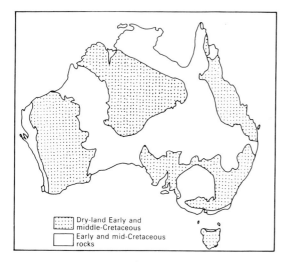

Figure 5.5 The extent of the marine inundation of the continent during the Cretaceous. (From Wasson, 1982.)

Cretaceous surface preserved throughout many of the major basins, particularly in the Eromanga, Officer and Canning Basins. Much of the Cretaceous surface is covered or partially covered by Cenozoic sediment, particularly in many of the eastern and southern basins. Each major basin is examined in turn and correlations are shown in Figure 5.6.

Post mid-Cretaceous deposition began in the Murray Basin during the Paleocene and continues today. Jones & Veevers (1982) suggested that sedimentation was initiated by the uplift of the Eastern Highlands and concomitant subsidence of the basin. Three major depositional cycles separated by erosional intervals occurred during the Paleocene–Early Oligocene, Oligocene–Middle Miocene, and Late Miocene–Pliocene. Marine incursions occurred during each cycle, with the most extensive during the Late Miocene. During each cycle a wide spectrum of environments existed, from shallow marine carbonate platforms through lagoonal, tidal and deltaic regions to riverine plains with lakes (Figure 5.7). The cause of these depositional cycles is not certain, but Brown (1983, 1985, 1989) attributed them to sea level changes accompanied by basinal isostatic adjustment. The effects of these sea level variations were restricted to the basin, but stream valleys on the margins of the Eastern Highlands eroded deeper during lower sea levels, were backfilled during high sea levels as a result of diminished erosion potential (Macumber, 1978). The final regression of the sea from the Murray Basin during the Late Miocene – Early Pliocene formed a prograding series of beach ridges and intervening fluvial and estuarine quartz sands (Parilla sand). These ridges form prominent landscape features in the western parts of the basin. Since the final regression of the sea, the basin has been dominated by alluvial, lacustrine and aeolian activity under oscillating cold and warm conditions that characterise the last 2.5 Ma. During this period the availability of water has varied, but not in any systematic way with regard to temperature (Bowler, 1978), except that at this time generally arid conditions prevailed in contrast to the generally humid conditions which had persisted since the Late Cretaceous.

The Eucla Basin was dominated by limestone deposition during the Eocene and Miocene. Rivers draining from the Yilgarn and Musgrave Blocks into the Eucla Basin were established after the withdrawal of the Cretaceous seas. During high sea levels in the Eocene, these channels were inundated by the sea and alluviated with the deposition of sands, lignite, spongolite and limestones across the southeastern Yilgarn (Jones, 1990). Across the Officer and southwestern Eromanga Basins these channels were filled with quartzose alluvium (Lampe Beds) and alluvium and marginal marine sediments (Pidinga Formation: Benbow *et al.*, 1982). These palaeochannels now form chains of playas on the Yilgarn and across the Officer and southwestern Eromanga Basins. After deposition of the Miocene limestone, the Eucla Basin was uplifted and the sea withdrew, leaving the flat surface of the Nullarbor Plain exposed. Minor alluvial and lake deposits in shallow drainage lines separated by low hills of deeply weathered Cretaceous rocks form the only Tertiary record in the Canning Basin. Longitudinal dunes, formed during the Quaternary, overlie these old drainage systems. The Eromanga Basin, the largest of the flanking basins, saw the deposition of the Winton Formation after the retreat

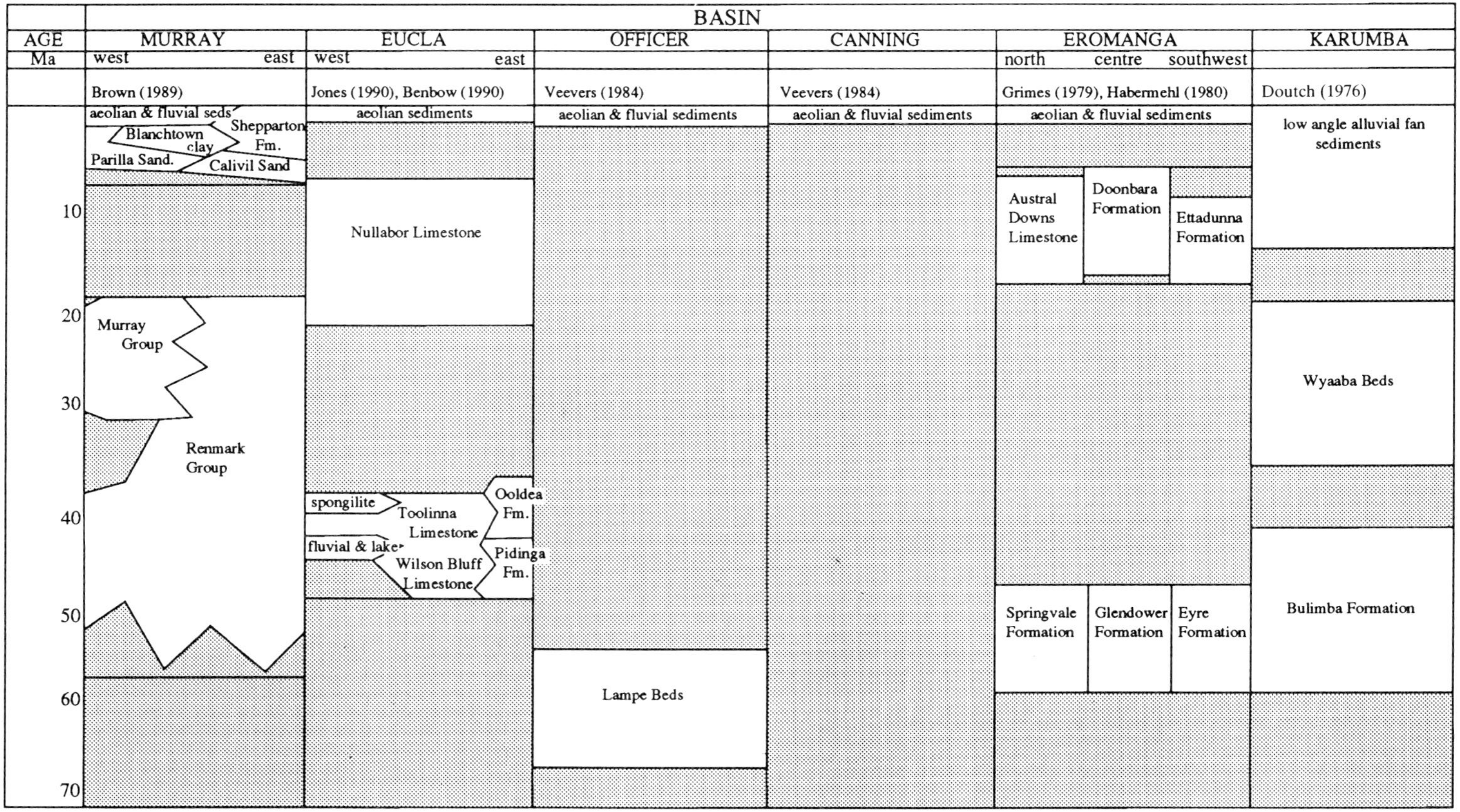

BASIN

AGE Ma	MURRAY		EUCLA		OFFICER	CANNING	EROMANGA			KARUMBA
	west	east	west	east			north	centre	southwest	
	Brown (1989)		Jones (1990), Benbow (1990)		Veevers (1984)	Veevers (1984)	Grimes (1979), Habermehl (1980)			Doutch (1976)

Murray – west/east: aeolian & fluvial seds; Parilla Sand., Blanchtown clay, Shepparton Fm., Calivil Sand; Murray Group; Renmark Group

Eucla – west/east: aeolian sediments; Nullabor Limestone; Ooldea Fm.; Toolinna Limestone; spongilite; Wilson Bluff Limestone; fluvial & lake; Pidinga Fm.

Officer: aeolian & fluvial sediments; Lampe Beds

Canning: aeolian & fluvial sediments

Eromanga: aeolian & fluvial sediments; Austral Downs Limestone; Doonbara Formation; Etadunna Formation; Springvale Formation; Glendower Formation; Eyre Formation

Karumba: low angle alluvial fan sediments; Wyaaba Beds; Bulimba Formation

AGE scale (Ma): 10, 20, 30, 40, 50, 60, 70

Figure 5.6 Correlation of formations, weathering events etc. acrosss the Murray, Eucla, Officer, Canning, Eromanga and Karumba Basins. Fm, Formation. (Karumba Basin from Doutch, 1976; Grimes, 1979; Veevers, 1984.)

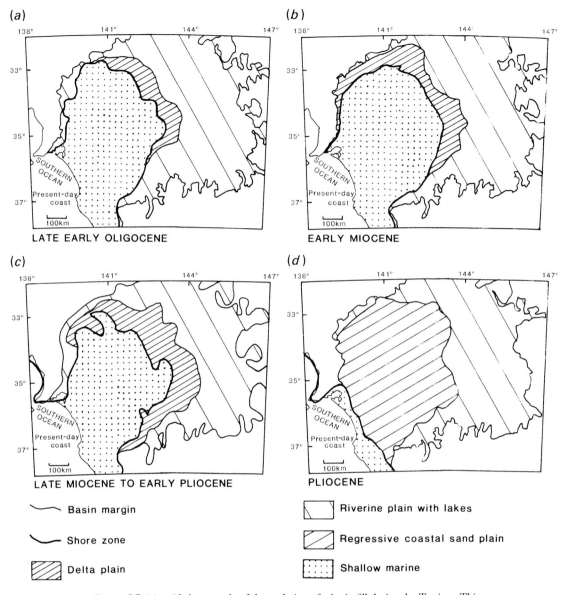

Figure 5.7 (*a*) to (*d*) An example of the evolution of a basin fill during the Tertiary. This much simplified example is for the Murray Basin in southeastern Australia. (Modified from Brown, 1989.)

of the sea during the Albian (95 Ma). This is a terrestrial sheet of volcanogenic sandstone, shale and coal deposited between about 95 and 90 Ma. This low relief surface was deeply weathered during the ensuing 20–30 Ma (the Mornay Profile of Idnurm & Senior, 1978). This phase of weathering was terminated by the deposition of the

quartz-rich fluvial Eyre Formation and its equivalents during the Paleocene and Eocene. Another break in deposition allowed time for the development of further weathering and the formation of silcrete (the Cordillo Silcrete of Wopfner, 1974; and the Cordillo Surface of Doutch, 1976). Post Eyre Formation warping and uplift during the

Oligocene produced the major landscape elements within the present basin. It also caused subsidence of the Lake Eyre and Lake Frome depressions with sedimentation in substantial lakes during the Miocene. The depressions elsewhere in the basin were the locus of deposition of the Etadunna Formation (and its equivalents). The Etadunna is a sequence of muds, carbonates and evaporites deposited in streams and lakes. Warping continued after the deposition of the Etadunna, and the surface was once again deeply weathered (the Canaway Profile and its equivalents, Curalie and Strathgordon; Doutch, 1976). The warped areas continued to be the locus of fluvial and lacustrine sedimentation. The Cooper Syncline beneath the channel country of Cooper Creek and Lake Yama Yama are examples of this, as are many other drainage lines of southern Queensland (Senior et al., 1978). Sedimentation in these loci consists of up to 160 m of quartz sandstones and conglomerates, mudrocks and thin gypsum layers, above the Etadunna Formation. These drainage lines are still the centres of fluvial and lacustrine sedimentation. They are separated by low ridges and mesas of deeply weathered and silcreted Cretaceous rocks and Eyre Formation. As fluvial and lake sedimentation continued through the Quaternary, dune fields swept across the interfluves, particularly in the southwest in the Simpson and Strezlecki Deserts. Evaporites and clastics accumulated in the lakes and muds and sands in the rivers during drier and wetter phases, respectively.

Other similar flanking basins such as the Gippsland, Otway and Perth Basins have post-Cretaceous records of marine and terrestrial sedimentation punctuated by tectonically or eustatically induced breaks.

REGOLITH AND SOILS OF THE CONTINENT

Regolith is a term used to encompass all the fragmental earth materials, residual or transported, that overlie bedrock and usually forms the surface of the land (Ollier, 1988). Soil forms the uppermost part of the regolith.

From the earlier parts of this chapter it is clear that much of the continent has elements of its landscape that are very ancient whereas other elements are comparatively young. The nature of the regolith depends primarily on climate, time, rate of stripping (erosion) and the type of rock or sediment on which it has developed.

The last major glaciation to occur in Australia was during the late Carboniferous and early Permian. These glaciers produced pavements and valleys still preserved in some of the old plateaux of Western Australia and in the Eastern Highlands of Victoria (Craig & Brown, 1984). After this glacial episode, conditions suitable for abundant plant growth persisted until the Jurassic, even though southern Australia was in a high latitude position. The climate began to dry about 160 Ma ago, but at about 140 Ma it again became humid. This humidity persisted and with it the landscapes became forested (Frakes et al., 1987). There is some debate about temperatures during the Mesozoic. Frakes et al. (1987) suggested that the Jurassic was generally temperate but that in the Early Cretaceous there were 'cooler than expected' temperatures. This is supported by Francis (1990) for the southwestern Eromanga Basin. This is not surprising, since between 175 and 40 Ma much of Australia lay within the Antarctic Circle (Wilford & Brown, Chapter 2, this volume). Taylor et al. (1990b) also provided evidence for cool climates in the southern sector of the Eastern Highlands during the Late Paleocene.

Frakes et al. (1987) noted three warm to cool cycles throughout the Tertiary: Late Cretaceous–Middle Eocene, Middle Eocene–Early Oligocene, and Early Oligocene–Late Miocene. These broad climatic cycles gave way to more arid climates during the period between mid-Pliocene and the start of the Quaternary, when temperatures oscillated between warm and cold on an about 100 ka cycle, during which time water availability also varied, but not necessarily in consort with temperature (Frakes et al., 1987). Throughout this time the history of groundwater is intimately related to regolith development (see e.g. Butt, 1989; Brown, 1989). The groundwater conditions are largely related to climate and geology.

Variations in rock type across the continent

have dramatically affected the nature of the landscape and regolith. Regions with rocks containing labile minerals (igneous rocks and volcanogenic or feldspathic sedimentary rocks) tend to have generally deep regoliths, whereas those on quartzose sedimentary rocks or limestones tend to have thin regoliths and soils.

Minor tectonic activity has occurred in parts of the continent from the Late Cretaceous. When combined with sea level changes during this period (Frakes *et al.*, 1987) their effect has been to cause much of the regolith to be stripped from the cratonic blocks and the Eastern Highlands, or to promote deposition in the flanking basins and on to the margins of the cratonic blocks.

While there are undoubtedly regolithic materials which pre-date the Late Cretaceous (e.g. Permian glacial materials and pavements), well-documented examples are rare. The majority date from the time of the regression of Cretaceous seas. Two major regolith-forming episodes can be identified since this time. The first, from the Late Cretaceous to the Late Miocene, gave rise to deep lateritic weathering under humid and most would argue warm (e.g. Frakes *et al.*, 1987; Butt, 1988; Ollier, 1988), conditions, although cooler phases are recognised (Frakes *et al.* (1987) summarised the evidence) based on micropalaeontological records both onshore and offshore and on $\delta^{18}O$ values. The second began about 10–6.5 Ma, when there was a rapid reduction in sea level brought about by expansion in the Antarctic ice sheets. By 2.5 Ma the Tertiary seas had retreated and the climate was altered to cyclical repetitions of cold (glacial) and warm (interglacial) periods in which warm periods were wetter than the cold periods and winds periodically became more intense. During this last 2.5 Ma, the acidic deep weathering environment gave way, at least in central and southern Australia, to alkaline weathering conditions with the production of widespread evaporites and calcareous regoliths.

Deep weathering

Many Australian land surfaces are covered by deeply weathered *in situ* or sedimentary materials.

Doutch (1976) showed multiple 'periods' of ferruginisation in the basins of Eastern Australia as did Frakes *et al.* (1987) and Grimes (1980).

Such 'weathering events' are easily recognised in depositional terrains, where the various 'events' are, at least in part, separated by intervening sedimentary events. It would be reasonable to assume that under the generally tectonically stable conditions that existed through the period between 90 and 6.5 Ma, intense weathering would be continuously occurring, but Frakes *et al.* (1987) correlated the deep weathering events with the warmer parts of their three Tertiary climatic cycles, which also happen to coincide with the uplift pulses of Jones & Veevers (1982) in the Eastern Highlands. The evidence of Francis (1990), Taylor *et al.* (1990a,b), Bird & Chivas (1988) and the postulation of cooler Late Cretaceous climate in Frakes *et al.*'s paper suggest, however, that the widely accepted belief that deep weathering needs warm climates may not be true. Recent evidence from Iceland (Gislason *et al.*, 1990) shows clearly that very rapid chemical denudation is possible in cool to cold climates as also suggested by Reynolds (1971). The idea of continuous weathering is certainly supported when the ages of deep weathering for the whole continent are plotted (Figure 5.8). Even considering the inaccuracies in the dating of weathering profiles, the stratigraphic and palaeomagnetic control is sufficient to place some value on the data.

The deeply weathered profiles generally grade upwards from parent rock through a zone of increasingly kaolinite-rich saprolite (rock weathered *in situ* that retains the original rock fabric), which becomes increasingly palid upwards. This is overlain by a mottled zone consisting of a white or pale-coloured matrix containing ferruginous red mottling and an uppermost ferruginous crust (ferricrete, laterite or bauxite). This is the classical 'laterite' profile originally described in Western Australia by Walther (1915).

These deeply weathered profiles show considerable variability. Many have transported upper horizons that are not related to the underlying saprolite (e.g. Churchward & Bettenay, 1973; Milnes *et al.*, 1985; Taylor & Ruxton, 1987).

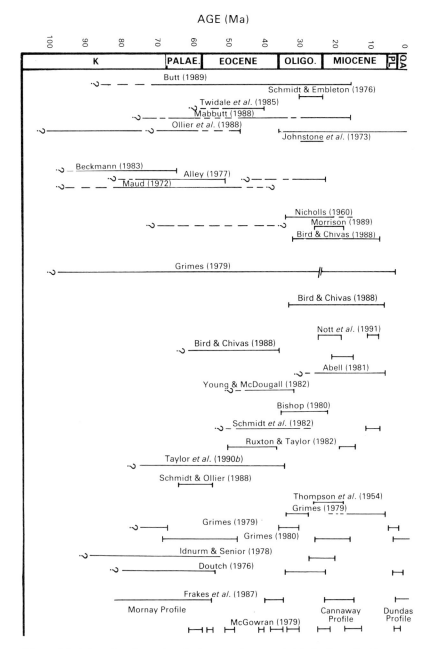

Figure 5.8 A time plot of some 'weathering events' as recorded in the Australian literature on weathering. Clearly, this plot shows there to be little cyclicity in these events, but it should be added that the data presented here are shown without critical review of their reliability. PALAE, Palaeocene; OLIGO., Oligocene; K, Cretaceous; PL, Pliocene; QA, Quaternary.

Some are stripped, with the upper zones removed (e.g. Senior, 1979; Butt, 1989; Ollier *et al.*, 1988), others have silicified upper zones (e.g. Ollier *et al.*, 1988) or are partially stripped with a silcrete in the pallid zone (e.g. Senior, 1979; Ollier, 1988) or capped by silcreted sediments unconformably overlying stripped profiles (e.g. Senior, 1979; Taylor & Ruxton, 1987; Ollier *et al.*, 1988).

These highly leached deep weathering profiles cover much of Australia, from east coast to west and from Tasmania (where they are less common) to Cape York. In places they are relatively thin (e.g. Monaro, where they vary from 2 m on basaltic lavas to tens of metres on surfaces exhumed from beneath early Tertiary basalts); in others very thick (e.g. southwest Queensland where they exceed 100 m; Senior, 1979). Most of these deeply weathered landscapes have been eroded into a series of mesas and low plateaux, protected somewhat by caps of ferricrete and/or silcrete, separating broad valleys cut into the weathered rocks. The cause of this erosion is tectonic, eustatic or due to climatic change or a combination of these. The products of the erosion (mainly kaolinite and quartz) are deposited in the lower parts of the landscape, particularly in the valleys, but some products occur as widely distributed sheets of alluvial sediments (e.g. Eyre and Glendower Formations). Some areas of the cratonic blocks, most notably the Yilgarn, are covered by sandplains. These are thought to originate by solution of kaolinite from the deep weathering profiles, leaving a lag of the solution-resistant quartz sand (Butt, 1985).

Soils formed on the deeply weathered materials and on sediments derived from them are highly leached and acid, and consist largely of kaolinite, quartz and sequioxides. The types of soil commonly associated with these parent materials are shown in Table 5.1. The parent materials for these soils are, in general, old (Figure 5.8), and it is likely that many of these leached acid soils have formed the substrate for vegetation in Australia since the early Tertiary. Indeed because of their significant erosion during the later Tertiary and Quaternary and their partial burial during the Quaternary they were more widespread than at

Table 5.1. *Soil parent materials and soils resulting largely from Tertiary deep weathering*

Soil parent materials	Soils
Kaolinite, laterite, bauxite, silcrete, mottled and pallid zone materials, and deeply kaolinitised rock	Massive red and yellow earths (Gn2) and the associates the earthy sands (Uc5.2), earthy loams (Um5.3, Um5.5), the gravels and possibly the red siliceous sands (Ucl.23), red structured earths (Gn3)

The notations in brackets are Northcote's soil forms.

present and as a consequence would have played a major part in the evolution of our flora.

Deep weathering in landscapes underlain by quartzose or limey sediments has not produced a thick regolith. In quartz sandstone terrains (e.g. eastern Kimberley, Arnhemland Plateau) the predominant form of weathering has been solution, resulting in the development of karstic landscapes dominated by solution channels along structural weaknesses in the rocks. In extreme cases, such as the Bungle Bungle Ranges (Young, 1986) and Ruined City of Arnhemland (Jennings, 1983), tower karst landscapes have been produced. Soils are minimal in these landscapes, except in some valleys that contain infertile quartz sand soils. The Nullarbor Plain is underlain by Miocene marine limestones, and, although it has only been exposed to terrestrial weathering for a comparatively short time, it has karstic landforms and discontinuous drainage typical of limestone terrains. Soils here are thin calcareous loams with extensive calcrete development (Northcote & Wright, 1982).

Regoliths and soils of the Quaternary

The growth of the Antarctic ice sheet near the end of the Miocene and the associated sudden decrease in sea surface temperatures resulted in an intensification of atmospheric circulation, initiating conditions, similar to those of the present day, that control the distribution of the mid-latitude dry deserts. For the next 3 Ma, landscapes and vegetation underwent the transition from those of the humid Tertiary to those of the more arid Quaternary (Frakes *et al.*, 1987). There is widespread evidence of predominantly alkaline weathering, associated with the post-Tertiary drying at about 2.5 Ma. Since then, oscillatory warm and cool climatic phases have been associated with widespread alluviation, lacustrine sedimentation and aeolian activity, with hillslope instability, erosion and valley alluviation in highland areas.

The Quaternary record for Australia prior to the last glacial–interglacial cycle, which began about 120 ka ago, is poor, but a good record for the last cycle is preserved. The development of the landscape and regolith during this phase was well described by Bowler (1982), Wasson & Clark (1987) and Frakes *et al.* (1987), so only a brief summary is included here.

Between 120 and 60 ka, at the start of the last cycle, landscapes and climates were very similar to those of the present. Inland Australia was dominated by stabilised longitudinal dunes (Frakes *et al.*, 1987). Major westerly flowing streams drained the Eastern Highlands, carrying fine sediments out across the Murray Basin. Dry lake beds were scattered across the landscape. About 60 to 36 ka, as the build-up of ice began in the northern hemisphere, runoff from the Eastern Highlands increased, rivers transported coarser sandy sediment across the Murray Basin, and the inland lakes filled (Bowler, 1978). A brief drying at this time reactivated the longitudinal dunes (Bowler & Wasson, 1983) but wetter conditions rapidly returned and persisted until 25 ka, when cooler, dryer conditions became dominant.

During the glacial maximum from about 30 to 15 ka, glaciers covered much of central and western Tasmania and a small area near Mt Kosciusko, leaving behind them glacial landscapes and moraine when they retreated. In the inland areas available precipitation decreased (Bowler & Wasson, 1983). High wind velocities enabled longitudinal dune fields to be reactivated and develop to their present dimensions. Source-bordering dune fields developed downwind of major streams (even in the highlands, e.g. Canberra), and lunettes of clay-rich materials formed on the leeward side of inland lakes and in Tas-

mania (Frakes *et al.*, 1987). The dune fields in many coastal regions also developed during this time in Tasmania. Inland lake systems dried, depositing increasing amounts of salt and gypsum, which, with clays, were deflated and deposited in the lunettes. During this same arid phase, dust and salts were carried downwind and deposited as blankets across much of Australia. Many of the desert loams and parna (loess) were formed during these arid climatic episodes and in the highlands, where they accreted in soil profiles (Walker *et al.*, 1988).

In the Eastern Highlands the cool, dry periods were marked by widespread hillslope erosion, formation of alluvial fans and valley alluviation with gravels and sands (between about 30 and 15 ka), followed by relative hillslope stability and the accumulation of fine alluvium in valley bottoms (Walker & Butler, 1983). The sand and gravel and the fine alluvium typically have red to yellow podzolic and earth soils and prairie soils, respectively developed on them.

This arid phase (and those of previous Quaternary cycles) generally produced alkaline weathering conditions in the arid areas. Increased quantities of expanding clays with high cation exchange capacity, eroded from weathering profiles formed under changed conditions in the Eastern Highlands, were deposited on the adjacent lowlands. These produced soils such as the alkaline, saline and cracking clay soils so typical of much of inland Australia (Table 5.2). The dune sands produced uniform red siliceous sands. This dramatic change from leached acid soils to the more alkaline, carbonate- and salt-rich soil of the Quaternary must have had a significant impact on vegetation, particularly when the climatic instability and changes in availability of water are considered.

Associated with the Quaternary climatic oscillations were sea level oscillations of 100 to 120 m. These changes in sea level resulted in the complex set of coastal environments and landscapes of the present. Many of the large coastal plains were built during this period, indeed many since the last glacial maximum. Complex coastal dune systems with complex soil landscape patterns

Table 5.2. *Soil parent materials and soils resulting from Quaternary arid weathering*

Soil parent materials	Soils
Sediments of Quaternary age, and all fresh rocks and sediments exposed by erosion of the deep weathering profile, or else always protruding above it	Calcareous earths (Gc soils), crusty and hard red duplex soils (Dr1, Dr2), saline clays and cracking clays (Uf1, Ug5), shallow sands (Uc), shallow loams (Um), shallow calcareous loams (Um5.11) and saline loams (Um1)

The notations in brackets are Northcote's soil forms.

developed during this period. Land-bridges between the mainland and Tasmania and New Guinea have formed and been severed on numerous occasions and the coastal plains around much of the continent have expanded and contracted with sea level change.

The cool to cold, windy and dry conditions which characterised the last glacial maximum gave way to present conditions *ca* 10 ka ago, and sea levels stabilised at about their present level *ca* 6 ka ago.

CONCLUSION

The interdependence of climate, vegetation and landscape is well known. This chapter has briefly examined the evolution of landscapes and the regoliths and the soils developed on them. Many of the landscapes of the Great Western Plateau have developed their present character since the region was swept by Permian glaciers. The retreat of the Cretaceous seas shaped the landscapes of the basins in this region. The Eastern Highlands, although their age is uncertain, were uplifted before the late Mesozoic. Hence all the major landscape elements have existed for at least 90 Ma, and are certainly ancient compared to those in most other continents.

Since the retreat of the Cretaceous seas, the continent has been tectonically quiet, with only minor uplift and sagging having any influence on landscapes. Climatic changes during this interval have, however, had a profound effect on the landscapes and their regoliths. The period from the Cretaceous to the Pliocene was predominantly humid, causing widespread deep weathering and leached acid soils. The onset of arid conditions about 2.5 Ma dramatically altered the weathering regimes to predominantly alkaline, leading to salt- and carbonate-rich soils. Intensified winds during this phase built the extensive aeolian landscapes that characterise much of the interior and coastal landscapes.

These dramatic shifts in landscape and soil development over a long time have caused equally dramatic shifts in vegetation, from humid forests during much of the Tertiary to the flora of today. Equally it is likely that earlier changes in vegetation can be viewed in the same way, although data on earlier landscape evolution and soil development are sparse.

REFERENCES

ABELL, R. S. (1981). Notes, to accompany the 1 : 25000 scale geological field compilation sheets of Canberra 1 : 100 000 sheet area. *Bureau of Mineral Resources, Australia. Record* 1981/6.

ALLEY, N. F. (1977). Age and origin of laterite and silcrete duricrusts and their relationship to episodic tectonism in the mid-north of South Australia. *Journal of the Geological Society of Australia*, 24, 107–16.

ANDREWS, E. C. (1911). Geographical unity of eastern Australia in late and post-Tertiary time, with application to biological problems. *Journal and Proceedings of the Royal Society of New South Wales*, 44, 420–80.

BECKMANN, G. G. (1983). Development of old landscapes and soils. In *Soils: An Australian Viewpoint*, pp. 51–72. Melbourne: CSIRO.

BENBOW, M. L. (1990). Tertiary coastal dunes of the Eucla Basin, Australia. *Geomorphology*, 3, 9–29.

BENBOW, M. L., LINDSAY, J. M., HARRIS, W. K. & COOPER, B. J. (1982). Latest Eocene marine incursion, northeast margin of the Eucla Basin. *Geological Survey of South Australia, Quarterly Geological Notes*, 81, 2–9.

BIRD, M. I. & CHIVAS, A. R. (1988). Stable-isotope evidence for low-temperature kaolinitic weathering and post formational hydrogen-isotope exchange in Permian kaolinites. *Chemical Geology (Isotope Geoscience Section)*, 77, 249–65.

BISHOP, P. (1980). Tertiary drainage on the southern tablelands of NSW. *Bureau of Mineral Resources, Australia, Record*, 1980/67, 7–8.

BISHOP, P. (1988). The eastern highlands of Australia: the evolution of an intraplate highland belt. *Progress in Physical Geography*, **12**, 159–82.

BOWLER, J. M. (1978). Quaternary climate and tectonics in the evolution of the Riverine Plain, southeastern Australia. In *Landform Evolution in Australia*, ed. J. L. Davies & M. A. J. Williams, pp. 35–46. Canberra: Australian National University Press.

BOWLER, J. M. (1982). Age, origin and landform expression of aridity in Australia. In *Evolution of Flora and Fauna of Arid Australia*, ed. W. R. Barker & P. J. M. Greenslade, pp. 35–45. Frewville: Peacock Publications.

BOWLER, J. M. & WASSON, R. J. (1983). Glacial age environments of inland Australia. In *Late Cainozoic Palaeoclimates of the Southern Hemisphere*, ed. J. C. Vogel, pp. 183–208. Rotterdam: Balkema.

BROWN, C. M. (1983). Discussion: a Cainozoic history of Australia's southeastern highlands. *Journal of the Geological Society of Australia*, **30**, 483–6.

BROWN, C. M. (1985). Murray Basin, southeastern Australia: stratigraphy and resource potential – a synopsis. *Bureau of Mineral Resources, Australia, Report* **264**.

BROWN, C. M. (1989). Structural and stratigraphic framework of groundwater occurrence and surface discharge in the Murray Basin, southeastern Australia. *BMR Journal of Australian Geology and Geophysics*, **11**, 127–46.

BROWNE, W. R. (1969). Geomorphology. General notes. *Journal of the Geological Society of Australia*, **16**, 559–69.

BUTT, C. R. M. (1985). Granite weathering and silcrete formation on the Yilgarn Block, Western Australia. *Australian Journal of Earth Sciences*, **32**, 425–33.

BUTT, C. R. M. (1989). Genesis of supergene gold deposits in the lateritic regolith of the Yilgarn Block, Western Australia. In *The geology of Gold Deposits: The Perspective in 88*, ed. R. R. Keys, W. R. H. Ramsay & D. I. Groves, *Economic Geology Monograph*, **6**, 460–70.

CALLEN, R. A. & TELFORD, R. H. (1976). New Late Cainozoic rock units and depositional environments, Lake Frome area, South Australia. *Transactions of the Royal Society of South Australia*, **100**, 125–67.

CHURCHWARD, H. M. & BETTENY, E. (1973). The physiographic significance of conglomeratic sediments and associated laterites in valleys of the Darling Plateau, near Harvey, Western Australia. *Journal of the Geological Society of Australia*, **20**, 309–18.

CRAIG, M. A. & BROWN, M. C. (1984). Permian glacial pavements and ice movement near Moyhu, north-east Victoria. *Australian Journal of Earth Sciences*, **31**, 439–44.

DOUTCH, H. F. (1976). The Karumba Basin, north-eastern Australia and southern New Guinea. *BMR Journal of Australian Geology and Geophysics*, **1**, 131–40.

FRAKES, L. A., McGOWRAN, B. & BOWLER, J. M. (1987). Evolution of Australian Environments. In *Fauna of Australia*, ed. G. R. Dyne & D. W. Walton, pp. 1–16. Canberra: Australian Government Publishing Service.

FRANCIS, J. E. (1990). Cooling the Cretaceous: evidence for Early Cretaceous cool climates in Australia from growth rings in fossil wood. In *Gondwana: Terranes and Resources, Tenth Australian Geological Convention, Geological Society of Australia, Abstracts*, **25**, 55.

GISLASON, S. R., ARNORSSON, S. & ARMANNSSON, H. (1990). Chemical denudation rates in SW-Iceland. *Chemical Geology*, **84**, 64–7.

GLEADOW, A. W. & OLLIER, C. D. (1987). The age of Gabbro at The Crescent, New South Wales. *Australian Journal of Earth Sciences*, **34**, 209–12.

GRAAF, W. J. W. van de (1981). Palaeogeographic evolution of a rifted cratonic margin: S. W. Australia – a discussion. *Palaeogeography, Palaeoclimatology, Palaeocology*, **34**, 163–72.

GRIMES, K. G. (1979). The stratigraphic sequence of old land-surfaces in northern Queensland. *BMR Journal of Australian Geology and Geophysics*, **4**, 33–46.

GRIMES, K. G. (1980). The Tertiary Geology of north Queensland. In *The Geology and Geophysics of Northeastern Australia*, ed. R. A. Henderson & P. J. Stephenson, pp. 329–47. Brisbane: Queensland Division of the Geological Society of Australia.

HABERMEHL, M. A. (1980). The Great Artesial Basin, Australia. *BMR Journal of Australian Geology and Geophysics*, **5**, 9–38.

HILL, E. S. (1975). *The Physiography of Victoria*. Melbourne: Whitcombe and Tombes.

IDNURM, M. & SENIOR, B. R. (1978). Palaeomagnetic ages of Late Cretaceous and Tertiary weathered profiles in the Eromanga Basin, Queensland. *Palaeogeography, Palaeoclimatology, Palaeoecology*, **24**, 263–77.

JACKSON, M. J. & VAN DE GRAAF, W. J. E. (1981). Geology of the Officer Basin. *Bureau of Mineral Resources, Geology and Geophysics, Australia, Bulletin* **206**.

JENNINGS, J. N. (1983). Sandstone pseudokarst or karst? In *Aspects of Australian Sandstone Landscapes*, ed. R. W. Young & G. C. Nanson, pp. 21–30. Wollongong: Australian and New Zealand Geomorphology Group.

JENNINGS, J. N. & MARBUTT, J. A. (1977). Physiographic outlines and regions. In *Australia: A geography*, ed. D. M. Jeans, pp. 38–52. Sydney: Sydney University Press.

JOHNSTON, M. H., LOWRY, D. C. & QUILTY, P. G. (1973). The geology of south-western Australia. *Journal of the Royal Society of Western Australia*, **56**, 5–15.

JONES, B. G. (1990). Cretaceous and Tertiary sedimentation on the western margin of the Eucla Basin. *Australian Journal of Earth Sciences*, **37**, 317–30.

JONES, J. G. & VEEVERS, J. J. (1982). A Cainozoic history of Australia's southeast highlands. *Journal of the Geological Society of Australia*, **29**, 1–12.

LAMBECK, K. & STEPHENSON, R. (1986). The post-Palaeozoic uplift history of south-eastern Australia. *Australian Journal of Earth Sciences*, **33**, 253–70.

MACUMBER, P. G. (1978). Evolution of the Murray River during the Tertiary Period: evidence from northern Victoria. *Proceedings of the Royal Society of Victoria*, **90**, 43–52.

MARBUTT, J. A. (1988). Landsurface evolution at the continental time-scale: an example from the interior of Western Australia. *Earth Science Reviews*, **25**, 457–66.

MAUD, R. R. (1972). Geology, geomorphology and soils of Central County Hindmarsh (Mount Compass-Milang), South Australia. *CSIRO Division of Soils, Publication*, **29**.

McGOWRAN, B. (1979). Comments on Early Tertiary tectonism and lateritization. *Geological Magazine*, **116**, 227–30.

MILNES, A. R., BOURMAN, R. P. & NORTHCOTE, K. H. (1985). Field relationships of ferricrete and weathered zones in southern Australia; a contribution to laterite studies in Australia. *Australian Journal of Soil Research*, **23**, 441–65.

MORGAN, R. (1980). Eustacy in the Australian Early and Middle Cretaceous. *Bulletin of the Geological Survey of New South Wales*, **27**.

MORRISON, K. C. (1989). Tin fields of northeast Tasmania. In *Geology and Mineral Resources of Tasmania*, ed. C. F. Burrett & E. L. Martin. *Geological Society of Australia Special Publication*, **15**, 369–70.

NICHOLLS, K. D. (1960). Erosion surfaces, river terraces and river capture in the Launceston Tertiary basin. *Papers and Proceedings of the Royal Society of Tasmania*, **94**, 1–12.

NORTHCOTE, K. H. & WRIGHT, M. J. (1982). Soil landscapes of Arid Australia. In *Evolution of the Flora and Fauna of Arid Australia*, ed. W. R. Barker & P. J. M. Greenslade, pp. 15–21. Frewville: Peacock Publications.

NOTT, J. F., IDNURM, M. & YOUNG, R. W. (1991). Sedimentology, weathering and geomorphological significance of Tertiary sediments on the NSW far south coast. *Australian Journal of Earth Sciences*, **38**, 357–72.

OLLIER, C. D. (1977). Early landform evolution. In *Australia: A Geography*, ed. D. N. Jeans, pp. 85–98. Sydney: Sydney University Press.

OLLIER, C. D. (1982). Geomorphology and tectonics of the Dorrigo Plateau, NSW. *Journal of the Geological Society of Australia*, **29**, 431–5.

OLLIER, C. D. (1988). The regolith in Australia. *Earth Science Reviews*, **25**, 355–61.

OLLIER, C. D., CHAN, R. A., CRAIG, M. A. & GIBSON, D. L. (1988). Aspects of landscape history and regolith in the Kalgoolie region, Western Australia. *BMR Journal of Australian Geology and Geophysics*, **10**, 309–21.

REYNOLDS, R. C. (1971). Clay mineral formation in an alpine environment. *Clays and Clay Minerals*, **19**, 361–74.

RUXTON, B. P. & TAYLOR, G. (1982). The Cainozoic geology of the Middle Shoalhaven Plain. *Journal of the Geological Society of Australia*, **29**, 239–46.

SCHMIDT, P. W. & EMBLETON, B. J. J. (1976). Palaeomagnetic results from sediments of the Perth Basin, Western Australia and their bearing on the timing of lateritization. *Palaeogeography, Palaeoclimatology, Palaeoecology*, **19**, 257–73.

SCHMIDT, P. W. & EMBLETON, B. J. J. (1987). Magnetic over-printing in southeastern Australia and the thermal history of its rifted margin. *Journal of Geophysical Research*, **86**, 3998–4008.

SCHMIDT, P. W. & OLLIER, C. D. (1988). Palaeomagnetic dating of Late Cretaceous to Early Tertiary weathering in New England, NSW, Australia. *Earth Science Reviews*, **25**, 363–71.

SCHMIDT, P. W., TAYLOR, G. & WALKER, P. H. (1982). Palaeo-magnetic dating and stratigraphy of a Cainozoic lake near Cooma, NSW. *Journal of the Geological Society of Australia*, **29**, 49–53.

SENIOR, B. R. (1979). Mineralogy and chemistry of weathered and parent sedimentary rocks in southwest Queensland. *BMR Journal of Australian Geology and Geophysics*, **4**, 111–24.

SENIOR, B. R., MOND, A. & HARRISON, P. L. (1978). Geology of the Eromanga Bagin. *Bureau of Mineral Resources, Geology and Geophysics, Australia, Bulletin*, **167**.

SKWARKO, W. K. (1966). Cretaceous stratigraphy and Palaeontology of the Northern Territory. *Bureau of Mineral Resources, Geology and Geophysics, Australia, Bulletin*, **73**.

TAYLOR, G. & RUXTON, B. P. (1987) A duricrust catens in southern Australia. *Zeitschrift für Geomorphologie*, **31**, 385–410.

TAYLOR, G., TAYLOR, G. R., BINK, M., FOUDOULIS, C., GORDON, I., HEDSTROM, J., MINELLO, J. & WHIPPY, F. (1985). Pre-basaltic topography of the northern Monaro and its implications. *Australian Journal of Earth Sciences*, **32**, 65–71.

TAYLOR, G., TRUSWELL, E. M., EGGLETON, R. A. & MUSGRAVE, R. (1990a). Cool climate bauxite. *Chemical Geology*, **84**, 183–4.

TAYLOR, G., TRUSWELL, E. M., McQUEEN, K. G. & BROWN, M. C. (1990b). Early Tertiary palaeogeography, landform evolution, and palaeoclimates of the Southern Monaro, NSW, Australia. *Palaeogeography, Palaeoclimatology, Palaeoecology*, **78**, 109–34.

THOMPSON, C. H., HUBBLE, G. D. & BECKMANN, G. G. (1954). A soil survey of the Middle Ridge Area, Parish of Drayton, Darling Downs, Queensland. *CSIRO, Division of Soils Report* **5/54**.

TWIDALE, C. R. & HARRIS, W. K. (1977). The age of Ayres Rock and the Olgas, central Australia. *Transactions of the Royal Society of South Australia*, **101**, 45–50.

TWIDALE, C. R. HORWITZ, R. C. & CAMPBELL, E. M. (1985). Hammersley landscapes of northwest Western Australia. *Revue de Géologie Dynamique et de Géographie Physique*, **26**, 32–50.

VEEVERS, J. J. (ed.) (1984). *Phanerozoic Earth History of Australia*. Oxford: Oxford Scientific Publications.

VEEVERS, J. J. & WELLS, A. T. (1962). Geology of the Canning Basin. *Bureau of Mineral Resources, Geology and Geophysics, Australia, Bulletin*, **60**.

WALKER, P. H. & BUTLER, B. E. (1983). Fluvial process. In *Soils*: An Australian viewpoint, pp. 83–90. Melbourne: CSIRO.

WALKER, P. H., CHARTRES, C. J. & HUTKA, J. (1988). The effects of aeolian accession on soil development on granitic rocks in southeastern Australia. I. Soil morphology and particle size distributions. *Australian Journal of Soil Research*, **26**, 1–16.

WALTHER, J. (1915). Laterit ein West Australian. *Zeitschrift-Deutsch Geologie*, **67B**, 113–40.

WASSON, R. J. (1982). Landform development in Australia. In *Evolution of the Flora and Fauna of Arid Australia*, ed. W. R. Barker & P. J. M. Greenslade, pp. 23–33. Frewville: Peacock Publications.

WASSON, R. J. & CLARK, R. L. (1987). The Quaternary in Australia – past, present and future. Australian Geoscience. *Bureau of Mineral Resources Australia, Report*, **282**, 29–34.

WELLMAN, P. (1987). Eastern Highlands of Australia: their uplift and erosion. *BMR Journal of Australian Geology and Geophysics*, **10**, 277–86.

WELLMAN, P. & McDOUGALL, I. (1974). Potassium-argon ages on the Cainozoic volcanic rocks of New South Wales. *Journal of the Geological Society of Australia*, **21**, 247–72.

WILLIAMS, E. (1989). Summary and synthesis. In *Geology and Mineral Resources of Tasmania*, ed. C. C. Burrett & E. L. Martin. *Geological Society of Australia Special Publication*, **15**, 468–99.

WOOLEY, D. R. (1978). Cainozoic sedimentation in the Murray drainage system, New South Wales section. *Proceedings of the Royal Society of Victoria*, **90**, 61–5.

WOPFNER, H. (1974). Post-Eocene history and stratigraphy of northeastern South Australia. *Transactions of the Royal Society of South Australia*, **98**, 1–12.

YOUNG, R. W. (1986). Tower karst in sandstone: Bungle Bungle Massif, northwestern Australia. *Zeitschrift für Geomorphologie*, **30**, 189–202.

YOUNG, R. W. & McDOUGALL, I. (1982). Basalts and silcretes on the south coast near Ulladulla, southern New South Wales. *Journal of the Geological Society of Australia*, **29**, 425–30.

6 Patterns in the history of Australia's mammals and inferences about palaeohabitats

M. ARCHER, S. J. HAND & H. GODTHELP

During the last two decades, vertebrate palaeontological research in Australia has entered a new
phase of development, with more investigators
backed by a significant increase in financial support from government and private financial
sources. The consequences of this accelerated
phase of investigation has been rapid growth in
information about vertebrate diversity, phylogenetic relationships, biocorrelation, palaeobio-
geography and palaeoecology. In this review, we
consider highlights of the developing late Mesozoic–late Cenozoic record of Australian terrestrial mammals, in part because the Cenozoic
record of these is better known than that for any
other group of vertebrates and in part because
the ability to infer aspects of palaeohabitats from
anatomical features is perhaps greatest for this
group.

Most modern orders of mammals underwent
adaptive radiations between the Late Cretaceous
and late Paleogene subsequent to the Early to
mid-Cretaceous diversification of angiosperms.
For this reason many aspects of the history and
structure of Australia's mammalian herbivores
reflect the requirements of harvesting and con-
suming particular groups of flowering plants. In
so far as this correlation holds, it is possible to
infer from the structure of the dentition of extinct
herbivores aspects of the vegetation upon which

they fed. Although experimental studies (e.g. Sanson, 1989) of the function of the teeth of living
Australian herbivores are few, deductive analysis
of the diets of extinct forms based on diets of
living species enables hypotheses about the timing
of key mid–late Tertiary changes in the structure
of Australia's terrestrial communities.

Higher-level systematic nomenclatures used
here follow those of Aplin & Archer (1987; marsupials), Watts & Aslin (1981; rodents) and
Walton & Richardson (1989; bats and other mammal groups). Biostratigraphic nomenclature,
unless otherwise indicated, follows those of
Woodburne et al. (1985) and Archer et al. (1989,
1991). The positions of the major fossil sites discussed in this chapter are shown in Figure 6.1
and the current understanding of the ages of the
sites is shown in Figure 6.2.

AUSTRALIAN MAMMAL DIVERSITY

There are 12 groups of ordinally distinct endemic
Australian mammals. Ten have still-living representatives: the egg-laying monotremes (platypuses and echidnas; since Early Cretaceous),
marsupial dasyuromorphians (dasyures, numbats
and thylacines; from Early Eocene), bandicoots
(bilbies, ordinary and forest bandicoots; from

Figure 6.1 Map showing the locations of the major fossil sites mentioned in this chapter.

Early Eocene), marsupial moles (from Early Miocene), diprotodontians (koalas, wombats, possums and kangaroos; from Late Oligocene), bats (from Early Eocene), whales (from Oligocene), rodents (from Early Pliocene), sirenians (dugongs; from late Tertiary), carnivores (dingoes from mid-Holocene and seals from Late Pleistocene) and primates (humans; from Late Pleistocene) (Archer, 1984; Rich, 1991). Of the two extinct ordinal-level groups known, yalkaparidontians

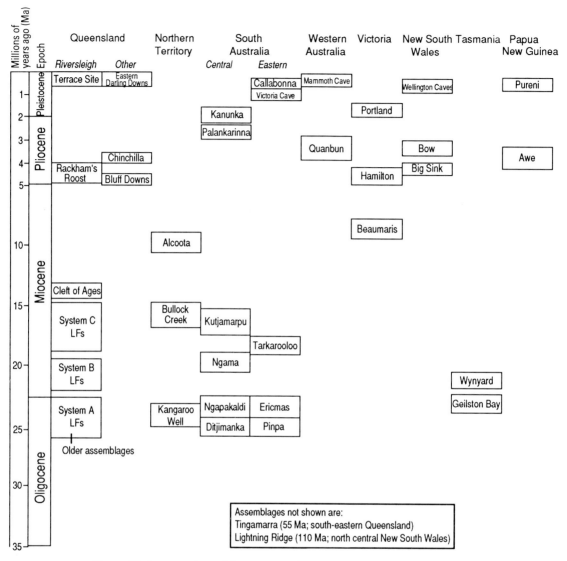

Figure 6.2 Summary chart of the current understanding of the ages of key Australian fossil mammal assemblages. The site names represent local faunas (LF singular, LFs plural).

(Archer *et al.*, 1988a) appear to have been a relatively minor radiation of highly autapomorphic marsupials that vanished between the Middle Miocene and Early Pliocene while Australian '?condylarthrans' (Godthelp *et al.*, 1992) appear to have been nonvolant placental terrestrial mammals that died out on this continent between the Early Eocene and Late Oligocene.

ORIGINS OF AUSTRALIA'S MAMMALS: SOURCE AREAS AND DISPERSAL ROUTES

Until recently, monotremes were presumed to be unique to Australia. This impression was furthered with discovery of the Early Cretaceous monotreme from Lightning Ridge (Archer *et al.*,

1985), Australia's oldest and only Mesozoic mammal. However, discovery of a platypus-like monotreme from 61–63 Ma old sediments in Patagonia (Pascual *et al.*, 1992*a,b*) demonstrates that the group had a Gondwanan distribution prior to the vicariated isolation of monotremes in Australia between Middle Eocene and Late Oligocene time. The possibility of extinct African and New Zealand monotremes should not be dismissed.

Australian marsupials have been presumed to have had a long, isolated history, one reflected by endemic ordinal- and family-level diversity. Recent discoveries in South America, however, have led to a revision of this view. Bonaparté (1990) demonstrated that the Late Cretaceous Los Alamitos faunal assemblage of South America contains no marsupials. Pascual & Ortez-Jaureguizar (1993) have further suggested that marsupials did not disperse to South America from North America until the latest Cretaceous or Early Paleocene because there are no undoubted marsupial fossils from South America prior to 61–63 Ma ago. Given current understanding that Australia's marsupials are monophyletic and descendants of microbiotheriids that are known from only the Cenozoic of South America and Antarctica (Aplin & Archer, 1987), the late appearance in South America sets constraints on the time for dispersal of marsupials from there, through Antartica (prior to Late Eocene; Woodburne & Zinsmeister, 1984), to Australia, which presumably occurred between 70 and 63 Ma ago. The oldest known Australian marsupials, at 55 Ma (from Tingamarra, southeastern Queensland: Godthelp *et al.* 1992), do not contradict this hypothesis.

The source for the ?condylarth from Tingamarra, the only nonvolant placental from Australia, is uncertain. Condylarths are known from the early Tertiary of South America (Marshall & de Muizon, 1988) and the Cretaceous of all northern continents. The affinities of the Australian ?condylarth are unclear but it at least resembles some condylarths known from the Cenomanian of Asia and some from the early Cenozoic of South America.

Australian microchiropteran bats were assumed, until recently, to have had a Eurasian origin, probably arriving sometime in the Oligocene. Discovery of bat-rich Oligo-Miocene sediments at Riversleigh, northwestern Queensland (Hand, 1987, 1989) and a single microchiropteran tooth from Late Oligocene sediments in the Lake Eyre Basin (Archer, 1978) suggested that rhinolophoid bats arrived sometime prior to the Late Oligocene. Now, however, discovery of an archaic (palaeochiropterygoid) bat from the Early Eocene Tingamarra Local Fauna (F) (Hand *et al.*, 1993) makes it clear that plesiomorphic bats were in Australia at least 10 Ma prior to separation of the continent from Antarctica. The source for these oldest Australian bats is unclear but similar bats are known from the Early and Middle Eocene of Europe and North America. Considering palaeogeographical patterns of the continents in the Early Eocene, there are two possible dispersal routes for Australia's microchiropteran bats: a rather circuitous route through North and South America and Antarctica, or a more direct route from southeastern Asia, via an archipelago, to Australia. There are no early Tertiary South American or Antarctic bats despite faunas from South America, at least, being reasonably well-known. In contrast, evidence for an Asian/archipelagic source is supported by the palaeogeographic studies summarised by Audley-Charles (1987) and by palaeobotanical studies by Truswell and others (e.g. Truswell *et al.*, 1987), which suggest that good dispersers among plant groups have used an archipelagic route to exchange in both directions between Asia and Australia since at least the Early Cretaceous.

Megachiropteran ('fruit') bats, common today in areas of northern and eastern Australia, are unrepresented in the Tertiary deposits of Australia; hence the time of their dispersal to this continent is unclear. This group of bats is represented in the mid-Tertiary of Africa and Europe and today is largely restricted to the Equatorial regions of the Old World. There can be little doubt that the source area for Australia's megachiropterans was southeastern Asia.

Australian cetaceans, sirenians and seals are in the main a subset of populations widespread in the southern oceans (Fordyce, 1991). The only probable exception to this cosmopolitan association is an endemic freshwater river dolphin (rhabdosteid) recovered from the Late Oligocene Namba Formation of the Lake Frome area (Fordyce, 1991). It is likely that the prior source for this freshwater lineage was the oceans bordering the southern margin of the continent, presumably via the ancestral Spencer Gulf.

Australian rodents, all of which are members of the family Muridae, are undoubtedly of Eurasian origin and probably arrived via island-hopping through archipelagoes linking Australia to southeastern Asia, sometime after the Middle Miocene and prior to the Early Pliocene (Godthelp, 1990).

Endemic Australian terrestrial carnivores include only the dingo (*Canis familiaris*). Newsome & Coman (1989) suggested that the earliest known specimens, approximately 3.5 ka old, exhibit attributes indicating that they were brought into Australia as domesticates of Aborigines.

Humans appear to have arrived sometime between 60 and 140 ka ago. There is argument about the timing, nature and number of human invasions, although considerable evidence suggests that there were at least two, an early one from southeastern Asia and a more recent one from mainland China.

THE STRATIGRAPHIC SOURCE FOR AUSTRALIA'S MAMMAL RECORD

As a continent, Australia has the second most poorly documented record of early mammals (exceeded only by Antarctica; Savage & Russell, 1983). Although mammals are distinguishable from derived synapsids by the Late Triassic on other continents and are diverse by the late Jurassic of the northern hemisphere, Australia's record for this class is mainly confined to the Cenozoic. A single monotreme (*Steropodon galmani*; Archer *et al.*, 1985) from the Early Cretaceous Griman Creek Formation at Lightning Ridge, northwestern New South Wales, is the only undoubted Australian Mesozoic mammal, although a second

possible mammal has been reported from the same deposit (Rich *et al.*, 1989).

The only early Tertiary mammal-containing assemblage from Australia is the as yet little-known Tingamarra LF from Murgon, southeastern Queensland (Godthelp *et al.*, 1992). Marsupial and placental fossils from Tingamarra were collected from an unnamed lacustrine clay that contained illites dated (K–Ar) at 55 Ma old (Godthelp *et al.*, 1992).

The earliest mid-Tertiary records for terrestrial mammals are arguably Late Oligocene or Early to even Middle Miocene in age. Doubts about the ages of the Oligo-Miocene central Australian mammal assemblages hinge primarily on the reliability of foraminiferal biocorrelation (Lindsay, 1987; Archer *et al.*, 1989, 1991; Rich *et al.*, 1991). One species (*Buliminoides* sp. *cf. B. chattonensis*), recovered in abundance from the Etadunna Formation, is otherwise known from only Late Oligocene deposits in other areas of the world. The vertebrates from central and northern Australia were derived from lacustrine and fluviatile clays and conglomerates of the Tirari Desert (the Etadunna and Wipajiri Formations) and the Lake Frome Embayment, South Australia (the Namba Formation – which is evidently significantly time transgressive; Martin, 1990) and freshwater limestones at Kangaroo Well, southern Northern Territory, Bullock Creek of central Northern Territory (Camfield Beds) and Riversleigh, northwestern Queensland (Plane & Gatehouse, 1968; Stirton *et al.*, 1968; Archer *et al.*, 1989, 1991). The central Australian deposits appear to span the Late Oligocene to Middle Miocene, ?Late Pliocene and Late Pleistocene (Callen & Tedford, 1976; Woodburne *et al.*, 1985; Archer *et al.*, 1989, 1991; Rich *et al.*, 1991). Fossil plants are known from the Middle Miocene Leaf Locality in the Wipajiri Formation (Stirton *et al.*, 1967) but these have not yet been comprehensively studied. The Riversleigh deposits (Figures 6.3, 6.4) appear to span the Late Oligocene to Middle Miocene (or possibly early Late Miocene) and Early Pliocene to Holocene. Fossil plants have just been recovered from the ?Middle to early Late Miocene Dunsinane Site at Riversleigh (Archer *et al.*,

Figure 6.3 The Oligo-Miocene (foreground), Pliocene, Pleistocene and Holocene limestones of Riversleigh, northwestern Queensland, have produced many unique fossil mammals as well as hundreds of other fossil vertebrates. (Photo by R. Arnett.)

1991), but most of these have not yet been studied. A small, probably Early Miocene, mammal assemblage is known from Geilston Bay, Tasmania (an unnamed clay deposit; Tedford et al., 1975). A single mammalian taxon, dated with certainty as Early Miocene, is known from Wynyard, Tasmania (Fossil Bluff Sandstone; Woodburne et al., 1985). Undoubted Middle Miocene mammal-bearing sites, such as the Batesford Quarry in Victoria, are very few in number and poor in fossil material.

Undoubted Late Miocene mammals are known from lacustrine clays of Alcoota in the southern Northern Territory (Woodburne, 1967; Woodburne et al., 1985) and near-shore marine deposits at Beaumaris, southern Victoria (Woodburne et al., 1985). Pliocene mammals are known, (for example, from: Hamilton, southwestern Victoria (Woodburne et al., 1985; Rich et al., 1991); Bluff Downs, northeastern Queensland (Archer &

Wade, 1976; Woodburne et al., 1985; Rich et al., 1991); Chinchilla, southeastern Queensland (Woodburne et al., 1985; Rich et al., 1991); South Australia (Pledge, 1984; Woodburne et al., 1985; Rich et al., 1991); and possibly northwestern Western Australia (Flannery, 1984). Pleistocene mammal deposits are very common and found in all states of Australia (Archer & Hand, 1984).

An attempt to correlate faunistically Australia's Tertiary mammal assemblages using Simpson's coefficient of faunal similarity was made by Rich et al. (1991). Although possibly useful at a superficial level, as noted by Rich et al. (1991), it suffers from a significant disparity in information about the assemblages involved in the analysis. Thus, the various Riversleigh Oligo-Miocene assemblages correlate with each other in large part because of their greater approximation to natural diversity, in contrast with those from other areas of Australia. Possibly significant correlations

Figure 6.4 Vertical sequences of Oligo-Miocene limestones on Riversleigh have produced a vast array of taxa that accumulated in carbonate-rich pools within lowland rainforests.

between specific Riversleigh assemblages and others from central Australia (e.g. the rat-kangaroo *Wakiewakie lawsoni* and the marsupial lions *Wakaleo oldfieldi* and *W.* cf. *W. oldfieldi* shared only between Riversleigh's Upper Site LF and the Kutjumarpu LF of central Australia) are thereby obscured.

SYNOPSIS OF THE PALAEOHABITATS FOR AUSTRALIA'S MAMMALS

Early Cretaceous

There is very little that can be deduced with confidence about the environment of the poorly

known Early Cretaceous mammals of Australia. Because the holotype and only known specimen of *Steropodon galmani* is a jaw fragment, we cannot determine much about its habits other than that it was probably a generalised carnivore/omnivore. Little can be concluded with confidence about palaeoclimate on the basis of the rest of the Lightning Ridge LF because there are arguments about the temperature tolerance of most groups involved (Molnar, 1991). The size of *S. galmani*, about that of a rabbit, is of interest because it is equal to or larger than the largest Mesozoic mammal known, even overlapping in size the smaller dinosaurs (Archer *et al.*, 1985). Large size among mammals is commonly a function of latitude, the larger the mammal the smaller its surface area:volume ratio and the easier it is for it to maintain thermal homeostasis. Considering recent suggestions (Rich & Rich, 1989) that during the Early Cretaceous, dinosaurs in nearby Victoria were living within the Antarctic Circle in climates that involved near 0 °C temperatures, the Lightning Ridge area of New South Wales was probably, though not certainly, cold. Modern echidnas endure subzero periods in the Alps of Australia by adjusting their metabolic rate through hibernation (Augee, 1991).

Late Paleocene/Early Eocene

No undoubted Paleocene mammal-bearing deposit is known from Australia but the Tingamarra LF from Murgon, southeastern Queensland, is arguably older than earliest Eocene, the illite date of 55 Ma possibly being a minimal age (Godthelp *et al.*, 1992). Tingamarra mammals so far recovered include only insectivorous and omnivorous marsupials, an omnivorous ?condylarth-like placental and a small insectivorous bat. There are no undoubted folivores and hence no indications of the nature of the vegetation that surrounded the Tingamarra depositional basin. Preponderance of insectivorous and omnivorous marsupials is a feature also of the Early Eocene Casamayoran mammals in Patagonia, one group of which (caroloameghiniids) may have had a representative in the Tingamarra LF (Godthelp *et al.*,

1992). Assemblages of this kind, dominated by small omnivores, suggest that at least some of these mammals were arboreal frugivores. In modern Australia, most comparably small omnivorous marsupials are arboreal possums. Hence we could suggest that the vegetation surrounding the Tingamarra lake included fruiting angiosperms. Other vertebrates in this deposit include unidentified small teleost fish, birds, frogs and a variety of reptiles including a madtsoiid snake (J. Scanlon, personal communication) and abundant large crocodylids (P. Willis, personal communication) and trionychid turtles (Gaffney & Bartholomai, 1979). Because large reptiles dominate this vertebrate fauna, we conclude that the Early Eocene climate of this area of southeastern Queensland was mild to warm throughout the year.

Late Oligocene to Early Miocene

No mammals are known from Australia between 55 and 25 million years of age; this 'dark age' includes the time when Australia separated from Antarctica, sometime between 45 and 38 Ma ago (Veevers, 1984). It also spans the Late Eocene–Early Oligocene, when on all other continents there was a major decline and turnover in mammalian faunas (e.g. the 'Grand Coupure' in Europe and the 'Terminal Eocene Event' of North America).

Many fossil assemblages span the interval between Late Oligocene and Early Miocene (Woodburne et al., 1985; Archer et al., 1991; Rich et al. 1991). Some of these provide reasonable indications of the structure of Australia's Tertiary mammal communities.

In central Australia, it is clear that biotas of this age from the Tirari Desert of the Lake Eyre Basin and those from the Lake Frome Embayment east of the Flinders Ranges share much in common, including the only representatives of a number of unique marsupial genera (e.g. Ilaria and Pilkipildra). Differences in the mammalian biotas of these two areas are generally at the species level except for the occurrence of freshwater river dolphins in the Namba Formation of Lake Pinpa, Frome Downs Station (Fordyce, 1991). Most if not all of these central Australian LFs have been recovered from fluvio-lacustrine deposits and are dominated by teleost fish, chelid turtles, crocodiles and sometimes water birds. It is presumed that the mammals were either washed into the depositional basins by streams (e.g. the Tarkarooloo LF; Rich et al., 1991), accumulated from the edges of the lakes possibly by aquatic predators (e.g. the Ditjimanka LF; Rich et al., 1991) or, in the case of larger animals sometimes found as articulated skeletons, possibly trapped along the muddy margins of drying lakes (e.g. the Pinpa LF; Tedford et al., 1977). Arguments that pollens from the Etadunna clays of the Tirari Desert indicate abundant grasslands interspersed with rainforest (Nothofagus subgenus Brassospora, W. Harris personal communication to Woodburne et al., 1985) during the Late Oligocene to Early Miocene have recently been challenged. Following reanalysis of Harris's material, Martin (1990) suggested that the grasses in the samples may represent aquatic species rather than evidence for mid-Tertiary central Australian grasslands. The abundant browsing marsupials and arboreal possums but lack of grazing mammals in the central Australian biotas corroborates Martin's interpretation.

In northern Australia, as evidenced particularly by the diverse Riversleigh assemblages, the balance of evidence from the older assemblages suggests that during the Oligo-Miocene, the area was covered in species-rich, lowland rainforest. Evidence for this conclusion was summarised by Archer et al. (1989, 1991) and includes a high diversity of sympatric arboreal folivorous marsupials and abundance of vertebrate groups whose living descendants inhabit only rainforest (e.g. log-runner birds, musky rat-kangaroos and Pseudochirops-type ringtail possums). Further, the structure of the Riversleigh mammal communities, with high taxonomic diversity but lack of markedly dominant species, resembles modern rainforest rather than mesic communities. There is some geological evidence for more open forest communities beyond the depositional basins containing the oldest of Riversleigh's Oligo-Miocene deposits (the 'System A' deposits; D. Megirian, personal communication), but there is no

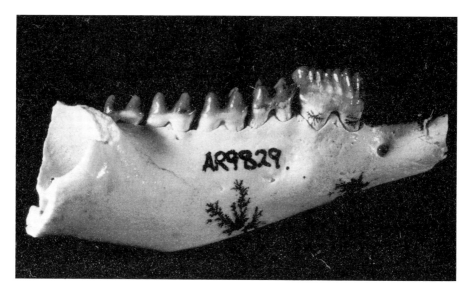

Figure 6.5 Riversleigh has a plethora of fossil macropodid kangaroos but all, like this undescribed species, are browsers with low-crowned molars and long, vertically ridged third premolars. (Photo courtesy of B. Brown.)

biological evidence that grasslands surrounded these basins, there being no sign of grazing adaptations in the dentitions of, for example, any of the kangaroos (Figure 6.5) so far recovered; many, however, show browsing adaptations.

The decline of the Early – Middle Miocene Riversleigh warm and everwet closed forests may be indicated by a single leaf fragment from the ?early Late Miocene Dunsinane site of Riversleigh, which is relatively small and has a dentate margin, features more common in temperate rainforest (which is unlikely at Riversleigh) or seasonal vegetation (R. Hill, personal communication).

Considering Late Oligocene – Middle Miocene climates, the rainforest faunas of northern Australia suggest a warm, wet and possibly seasonal climate. Contemporaneous climates of central Australia are less clear. Abundant crocodiles and large turtles suggest warm temperatures. *Nothofagus* subgenus *Brassospora* pollen suggests high rainfall. However, compared with the Riversleigh biotas of northern Australia, the central Australia assemblages are relatively depauperate in arboreal species with, for example, no more than three folivorous pseudocheirids per LF (cf. nine in

some Riversleigh assemblages). Further, in some areas (e.g. 'Discovery Basin' of Lake Pitikantia) there appears to have been a decided faunal dominance by the terrestrial palorchestid marsupial *Ngapakaldia tedfordi*. This sheep-sized marsupial appears to have been present in relatively large numbers, possibly as 'mobs'. Numerical dominance by this generalised browser suggests that, although the local region supported rainforest, areas of open forest or woodland may also have been present.

The Early Miocene environments of Tasmania (Geilston Bay LF, Wynyard LF) contained a subset of the variety of marsupials present at the same time in central and/or northern Australia but none that gives any clear indication of local climates or habitats.

Middle–Late Miocene

What are interpreted to be diverse Middle Miocene assemblages from central Australia (Kutjamarpu LF from the Tirari Desert of South Australia; Woodburne *et al.*, 1985; Rich *et al.*, 1991) and northern Australia (e.g. 'System C' LFs of Archer *et al.*, 1989, 1991, and the Bullock

Creek LF of the Northern Territory *fide* Woodburne *et al.*, 1985) are reasonably well known and indicate at least local wet forest environments characterised by browsing herbivores. It is possible that these are riparian assemblages and that drier, more open habitats existed further from these sites of deposition.

The earliest evidence for possible grazing adaptations in Australian Oligo-Miocene mammals occurs in the Bullock Creek LF of the Northern Territory and a few of the stratigraphically highest of the 'System C' LFs at Riversleigh. In the former LF, a single, relatively high-crowned macropodid kangaroo molar appears to represent an animal adapted to feeding on abrasive plant materials (B. Cooke, personal communication), although these need not have been grasses. In the latter LF, the presence of a limited number of high-crowned 'proto-wombat' teeth suggest that these marsupials were already specialising on abrasive grasses or were pre-adapted for this diet by specialising on some other abrasive food available within the late Middle Miocene Riversleigh forests. Modern wombats are all grazers but these forms also have far more specialised, rootless and high-crowned teeth than their Riversleigh antecedents.

Late Miocene

There are only two significant Late Miocene mammal-producing sites in Australia: Beaumaris in southern Victoria, and Alcoota in the southern Northern Territory. The Beaumaris LF, obtained from a near-shore marine deposit, is depauperate and provides little information relevant to interpreting palaeoclimates or palaeohabitats.

Although the Alcoota LF is the only reasonably diverse Australian Late Miocene assemblage, it lacks almost all representation of arboreal mammals. Even the single possibly arboreal animal, a species of *Pseudochirops*-like possum known from a single dentary, is not certainly arboreal. A close living relative, the folivorous rock-haunting ringtail possum (*Petropseudes dahli*) inhabits rock piles in areas of northern Australia and is only partly arboreal. Evidently, by Late Miocene time, the

rainforests of at least the southern Northern Territory had given way to open sclerophyllous forest or, more probably, woodland. Yet, in contrast to modern woodland herbivores, the few kangaroos and diverse diprotodontoid marsupials present in the Alcoota LF were mostly browsers, suggesting that although the forests of that time and place were no longer closed, abrasive grasses were not a significant component of the emergent understorey.

Pliocene

There seems to be a significant evolutionary gap in the history of Australia's mammals between Late Miocene assemblages such as the Alcoota LF and the Early Pliocene assemblages of northern and southeastern Australia. The latter are characterised by grazing as well as browsing kangaroos, larger but less diverse browsing diprotodontoids and a variety of specialised grazing wombats. This distinction increases through the Middle and Late Pliocene assemblages of Australia. The generalisation, however, may not hold for New Guinea where some of the ?Middle Pliocene Awe mammals (e.g. *Kolopsis rotundus*; Plane, 1967; Rich *et al.*, 1991) more closely resemble Australian Late Miocene forms. A possible explanation for this may be that the Awe area of New Guinea was ecologically more stable during the Late Tertiary and as a consequence provided a refuge for lineages that became extinct earlier in Australia.

Early Pliocene assemblages such as the Bluff Downs LF of northeastern Queensland, the Bow LF of eastern New South Wales and the Sunlands LF of South Australia are also the first to exhibit a high percentage of mammals representing still-living genera. The Late Miocene Alcoota LF has only one taxon: the *Pseudochirops*-like possum. In contrast, the Bluff Downs LF (for example) contains species of *Macropus (Macropus)*, *M. (Osphranter)*, *Petrogale*, *Pseudochirops*, *Perameles* and *Planigale*. One Early Pliocene assemblage, the Hamilton LF of southwestern Victoria (Rich *et al.*, 1991) is the only one that appears to have accumulated in a rainforest. It contains generic-level taxa

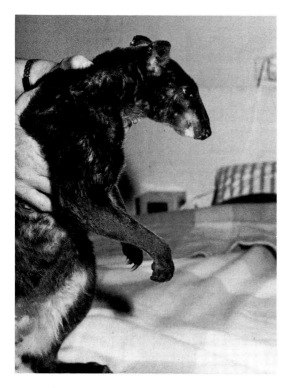

Figure 6.6 A living forest wallaby (*Dorcopsis*) from New Guinea. The dentition of this rainforest browser is similar in form to that which characterises the macropodids of Australia's Oligo-Miocene rainforests.

typical of modern rainforests in northeastern Queensland or New Guinea such as *Hypsiprymnodon*, *Strigocuscus*, *Thylogale*, cf. *Dendrolagus* and *Dorcopsis* (Figure 6.6). However, the presence of species of *Petaurus*, gliding possums that today are not normally found in rainforest suggests the proximal occurence of more open forests.

Although there are no undoubted diverse Late Pliocene mammal assemblages, the Kanunka LF, from the Tirari Desert of central South Australia, may be of this age (Tedford *et al.*, 1986). It contains an abundance of Pleistocene-like taxa as well as additional representatives of living genera including *Lagorchestes*, *Bettongia* and *Vombatus*. There is nothing here, however, to suggest unambiguously the desert conditions that now dominate the Tirari Desert. It is probable that perhaps as late as 2.5 Ma ago, this region was dominated by open sclerophyllous forest or woodlands rather than grasslands and dunes.

Another major area of uncertainty concerns the western half of the continent. There are no diverse Pliocene mammal assemblages from Western Australia, the only one being the faunistically impoverished Quanbun LF of the Kimberley Region (Flannery, 1984).

Quaternary

All states of Australia and New Guinea have an abundance of Pleistocene and Holocene mammal assemblages (Archer & Hand, 1984; Murray, 1991). Although most lack radiometric dates, on the basis of faunal composition, the majority appear to be Late Pleistocene in age. The Fisherman's Cliff and Portland LFs of Early Pleistocene age represent distinctively different assemblages. The latter contains the last surviving representative of the otherwise mid–late Tertiary ?frugivorous ektopodontid possums (Figure 6.7) as well as many taxa not found in the later part of the Quaternary. The Quaternary of New Guinea is least well understood but from recently discovered deposits in Irian Jaya and Papua it is clear that some megafaunal groups such as zygomaturine diprotodontoids that became extinct in Australia approximately 30 ka ago survived in New Guinea at least until the latest Pleistocene.

Regional changes in habitat have been detected from changes in the balance of indicator species (e.g. local extinction of the arid/semiarid western barred bandicoot, *Perameles bougainville*, from the extreme southwest of Western Australia approximately 20 ka ago; Merrilees, 1979). On a broader scale, it is evident that several periods of aridity and lowered temperatures in Australia correspond to periods of glacial maxima in the northern hemisphere. For example, approximately 17 ka ago, severe aridity caused a major contraction in the rainforest biotas of northeastern Queensland (Kershaw, 1983). Although there is no evidence from the mammal record itself because of an almost total lack of undoubted Early–Middle Pleistocene deposits from eastern Australia, it is probable that Australia's wet forest mammals underwent significant declines in diversity and range during these periods of climatic rigour.

Gigantism in Australia's mammals as well as

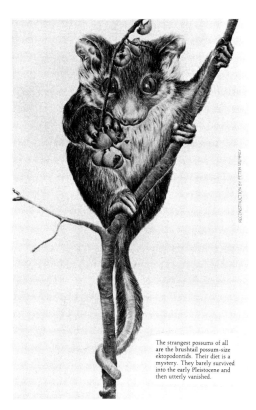

Figure 6.7 A suggested likeness of an Oligo-Miocene ektopodontid possum from central Australia. These once widespread rodent-like marsupial possums declined until their extinction in the Early Pleistocene. (Drawing by P. Murray.)

Figure 6.8 Life-size reconstructions in the Australian Museum's 'Dreamtime to Dust' gallery of megafaunal species. The giant Late Pleistocene varanid lizard, *Megalania prisca*, is shown scavenging a carcass of the largest marsupial known, the herbivorous diprotodontoid *Diprotodon optatum*. (Courtesy of the Australian Museum.)

other vertebrates is evident after Middle Miocene time and is particularly marked in herbivorous marsupial lineages that survived the Late Miocene (e.g. kangaroos, wombats and diprotodontoids). Whereas during the Oligo-Miocene, diprotodontoids, for example, were on average about the size of a calf, by mid-Pliocene to Early Pleistocene time, this mean had shifted to the size of a cow. By Late Pleistocene time, there were no small diprotodontoids left on mainland Australia, although calf-sized zygomaturines persisted in Irian Jaya (Plane, 1991) and slightly larger, panda-like zygomaturines persisted in Papua (Flannery & Plane, 1986). This gigantism, which also affected other vertebrates (e.g. the giant lizard *Megalania prisca* (Figure 6.8) and the giant madtsoiid python *Wonambi naracoortensis*), resulted in what has been described as the 'megafauna' (Murray, 1991). The

effects the megafauna may have had on Australia's plant communities are unclear, but presumably included the ability of terrestrial browsers to forage at progressively greater heights, the ability to transport large seeds for greater distances and the effects of graviportal trampling of surface plants and soils.

A converse trend for dwarfing is evident among Late Pleistocene–Holocene giants (Marshall & Corruccini, 1978), such that the mean size of some species (e.g. large grazing kangaroos in the genus *Macropus*) may have declined by more than 30%. Possible ecophysiological reasons for these declines have been considered (see e.g. Archer, 1984; Murray, 1991) and include the reproductive disadvantage larger mammals must have endured

Figure 6.9 An array of Late Pleistocene megafaunal species from Australia with a human to give scale. (Drawing by and courtesy of Peter Murray.)

in attempting to raise large, dependent young in ecosystems with an unpredictable resource base such as sometimes characterised more than 50% of Late Quaternary Australia.

Late Pleistocene mammal extinctions were profound (Figure 6.9), resulting in the loss of at least 40 species of mammal, including extinction of all species greater than 60 kg in weight (Murray, 1991). The timing and cause(s) of these extinctions are the subject of controversy but, whenever this loss began and for however long it lasted, it had largely run its course by 25 ka ago, after which time most species present survived into the modern world. Climatic change and the direct and indirect effects of humans were almost certainly primary causes for these extinctions, which appear to have focussed on large, more conspicuous and perhaps less adaptable species such as marsupial lions, giant short-faced kangaroos, diprotodontoids and giant wombats. However, a few giant grazing kangaroos, such as the eastern grey kangaroo, red kangaroo and euro, survived. Following European interference in land practices and the introduction of nonendemic mammals, 20 small- to medium-sized mammals have become extinct.

Although it has commonly been presumed that extinctions of small mammals, such as the crescent nailtail wallaby (*Onychogalea lunata*, Figure 6.10) and pig-footed bandicoot (*Chaeropus ecaudatus*), after European settlement were the result of introduced animals or habitat clearance, Flannery (1990) has suggested an alternative explanation. Presuming, as seems reasonable, that some of the smaller species in pre-human Pleistocene communities of Australia depended on regrowth periodically revitalised by the browsing activities of now extinct megafaunal herbivores, extinction of these herbivores should have precipitated extinction of the dependent species. If, however, a concomitant change was an increase in frequency of burning following human arrival ('firestick farming'), cycling of the necessary seral stages may have continued despite loss of the megafaunal herbivores. Modern extinctions of small mammals may in part, then, be explained by European-induced changes to Aboriginal activities. With decline of Aboriginal land 'management' activities and the consequent simplification of formerly diverse ecosystems, small mammals that serendipitously escaped Late Pleistocene extinction after decline of the megafaunal

Figure 6.10 The crescent nailtail wallaby, *Onychogalea lunata*, one of the medium-sized wallabies that became extinct soon after Europeans settled Australia. (Painting by J. Gould.)

herbivores may have only now begun dropping out.

Studies of Pleistocene ecosystems on other continents (e.g. Diamond, 1990) have led to recognition of 'ghosts' in modern ecosystems, such as declines in distribution and diversity of large-seeded trees whose dispersal and perhaps seed germination depended on the activities of megafaunal species. Similarly, defensive features of some plants (e.g. spines and toxins) may be inexplicable in terms of modern herbivores and reflect, instead, defences against now extinct megafaunal herbivores. In Australian ecosystems, most plants that depended for seed dispersal on megafaunal species would by now, at least 25 ka after the extinction of most of these giant herbivores, have become extinct. Some, however, might have survived because of the seed-transporting activities of large animals such as cassowaries (*Casuarius casuarius*). Anachronous defensive structures or chemicals (e.g. possibly the toxins of the stinging trees) (*Dendrocnide* spp.) may similarly have survived, provided they did not disadvantage the plants in which they occurred.

KEY ASPECTS OF THE HISTORY OF PARTICULAR MAMMAL LINEAGES

Here we note key aspects of the history of families of Australian mammals that appear to reflect broader changes in Australia's mammal communities. Unless otherwise noted or already given above, references for these data can be retrieved from Archer (1984), Archer *et al.* (1991), Rich (1991) and Rich *et al.* (1991).

Monotremes (egg-laying mammals)

Ornithorhynchidae (platypuses)

The record starts in the Early Cretaceous with the possible ornithorhynchid *Steropodon galmani* from Lightning Ridge, New South Wales. One genus has recently been found in 61–63 Ma old sediments in Patagonia (Pascual *et al.*, 1992*a,b*) demonstrating a Gondwanan vicariance pattern. No post-Early Paleocene monotremes are known from anywhere other than the Australian region. *Obdurodon* spp. occur in the Oligo-Miocene sediments of central Australia and Riversleigh, where they reflect the presence of permanent fresh water. Loss of inland waterways around the Lake Eyre and Lake Frome areas meant the loss of platypuses from central Australia. *Ornithorhynchus anatinus*, the modern platypus, is the only survivor of a lineage that appears to have undergone geographical and morphological decline since the Middle Miocene. There is no evidence that platypuses ever existed in New Guinea.

Tachyglossidae (echidnas)

The record starts in the Late Miocene of New South Wales. Three genera are known: *Tachyglossus* (short-beaked, termite-eating echidnas) from Australia and New Guinea; *Megalibgwilia* (giant long-beaked, ?large insect-eating echidnas) from the Late Miocene to Pleistocene of Australia; *Zaglossus* (long-beaked worm-eating echidnas) from the Pleistocene of Australia and Holocene of New Guinea (Griffiths *et al.*, 1991). The reasons for the Pleistocene extinction of *Zaglossus* and *Megalibgwilia* from the Australian mainland are unclear but two points may be relevant: (1) modern *Zaglossus bruijnii* persists, albeit rarely, in New Guinean rainforest; (2) it is a favourite food item of the Papuans. Late Pleistocene declines in Australia's rainforests and the arrival of humans may, therefore, have contributed to its decline on the mainland.

Marsupials ('pouched' mammals)

Dasyuridae (dasyures), Thylacinidae (thylacines), Myrmecobiidae (numbats)

These were the common carnivores and insectivores of the late Tertiary and Quaternary. Dasyure diversity expands in northern Australia concomitant with a decline in bandicoot (perameloid) diversity. Thylacine diversity declined from at least five species in the Late Oligocene–Middle Miocene, ranging from small- to large-dog in size, to one large species in the Late Miocene (Alcoota) and the single modern species from the Early Pliocene–Holocene. Geographical decline included loss from New Guinea approximately 10 ka ago, loss from mainland Australia at the time dingoes were introduced by Aborigines about 4 ka ago and extinction from Tasmania in the 1930s following intense persecution by Europeans. Numbats, like honey possums and fruit bats, have not yet been found in the Tertiary.

Notoryctidae (marsupial moles), yalkaparidontians ('thingodontans'), yingabalanarids ('weirdodontans')

Marsupial moles are represented today by the single central desert species *Notoryctes typhlops*. Discovery of the only known and relatively plesiomorphic fossil member of this family in Oligo-Miocene rainforest deposits of Riversleigh suggests that they pre-adapted within forests for a subterranean lifestyle in deserts. The notion that the Tertiary rainforests were the source for most of Australia's dry country animals has been called the 'Green Cradle Concept' (Archer *et al.*, 1988*b*; Archer, 1990). Thingodonts (Figure 6.11) and weirdodonts exemplify another aspect of Australia's mammal record – the persistence of strikingly distinctive, presumably archaic Gondwanan groups of mammals until the Middle Miocene, after which time they evidently vanished with the great inland rainforests.

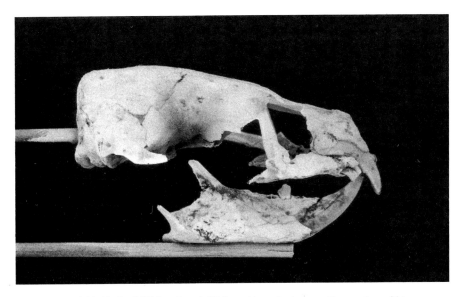

Figure 6.11 Skull of 'Thingodonta' (*Yalkaparidon coheni*), one of a number of bizarre, Gondwanan mammal groups that survived into the Oligo-Miocene of Riversleigh and then vanished, presumably in response to the Late Miocene collapse of Australia's continental rainforests.

Perameloids (bandicoots)

The perameloids, mouse- to cat-size marsupials, were present in the Early Eocene Tingamarra deposit and hence were distinct prior to Australia's separation from Antarctica. In the Oligo-Miocene rainforests of northern Australia insectivorous/omnivorous forms were among the most common type of mammal. Decline in bandicoot diversity presumably in the Late Miocene corresponded with an apparent rise in diversity of small insectivorous dasyurids. Arrival and diversification of the omnivorous/herbivorous rodents in the Late Miocene–Early Pliocene followed and may not have been, contrary to earlier speculations, the cause of the decline in bandicoot diversity. Today forest bandicoots (Peroryctidae) predominate in the rainforests of New Guinea, while common bandicoots (Peramelidae) predominate in the more open forests, woodlands and deserts of Australia. The burrowing bilbies (Thylacomyidae) are restricted to the arid/semi-arid habitats of Australia and have a fossil record no older than Pliocene. The rabbit-size pig-footed bandicoot (*Chaeropus ecaudatus*), a recently extinct peramelid from inland southern Australia,

was evidently a grazer and as such the smallest known from Australia (Wright *et al.*, 1991).

Phascolarctidae (koalas)

Koalas are uncommon elements in the Oligo-Miocene rainforests. Molar structure implies that throughout their history they have been arboreal browsers. Archer & Hand (1987) suggested that they may have specialised on the relatively rare eucalypts within Australia's Tertiary rainforests. If so, as the inland rainforests declined, koalas would have been advantaged by the spread of eucalypts, these trees being better adapted to poorer, nutrient-deficient soils. As a consequence, in modern eucalypt-dominated forests of eastern Australia, koalas appear to be far more abundant than they were in the rainforests of the Tertiary.

Vombatidae (wombats)

All living wombats are specialised grazers. Their more extraordinary adaptations include: rootless teeth that cannot be annihilated by abrasive foods; and their reshaping of ancestrally derived

Figure 6.12 Isolated molars of a 'proto-wombat' (*Rhizophascolonus crowcrofti*) from the Oligo-Miocene Kutjamarpu LF of central Australia. It may have been specialising on abrasive foods but, unlike modern grazing wombats, the molars of this taxon had roots.

subselenodont molar crowns into transverse ridges suitable for masticating grass, while still in their mother's pouch, using a tooth-sharpening process (thegosis). The fossil record indicates that Oligo-Miocene wombats were not as specialised (Figure 6.12), although high-crowned molars have been found as rare items in newly discovered ?late Middle Miocene sediments at Riversleigh (Archer *et al.*, 1991). By the Early Pliocene, at least two different kinds of grazing wombat were present in eastern Australia. By the Late Pleistocene there were many different kinds throughout the continent, including pygmy species as well as cow-sized giants. This diversity declined as part of the Late Pleistocene megafaunal extinctions, to leave three modern species.

Diprotodontoidea (diprotodontoids)

Holocene Australia is unique within at least the last 25 Ma because of the absence of the herbivorous diprotodontoids. Until at least the Late Miocene the two families in this group dominated most terrestrial herbivorous niches in central Aus-

tralia. As a generalisation (see above), progressively younger representatives of this superfamily became larger until, by the Pleistocene, some (e.g. *Diprotodon optatum*, the world's largest marsupial) were the size of a rhinoceros (Figure 6.13). In the Oligo-Miocene habitats of central and northern Australia, diprotodontoids appear to have been as common as large kangaroos are today. However, as the forests of these areas opened up in the Late Miocene–Early Pliocene, their numbers gave way to a steep rise in diversity of grazing as well as browsing kangaroos. Unfortunately, the actual diets of most diprotodontoids are unknown. Judging from molar morphology, however, late Tertiary and Pleistocene palorchestid diprotodontoids may have been cow- to horse-sized grazers.

Macropodoidea (kangaroos)

Almost all of the palaeocommunity trends evident in other groups are mirrored in one or another of the diverse kangaroo lineages. There are two families of kangaroos:

Figure 6.13 Palaeontologists from the Australian Museum excavating a skeleton of the Pleistocene *Diprotodon optatum* from northern New South Wales. Some of the bones recovered have marks that suggest that the animal was butchered or scavenged by humans. (Photo courtesy of the Australian Museum.)

1. The Potoroidae (rat-kangaroos) with browsing (Bulungamayinae), omnivorous (Hypsiprymnodontinae, Potoroinae) and carnivorous (Propleopinae) lineages.
2. The Macropodidae (ordinary kangaroos) with browsing (Macropodinae and Sthenurinae) and grazing forms (same groups).

None is known from the Early Eocene Tingamarra LF. In the Oligo-Miocene forest assemblages, there are no grazers (a single tooth being a possible exception, see above), reasonably abundant omnivorous (both subfamilies) and browsing potoroids, very abundant balbarine macropodids, and rare (as befits a meat-eater) carnivorous potoroids (Figure 6.14). In the Late Miocene faunas only browsing macropodids (macropodines and ?sthenurines) are known. In the Early Pliocene, there are rare hypsiprymnodontines, very abundant browsing (sthenurine, macropodine) and abundant grazing (macropodine) macropodids, and rare propleopines. By the Late Pleistocene browsers were very rare, whereas the grazing macropodines and sthenurines had undergone an explosive rate of speciation. Following Late Pleistocene megafaunal extinction, very few browsers (swamp wallaby (*Wallabia*), pademelons (*Thylogale*), quokka (*Setonix*), tree kangaroos (*Dendrolagus*)) survived on mainland Australia. In contrast, today, browsers (*Dorcopsis, Dorcopsulus, Dendrolagus, Thylogale*) dominate modern New Guinea, reflecting the widespread persistence there of wet forests. The late appearance (by Late Miocene), early diversification (by Early Pliocene) and even later major diversification (Pleistocene) of grazing kangaroos must be a reflection of the timing of the development of Australia's now extensive grasslands.

Pseudocheiridae (ringtail possums)

The history of this group is almost the inverse of that of the grazing kangaroos. Ringtail possums are arboreal folivores. They were very common in the Oligo-Miocene of Riversleigh with up to nine

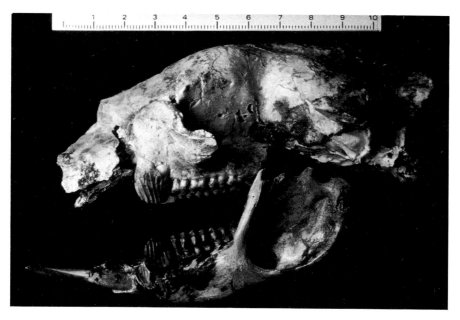

Figure 6.14 Skull and lower jaw of an Oligo-Miocene propleopine rat-kangaroo (*Ekaltadeta ima*). The teeth and cranial morphology of this large kangaroo indicate that it was carnivorous. (Photograph courtesy of R. Arnett.)

Figure 6.15 The green ringtail (*Pseudochirops* sp.) of Australia's wet tropics. All ringtail possums of this particular genus live in rainforest, one species in Australia, several in New Guinea. They are all arboreal folivores. The Oligo-Miocene rainforests of Riversleigh supported a wide variety of closely related as well as other genera. (Photo courtesy of the Queensland Museum.)

species in single LFs. Even in the Oligo-Miocene assemblages of central Australia there were as many as three sympatric species, albeit representing extinct groups. However, in the Oligo-Miocene sediments of Riversleigh there are diverse *Pseudochirops*-like and *Pseudochirops* ringtails, taxa that today exist only in the rainforests of New Guinea and Australia (Figure 6.15). Because of the rarity of Tertiary rainforest assemblages apart from Riversleigh, it is not until the Early Pliocene Hamilton LF that rainforest ringtails are again visible in this remnant of declining continental rainforest. In the drier parts of modern Australia, only the common ringtail (a decidedly autapomorphic taxon) is widespread, the others being confined either to wet forests in eastern Australia or to northern rocky habitats of the Northern Territory, Queensland and Western Australia. Ringtails also provide one (greater gliders, *Petauroides*) of three lineages of possums (pseudocheirids, petaurids, acrobatids) that have developed gliding capacities presumably in response to the late Tertiary widespread opening of the forest canopy. The first sign in the record of gliding possum *lineages* comes from the Hamilton LF of southwestern Victoria, where possible ancestors or close relatives of the gliding ringtails and gliding petaurids occur. Although it is far from clear that these Early Pliocene species (*Petauroides stirtoni*, *P. marshalli* and *Petaurus* spp.) could glide, the possibility qualifies assumptions that the Hamilton palaeohabitat was exclusively rainforest.

Placental mammals

Palaeochiropterygoidea (archaic bats)

The presence of a 'palaeochiropterygoid' in the Early Eocene Tingamarra LF of southeastern Queensland may be a palaeoclimatic indicator. Elsewhere in the world, microchiropteran bats of this type are known only from warm, tropical–subtropical palaeoenvironments such as the Early and Middle Eocene of southern Europe and central North America.

Hipposideridae (leaf-nosed bats)

Oligo-Miocene bat assemblages of tropical and subtropical middle Europe contained abundant hipposiderid bats. In the Australian fossil record, the bat-rich Oligo-Miocene deposits of Riversleigh are similarly dominated by a wide variety of hipposiderids (Figure 6.16), some of which persist in the riparian Pliocene Rackham's Roost LF (Hand, 1989). Presumably by Pleistocene and certainly by Holocene time, hipposiderid diversity in this region had steeply declined, with a concomitant rise in the diversity and abundance of vespertilionids, microchiropterans that today dominate the Australian as well as world faunas, modern hipposiderids having their centres of abundance mainly in the wet, Old World tropics (e.g. the lowland rainforests of Sarawak). Although hipposiderids no longer dominate the modern Australian bat fauna, they remain common in parts of the tropics. The overall phylogenetic and geographical decline in the Australian record presumably reflects the decline in extent and diversity of Australia's tropical forests.

Muridae (rodents)

At least one uncommon mouse-sized rodent (*Pseudomys vandyckii* Godthelp, 1990) is known from the Early–Middle Pliocene Chinchilla Sand of southeastern Queensland and a single undetermined murid is represented in the Bluff Downs LF (Archer & Wade, 1976). However, in the ?Early–Middle Pliocene Rackham's Roost LF of Riversleigh, northwestern Queensland, there is a plethora of plesiomorphic as well as derived endemic murid taxa, suggesting an earlier arrival and evolution in isolation for this group in Australia. Modern species of the genera represented in the Rackham's Roost deposit are only known from nonrainforest habitats. Godthelp (1990) suggested that murids first dispersed to Australia sometime in the Late Miocene or earliest Pliocene via sclerophyllous forest or woodland archipelagic corridors linking northern Australia and southeastern Asia. Lineages that today inhabit Australia's rainforest (*Uromys*, *Melomys* and some

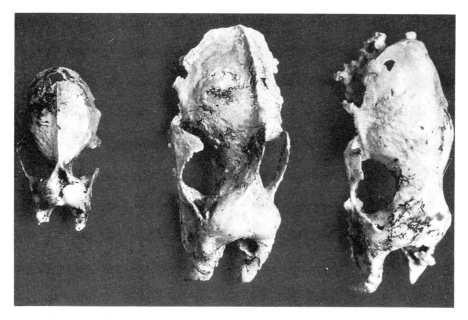

Figure 6.16 Before the collapse of the rainforests of northern Australia, hipposiderid microchiropteran bats, such as these from the Oligo-Miocene deposits at Riversleigh, were very diverse. (Photograph courtesy of R. Arnett.)

Rattus) do not make an appearance in the record prior to the Pleistocene and appear to have been Pleistocene immigrants via the rainforests of northeastern Queensland.

SUMMARY

The limited Early Cretaceous mammal record from central northern New South Wales tells us nothing about palaeoclimates but does corroborate the biological integrity of Gondwana. The better-known, but still limited Early Eocene record, from southeastern Queensland perhaps suggests a complex, fruit- and insect-rich forest. The mammal records of the Late Oligocene–Middle Miocene of central and northern Australia argue for at least regionally diverse lowland rainforests. Australia's bats dispersed to Australia, probably from Eurasia via an archipelago that had linked Australia to Asia since the Early Eocene. Most of the trends we have discussed above reflect the evident climatic crisis(es) that overtook Australia between the Middle Miocene and Early Pliocene. Most involve declines and extinctions of forest groups (e.g. ringtail possums) and the rise of open country groups (e.g. grazing kangaroos), with the Early Pliocene being the first period characterised by an abundance of grazing mammals. New Guinea and small areas of the wet tropics of northeastern Australia appear to have provided a refuge for specific forest groups that vanished from the rest of mainland Australia. Triple convergence in gliding among Australian possums appears to be a measure of the late Tertiary (Middle–Late Miocene) opening of the forests. The first diverse rodent assemblages suggest the group invaded in the latest Miocene–earliest Pliocene via relatively dry archipelagic corridors linking southeastern Asia and north central Australia. Climatic and/or human-induced crises appear to have precipitated a profound loss of mammal diversity in the Late Pleistocene. Modern extinctions may be a consequence of European-induced revitalisation of processes first set in train by Late Pleistocene megafaunal extinctions.

We acknowledge the vital support our research work on Tertiary mammals has had from (among others): the Australian Research Grant Scheme; the National Estate Grants Scheme (Queensland); the Department of Arts, Sport, the Environment, Tourism and Territories; the Queensland National Parks and Wildlife Service; the University of New South Wales; IBM Australia Pty Ltd; ICI Australia Pty Ltd; the Australian Geographic Society; Wang Australia Pty Ltd; the Queensland Museum; the Australian Museum; Mount Isa Mines Pty Ltd; Surrey Beatty & Sons Pty Limited; the Riversleigh Society; private supporters including Elaine Clark, Martin Dickson, Sue and Jim Lavarack, Sue and Don Scott-Orr and many enthusiastic students (notably Bernie Cooke, Brian Mackness, Jeanette Muirhead, Neville Pledge, John Scanlon and Paul Willis), professional colleagues (notably Alan Bartholomai, Alex Baynes, Tim Flannery, Peter Murray, Tom Rich, David Ride, Richard Tedford and Michael Woodburne) and many enthusiastic workers (notably Anna Gillespie, Cathy Nock and Stephan Williams). Photographs were taken by Ross Arnett (University of New South Wales).

REFERENCES

APLIN, K. & ARCHER, M. (1987). Recent advances in marsupial systematics with a new syncretic classification. In *Possums and Opossums: Studies in Evolution*, pp. xv–lxxii. Sydney: Surrey Beatty & Sons Pty Ltd and the Royal Zoological Society of New South Wales.

ARCHER, M. (1978). Australia's oldest bat, a possible rhinolophid. *Proceedings of the Royal Society of Queensland*, 89, 23.

ARCHER, M. (1984). The Australian marsupial radiation. In *Vertebrate Zoogeography and Evolution in Australasia*, ed. M. Archer & G. Clayton, pp. 633–808. Perth: Hesperian Press.

ARCHER, M. (1990). Why are Australia's animals so distinctive? In *The Australian Environment: Taking Stock Looking Ahead*, ed. S. Neville, pp. 32–40. Melbourne: Australian Conservation Foundation.

ARCHER M., FLANNERY, T. F., RITCHIE, A. & MOLNAR, R. E. (1985). First Mesozoic mammal from Australia – an early Cretaceous monotreme. *Nature*, 318, 363–6.

ARCHER, M., GODTHELP, H., HAND, S. J. & MEGIRIAN, D. (1989). Fossil mammals of Riversleigh, northwestern Queensland: preliminary overview of biostratigraphy, correlation and environmental change. *Australian Zoologist*, 25, 29–65.

ARCHER, M. & HAND, S. J. (1984). Background to the search for Australia's oldest mammals. In *Vertebrate Zoogeography and Evolution in Australasia*, ed. M. Archer & G. Clayton, pp. 517–65. Perth: Hesperian Press.

ARCHER, M. & HAND, S. J. (1987). Evolutionary considerations. In *The Koala: Australia's Endearing Marsupial*, ed. L. Cronin, pp. 79–106. Sydney: Reed Books.

ARCHER, M., HAND, S. J. & GODTHELP, H. (1988a). A new order of Tertiary zalambdodont marsupials. *Science*, 239, 1528–31.

ARCHER, M., HAND, S. J. & GODTHELP, H. (1988b). Green cradle: the rainforest origins of Australia's marsupials. *Abstract, 7th International Palynological Congress*, Brisbane, p. 2.

ARCHER, M., HAND, S. J. & GODTHELP, H. (1991). *Riversleigh*. Sydney: Reed Books Pty Ltd.

ARCHER, M. & WADE, M. (1976). Results of the Ray Lemley Expeditions. Part 1. The Allingham Formation and a new Pliocene vertebrate fauna from northern Queensland. *Memoirs of the Queensland Museum*, 17, 379–97.

AUDLEY-CHARLES, M. G. (1987). Dispersal of Gondwanaland: relevance to the evolution of the angiosperms. In *Biogeographical Evolution of the Malay Archipelago*, ed. T. C. Whitmore, pp. 5–25. Oxford: Clarendon Press.

AUGEE, M. (1991). Echidnas and the evolution of hibernation. In *Abstracts, Symposium on Monotreme Biology*, p. 2. Sydney: Royal Zoological Society of New South Wales, University of New South Wales.

BONAPARTÉ, J. (1990). New late Cretaceous mammals from the Los Alamitos Formation, northern Patagonia, and their significance. *National Geographic Research*, 6, 63–93.

CALLEN, R. A. & TEDFORD, R. H. (1976). New late Cainozoic rock units and depositional environments, Lake Frome area, South Australia. *Transactions of the Royal Society of South Australia*, 100, 125–67.

DIAMOND, J. M. (1990). Biological effects of ghosts. *Nature*, 345, 769–70.

FLANNERY, T. F. (1984). Re-examination of the Quanbun Local Fauna, a late Cenozoic vertebrate fauna from Western Australia. *Records of the Western Australian Museum*, 11, 119–28.

FLANNERY, T. F. (1990). Pleistocene faunal loss: implications of the aftershock for Australia's past and future. *Archaeology in Oceania*, 25, 45–67.

FLANNERY, T. F. & PLANE, M. D. (1986). A new late Pleistocene diprotodontoid (Marsupialia) from Pureni, Southern Highlands Province, Papua New Guinea. *BMR Journal of Australian Geology and Geophysics*, 10, 65–76.

FORDYCE, E. (1991). The Australasian marine vertebrate record and its climatic and geographic implications. In *Vertebrate Palaeontology of Australasia*, ed. P. Vickers-Rich, J. M. Monaghan, R. F. Baird & T. H. Rich, pp. 1165–90. Melbourne: Pioneer Design Studio.

GAFFNEY, E. S. & BARTHOLOMAI, A. (1979). Fossil trionychids of Australia. *Journal of Paleontology*, 53, 1354–60.

GODTHELP, H. (1990). *Pseudomys vandycki*, a Tertiary murid from Australia. *Memoirs of the Queensland Museum*, 28, 171–3.

GODTHELP, H., ARCHER, M., CIFELLI, R.,
    HAND, S. J. & GILKESON, C. (1992). Australia's
    first early Tertiary mammals. *Nature*, **356**, 514–16.
GRIFFITHS, M., WELLS, R. T. & BARRIE, D. J.
    (1991). Observations on the skulls of fossil and extant ech-
    idnas (Monotremata: Tachyglossidae). *Australian Mam-
    malogy*, **14**, 87–101.
HAND, S. J. (1987). Phylogenetic studies of Australian
    Tertiary bats: summary of PhD thesis. *Macroderma*, **3**,
    9–12.
HAND, S. J. (1989). On the wings of fortune. The origin
    of Australia's bat fauna. *Australian Natural History*, **24**,
    130–8.
HAND, S. J., NOVACEK, M. J., GODTHELP, H. &
    ARCHER, M. (1993). First Eocene bat from Australia.
    Journal of Vertebrate Paleontology, in press.
KERSHAW, A. P. (1983). A Holocene pollen diagram
    from Lynch's Crater, north-eastern Queensland, Aus-
    tralia. *New Phytologist*, **94**, 669–82.
LINDSAY, J. M. (1987). Age and habitat of a monospecific
    foraminiferal fauna from near-type Etadunna Forma-
    tion, Lake Palankarinna, Lake Eyre Basin. *South Austra-
    lian Department of Mines Report*, 1987/93.
MARSHALL, L. G. & CORRUCCINI (1978). Variability,
    evolutionary rates, and allometry in dwarfing lineages.
    Paleobiology, **4**, 101–18.
MARSHALL, L. G. & MUIZON DE, C. (1988). The
    dawn of the Age of Mammals in South America. *National
    Geographic Research*, **4**, 23–55.
MARTIN, H. (1990). The palynology of the Namba For-
    mation in the Wooltana-1 bore, Callabonna Basin (Lake
    Frome), South Australia, and its relevance to Miocene
    grasslands in central Australia. *Alcheringa*, **14**, 247–56.
MERRILEES, D. (1979). The prehistoric environment in
    Western Australia. *Journal and Proceedings of the Royal
    Society of Western Australia*, **62**, 109–42.
MOLNAR, R. (1991). Fossil reptiles in Australia. In *Ver-
    tebrate Palaeontology of Australasia*, ed. P. Vickers-Rich,
    J. M. Monaghan, R. F. Baird & T. H. Rich, pp. 605–701.
    Melbourne: Pioneer Design Studio.
MURRAY, P. (1991). The Pleistocene megafauna of Aus-
    tralia. In *Vertebrate Palaeontology of Australasia*, ed.
    P. Vickers-Rich, J. M. Monaghan, R. F. Baird &
    T. H. Rich, pp. 1071–164. Melbourne: Pioneer Design
    Studio.
NEWSOME, A. E. & COMAN, B. J. (1989). Canidae. In
    Fauna of Australia, vol. 1B, ed. D. W. Walton &
    B. J. Richardson, pp. 993–1005. Canberra: Australian
    Government Publishing Service.
PASCUAL, R., ARCHER, M., ORTIZ JAUREGUIZAR,
    E., PRADO, J. L., GODTHELP, H. & HAND, S. J.
    (1992*a*). First discovery of monotremes in South
    America. *Nature*, **356**, 704–6.
PASCUAL, R., ARCHER, M., ORTIZ JAUREGUIZAR,
    E., PRADO, J. L., GODTHELP, H. & HAND, S. J.
    (1992*b*). The first non-Australian monotreme: an early
    Paleocene South American platypus (Monotremata,

Ornithorhynchidae). In *Platypus and Echidnas*, ed. M.
    Augee, pp. 1–14. Sydney: Royal Zoological Society of
    New South Wales.
PASCUAL, R. & ORTEZ JAUREGUIZAR, E. (1993).
    El Ciclo Faunistico Cochabambiano (Paleoceno Tem-
    prano): su incidencia en la historia biogeografica de los
    mamiferos Sudamericanos. In *Fosiles y facies de Bolivia
    vol. 1 Vertebrados*, ed. R. Suárez-Soruco, in press.
PLANE, M. D. (1967). Stratigraphy and vertebrate fauna
    of the Otibanda Formation, New Guinea. *Bureau of
    Mineral Resources Bulletin*, **86**, 1–64.
PLANE, M. D. (1991). Musings on New Guinea fossil
    vertebrate discoveries. In *Vertebrate Palaeontology of Austra-
    lasia*, ed. P. Vickers-Rich, J. M. Monaghan, R. F. Baird
    & T. H. Rich, pp. 85–109. Melbourne: Pioneer Design
    Studio.
PLANE, M. D. & GATEHOUSE, C. G. (1968). A new
    vertebrate fauna from the Tertiary of northern Australia.
    Australian Journal of Science, **30**, 272–3.
PLEDGE, N. S. (1984). A new Miocene vertebrate faunal
    assemblage from the Lake Eyre Basin: a preliminary report.
    Australian Zoologist, **21**, 345–55.
RICH, T. H. (1991). Monotremes, placentals, and mar-
    supials: their record in Australia and its biases. *In Vertebrate
    Palaeontology of Australasia*, ed. P. Vickers-Rich, J. M.
    Monaghan, R. F. Baird & T. H. Rich, pp. 893–1004.
    Melbourne: Pioneer Design Studio.
RICH, T. H., ARCHER, M., HAND, S. J., GODT-
    HELP, H., MUIRHEAD, J., PLEDGE, N. S., FLAN-
    NERY, T. F., WOODBURNE, M. O., CASE, J. A.,
    TEDFORD, R. H., TURNBULL, W. D., LUNDEL-
    IUS, E. L. Jr, RICH, L. S. V., WHITELAW, M. J.,
    KEMP, A. & RICH, P. V. (1991). Appendix 1: Austra-
    lian Mesozoic and Tertiary terrestrial mammal localities.
    In *Vertebrate Palaeontology of Australasia*, ed. P. Vickers-
    Rich, J. M. Monaghan, R. F. Baird & T. H. Rich, pp.
    1005–58. Melbourne: Pioneer Design Studio.
RICH, T. H., FLANNERY, T. F. & ARCHER, M.
    (1989). A second Cretaceous mammalian specimen from
    Lightning Ridge, New South Wales, Australia. *Alcheringa*,
    13, 85–8.
RICH, T. H. & RICH, P. V. (1989). Polar dinosaurs and
    biotas of the early Cretaceous of southeastern Australia.
    National Geographic Research, **1**, 15–53.
SANSON, G. (1989). Morphological adaptations of teeth
    to diets and feeling in the Macropodoidea. In *Kangaroos,
    Wallabies and Rat-kangaroos*, ed. G. Grigg, P. Jarman &
    I. Hume, pp. 151–68. Sydney: Surrey Beatty & Sons Pty
    Ltd.
SAVAGE, D. E. & RUSSELL, D. E. (1983). *Mammalian
    Paleofaunas of the World*. Reading, MA: Addison-Wesley.
STIRTON, R. A., TEDFORD, R. H. & WOOD-
    BURNE, M. O. (1967). A new Tertiary formation and
    fauna from the Tirari Desert, South Australia. *Records of
    the South Australian Museum*, **15**, 427–62.
STIRTON, R. A., TEDFORD, R. H. & WOOD-
    BURNE, M. O. (1968). Australian Tertiary deposits con-

taining terrestrial mammals. *University of California Publications in Geological Science*, 77, 1–30.

TEDFORD, R. H., ARCHER, M., BARTHOL-OMAI, A., PLANE, M., PLEDGE, N. S., RICH, T., RICH. P. & WELLS, R. T. (1977). The discovery of Miocene vertebrates, Lake Frome area, South Australia. *BMR Journal of Australian Geology and Geophysics*, 2, 53–7.

TEDFORD, R. H., BANKS, M. R., KEMP, N. R., McDOUGALL, I. & SUTHERLAND, F. L. (1975). Recognition of the oldest known fossil marsupials from Australia. *Nature*, 255, 141–42.

TEDFORD, R. H., WILLIAMS, D. & WELLS, R. T. (1986). Lake Eyre and Birdsville Basins: Late Cainozoic sediments and fossil vertebrates. In *The Lake Eyre Basin – Cainozoic Sediments, Fossil Vertebrates and Plants, Landforms, Silcretes and Climatic Implications*, ed. R. T. Wells & R. A. Callen. *Australasian Sedimentology Gp Field Guide Series*, 4, 42–72. Sydney: Geological Society of Australia.

TRUSWELL, E. M., KERSHAW, A. P. & SLUITER, I. R. (1987). The Australian–South-east Asian connection: evidence from the palaeobotanical record. In *Biogeographical Evolution of the Malay Archipelago*, ed. T. C. Whitmore, pp. 32–49. Oxford: Clarendon Press.

VEEVERS, J. J. (1984). *Phanerozoic Earth History of Australia* Oxford: Clarendon Press.

WALTON, D. W. & RICHARDSON, B. J. (eds.) (1989). *Fauna of Australia. Vol. 1B Mammalia* Canberra: Australian Government Publishing Service.

WATTS, C. H. S. & ASLIN, H. J. (1981). *The Rodents of Australia*. Sydney: Angus & Robertson.

WOODBURNE, M. O. (1967). The Alcoota Fauna, central Australia: an integrated palaeontological and geological study. *Bureau of Mineral Resources Bulletin*, 87, 1–187.

WOODBURNE, M. O., TEDFORD, R. H., ARCHER, M., TURNBULL, W. D., PLANE, M. D. & LUNDELIUS, E. L. Jr (1985). Biochronology of the continental mammal record of Australia and New Guinea. *South Australian Department of Mines and Energy Special Publication*, 5, 347–65.

WOODBURNE, M. O. & ZINSMEISTER, W. J. (1984). The first land mammal from Antarctica and its biogeographic implications. *Journal of Paleontology*, 58, 913–48.

WRIGHT, W., SANSON, G. D. & McARTHUR, C. (1991). The diet of the extinct bandicoot *Chaeropus ecaudatus*. In *Vertebrate Palaeontology of Australasia*, ed. P. Vickers-Rich, J. M. Monaghan, R. F. Baird & T. H. Rich, pp. 229–45. Melbourne: Pioneer Design Studies.

7 Australian Tertiary phytogeography: evidence from palynology

H. A. MARTIN

This chapter reconstructs Tertiary vegetation and phytogeography from the fossil spore and pollen record. Plants are constrained by their environment: they cannot grow outside of acceptable environmental limits, and climate is the most important of all environmental factors. For this reason, past distributions are considered in conjunction with the appropriate past climate and other environmental factors that may have been very different from those of today. Some well-established phytogeographical hypotheses are discussed in the light of the fossil record.

The palynological record is comprehensive and well suited to this purpose, but it has limitations. Not all pollen preserves as fossils. For example, the Lauraceae, an important Australian family and common in the macrofossil record, is absent from the fossil pollen record. When micro- (spores, pollen) and macrofossil (leaves, flowers, fruit) floras are compared, there is a core of common taxa, but some taxa are restricted to one or the other flora. For example, of the total flora identified in one deposit by Graham & Jarzen (1969), 27% were found as microfossils only and 17% as macrofossils only, with a core of 56% found as both. Furthermore, micro- and macrofossils from six Eocene lenses at Anglesea (Victoria) contain spore and pollen floras that vary somewhat, but are generally similar. However, macrofossil percentages are markedly different between lenses, and include taxa not identified as microfossils

(Christophel et al., 1987). The microfossil assemblage samples a broader area and gives a more general picture of the vegetation, whereas macrofossils present more localised and variable facets. Obviously, if both micro- and macrofossils are studied, a better picture of the vegetation emerges than from either one alone, but this has rarely occurred. It is not always possible to study both in the same deposit. Sediments deep underground are accessible by bores, which is a great advantage, especially in the Australian context where most of the landscape is so flat. Microfossils may be extracted from bore samples which are unsuited for macrofossil recovery.

Studies of pollen assemblages collected from the forest floor show that they are identifiable with, and may characterise the vegetation at, a general or formation level (Birks & Birks, 1980; Sluiter & Kershaw, 1982). Small patches of different kinds of vegetation, such as may be found in sheltered gullies or gorges, are palynologically invisible (Kershaw, 1973; Ladd, 1979). Such patches, however, are very important biologically. Thus, interpretations of the vegetation are those of the dominant type in the landscape and exclude small areas of different kinds of vegetation.

Most palynological sites are in southeastern Australia. Evidence from elsewhere in Australia is too sporadic for a detailed history, but is invaluable for comparisons with the southeast and for illustrations of geographical variation.

FIRST APPEARANCES

Fgure 7.1 shows the time ranges of taxa recorded from southeastern Australia. Note that the level of identification is not uniform. Some pollen types are sufficiently distinctive to be identified with species, e.g. *Lagarostrobos franklinii*, *Dodonaea triquetra* and *Gardenia chartacea*, but they are few. Others can be identified only to genus, family or order.

This range chart applies to southeastern Australia in a general sense. Pollen of some of the taxa are found only in inland areas, such as the Murray Basin and not in the coastal Gippsland Basin, e.g. *Alangium*, *Fuchsia*, Caesalpinoides, *Guettarda*, *Alyxia*, Malpighiaceae (Macphail & Truswell, 1989). Ranges may differ elsewhere in Australia. For example, *Anacolosa* pollen disappeared from southeastern Australia at about the end of the Eocene, but it was present in the Early Miocene in Queensland (Hekel, 1972). Those taxa not found in Australia today are given in Figure 7.1. Most of the taxa, however, are present in the modern vegetation, but their distributions have changed drastically. One group given in Figure 7.1 is now restricted to Tasmania but was once widespread in Australia. Others are now restricted to the north, e.g. *Ilex* (Martin, 1977) and *Nypa* (Lear & Turner, 1977). Many once found in inland areas are now restricted to the northern and eastern coastal strip.

Figure 7.1 shows that there is a group of gymnosperms that range throughout most of the Mesozoic. Angiosperm pollen types identifiable with extant taxa appear within the Late Cretaceous, but these are not the very first angiosperms because there are a number of extinct pollen forms that cannot be identified with extant taxa (Dettmann & Playford, 1969; Dettmann, 1973). Figure 7.1 lists those pollen taxa that can be identified with extant taxa, but there are many fossil types, especially among the tricolpate or tricolporate grains, that are not yet identifiable.

These first appearances may have been the result of evolution in southeastern Australia, or migration from elsewhere. Evolution and migration are discussed later.

Changes through time

The palynofloras through the Tertiary are described in detail in Chapters 10 and 13. A brief overview is necessary for the discussion here.

Pollen frequencies reflect the relative abundance of some taxa in the vegetation, but this reflection is biassed towards wind-pollinated taxa, which produce large quantities of pollen. Insect-pollinated plants usually produce smaller quantities of pollen with larger, heavier grains that may not be transported by the wind at all. For example, Ladd (1979) noted the absence of *Acacia* in the pollen assemblage from the floor of a forest in which it was classified as codominant. *Acacia* is regularly found in the fossil record, but only as a small percentage of the total assemblage, hence little can be said of its abundance and any possible variation in the vegetation. The Myrtaceae, though mainly insect/bird/bat-pollinated (Churchill, 1970; Crome & Irvine, 1986; McCoy, 1989), produce copious pollen, and high percentages may be found in fossil assemblages. Though the taxa with the highest pollen percentages may not have been the most abundant in the vegetation, when percentages change, sometimes dramatically, it indicates change in the vegetation. The frequencies in the pollen diagrams are a 'pollen signature' that characterises the vegetation.

In general, the spores of ferns and their allies show no particular variation with time and probably reflect ecological conditions. The group contains many taxa, but only a few may be abundant. *Cyathea*, a tree fern, is probably the most abundant taxon and is usually present in appreciable spore frequencies in the Late Miocene–Pliocene sequence. In the Late Pliocene–Early Pleistocene, the bryophyte Anthocerotae is an important spore type, although the composition of the group may not be temporally and spatially uniform.

The gymnosperms, as a group, were most abundant in the earliest Tertiary, and decreased in the Early Eocene. For the remainder of the Tertiary no particular pattern is observed, indicating geographical variation or a response to local ecological conditions.

Casuarinaceae pollen may be abundant at any

Geological periods (left to right): LATE CRETACEOUS | PALEOCENE | EOCENE | OLIGOCENE | MIOCENE | PLIOCENE

Taxa:

- Araucariaceae
- Microcachrys ◆
- Podocarpus
- Phyllocladus, 3-saccate ◆
- cf. Callitriche
- Ascarina ●
- Gunnera ◆
- Ilex
- Lagarostrobos ●
- Dacrydium ●
- Nothofagus, ancestral type
- Macadamia, Helicia/Orites complex
- Beauprea ●
- Gevuina/Hicksbeachia
- ?Xylomelum
- Winteraceae
- Nothofagus, brassii type ●
- Nothofagus, fusca type ◆
- Casuarinaceae
- Banksieae
- Austrobuxus/Dissiliaria
- Myrtaceae
- Phyllocladus, 2-saccate ◆
- Dacrycarpus ●
- Eucalyptus, bloodwood/Angophora type
- Anacolosa ●
- Sparganiaceae
- Restionaceae, Restio type
- Restionaceae, Hypolaena type
- Cyperaceae
- Cunoniaceae
- Santalum type
- Polygalaceae
- Liliaceae
- Cupaneae
- Nypa
- Ericales
- Bombax type ●
- Poaceae
- Sapotaceae
- Guettarda
- Nothofagus, menziesii type
- Merrimia
- Alyxia
- Loranthaceae
- Rhizophoraceae

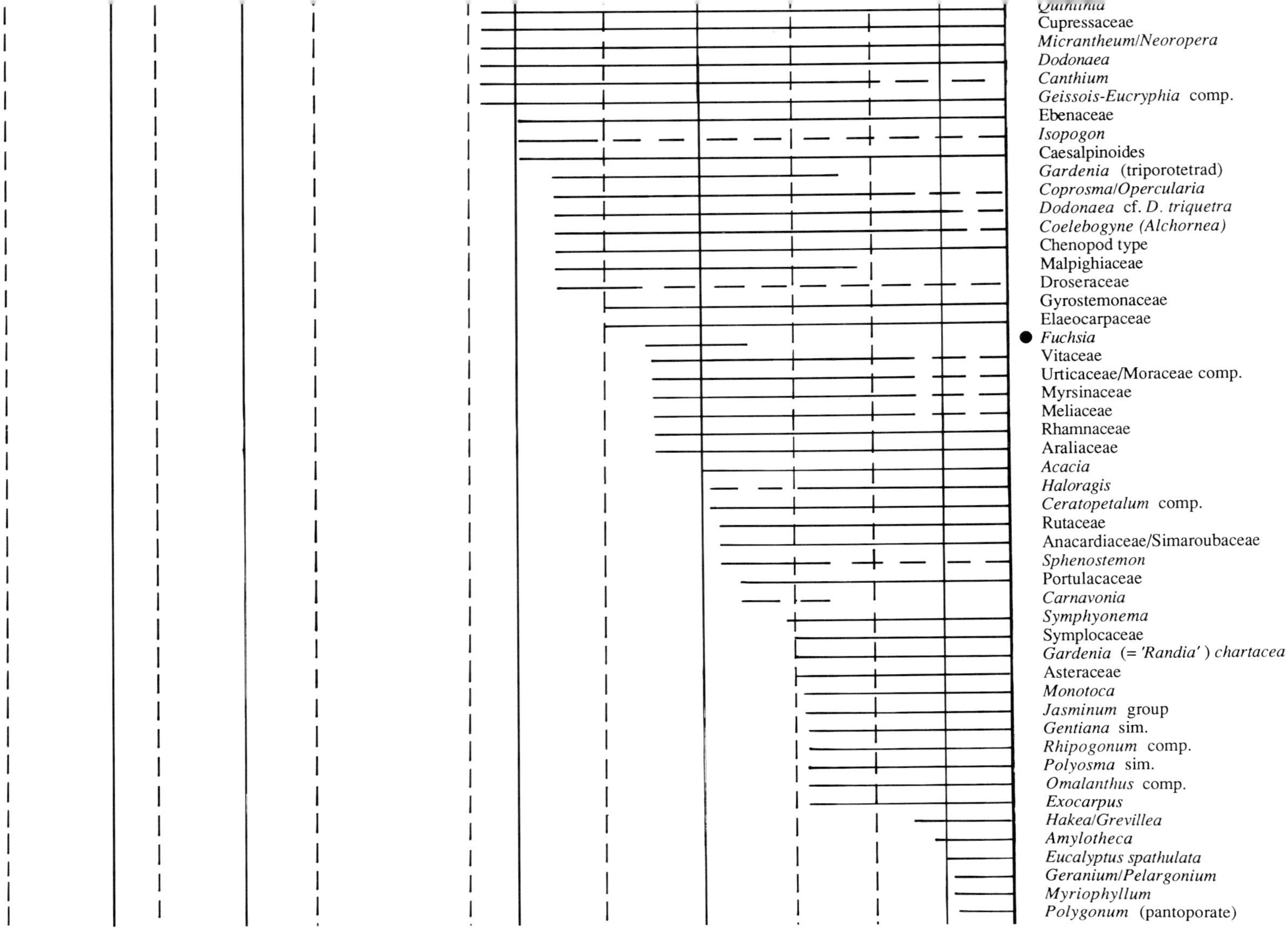

Figure 7.1 Range chart of first appearances of the taxa. For fossil names, and references where they may be found, see Appendix 7.1. Black dots indicate taxa not present in Australia today. Black diamonds denote taxa confined to Tasmania today.

time during the Tertiary, reflecting local ecological conditions. Pollen cannot be identified more precisely than the family (Kershaw, 1970a), but the cones and foliage of *Gymnostoma*, a rainforest taxon, are distinguishable from *Casuarina–Allocasuarina* of open forests, woodlands and shrublands. Macrofossil evidence is therefore required to determine the environmental significance of the presence of this family.

Nothofagus pollen first appeared in the Late Cretaceous and maintained relatively low percentages until the Late Eocene, when it increased dramatically. Higher percentages continued throughout the Late Eocene and Oligocene, decreased a little in the Early Miocene and drastically in the Middle–Late Miocene. *Nothofagus* subgenus *Brassospora* constitutes the bulk of the pollen. There was a minor resurgence of *Nothofagus* in the Early Pliocene, but this did not include subgenus *Brassospora*. *Nothofagus* disappeared from inland regions after the Early Pliocene, but persisted in highland regions until Plio-Pleistocene time (McEwen Mason, 1989) and in the most southeasterly part of Australia, east Gippsland, where relict populations are found today. *Nothofagus* is important in the extant vegetation of Tasmania.

The Myrtaceae contains fossil taxa with diverse ecological requirements. One eucalypt pollen type, *Myrtaceidites eucalyptoides*, conforms to the bloodwood eucalypt/*Angophora* morphology, but other eucalypt pollen types are also present. Eucalypts are well-known indicators of open sclerophyllous forests and woodlands, but some of the myrtaceous pollen is recognisable as *Backhousia*, *Syzygium*, *Tristania* (Martin, 1978), *Decaspermum*, *Austromyrtus*, *Rhodamnia* (Truswell *et al.*, 1985), *Acmena* and *Syzygium–Cleistocalyx* (Luly *et al.*, 1980), which are predominantly rainforest taxa. Fire history provides indirect evidence about the kind of vegetation this group represents. The charcoal content parallels the abundance of Myrtaceae (Martin, 1987). Rainforests rarely burn, except in drought periods when they are subjected to exceptional drying (Webb, 1970; Luke & McArthur, 1978), whereas eucalypt forests burn

on a regular basis (Ashton, 1981). Pollen evidence indicates that eucalypts became dominant in the vegetation at about Late Miocene time.

Proteaceae pollen was most abundant in the Paleocene–Middle Eocene, decreased in the Late Eocene and maintained relatively low percentages throughout the rest of the Tertiary. This group consists of up to 23 pollen types and its abundance is remarkable, given the low pollen productivity of modern Proteaceae (Martin, 1978). These ancient Proteaceae were probably rainforest trees, but in the extant rainforest of northeastern Australia Proteaceae, although relatively common, contributes only low percentages to the pollen rain (Kershaw, 1970b). Many of the early Tertiary pollen species became extinct at about the end of the Eocene. The geographical diversity of early Tertiary proteaceous pollen species is also remarkable, with whole suites described from Western Australia (Stover & Partridge, 1982), and Queensland (Dudgeon, 1983), as well as from southeastern Australia. Some species are common to each of these areas, but others appear to be restricted to the area of their description.

Asteraceae pollen, which first appeared in the Miocene, remained low until the Late Pliocene–Pleistocene, when it increased dramatically.

Poaceae pollen, which first appeared in the Eocene, was rare until the Late Miocene–Pliocene, when there were local increases. It increased dramatically in the Late Pliocene–Pleistocene.

THE VEGETATION

Tertiary vegetation is described in detail elsewhere in this book (Chapters 10 and 13). Briefly, it was rainforest, or closed forest throughout the early to-mid-Tertiary, when *Nothofagus* was common. In the mid-Late Miocene, the change to abundant Myrtaceae and the higher charcoal content, but with some rainforest taxa still present, may indicate the presence of wet sclerophyllous forest, which has a tall canopy of eucalypts, some rainforest taxa in the understorey and burns on a regular basis (Ashton, 1981). In the Early Pliocene, there was a minor resurgence of rainforest, which probably did not entirely displace the wet sclerophyll. This return of rainforest did not last long and wet sclerophyllous taxa became dominant in the mid–Late Pliocene. With the increase in Asteraceae and Poaceae, the vegetation was probably more open, with the development of woodlands and grasslands/herbfield in the Late Pliocene–Pleistocene. Within these general types of vegetation there is both temporal and spatial variation.

Paleocene

Podocarpaceous gymnosperms dominated and the proteaceous type was significant and diverse in the Paleocene vegetation. Myrtaceae, Casuarinaceae and *Nothofagus* were minor components and there was a diversity of other angiosperms. The vegetation at Princetown suggests a cool temperate climate, similar to that of Tasmania today (Harris, 1965).

Late Paleocene floras from Bombala in the Southeastern Highlands are probably slightly younger than those of Princetown. Bombala resembles Princetown in that podocarp pollen is dominant, and there is a diversity of angiosperms, including a low occurrence of Casuarinaceae, but it differs in the greater abundance of *Nothofagus* pollen, the absence of Myrtaceae and the greater restriction of the proteaceous type. Bombala vegetation is interpreted as cool temperate rainforest, with a suggested mean annual temperature of 14–20 °C and precipitation of 1200–2400 mm. Fossil tree ring analysis indicates that the trees grew in a favourable and uniform environment, growth was slow, the growing season ended abruptly and frosts and droughts did not occur. These forests grew at palaeolatitudes of 57° S and daylength would have had considerable seasonal variation, but total availability of light was not limiting. Growth increments were complacent, which is a feature of extensive and dense forests (Taylor *et al.*, 1990).

Mount Royal Range, another Late Paleocene highland site north of Bombala, also has gymnosperm pollen dominant but the angiosperm flora is less diverse and *Nothofagus* is absent (Martin *et al.*, 1987).

Early Eocene

The pollen floras are very diverse and have a wealth of angiosperms in the Early Eocene. When compared with the Paleocene, gymnosperms were less abundant and the proteaceous type remained significant. *Nothofagus* and Myrtaceae were minor elements in the vegetation.

Middle Eocene

Middle Eocene pollen floras are similar to those of the Early Eocene, with diverse angiosperms and a significant proteaceous component, but differ in that *Nothofagus* frequencies increased.

Late Eocene

Nothofagus was the dominant pollen group in the Late Eocene. Since *Nothofagus* is a high pollen producer, trees in the forest were probably not as abundant as suggested by the pollen diagrams. While it may have been no more abundant than some of the other taxa found in low frequencies, it was more common than in the Paleocene–Early Eocene. Many of the Early–Middle Eocene angiosperms continued into the Late Eocene and there was high diversity. By the end of the Eocene, however, many of these older angiosperms had disappeared.

Oligocene

Nothofagus pollen remained common in the Oligocene. In comparison with the Eocene, the floras are not as diverse, and the proteaceous type in particular is much reduced. There are, however, a number of angiosperms, currently of more tropical affinities, found in inland regions (Macphail & Truswell, 1989).

Early–Middle Miocene

Nothofagus pollen decreased somewhat in inland regions during the Early–Middle Miocene, while frequencies remained high in coastal and extreme southeastern localities. Myrtaceae increased and araucarians, Casuarinaceae or other angiosperms may have been locally abundant, especially in inland regions. The Early–Middle Miocene appears to be a time of maximum diversity of forest types (discussed further below).

Late Miocene

A drastic change occurred in Late Miocene, with the disappearance of *Nothofagus* pollen and the increase of Myrtaceae. Many of the rainforest flora also disappeared. The Late Miocene marks the end of widespread rainforest, although small pockets probably remained in suitable habitats such as sheltered gullies, etc. Wet sclerophyllous forest, with a tall, open eucalypt canopy and rainforest species in the understorey replaced the rainforest. In many respects, wet sclerophyllous may be regarded as intermediate between rainforest and dry sclerophyllous forests, or open eucalypt forests without rainforest taxa. Wet sclerophyllous forests require fire for regeneration. *Cyathea*, the tree fern, is common in sheltered gullies of extant wet sclerophyllous forests in southeastern Australia (Ashton, 1981). The herbaceous element, i.e. Asteraceae and Poaceae, is present but in limited frequencies in open forests. The fossil record is consistent with all of these features.

Early Pliocene

In the Early Pliocene there was a brief resurgence of rainforest, but it is only a fraction of that of the Early–Middle Miocene and it did not completely displace the myrtaceous forests. There was probably a mosaic, with rainforest in the moister habitats and wet sclerophyll in the drier parts of the landscape. The composition of the rainforest varied with locality.

In the Lachlan Valley, there is appreciable

Nothofagus pollen but very little at Jemalong Gap (Martin, 1987). The latter site, however, has Cupanieae, *Macaranga/Mallotus*, *Helicia/Orites/Macadamia* and *Symplocos*, rainforest taxa not found in the Lachlan Valley. Gymnosperms may have been more abundant in the Early Pliocene.

Mid–Late Pliocene

The resurgence of rainforest in the Early Pliocene was short lived and the vegetation returned to its former state, i.e. wet sclerophyllous forest. Towards the latest Pliocene, the rainforest taxa including gymnosperms disappeared and the myrtaceous forests became more dry sclerophyll. Pollen of herbaceous taxa also increased.

Latest Pliocene–Pleistocene

There was a dramatic change in the latest Pliocene–Pleistocene. The vegetation opened, and woodlands and grasslands/herbfields became dominant.

CLIMATE

Quilty (Chapter 3, this volume) discusses climate as deduced from oceanic evidence, but climate may also be deduced from the vegetation. The oceanic climatic record is one mainly of temperature, whereas precipitation may be deduced from the vegetation. Temperature and precipitation are not independent, and changes in oceanic temperature have repercussions far inland, and on precipitation, as this discussion shows. A reconstruction of the precipitation for the Late Eocene–Plio-Pleistocene from the Lachlan Valley region is shown on Figure 7.2. This reconstruction is based upon the present-day requirements of a number of taxa and the types of vegetation (see Martin, 1987).

In the Late Eocene–Oligocene of the Lachlan River region (Figure 7.2), rainfall was very high, probably above 1800 mm annually. High humidities would have been maintained throughout the year and the forests rarely dried out sufficiently to allow burning. There was a slight decrease in

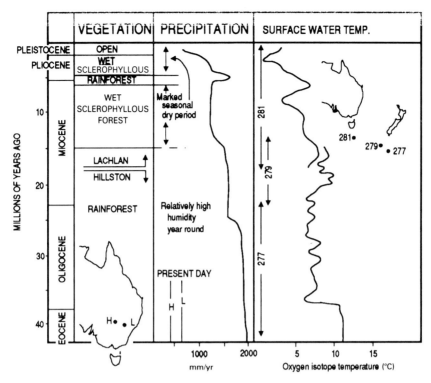

Figure 7.2 Precipitation deduced from the rainfall requirements of living taxa (see Martin, 1987) and the surface water oxygen isotope temperatures, from Shackleton & Kennett, (1975). H, Hillston. L, Lachlan.

precipitation at about the Late Oligocene–Early Miocene, but it was still above 1500 mm annually, the lower limit for widespread rainforest in New South Wales today (Baur, 1957). There would have been no definite dry period to allow burning on a regular basis.

In the Late Miocene, there was a further drop in precipitation, probably to 1500–1000 mm annually, between the lower limit for rainforest and the lower limit for wet sclerophyllous forest (Ashton, 1981). There was a definite dry period and burning became a regular event.

In the Early Pliocene precipitation increased, probably to just over 1500 mm annually for a brief interval. It decreased again to 1500–1000 mm annually in the mid–Late Pliocene. In the latest Plio-Pleistocene, precipitation decreased further to about 500–800 mm annually, the lower limit for the major eucalypt species in wet sclerophyl-

lous forest (Ashton, 1981; Boland et al., 1984) and for isolated pockets of rainforest across northern and northeastern Australia (Webb & Tracey, 1981). There were subsequent decreases during the Pleistocene, but they are not pursued here.

The nature of sediments is strongly influenced by the climate at the time of deposition. The climatic record indicated by sediments strongly supports evidence deduced from the vegetation. In the Late Eocene–Oligocene, carbonaceous clays and lignites containing wood are common. Such lignites would have been formed from swamp forests, which require very wet conditions. In the Early–Middle Miocene, pale grey clays with a lower carbonaceous content indicate that conditions of deposition were not as wet, but high humidities were maintained around the year. In the Late Miocene–Pliocene, grey clays are very restricted, showing that permanent swamps, bogs etc. were not common in the landscape. Most of the sediments contain variegated clays of reds, browns, yellows, fawns and whites. The red-brown colours indicate a well-marked dry season. There is one horizon in this Late Miocene–Pliocene sequence of the western slopes of New South Wales, however, where carbonaceous clays, some wood and lignite, are found and it is these sediments that indicate the Early Pliocene resurgence of rainforest. The Pleistocene sediments are almost entirely brown in colour and grey clays are rare. The climate had become too dry to support the permanently wet conditions required for pollen preservation, except in a few, rare localities (Martin, 1986).

The oxygen isotope sea surface temperature curve (Shackleton & Kennett, 1975) is shown in Figure 7.2. It depicts the decline of temperatures throughout the Tertiary, from possibly the warmest intervals in earth's history during the Cretaceous (Frakes & Francis, 1990), to the Pleistocene glaciations. The Oligocene–Pleistocene oxygen isotope temperatures roughly parallel the precipitation curve. Thus, a decline in temperature was accompanied by a decline in precipitation.

The light regime must also be considered. In the early Tertiary, southeastern Australia was much further south and would have experienced long summer days and short winter days with low light levels. Paleocene wood from the Monaro region (discussed previously) has clear and well-defined rings, showing that growth was controlled by well-defined seasons. The end of the growing season was abrupt, as shown by few late wood cells. These characteristics are found today in forests that are uniformly wet. The wood shows that frosts did not occur. Since this locality was at 57° S, these forests would have had the low winter light levels of the high latitudes (Taylor *et al.*, 1990), which would have been the cause of seasonality at this time.

GEOGRAPHICAL VARIATION

Where sufficient detail of the palynofloras has been reported, geographical variation can be described, both within southeastern Australia and between southeastern Australia and other regions of Australia. For the Paleocene–Middle Eocene, there are only a few sites, but even so, some similarities and differences emerge.

Late Paleocene–Early Eocene palynofloras in the Lake Eyre region of Central Australia are strikingly different from those in southeastern Australia. Here, the vegetation was dominated by Cunoniaceae and there was probably a significant gymnosperm content. In the Early Eocene there was a major change to Myrtaceae forests, and in the Middle Eocene *Nothofagus* subgenus *Brassospora* became significant (Sluiter, 1992). In southeastern Australia, Cunoniaceae pollen has not been recorded and Myrtaceae is only a minor component.

Middle Eocene palynofloras in the Eyre Formation of central Australia have a low frequency of *Nothofagus* and high Casuarinaceae, but one, from the late Middle Eocene, is dominated by *Nothofagus* (Wopfner *et al.*, 1974). A late Middle Eocene palynoflora from the St Vincent Basin, South Australia, has a diverse flora with a small amount of *Nothofagus* pollen (Hos, 1977), far less than that at Anglesea, which is a site of comparable age in southeastern Australia (Christophel *et al.*, 1987). The Casuarinaceae pollen content is high, similar to that of the basal sample at Anglesea. It has a remarkable 26% of proteaceous pollen types, the highest of any Australian locality, but Myrtaceae pollen is a minor component.

Middle–Late Eocene palynofloras of the Yaamba Basin in Queensland have a minor amount of *Nothofagus*, but a rich diversity. The most abundant pollen type is variously Casuarinaceae, Araucariaceae, ?palm (*Arecipites* sp.), or unidentified tricolpate or tricolporate types (Foster, 1982).

The Middle–Late Eocene is of special interest, as it is the time that *Nothofagus* increases in the vegetation, but not uniformly over the continent. There are a number of Middle–Late Eocene palynofloras outside of southeast Australia. In the Eucla Basin, Western Australia, a coastal site of Late Eocene–Early Oligocene age yielded a rich flora that has abundant *Nothofagus* pollen and a great diversity of proteaceous species, in keeping with south-eastern Australia. The most common gymnosperm pollen is *Lagarostrobos* (Milne, 1988). In the Officer Basin, pollen assemblages of latest Middle Eocene–early Late Eocene age have an appreciable content of *Nothofagus*, and Casuarinaceae may be abundant (Hos, 1978). Although the proteaceous content is not high, it is extremely diverse (Alley, 1985). These assemblages are thus similar to those of southeastern Australia.

Pollen assemblages at Napperby, Central Australia, reported as Middle Eocene (Kemp, 1976), have appreciable *Nothofagus*, and Casuarinaceae is sometimes abundant, similar to the palynofloras in southeastern Australia. There is, however, a significant proportion of the swamp taxa Sparganiaceae, Cyperaceae and Restionaceae, which indicate a swamp with open vegetation. *Micrantheum*, a heath taxon, is also present (Kemp, 1976) and this taxon is rarely seen in southeastern Australia until the Late Miocene–Pliocene (Martin, 1987). Thus, at this locality there were some open swamps and shrubby vegetation as well as rainforest.

Lagarostrobos franklinii, the huon pine, has a long history (Figure 7.1) and is often the most common gymnosperm in Late Eocene assemblages in the Lachlan and Murrumbidgee regions of New South Wales (Figure 7.3). In Late Eocene–Early Oligocene coal assemblages of the Gippsland Basin, Victoria, pollen of *Lagarostrobos* may constitute more than 80% of the assemblage, which is exceptionally high (Stover & Partridge, 1973). In the Early Oligocene, any abundance was restricted to the more southerly part of New South Wales, and by Late Oligocene–Early Miocene time it was rare. In Early–Middle Miocene assemblages of the Yallourn region, however, it may be the most common gymnosperm pollen. Thus *Lagarostrobos*, which was widespread and frequently common in the Late Eocene (and some earlier times), was restricted to the most southeasterly part of the Australian mainland by the Middle Miocene. Today, *Lagarostrobos* is endemic to southwest Tasmania, which has an annual precipitation of up to 2500 mm (Boland *et al.*, 1984). This pattern of change suggests that the distri-

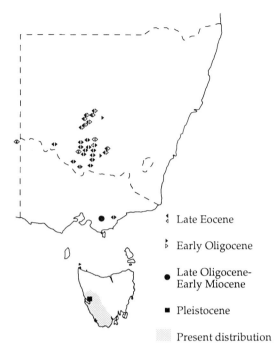

Figure 7.3 Occurence of *Lagarostrobos franklinii*. Solid symbols indicate where *L. franklinii* is the most abundant gymnosperm, open symbols where it is not. Greater abundance is restricted earlier in inland regions where the climate is drier. H, Hillston; L, Lachlan.

bution of *Lagarostrobos* became progressively restricted to the wetter, more southeasterly regions, as the climate became progressively drier.

There are many Early–Middle Miocene sites within southeastern Australia and the distribution of different abundances can be mapped (Figure 7.4). Vegetation with high Myrtaceae is found mainly in inland regions. Some of these high Myrtaceae assemblages have a considerable amount of the bloodwood eucalypt/*Angophora* pollen type, e.g. Lake Menindee (Figure 7.5). The northeast coastal site, Alstonville (Figure 7.5), however, is clearly different from the inland sites in that it has a high proportion of a different myrtaceous grain, which is more like the rainforest taxa *Rhodamnia/Archirhodomyrtus*. Vegetation with abundant *Nothofagus* was concentrated towards the wetter regions at the coast and to the south.

Assemblages with Araucariaceae as the most

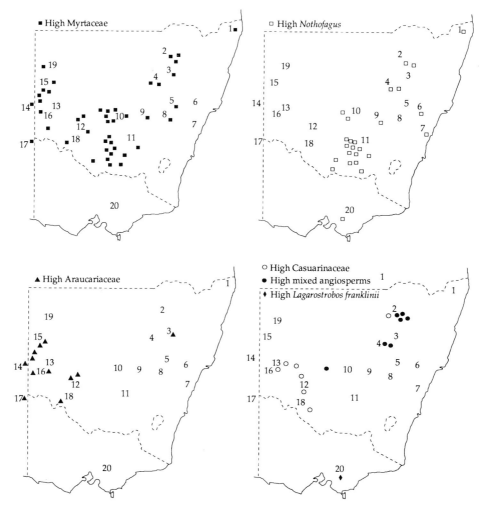

Figure 7.4 Distribution of greater abundance of selected taxa for the Early–Middle Miocene. 1, Alstonville, H. A. Martin (unpublished data). 2, Namoi River, Martin (1980). 3, The Warrumbungles, Holmes *et al.* (1983). 4, Castlereagh River, Martin (1981). 5, Mudgee, H. A. Martin (unpublished data). 6, Home Rule, McMinn (1981). 7, Little Bay, A. D. Partridge *et al.* (unpublished data). 8, Cadia, Owen (1975). 9, Lachlan River Valley, Martin (1987). 10, Lachlan region of Murray Basin, Martin (1984*b*). 11, Murrumbidgee area, Murray Basin, Martin (1984*a*). 12, Western Murray Basin, Truswell (1987). 13, Manilla, Macphail (1987). 14, Oakvale, Truswell *et al.* (1985). 15, Lake Menindee region, Martin (1988). 16, Canegrass, Martin (1986). 17, SADME MC63, Martin (1991*b*). 18, Mallee Cliffs, Martin (1993). 19, Glenmore, A. McMinn (unpublished data). 20, Latrobe Valley, Sluiter & Kershaw (1982) and Luly *et al.* (1980).

common gymnosperm pollen and high Casuarinaceae pollen frequencies are found in the western part of New South Wales. This region, the Murray Basin, was an inland sea during the Early–Middle Miocene (Brown, 1989). The highest frequencies of both these pollen types are found in the more marine sediments (Martin,

1991*b*) and were probably water transported. This distribution suggests that these taxa may have been associated with shorelines, deltas and river banks. There is one high Araucariaceae pollen record a long way from the Murray Basin (3 on Figure 7.4), but, as pollen identification is only to the family level, the Araucariaceae at this site need

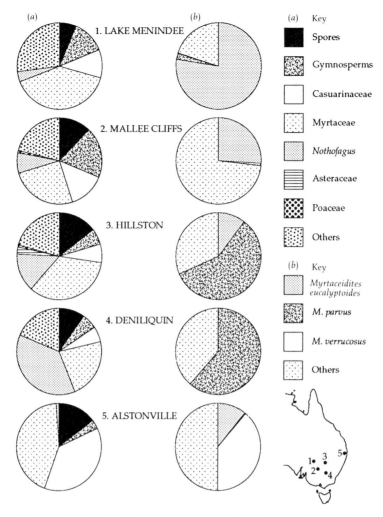

Figure 7.5 A comparison of the myrtaceous content of some Early Miocene pollen assemblages. When the Myrtaceae group is examined according to its component species, it shows that the myrtaceous part of the vegetation was not uniform. (a) The total assemblage. (b) The component pollen species of the Myrtaceae group.

not be the same species as that (those) in western New South Wales. Today, Araucariaceae is characteristic of drier kinds of rainforest, but it is not restricted to this vegetation type.

Late Oligocene–Early Miocene floras at Sandy Cape on Fraser Island, Queensland, have a moderate amount of *Nothofagus* pollen. Spores of the tree fern *Cyathea* dominate the assemblages and there is one with Casuarinaceae pollen dominant. Angiosperms are diverse (Wood, 1986). Hekel (1972) reported on other palynofloras

in Queensland, but the dating is questionable.

At Lake Frome, central Australia, a Late Oligocene–Early Miocene assemblage has pollen of podocarpaceous gymnosperms as dominants and a moderate amount of *Cyathea* spores. Microfossils of the swamp taxa Sparganiaceae and Restionaceae and the freshwater algae *Pediastrum* and *Botryococcus* are abundant (Martin, 1990)

Early–Middle Miocene assemblages from a deep sea core on the Northwest Shelf give some indication of the land flora. The palynoflora is very restricted and most of it is Casuarinaceae. *Cyathea* spores are present, but there are no taxa that would unequivocally indicate rainforest (McMinn & Martin, 1992).

The Early–Middle Miocene is of special interest as it immediately pre-dates the widespread decline of rainforest. The taxa requiring the highest precipitation, *Lagarostrobos* and *Nothofagus*, became rare and less common, respectively, in drier inland regions.

There are few Late Miocene palynofloras immediately postdating the disappearance of widespread rainforest, since the Late Miocene was a probably a time of nondeposition or erosion. The few palynofloras found in southeastern Australia (Martin, 1987) are dominated by Myrtaceae and are very like the mid–Late Pliocene wet sclerophyllous forest assemblages.

Late Miocene palynofloras of a deep-sea core of the Northwest Shelf are dominated by Casuarinaceae, *Acacia*, Gyrostemonaceae, Restionaceae and *Haloragis/Gonocarpus*. Small amounts of Asteraceae and Poaceae pollen are also present (McMinn & Martin, 1992).

The Early Pliocene resurgence of rainforest was extremely variable. *Nothofagus* was most significant in the southern part of the Western Slopes, closer to the highlands (Figure 7.6). Traces of *Nothofagus* pollen are found in the northern part (Martin, 1991*a*) and in one site to the west in South Australia. Higher pollen counts of gymnosperms, particularly *Podocarpus* and occasionally *Phyllocladus* (the 2-saccate form) and/or Cupressaceae, are frequently found at the same stratigraphic level. This pattern suggests a mosaic of rainforest communities, with *Nothofagus* and/or gymnosperms, and/or other rainforest angiosperms. The Late Miocene wet sclerophyllous forest was not entirely replaced, but was probably confined to drier habitats.

When *Nothofagus* disappeared from the inland areas in the Middle Miocene, it was most likely restricted to the Southeastern Highlands, which would have had a higher rainfall. With an increase in precipitation in the Early Pliocene, rainforest migrated westward. No doubt there was a rainfall gradient down the Western Slopes, just as there

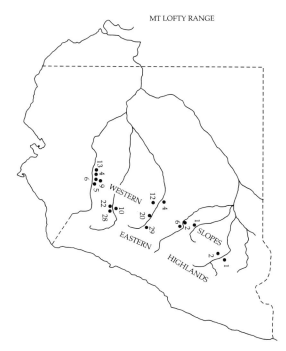

Figure 7.6 Percentages of *Nothofagus* in the Early Pliocene rainforest resurgence. The sites with low percentages often have rainforest taxa not found with the high percentages. For further explanation, see the text.

is today, and the decreasing frequencies of *Nothofagus* pollen with distance westward seen in the Lachlan River Valley (Figure 7.6) is an expression of this gradient. The low pollen frequencies (below ?5%) are more likely to be the result of being transported long distances than a reflection of *Nothofagus* growing at that locality. Thus, on the northern Western Slopes, *Nothofagus* was probably growing to the east, further into the highlands than at the localities sampled. The occurrence of *Nothofagus* in the South Australian locality (Figure 7.6) raises intriguing possibilities. Is it the result of long-distance transport from trees that grew in the Mount Lofty Ranges?

At Lake Tay, Western Australia, Early Pliocene assemblages are generally similar to those of the Late Miocene and mid–Late Pliocene of southeastern Australia. Myrtaceae or Casuarinaceae are the dominant pollen groups and there is a small amount of Asteraceae pollen. They differ in that the rainforest element is lacking, except for trace quantities of pollen of *Nothofagus* subgenus *Brassospora* and Podocarpaceae. These two taxa may have been blown in from refugia elsewhere (Bint, 1981), or they may have been reworked from Early Tertiary deposits in the lake catchment. Moreover, *Podocarpus* is not restricted to rainforest and there is one species in the sclerophyllous vegetation of Western Australia today. Apart from these two taxa, the assemblages indicate dry sclerophyllous forest.

In the southern river valleys of the Western Slopes, mid–Late Pliocene pollen floras are dominated by Myrtaceae. The rainforest element, and especially the gymnosperms, dwindles towards the top of the sequence (Martin, 1987, 1991*a*). In the northern river valleys, however, gymnosperm pollen may be conspicuous and Araucariaceae usually accounts for most of it. Casuarinaceae can also be abundant.

Rainforest persisted in the highlands in the late Tertiary–Early Pleistocene. Assemblages at Lake George, New South Wales, contain some *Nothofagus* subgenus *Brassospora* pollen immediately prior to the increase of herbs and grasses (McEwen Mason, 1989). In the Atherton Tablelands, pollen assemblages contain a rich diversity of rainforest taxa, including *Nothofagus* subgenus *Brassospora*. The dominant pollen group is either *Nothofagus*, or other rainforest taxa, or Myrtaceae, or Casuarinaceae (Kershaw & Sluiter, 1982).

All this variation suggests that there was a rainfall gradient similar to that of today, i.e. it was wetter on the coast and in the southeast of Australia, and drier inland. A similar gradient is to be expected if the landforms of Australia were much the same during the Tertiary as they are today, and all evidence indicates that this was so.

Grasslands, herbfields and woodlands became widespread in the latest Pliocene–earliest Pleistocene. There are two sites, of about the same age, where this event has been dated, one at Lake George in the Southeastern Highlands (McEwen Mason, 1989) and the other on the Northwest Shelf (McMinn & Martin, 1992). Available evidence therefore suggests that grasslands became widespread at about the same time throughout Australia.

COMPARISONS WITH AUSTRALIA'S NEIGHBOURS

Antarctica

Antarctica was forested during the Cretaceous (Jefferson, 1982; Francis, 1990). Studies of fossil tree rings indicate strong seasonal growth and a lack of drought and frosts during the growing period. Seasonality may have resulted from colder winters when temperatures may have been below freezing, but the light regime, with the long polar winter, would certainly have determined seasonality. In the Late Cretaceous, gymnosperm pollen of Araucariaceae, *Lagarostrobos*, *Dacrycarpus*, *Dacrydium* and *Podocarpus* was present. Pollen of the angiosperms *Ilex*, *Gunnera*, *Nothofagus* (all four subgenera), Myrtaceae, *?Macadamia*, *Gevuina/Hicksbeachia*, *?Xylomelum* and Winteraceae were present, and there were 12 taxa of ferns and their allies and also some bryophytes (Dettmann, 1989).

Early Tertiary pollen is found in erratics and in marine sediments, having been reworked from the site of deposition. Since *in situ* pollen deposits are difficult to locate in Antarctica, these other sources yield important information about the vegetation. The assemblages are not as diverse as those of southeastern Australia. The floral composition, however, is very similar in its gymnosperm, *Nothofagus*, proteaceous and myrtaceous content (Kemp, 1972; Kemp & Barrett, 1975; Truswell, 1982). Antarctica was vegetated until the Eocene–Oligocene, when glaciation reached major proportions. Pliocene wood and leaves of *Nothofagus* have been found (Hill *et al.*, 1991), and the species must have survived until that time, at least as a small alpine shrub (Francis, 1990). It is not known when tundra shrublands developed. There is no evidence of the herbaceous element that is so common in modern tundras (Kemp & Barrett, 1975).

New Zealand

New Zealand vegetation throughout most of the Tertiary bears a striking resemblance to that of

Australia. The floral lists show some differences and the first appearance of taxa common to both may be different (Martin, 1982; Macphail *et al.*, Chapter 10, this volume), but the dominant pollen groups are similar.

From the Paleocene to the Middle Eocene, podocarps were very common. Casuarinaceae and Proteaceae were probably the most abundant pollen groups. *Nothofagus* pollen was present throughout, but scarce, just as it was in Australia (Mildenhall, 1980; Pocknall, 1989).

During the Eocene, *Nothofagus* became the dominant pollen group, subgenus *Fuscospora* first and then, in the Late Eocene, subgenus *Brassospora*. Proteaceae was well represented and included the tribe Banksieae, which is not native to New Zealand today. Myrtaceae was sometimes prominent. There are a number of tropical taxa present in the pollen floras, although only in low frequencies. Thus, in its general aspects, the Eocene vegetation must have been remarkably similar to that of Australia.

The Oligocene forests were dominated by *Nothofagus* subgenus *Brassospora*, with Casuarinaceae often abundant and Myrtaceae, Palmae, podocarps and *Nothofagus* subgenus *Fuscospora* prominent. Tropical elements, such as Caesalpineae, were also present. There was a variety of Proteaceae, but never as abundant as in the Late Eocene palynofloras, just as it was in Australia. In the Late Oligocene, araucarians were prominent in some localities.

Early–Middle Miocene New Zealand palynofloras are variable, as they are in southeastern Australia. Assemblages are dominated by *Nothofagus* subgenus *Brassospora*, Myrtaceae, Podocarpaceae, Palmae, *Macaranga/Mallotus*, Casuarinaceae, Sparganiaceae and occasionally tree ferns (Mildenhall, 1980; Pocknall, 1989).

There was a dramatic change in the Late Miocene similar to what happened in southeastern Australia. Many herbaceous taxa appeared and temperate taxa became prominent. *Nothofagus* subgenus *Fuscospora* and podocarps were usually dominant in pollen assemblages, but *Nothofagus* subgenus *Brassospora* and Casuarinaceae were still

common and occasionally dominant. There were a number of important extinctions, including *Lagarostrobos*, *Gardinia* ('*Randia*') *chartacea* type and *Micrantheum* (Mildenhall, 1980).

When compared with Australia, the Pliocene marks a divergence in the types of environment. Mountain forming commenced in New Zealand in the Miocene and glaciations reduced some forested areas to shrubland/grassland (Mildenhall, 1980), whereas developing aridity reduced forest cover in Australia. In the Pliocene New Zealand vegetation, the dominant pollen types were Myrtaceae, *Dacrydium cupressinum* and *Nothofagus* subgenus *Fuscospora*. *Cyathea* spores were more abundant than today. The palynofloras were extremely variable, even sequentially, indicating rapid change in the environment. Grasslands were present in some localities (Mildenhall, 1980).

The current New Zealand vegetation is substantially different from that of Australia, but this difference is a relatively late development. It should be noted that times of major change in Australia were also times of major change in New Zealand, e.g. the rise to dominance of *Nothofagus* in the Middle–Late Eocene and its decline in the Middle Miocene.

Southeast Asia and New Guinea

Records from Australia's nearest neighbour, New Guinea, are limited to the Neogene (Khan, 1974; Playford, 1982). Palynofloras from the Central Delta region of New Guinea extend back to the Late Miocene and consist mainly of contemporaneous taxa. The spores of ferns and their allies dominate. Gymnosperms form a very minor component, with *Microcachrys*, *Dacrydium*, *Podocarpus*, *Dacrycarpus* and traces of Araucariaceae present. The composition of the gymnosperm flora is thus very like that of Australia. *Nothofagus* pollen is usually present in appreciable amounts, especially in the Pliocene, and almost all of it belongs to subgenus *Brassospora*. There are rare occurrences of subgenus *Fuscospora*, but it is not known if they are the result of long-distance transport or indicate that it was growing in New

Guinea. Relatively minor amounts of Proteaceae, Casuarinaceae and Myrtaceae are present. These aspects of the assemblages are also like Australia. There is, however, a substantial mangrove component of Rhizophoraceae, Sonneratiaceae (Khan, 1974) and *Barringtonia* (Playford, 1982) in the New Guinea pollen assemblages, which is unlike the Australian record.

Overall, the New Guinea Neogene flora is substantially like that of Australia and the major differences, the fern spore and mangrove component, may be attributed to specific environments. These Neogene reports are from the section of New Guinea that has always been part of the Australian plate.

Borneo has a good Tertiary record. In the Paleocene–Eocene, none of the dominant Australian pollen groups, such as Proteaceae, *Notho-fagus*, or the austral gymnosperms, was present and Myrtaceae were rare (Morley, 1977). There were some taxa common to both, e.g. *Ilex, Anacolosa, Nypa*, Rhizophoraceae and Sapotaceae.

In the Oligocene–Miocene of Borneo, the flora became more modern and most taxa are related to the present flora. The affinities are essentially Indonesian-Malaysian and the mangrove flora is an important element. Asiatic montane taxa, such as *Abies, Alnus, Ephedra, Pinus, Picea* and *Tsuga* also became common in the Oligocene–Miocene (Morley, 1977). These features of the vegetation are very different to that of the contemporaneous Australian vegetation.

There are also some similarities with Australia. Myrtaceae, rare in pre-Miocene time, increased markedly in the Early Miocene and remained abundant thereafter. Casuarinaceae and *Dacrydium*, found in the unique heath forests today, became more abundant in the Miocene (Muller, 1972). These heath forests are now found on infertile sands and acid peats (Specht & Womersley, 1979), environments similar to the poor and acid soils of Australia (Bowen, 1981). *Podocarpus* and *Phyllocladus* are found from the Plio-Pleistocene onwards (Muller, 1966). Several other taxa are found in both and these were fully discussed by Truswell *et al.* (1987).

PHYTOGEOGRAPHY

Barlow (1981) presented an excellent review of the development of Australian phytogeographical thought. He divided phytogeography into the 'old', from Hooker (1860) to Burbidge (1960), developed before plate tectonics became accepted, and the 'new', based on a framework of modern palaeogeography and cytogenetics. This section examines the main tenets of both old and new in the light of the current fossil record.

In her analysis of the Australian flora, Burbidge (1960) delimited the principal floristic zones (Figure 7.7) as:

1. The Tropical Zone, extending across northern Australia and half way down the east coast.
2. The Temperate Zone, across southern Australia.
3. The Eremaean Zone, covering the whole of arid Australia (approximates the Eyrian faunal province). This zone is the largest in area.

There are interzones between the Tropical/Eremaean and Temperate/Eremaean. In these interzones, the wetter habitats along the watercourses, floodplains, sheltered gullies etc., support the tropical or temperate taxa, while the drier

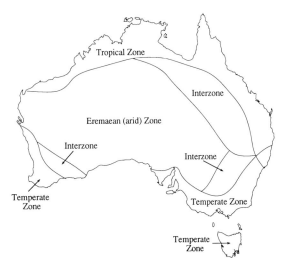

Figure 7.7 The phytogeographical zones of Australia. From Burbidge (1960).

slopes and ridgetops harbour the Eremaean taxa. These phytogeographical zones are a reasonable representation of the present-day flora and are useful as such.

It is customary to consider the Australian flora in terms of 'elements' or subfloras, i.e.:

1. The tropical, Malayan, Malaysian or Indo-Melanesian element, whose affinities are with Indonesia, Malaysia and the tropics in general.
2. The Australian or autochthonous element, which developed in Australia.
3. The Antarctic element, with affinities to other southern land masses. Alpine elements have, at times, been included here.

There are other elements of a minor nature, e.g. the cosmopolitan element and an element with affinities to the northern hemisphere (Burbidge, 1960; Barlow, 1981).

The use of these elements as an expression of present-day affinities is reasonable, provided that origins are not implied. To use present distributions as a measure of origins, migrations or evolution is fundamentally unsound, because spatial patterns cannot measure the time dimension. They may, however, suggest hypotheses to be tested. If the hypothesis is about origins or migrations, i.e. the time dimension, then the fossil record provides the test (Martin, 1984c). The theory that a taxon evolves somewhere and expands by dispersal from its area of origin is a sound working hypothesis. Expansion may be channelled by barriers or migration routes. Widespread distributions may become fragmented (vicariance biogeography). While such hypotheses assist our understanding of distributions, present-day distributions cannot be extrapolated backwards, along migration routes, to a centre of origin. This might be possible if *all other factors* were equal, but they are not. Palaeogeography, climate and many other factors have all changed, or rather change continuously, through time, in such complex patterns that a backwards extrapolation is impossible (Martin, 1984c), as is shown in the following assessment of such attempts in the light of the fossil record.

The Tropical element and the invasion theory

Taxa in the Australian Tropical Zone show affinities with floras of the Indo-Malaysian region. As they are limited to the north and northeast of Australia, it was assumed that they were recent immigrants into Australia. This assumption ignores the fact that the only environment in Australia with a climate both warm enough and wet enough for tropical taxa (more or less equivalent to tropical rainforest) is found in the north and northeast. The fossil record shows very clearly that many of these taxa have been in Australia for a very long time.

Truswell *et al.* (1987) examined the invasion theory in depth and showed from the fossil record that there has been almost continuous interchange throughout the Tertiary. Macphail & Truswell (1989) documented a number of tropical taxa in inland Australia from the Middle Eocene onwards. How this interchange was achieved is not clear, for the Australian Plate and the Indonesian–Malaysian island chain to the north were separated by 3000 km of deep ocean at the beginning of the Tertiary. Australia moved northward to come into contact with the island chain in the Middle Miocene. There may have been microcontinents that acted as 'stepping stones' between the two land masses. An examination of pollen assemblages before and after contact of the two land masses shows minimal interchange, which might be attributed to immigrant taxa (Truswell *et al.*, 1987).

Quantitatively, taxa that entered Australia from the north during the Tertiary account for only minor proportions of the pollen assemblages. It is difficult to assess their likely importance in the vegetation, given that most of them were probably pollinated by insects or other animals and thus would be underrepresented in pollen assemblages. The dominant pollen groups, however, remain essentially Australian and are not affected by immigrants from the north. It is important to consider the environment here. For the most part, Australia has very poor soils (Martin, 1982), unlike Indonesia–Malaysia. If the latitudinal difference, which would have been large in the early Tertiary, is added, then for the most part the environment of the two land masses would have been very different. This environmental difference constitutes a substantial barrier to any large-scale migration, whether in the early Tertiary, Middle Miocene or the present. Webb & Tracey (1972) showed that structurally similar communities, in similar habitats, have similar floristics, whether in northeast Australia or in New Guinea. The rainforest communities on rich soils, however, are vastly different to adjacent sclerophyllous communities on poor soils, and the difference is far greater than between similar rainforest communities in Australia and in New Guinea.

Mangroves are tropical–subtropical and it is puzzling that they have so few records in the Australian Tertiary. Macphail *et al.* (Chapter 10, this volume) have detailed the fossil pollen record for the early Tertiary. The reason for the lack of any coherent history of mangroves may be that there are no studies along the northern coastline, the region where a history is most likely to be found. Another reason may be that, during the Tertiary, Australia was in higher latitudes and it may not have been truly tropical, although the presence of large quantities of the dinoflagellate *Polysphaeridium zoharyi* in the mid-Tertiary of the river estuaries of the Murray Basin show that temperatures were comparable with those of Port Morseby and the Gulf of Carpentaria today (H. A. Martin, unpublished data). It may be that this tropical interval in the Middle Miocene of southeastern Australia was too brief for mangroves to migrate all the way from tropical regions. If the mangrove habitat existed, as it probably did in the Murray Basin, it is unlikely that it was not colonised by something. The mangrove habit has evolved many times, and it may be that mangroves existed in southeastern Australia but have not been recognised.

The Australian or autochthonous element

By definition, the taxa in this element originated within Australia, and the term applies particularly

to sclerophyllous plants, although it is not restricted to this group (Burbidge, 1960).

Eucalyptus, with approximately 800 species (L. A. S. Johson, personal communication) is the most significant genus. It is found in all vegetation types, though rarely in rainforest, and is dominant in all except the Eremaean Zone. The bloodwood eucalypt/*Angophora* pollen type (*Myrtaceidites eucalyptoides*) is distinctive and, although other pollen types within the genus may be distinguished with detailed study (Churchill, 1968), such studies have not been carried out on Tertiary pollen. *Myrtaceidites eucalyptoides* occurs in Late Paleocene deposits in the Lake Eyre Basin of inland Australia (I. R. K. Sluiter & N. F. Alley, unpublished data). When the Late Eocene–Early Oligocene of the Port Kenny district is compared with several typical sites in southeastern Australia (Figure 7.8), the Myrtaceae group is more common in the former, except for one site in the latter (Hillston), where it is about equal. When the Myrtaceae group is divided into its component species (Figure 7.9), however, *M. eucalyptoides* is much higher at Port Kenny than at Hillston during both the Late Eocene and Early Oligocene (Martin, unpublished data). Assemblages with a similar content of *M. eucalyptoides* are not found in southeastern Australia until the Early Miocene (Martin, 1988), see Figure 7.5.

Fruit and leaf remains of Middle Miocene eucalypts show some advanced states, comparable with extant taxa (Holmes *et al.*, 1983). A suite of silicified moulds and casts of fruits like those of *Eucalyptus*, *Leptospermum*, *Melaleuca*, *Callistemon* and *Angophora* from inland Australia suggest that the Leptospermoideae originated or diversified in inland Australia (Lange, 1978). Unfortunately, the age of these silicified fossils is equivocal, either Eocene–Oligocene or Miocene (Ambrose *et al.*, 1979) with a Miocene age more likely (Truswell & Harris, 1982). This topic is discussed in detail by Hill (Chapter 16).

The family Myrtaceae did not originate in Australia. Late Cretaceous Myrtaceae are found in Africa (Santonian), Borneo (early Senonian) and South America (Maastrichtian) (Muller, 1981). Myrtaceae pollen similar to *Syzygium* and *Eugenia*

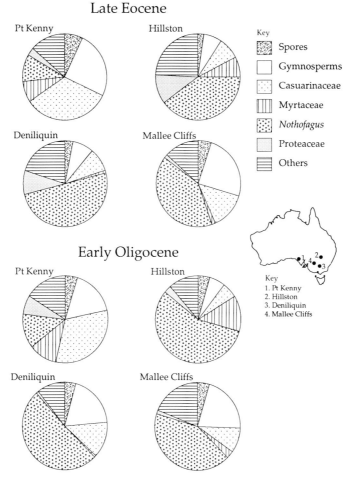

Figure 7.8 Comparison of the Late Eocene and Early Oligocene pollen assemblages at Port Kenny with those from several selected sites in southeastern Australia. For further explanation, see the text.

are found in Campanian and Maastrichtian sediments of the Antarctic Peninsula. These records are younger than those from Africa and Borneo and are about the same age as that of South America. The Antarctic records pre-date the first appearance of Myrtaceae in the Paleocene in both Australia and New Zealand. The fossil record suggests that Myrtaceae originated elsewhere in the world and migrated into Australia via South America and Antarctica. Australia has since become an important centre of diversity (Dettmann, 1989).

Acacia, with many more species than *Eucalyptus*, is common in all types of Australian vegetation, but very few species are found in rainforest. It may be dominant in the Eremaean Zone. *Acacia* pollen first appears in the Eocene in Australia (see Macphail *et al.*, Chapter 10), and there are Late Eocene records in Africa and Oligocene records in South America (Muller, 1981). It is therefore possible that *Acacia* originated in Australia, and the subcontinent has become an important centre of diversity.

Proteaceae is an important element in the

Late Eocene

Pt Kenny   Hillston

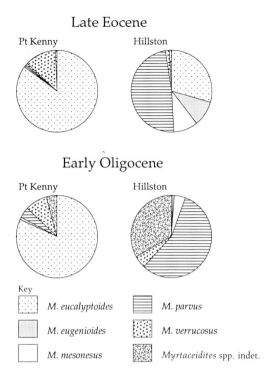

Early Oligocene

Pt Kenny   Hillston

Key

▨	M. eucalyptoides	▤ M. parvus
▨	M. eugenioides	▨ M. verrucosus
☐	M. mesonesus	▨ Myrtaceidites spp. indet.

Figure 7.9 Comparison of the component pollen species of the Myrtaceae group at Port Kenny and Hillston. The bloodwood eucalypt/*Angophora* type (*Myrtaceidites eucalyptoides*) is more abundant at Port Kenny in both the Late Eocene and Early Oligocene.

sclerophyllous vegetation. The *Xylomelum* pollen type first appeared in Australia and New Zealand in the Paleocene, but it was present in the Antarctic Peninsula in the Campanian, suggesting that its origin was not Australian. *Knightia* and *Gevuina/ Hicksbeachia* (rainforest taxa) probably originated in southeastern Australia. The place of origin of ?*Macadamia* (also a rainforest taxon) is uncertain, as pollen is found in both Antarctica and southeastern Australia in Campanian sediments (Dettmann, 1989). Unfortunately there are no Tertiary Antarctic records suitable for tracing origins and migrations.

The tribe Banksieae first appeared in the Paleocene. The tribe contains the rainforest genera *Austromuellera* and *Musgravea*, whose pollen cannot be distinguished from *Banksia* and *Dryandra*. *Banksia* cones are found in lignite at Yallourn (?Oligocene–Middle Miocene), where the parent

plants grew in a swamp (Cookson & Duigan, 1950). There are swamp species of *Banksia* today. Fossil *Banksia* fruits have also been found in the Eocene of Western Australia (McNamara & Scott, 1983). Pollen of the distinctively sclerophyllous *Grevillea/Hakea* complex (excluding *Grevillea robusta*, which is a rainforest species) is not found until the Late Miocene–Pliocene. It should be remembered that an environment suitable for sclerophyllous vegetation on a grand scale only became available in the Middle–Late Miocene in southeastern Australia.

Casuarinaceae are also distinctive in the Australian vegetation. Casuarinaceae pollen first appears in the Paleocene in Australia (Stover & Partridge, 1973), New Zealand (Mildenhall, 1980) and South America (Muller, 1981; Romero, 1986). Its origin is thus equivocal. Casuarinaceae has also been reported from South Africa, but these are Miocene in age (Coetzee & Muller, 1984; Coetzee & Praglowski, 1984).

The 'Pan Australian' flora and east–west disjunctions

The sclerophyllous vegetation in western and eastern Australia show many disjunct distributions. Today, these distributions are disrupted by the arid centre and the Nullarbor Plain, a limestone plateau unsuited to most sclerophyllous species. Such distributions led to the notion of a Pan Australian flora that has been disrupted (Burbidge, 1960). Barlow (1981), however, pointed out that the only uniformity was structural, i.e. southern Australia was once covered by closed forest, from east to west. Some superficial interpretations of the fossil record have continued a popular notion of a once uniform flora.

The fossil record clearly shows that the flora was not uniform at any time, as discussed above under Geographical variation. One more illustration may be seen amongst the Eocene proteaceous types. Descriptions of suites of pollen types in the west (Stover & Partridge, 1982), southeast (Stover & Partridge, 1973), central (Harris, 1972) and northeast (Dudgeon, 1983) of Australia show species in common but some of the species are not found, or rarely found, outside of the region where they were described.

The vegetation may not have been structurally uniform from east to west either. Casuarinaceae and the bloodwood eucalypt/*Angophora* pollen types are frequently more common in central Australian regions, suggesting that there may have been some sclerophyllous vegetation in the landscape. Moreover, some leaf assemblages are dominated by quite small leaves, or there is a mixture of small and larger leaves, suggesting that the well-watered river flats supported rainforest types, whereas the drier interfluvial areas may have been dominated by sclerophyllous vegetation or stunted vine forest, as seen in Arnhemland today (Greenwood *et al.*, 1990). Some pollen assemblages from Port Kenny, at the eastern end of the Nullarbor Plain (discussed above) are in accord with the macrofossil evidence. These assemblages have a higher Casuarinaceae and Myrtaceae pollen content than most in southeastern Australia (Figure 7.8), and the bloodwood eucalypt/*Angophora* type component of the Myrtaceae group is much higher at Port Kenny (Figure 7.9). The Casuarinaceae and bloodwood eucalypt/*Angophora* pollen type together suggest sclerophyllous forest, at least in parts of the landscape.

All the fossil evidence indicates therefore that there has never been a uniform flora across Australia. Probably there was just as much variation during the Tertiary as is seen today.

THE ANTARCTIC/SOUTHERN/ 'GONDWANAN' ELEMENT

Taxa in this element show affinities to New Zealand, South America and Africa. This element has been redefined and renamed the Gondwanan element or flora (Barlow, 1981; Nelson, 1981), unfortunately causing confusion. The Permian *Glossopteris* and the Triassic *Dicroidium* floras, which are found throughout the supercontinent of Gondwana (Gould, 1983) and are extinct, are also known as Gondwanan floras. Fragmentation of the supercontinent into the present-day landmasses was largely complete by the Early Cre-

taceous, when angiosperms first evolved. Thus there are no Gondwanan angiosperms living today.

When the origin of a taxon can be located, it is on one or another, or several of the southern landmasses (see Dettmann, 1989). Dettman's (1989; Chapter 8, this volume) excellent reviews of Cretaceous austral temperate forests show that Antarctica and adjacent land masses of Australia, New Zealand and South America were almost certainly the centres of origin for *Lagarostrobos*, certain *Dacrydium*, *Nothofagus*, Gunneraceae and the proteaceous taxa *Beauprea*, ?*Macadamia* and *Gevuina/Hicksbeachia*. *Dacrycarpus* probably originated in Antarctica/southern Australia and *Ilex* and Chloranthaceae in southern Australia. ?*Xylomelum* probably originated in the Antarctic Peninsula, adjacent to South America. *Microcachrys*, Winteraceae and Myrtaceae almost certainly originated outside of Antarctica and adjacent land masses, but migrated into Australia through the Antarctic pathway (Dettmann, 1989). Unfortunately, the Tertiary record is fragmentary, but Antarctica would have continued to function as a corridor for interchange during the early Tertiary, before extensive icecap development, and/or development of a seaway, disrupted this avenue. Interchange between the southern landmasses continued throughout the Tertiary, particularly between Australia and New Zealand (see discussion below).

Alpine/subalpine floras

The alpine/subalpine environment is today very limited on the Australian mainland but is more extensive in Tasmania (Galloway, 1986). In Australia, the mountains are old though not very high (Taylor, Chapter 5, this volume) and cold climates are a relatively recent development. Alpine–subalpine environments are only a few million years old, having been shaped by Pleistocene glaciations (Ollier, 1986).

Alpine–subalpine floras, with their preponderance of shrubby and herbaceous taxa, show most affinities with each other (Smith, 1986) and the cosmopolitan element is high. These floras, isolated on mountains, are regarded as 'islands' that would have expanded and contracted with the glacial/interglacial cycles. Migration between the 'islands' would have been by long-distance dispersal (Smith, 1986), though this is controversial (see below).

The alpine–subalpine floras have sometimes been included in the Antarctic element (see e.g. Hooker, 1860). This is inadvisable, as both the environment and floristics are very different. Macphail *et al.* (Chapter 10, this volume) and Carpenter *et al.* (Chapter 12, this volume) provide a detailed history of the origin of cool climate floras in Australia.

The development of an arid zone flora

A review of early to mid-Tertiary assemblages in the arid zone (Truswell & Harris, 1982) shows that rainforest taxa were found throughout. Some Pliocene assemblages are interpreted as open vegetation, but *Dacrydium*, *Podocarpus*, *Phyllocladus* and rare *Nothofagus* are present (Truswell & Harris, 1982). At Lake Frome, a Late Oligocene–Early Miocene pollen assemblage has rainforest taxa, including *Nothofagus* and abundant swamp taxa of Restionaceae, Cyperaceae and Sparganiaceae. The freshwater algae *Pediastrum* and *Botryococcus* are also abundant. A Late Miocene–Pliocene pollen assemblage has the same algae and swamp taxa, but rainforest taxa are absent. Casuarinaceae dominate the assemblage with the bloodwood eucalypt/*Angophora* subdominant and *Acacia*, *Dodonaea* and *Hakea/Grevillea* are also present. This assemblage is interpreted as dry sclerophyllous forest (Martin, 1990). In most of these inland assemblages, Casuarinaceae is more abundant than in southeast Australia, suggesting a drier environment inland, in keeping with the gradient seen in the southeast. However, in all of these Tertiary records there are no assemblages that could be classified as arid.

That inland Australia was well watered during the Tertiary is attested by the permanent freshwater lakes that existed during that time. Krieg

et al. (1990) described a giant lake phase in the Miocene of the Lake Eyre–Lake Frome region. The lakes contracted through the Miocene–Pliocene, but ancestral Lake Frome in ?Plio-Pleistocene time was larger, and precipitation higher, than today. Fish fossils of this time have a rotted appearance reminscent of the present situation on the shores of Lake Eyre, when fish are killed by increasing salinity, float, are washed onshore and decay (Callan, 1977), on the occasions when Lake Eyre, normally a dry salina, becomes filled with fresh water and dries out once again. Thus, whereas the climate was becoming drier from about the mid-Tertiary, it was wetter than today at the beginning of the Pleistocene and aridity reached its present proportions about 500 ka (Bowler, 1982). Unfortunately, there are no palynological records of this crucial period.

Burbidge (1960) suggested that many arid zone taxa may have developed from species of coastal habitats. This suggestion was based upon the fact that taxa such as Chenopodiaceae, common in coastal saltmarshes and sand dunes, are common in the arid zone as well. This has led to the notion that shorelines of Cretaceous seas that covered inland Australia were the cradle of the arid flora. The chenopod pollen type, and other angiosperm taxa in the arid flora, evolved long after the Cretaceous seas had disappeared (see Figure 7.1), at a time when freshwater lakes were present in inland Australia. Saltmarshes, however, are physiologically dry and sand dunes may be a desiccating medium for plant growth; taxa that can tolerate these environments are therefore able to grow in the arid interior. Coastal and arid taxa may have common ancestors, but they were not growing on the Cretaceous shorelines of inland seas. Using phylogenetic studies of *Atriplex*, Parr-Smith (1982) defined five major lines of evolution, each of which is either coastal or arid. The postulated precursor of the evolutionary sequence is an edge of the arid zone, near coastal species (Parr-Smith, 1982) growing today where these two environments are close to each other. There is no support for an hypothesis of a coastal origin of *Atriplex* in this study.

The decline of rainforest as the dominant vegetation

A striking feature of the palynological record is the demise of rainforest as a major part of the vegetation in southeastern Australia in the Middle–Late Miocene. Did rainforest decline all over Australia at the same time?

As discussed previously, Late Oligocene–Early Miocene microfloras at Lake Frome, in inland Australia, clearly indicate rainforest, whereas the Late Miocene–Pliocene microfloras represent dry sclerophyllous forest (Martin, 1990). Other Miocene sites reported very briefly by Truswell & Harris (1982) contain at least some rainforest taxa, but these sites will have to be re-examined before their precise nature can be assessed. Pliocene sites similarly reported by these authors contain a few rainforest taxa but pollen of Casuarinaceae, the eucalypt-type, Chenopodiaceae, Asteraceae, and Cyperaceae are locally abundant. These sites require revision, particularly the grass component (discussed below). Dating of these assemblages is somewhat uncertain (Truswell & Harris, 1982). What evidence exists suggests therefore that rainforest may have declined at the same time in inland as in southeastern Australia.

In the offshore Capricorn Basin, Queensland, Early Miocene pollen assemblages are dominated by gymnosperms and *Nothofagus*, with lesser amounts of Myrtaceae and Casuarinaceae. Chenopodiaceae, Asteraceae and Poaceae pollen are present but in very low quantities. In the Middle Miocene, *Nothofagus* became rare and in the Late Miocene–Early Pliocene, Myrtaceae and gymnosperms were dominant. Asteraceae and Chenopodiaceae increased in the Late Pliocene (Hekel, 1972). Thus, the major changes in the vegetation appear to be synchronous with those in southeastern Australia and there was a significant decline in rainforest about the Middle-Late Miocene.

In a deep-sea core offshore of northwest Australia, Early–Middle Miocene microfossil assemblages are restricted and contain only Casuarinaceae, *Cyathea*, Restionaceae and Asteraceae. There are no unequivocal rainforest taxa present. Late Miocene–Pliocene microfossil assemblages are more diverse and contain Casuarinaceae, Gyrostemonaceae, *Acacia* and rare *Podocarpus*, *Dodonaea* and *Cyathea*. Restionaceae, Chenopodiaceae, Asteraceae and Poaceae pollen are present in relatively low quantities, but Poaceae increased towards the top of the Late Pliocene. There are no unequivocal rainforest taxa throughout the sequence (McMinn & Martin, 1992). An Early and a Late Eocene site in the northwest, reported briefly by Truswell & Harris (1982), both contain pollen of rainforest taxa generally typical of the period, including *Nothofagus* in the Late Eocene site. Rainforest was therefore present in the northwest at that time, but appears to have declined before the Miocene, suggesting an earlier decline here than elsewhere.

In all these sites the change was from rainforest dominance to another, drier, kind of forest. As far as can be judged, therefore, practically all of Australia was forested through the Tertiary, including most of the Pliocene.

The development of grasslands

Poaceae first appeared in the Eocene (Figure 7.1), but it was rare throughout the Eocene–Middle Miocene. Under a closed forest canopy, as would have existed then, insufficient light reaches the ground to support grasses and other herbaceous plants. Older reports of Early Miocene grasslands in central Australia (Martin, 1978; Truswell & Harris, 1982) are based on a misidentification of the *Restio*-type of Restionaceae, a swamp taxon (Martin, 1990). Grasses increased somewhat in the Late Miocene–Pliocene of southeastern Australia, but it is uncertain whether they were dryland or swamp grasses. Some grasses are found today under an open eucalypt canopy as would have existed in the Late Miocene–Pliocene. It is doubtful whether true grasslands existed until late in the Pliocene.

The increase in grass pollen in the sequence of assemblages is a recognisable horizon (Martin, 1979). Did this increase occur synchronously all over Australia? McEwen Mason (1990), using palaeomagnetic techniques, dated this increase at 2.5–3 Ma, i.e. Late Pliocene, at Lake George, in the highlands of southeast Australia. In northwest Australia, a well-dated deep-sea core shows an initial, small increase in grass pollen during the Late Miocene–Middle Pliocene sequence, continuing into the Late Pliocene, with a further large increase during the Late Pliocene, continuing into the Pleistocene. Thus this northwest sequence shows essentially the same pattern as that of the southeast. In the offshore Capricorn Basin of Queensland, grass pollen is minimal and Asteraceae increased during the Late Pliocene. Chenopodiaceae, however, increased during the Late Miocene–Early Pliocene, to about 6% of total pollen (a very modest increase) in the Late Pliocene (Hekel, 1972). It is not known whether the Chenopodiaceae pollen indicates dryland or saltmarsh taxa. Herbfields rather than grasslands therefore developed in the northeast of Australia during the Late Pliocene.

Available evidence suggests that grasslands and herbfields developed at about the same time across Australia. Concomitant with this increase in herbaceous vegetation was forest decline.

The development of fire tolerance/ dependency

The ability to cope with burning on a regular basis, even to the extent that many species are dependent on fire for germination, probably evolved throughout millions of years. Ridgetops, the driest habitat in the landscape, would have been subjected to lightning strikes and small, limited fires, even while the major part of the landscape was covered in rainforest. When the climate became too dry for rainforest, and burning became an integral part of the environment, these taxa rapidly expanded their distribution (see A model for change, p. 131).

It is popularly believed that Aborigines, who must have been the original (but purposeful) pyromaniacs, drastically modified the vegetation through the use of fire. Any such modification would have been limited to fire-prone vegetation, where burning on short-term cycles, such as three

or five years, maintains a greater diversity of species than burning on longer-term cycles such as 20 or 30 years, when all the species with life cycles shorter than the time between fires disappear from the community (Specht et al., 1958).

Aborigines could not have caused the widespread decline of rainforest by burning. Under normal circumstances rainforest is too damp to burn, and to do so requires exceptional drying, such as a drought (Webb, 1970; Luke & McArthur, 1978). Today, the frequent fires in sclerophyllous vegetation bordering rainforest usually stop when then they reach the rainforest, and only burn the ecotone between the two, modifying the boundary between them. Rainforest may be eroded at the edges by repeated fires (Barker, 1990), but it would require the assistance of a change to a drier climate before wholesale destruction by fire would be possible.

The fossil record, with the change from rainforest to wet sclerophyllous forest and the concomitant increase in charcoal particles, shows that fire had become an integral part of the environment in the Late Miocene, some ?10 Ma before Aborigines arrived in Australia. The decline in rainforest was caused by climatic change.

ORIGINS AND MIGRATIONS

The fossil record may reveal the origins and likely migrations of taxa, if there is sufficient evidence, and this has been discussed where appropriate. For many taxa with only a few fossil records, however, their origin is obscure.

Even if origins are obscure, the fossil record may reveal some surprising distributions. Many rainforest taxa with a restricted distribution today once had widespread distributions in Australia. Nonrainforest taxa may also have had a wider past distribution. For example, Eucalyptus spathulata, which has a distinctive pollen type and is now endemic to southwest Australia, first appears in the Late Oligocene (N. F. Alley, personal communication), was once also in southeastern Australia in the Early Pliocene (Martin & Gadek, 1988) and has been found off the Great Barrier Reef (H. A. Martin, unpublished data).

Figure 7.10 The distribution of the *Dodonaea triquetra* pollen type, fossil and modern. 1, As *Tricolporites* sp. aff. *Diplopeltis*; Bint (1981), Early Pliocene. 2, As *Nuxpollenites* aff. *D. triquetra*; Martin (1991b), Late Eocene. 3, As *Tricolporites* sp. aff. *Diplopeltis*; Kemp (1976), Middle–?Late Eocene. 4, As *Nuxpollenites* sp.; Macphail & Truswell (1989), Late Eocene–Middle Miocene. 5, As *Nuxpollenites* sp.; Truswell (1987), Middle Miocene. 6, As *Nuxpollenites* aff. *Dodonaea triquetra*; Martin (1991b), Early–Middle Miocene. 7, Martin (unpublished data), Early–Middle Miocene. 8, Martin (unpubl.), Pliocene. 9, As *Nuxpollenites* sp. aff. *Diplopeltis*; Mildenhall & Pocknall (1989), Miocene. 10, H. A. Martin (unpublished), Early Pliocene. Shaded area, present distribution of *Dodonaea triquetra*. (From West, 1984).

Figure 7.11 Photos of *Dodonae triquetra* pollen. (*a*)–(*k*) Fossil specimens. (*a*)–(*c*) From Port Kenny (site 2 on Fig. 7.10). (*d*)–(*f*) From the Western Murray Basin (site 6). (*g*)–(*h*) From the Macquarie River Valley (site 8). (*i*) From a deep-sea core off the Great Barrier Reef (site 10), a lightly thickened form. (*j*)–(*k*) From New Zealand (site 9). (with kind permission from D. C. Mildenhall.)

Modern pollen of *Dodonaea triquetra* is tricolporate, with heavily thickened pores. It is distinguished from all other species of *Dodonaea* by unique, thickened bands. These bands are intercolpal, joining over the poles. The intercolpal bands may be less thickened or unthickened in the equatorial region. The variable aspect viewed and expansion/contraction of the grain adds to the observed variability.

(*l*–*v*) Modern specimens. (*l*), (*m*) From Sydney. The protruding pore membrane (*m*) is not seen on fossil specimens. (*n*) From Tilba Tilba, New South Wales. (*o*), (*p*) From Bundanoon, New South Wales, oblique polar view of a heavily thickened grain showing thickened bands in the equatorial region (arrows). (*q*)–(*t*) From Mt Coolum, Queensland. (*q*), (*r*) Polar view showing the outline of thickening over the pole (*q*) but no thickening in the equatorial region (*r*). (*s*), (*t*) Equatorial view. Arrows in (*s*) indicate thickening over poles, and in (*t*) thinning of the exine towards the equator. (*u*), (*v*) From Glenbrook, New South Wales. The oblique view produces a variable outline. The variability seen amongst the fossil specimens thus falls within intraspecific variability observed in *D. triquetra*. Scale bars represent 10 μm.

Note: Different chemical treatments cause differences in swelling of the grains. All modern specimens and some fossils (e.g. (*a*)–(*c*)) have been acetolysed, which is known to cause swelling. Compression of fossils may alter their look also.

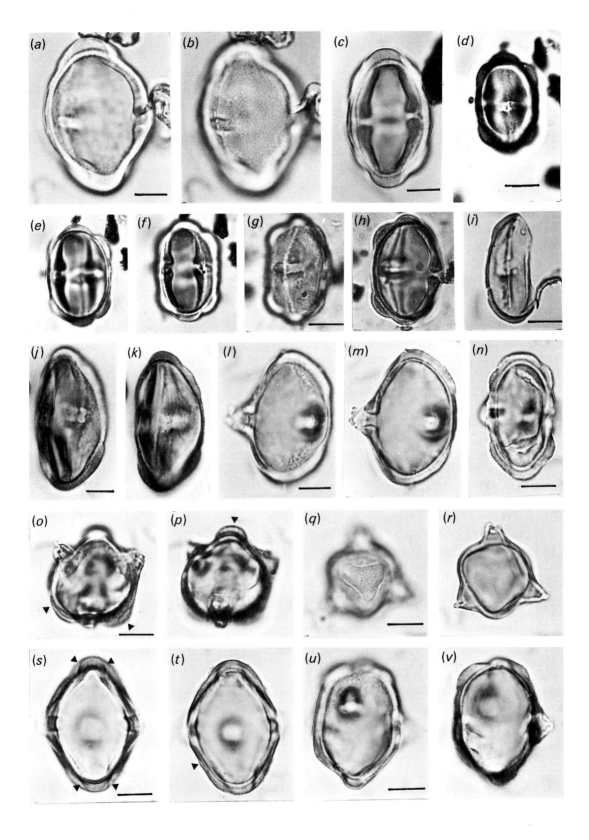

Another example with a distinctive pollen type, *Dodonaea triquetra*, now has a restricted distribution along the southeast coastal strip (Figure 7.10) and is found in wet and dry sclerophyllous forests. This species has been classified as a fire weed because of its high rate of germination after a hot fire (West, 1984). *Dodonaea triquetra* first appeared in the Late Eocene, although the morphology of the pollen grain is not exactly the same as that of today (Figure 7.11). These differences are thought to indicate an evolving lineage, such that the Eocene species producing *D. triquetra* pollen, and possibly the late Tertiary species as well, would not have been the same as that of today. Fossil distributions (Figure 7.10) show that it was present in northeastern, western and central Australia, inland southeastern Australia and New Zealand (Mildenhall & Pocknall, 1989). The morphology of the New Zealand specimens, however, is different from any seen in Australia, but given the range of variation amongst modern specimens it is reasonable to include it in the lineage.

Acacia is widespread in Australia but is absent from the modern flora of New Zealand, although it was present there in the past. In the southern part of Southland, *Acacia* pollen occurs in Early Miocene assemblages that are typically rainforest, with abundant *Nothofagus* (Pocknall, 1981; Mildenhall & Pocknall, 1989). In the southern part of the North Island, *Acacia* pollen is found in Pliocene–Early Pleistocene assemblages from what was coastal swamp/scrub/grassland with few trees (Mildenhall, 1975). There was probably more than one introduction of *Acacia* into New Zealand from Australia.

Fuchsia, native to New Zealand, Tahiti and Central and South America, but not Australia, is found in the Australian fossil record from the Late Oligocene to Early Miocene. It also reached New Zealand in the Late Oligocene and has remained there to the present day (Berry *et al.*, 1990). These are only a few examples of former distributions that would never be suspected from extant distributions. They illustrate the complexity of past migrations.

Plant migration is normally achieved through seed or spore dispersal. Different strategies for transport, longevity and germination will affect migrations, and this may be illustrated with *Nothofagus* and *Ilex*. Both originated in the southern Australian and/or Antarctic region (Dettmann, 1989) in the Late Cretaceous (Figure 7.1), yet *Nothofagus* has a strictly southern distribution whereas *Ilex* is cosmopolitan and had achieved a world-wide distribution in the Late Cretaceous (Martin, 1977). The seeds of *Nothofagus* do not remain viable for more than one year and rarely disperse more than a few metres from the parent tree. If birds and animals eat them, the kernel is digested, and if they dry out they will not germinate, all factors which slow down migrations. On the other hand, the fruits of *Ilex* are eaten by birds and the seeds are voided unharmed. The seed requires an after-ripening period before germination, which may be two years but in one species is eight years (Martin, 1977); the seed is therefore effectively much longer-lived than that of *Nothofagus*. All these factors enabled *Ilex* to disperse widely and migrate rapidly.

The fossil record shows conclusively that long-distance dispersal occurs, and would have occurred, on a regular (geological) basis. How this would have been originally achieved is unknown. Much speculation has gone into devising island stepping stones, land-bridges, etc., most of which are impossible on current palaeogeography. It is likely that long-distance dispersal has occurred by normal methods, such as birds, and cyclonic and gale-force winds, which may transport quite large objects and which occur on a regular basis. Ocean currents may transport plant propagules for long distances. Even if such events are rarely seen today, they were common in geological time.

Transport is not the only criterion for migration. Once transported, the seed must germinate, establish and compete successfully with the species already in possession of the site. Disturbances caused by cyclones and gales may well create openings in the vegetation that would allow migrants to become established.

Migrants, once established, may not find the environment conducive to permanent residency. As examples, there have been at least two introductions of *Acacia* into New Zealand, but both eventually died out. The introduction of *Fuchsia* into Australia did not become permanent.

It is very clear from the fossil record that taxa, not communities, migrate.

A MODEL FOR CHANGE

Two kinds of change may be distinguished:

1. Changes in the flora occur more or less constantly, as seen from records of first appearances (Figure 7.1).
2. Changes through time in the vegetation, however, appear relatively slight for most of the time, with some periods of drastic change.

The Oligocene was a time of slow and gradual change, with the dominant features of the vegetation remaining much the same. When assemblages of earliest and latest Oligocene age in inland New South Wales are compared, there are considerable differences. Taxa that were relatively abundant and usually present in the earliest Oligocene were rare by the end of the Oligocene (e.g. *Lagarostrobos* (Figure 7.3) and some of the proteaceous species). It may be inferred that the climate was slowly becoming drier. In latest Oligocene–earliest Miocene time, in a more abrupt change, *Nothofagus* declined, especially in the more westerly localities and Myrtaceae, Araucariaceae and/or Casuarinaceae became dominant, indicative of a further step to a drier climate. It is difficult to assess just how big this change was, as major kinds of vegetation are found within a climatic range and only change when some threshold is crossed.

In the Middle–Late Miocene, there was a drastic change and widespread rainforest disappeared from inland New South Wales, though it would have remained in the highlands and coastal regions where the climate was wetter than that of the inland. Taxa that required wetter climates were restricted to refugia in the highlands or to small, favourable habitats such as sheltered gullies and gorges. When the climate became wetter for a brief period in the Early Pliocene, these rainforest taxa expanded their distribution.

The Late Pliocene–Early Pleistocene was another time of drastic change, when the forests opened up and grasslands/herbfields with some woodlands developed.

These changes are undoubtedly driven by climatic change. How does the vegetation respond to these changes? In any given landscape, there is a variety of habitats, from well-watered valley bottoms to drier hillslopes and ridgetops that are subjected to the most desiccation. Evolution is occurring continuously in these different habitats. For example, tolerance to burning was probably developing on ridgetops subjected to lightning strikes and small, limited fires for millions of years, so that small, isolated communities, 'nuclei for change' (Herbert, 1933), existed long before the climate allowed burning on a regular basis. When change comes, be it wetter or drier, the vegetation responds rapidly with an expansion/contraction of the different habitats. The fossil record supports this model of change. Taxa, when they first appear, are usually not common and become abundant at some later time. As far as can be assessed, all the major changes in the vegetation occur at a time of climatic change, and at the same time across Australia, although the precise nature of the change differs with the geographical region.

Plants thus respond rapidly to change, either expanding or contracting their distribution as demanded by the climate. The little pockets of rainforest taxa in favourable niches in an otherwise unfavourable terrain, seen across northern Australia today, will probably expand, should the climate become wetter, as it may, if predictions of the greenhouse effect are accurate. Most plants are hermaphrodites but there are many mechanisms to promote outcrossing. Some taxa, however, are able to self-pollinate if pollination by normal outcrossing has not occurred, thus enhancing the chances of survival (Crome & Irvine, 1986). Experiments show that self-pollination results in inbreeding depression in growth and competitive ability, which, however, has not been observed in the field, even though self-fertilisation may be common. It is thought that the variance in habitat is much greater than the variance in plant fitness (Crome & Irvine, 1986). This ability to self-fertilise is advantageous for survival in times of drastic change, should the population fall to very low levels or be fragmented to isolated patches.

All animals are dependent, ultimately, on plant life. How are the times of change likely to affect animal populations? If change is slow, animals may be able to adapt, but change may be too rapid for this to happen. Today, many animal populations suffer a shrinking habitat, the result of pressures from the human population, and the deleterious effects may be observed. Plants may be able to survive a drastically fragmented distribution, which, however, may be insufficient to support the animal populations dependent on them. Moreover, animals do not have recourse to self-fertilisation, as do plants. The Late Miocene demise of rainforest on a grand scale must have had a devastating effect on the animal populations. At Riversleigh in northeastern Australia, species diversity dropped markedly from a high in the Late Oligocene–Early Miocene, to a low in the Pliocene–Quaternary (Archer *et al.*, 1989*a*). Prior to this Late Miocene devastation, all the species were rainforest inhabitants (Archer *et al.*, 1989*b*).

If this model for change is accepted, then the implications for conservation are clear. A maximum of diversity should be maintained as the nucleus for future change.

CONCLUSIONS

The landscape was almost entirely forested throughout the Tertiary. The vegetation was mainly rainforest prior to the Middle Miocene, but afterwards drier kinds of forest dominated. In the Late Pliocene–Pleistocene, the herbaceous element increased and the forest opened up. Woodlands/grasslands/herbfields were dominant.

There was almost continuous change in the flora with time. Vegetation changes were small for most of the time, but there were several periods of drastic vegetation change. The general trends in the vegetation and times of change are recognisable in different parts of Australia, but there was variation in the vegetation, comparable with that seen today. The times of major change correspond with climate changes. A climatic gradient parallel to that of today, i.e. wetter on the coast and drier inland, existed throughout the Tertiary.

Comparisons of Australia with neighbouring land masses generally show differences commensurate with environmental differences. Thus, the Tertiary vegetation of Australia and New Zealand was generally similar until the Miocene, after which considerable differences emerged. The environment diverged during the late Tertiary: aridity developed in Australia, whereas mountain building and glaciation shaped New Zealand's environment. The flora and vegetation of the two are very different today.

When the long-established phytogeographical hypotheses are tested via the fossil record, they are found to be almost totally inadequate. Present distributions, taxonomy and cytogenetics cannot be read backwards to deduce the origin and migration routes of plant taxa. Evolution of taxa would have occurred almost continuously. Each taxon occupied its favoured niche in the landscape. Small patches of vegetation different from the dominant type(s), such as that in protected gorges or on ridgetops subjected to lightning strikes, were 'nuclei for change'. When climatic change comes, an expansion or contraction of the different communities rapidly accommodates the change.

Appendix 7.1. *Taxa in the range chart (Figure 7.1) with their fossil name(s)*

Botanical name	Fossil name	Reference(s)
Acacia (Mimosaceae)	*Polyadopollenites myriosporites* (Cookson) Stover & Patridge *Acaciapollenites myriosporites* (Cookson) Mildenhall	Cookson (1954), Stover & Partridge (1973)
Alangium (Alangiaceae)	*Alangiopollis* sp.	Foster (1982), Macphail & Truswell (1989), Reitsma (1970)
Alyxia (Apocynanceae)	*Psilodiporites* sp.	Macphail & Truswell (1989)
Amylotheca (Loranthaceae)	*Amylotheca*	Martin (1978), Cookson (1956)
Anacardiaceae/Simaroubaceae	?*Striatocolporites minor* Muller	Sluiter (1984)
Anacolosa (Olacaceae)	*Anacolosidites* spp.	Cookson & Pike (1954*b*), Stover & Partridge (1973)
Araliaceae	Araliaceae	Luly *et al.* (1980), Sluiter (1984), Truswell *et al.* (1987)
Araucariaceae	*Araucariacites australis* Cookson, ?*Dilwynites granulatus* Harris	Martin (1978), Harris (1965), Dettmann (1989)
Ascarina (Chloranthaceae)	*Clavatipollenites* spp.	Dettmann (1989), Muller (1981), Macphail & Truswell (1989)
Asteraceae	*Tubulifloridites* spp.	Stover & Partridge (1973), Martin (1973*a*)
Austrobuxus (= *Longetia*) *Dissiliaria* (Euphorbiaceae)	*Malvacipollis* spp. *Polyorificites oblatus* Martin = *Heliciporites astrus* Stover & Partridge	Martin (1974), Stover & Partridge (1973)
Banksieae (Proteaceae)	*Banksieaeidites* sp.	Cookson (1950), Stover & Partridge (1973)
Beauprea (Proteaceae)	*Beaupreaidites* spp. Cookson	Cookson (1950), Dettmann (1989)
Bombax type (Bombacaceae)	*Bombacidites bombaxoides* Couper	Stover & Partridge (1973), Muller (1981)
Caesalpinoides (Caesalpiniaceae)	*Margocolporites vanwijhei* Germeraad *et al.*	Macphail & Truswell (1989)
Calamus (Arecaceae)	*Calamus* Truswell *et al.*	Truswell *et al.* (1987)
cf. *Callitriche* (Callitrichaceae)	*Australopollis obscurus* (Harris) Krutzsch emend. Stover & Partridge	Macphail (1990)
Canthium (Rubiaceae)	*Canthiumidites oblatus* Pocknall & Mildenhall	Macphail & Truswell (1989)
Carnarvonia (Proteaceae)	*Proteacidites stratosus* Pocknall & Mildenhall	Sluiter (1984)
Casuarinaceae	*Haloragacidites harrisii* (Couper) Harris, *Casuarinidites cainozoicus* Cookson & Pike	Cookson & Pike (1954*b*), Stover & Partridge (1973), Kershaw (1970*a*)
Celtis (Ulmaceae)	*Celtis* sim. Sluiter	Sluiter (1984), Truswell *et al.* (1987)
Ceratopetalum (Cunoniaceae)	*Ceratopetalum* comp.	Luly *et al.* (1980), Sluiter (1984)
Chenopod type (Chenopodiaceae/ Amaranthaceae)	*Chenopodipollis chenopodiaceoides* (Martin) Truswell	Truswell *et al.* (1985), Macphail & Truswell (1989)
Coelebogyne (= *Alchornea*) (Euphorbiaceae)	*Psilatricolporites operculatus* van der Hammen & Wymystra	Germeraad *et al.* (1968), Martin (1974)
Coprosma/Opercularia (Rubiaceae)	*Palaeocoprosmadites* Macphail & Truswell, *Coprosma* sim. Sluiter	Macphail & Truswell (1989), Sluiter (1984)
Cunoniaceae	Cunoniaceae	Luly *et al.* (1980), Sluiter (1988)

Appendix 7.1. (*cont.*)

Botanical name	Fossil name	Reference(s)
Cupaneae (Sapindaceae)	*Cupaneidites* spp.	Cookson & Pike (1954*b*), Stover & Partridge (1973)
Cupressaceae	Cupressaceae	Martin (1973*a*, 1978)
Cyperaceae	Cyperaceae, *Cyperaceaepollis* spp.	Martin (1973*a*, 1978), Sluiter (1988), Truswell *et al.* (1985)
Dacrycarpus (Podocarpaceae)	*Dacrycarpites australiensis* Cookson & Pike	Cookson & Pike (1953), Taylor *et al.* (1990)
Dacrydium (Podocarpaceae)	*Lygistepollenites florinii* (Cookson & Pike) Stover & Evans	Martin (1978), Stover & Partridge (1973)
Dodonaea (Sapindaceae)	*Dodonaea sphaerica* Martin	Martin (1973*a*)
Dodonaea cf. *D. triquetra* (Sapindaceae)	*Nuxpollenites* sp. Truswell *et al.*	Truswell *et al.* (1985), Macphail & Truswell (1989)
Droseraceae	Droseraceae, *Droserapollis*	Sluiter (1984), Macphail & Truswell (1989), Truswell & Marchant (1986)
Ebenaceae	Ebenaceae, *Diospyros* sim. Sluiter	Sluiter (1984), Truswell *et al.* (1987)
Elaeocarpaceae	*Elaeocarpus* com. Elaeocarpaceae	Luly *et al.* (1980), Truswell *et al.* (1987)
Ericales	*Ericipites* spp.	Harris (1965), Stover & Partridge (1973)
Eucalyptus, bloodwood/*Angophora* type (Myrtaceae)	*Myrtaceidites eucalyptoides* Cookson & Pike, *Eucalyptus* comp. Sluiter	Cookson & Pike (1954*b*), Martin (1978), Sluiter (1984)
Eucalyptus spathulata	*Eucalyptus spathulata* Martin & Gadek, *Myrtaceidites* spp. C, Bint	Martin & Gadek (1988), Bint (1981)
Exocarpus (Santalaceae)	*Exocarpus* sim. Sluiter	Sluiter (1984)
Fuchsia (Onagraceae)	*Diporites aspis* Pocknall & Mildenhall	Macphail & Truswell (1989), Berry *et al.* (1990)
Gardenia = '*Randia*' *chartacea* (Rubiaceae)	*Triporopollenites bellus* Stover & Partridge	Stover & Partridge (1973), Martin (1978)
Gardenia (triporotetrad) (Rubiaceae)	*Triporotetradites* sp. Truswell *et al.*	Truswell *et al.* (1985), Macphail & Truswell (1989)
Geissois/Eucryphia (Cunoniaceae/ Eucryphiaceae)	*Geissois/Eucryphia* comp.	Luly *et al.* (1980), Martin (unpubl.), Sluiter (1984)
Gentiana (Gentianaceae)	*Gentiana* sim.	Luly *et al.* (1980)
Geranium/Pelargonium (Geraniaceae)	'*Tricolporites geranioides*' Couper, *Tricolporopollenites pelargonioides* Martin	Martin (1973*b*)
Gevuina/Hicksbeachia (Proteaceae)	*Propylipollis* (*Proteacidites*) *reticuloscabratus* (Harris) Dettmann	Dettmann (1989), Martin (1987)
Guettarda (Rubiaceae)	cf. *Guettardidites*	Macphail & Truswell (1989)
Gunnera (Gunneraceae)	*Rhoipites alveolatus* (Couper) Pocknall & Crosbie, *Gunnerites reticulatus* Cookson & Pike	Dettman (1989), Pocknall & Crosbie (1982), Muller (1981)
Gyrostemonaceae	Gyrostemonaceae	Truswell *et al.* (1985), Macphail & Truswell (1989)
Hakea/Grevillea (Proteaceae)	Proteaceae cf. *Grevillea* Martin	Martin (1973*a*, unpubl.)
Haloragis (Haloragaceae)	*Haloragacidites haloragoides* Cookson & Pike	Cookson & Pike (1954*b*), Stover & Partridge (1973)
Ilex (Aquifoliaceae)	*Ilexpollenites* spp.	Martin (1977)
Isopogon (Proteaceae)	*Proteacidites isopogoniformis* Couper	Macphail & Truswell (1989)

Appendix 7.1. (*cont.*)

Botanical name	Fossil name	Reference(s)
Jasminum (Rubiaceae)	*Jasminum* comp.	Luly *et al.* (1980)
Lagarostrobos (Podocarpaceae)	*Phyllocladidites mawsonii* Cookson ex Couper	Martin (1978), Playford & Dettmann (1979)
?Liliaceae	*Liliacidites* spp.	Stover & Partridge (1973)
Loranthaceae	*Gothanipollis bassensis* Stover *Tricolpites simatus* Partridge	Stover & Partridge (1973), Martin (1978), Sluiter (1984)
Macadamia/Helicia/Orites comp. (Proteaceae)	*Propylipollis* cf. *scaboratus* (Couper) Dettmann, *Proteacidites ivanhoensis* Martin	Dettmann (1989), Martin (1973a, 1978)
Macaranga/Mallotus (Euphorbiaceae)	*Triporopollenites endobalteus* McIntyre	Martin (1974)
Malpighiaceae	*Perisyncolporites pokornyi* Germeraad, Hopping & Muller	Macphail & Truswell (1989), Germeraad *et al.* (1968)
Malvaceae	*Malvacearumpollis* sp.	Macphail & Truswell (1989)
Meliaceae	Meliaceae, *Dysoxylum* sim.	Sluiter (1984), Truswell *et al.* (1987)
Merrimia (Convolvulaceae)	*Perfotricolporites digitatus* Gonzalez	Ferguson *et al.* (1977), Germeraad *et al.* (1968), Macphail & Truswell (1989)
Micrantheum/Neoropera (Euphorbiaceae)	*Malvacipollis spinyspora*	Martin (1973a, 1974, 1978), Kemp (1976)
Microcachrys (Podocarpaceae)	*Microcachrydites antarcitcus* Cookson	Martin (1978), Dettmann (1989)
Monotoca (Epacridaceae)	*Monotoca* sp.	Hill & Macphail (1985), Martin (unpubl.)
Myriophyllum (Haloragaceae)	*Haloragacidites myriophylloides* Cookson & Pike	Cookson & Pike (1954b), Martin (1978)
Myrsinaceae	Myrsinaceae, *Rapanea* sim.	Sluiter (1984), Truswell *et al.* (1987)
Myrtaceae	*Myrtaceidites* spp. Cookson & Pike, *Acmena* comp., *Baeckea/Leptospermum* comp., *Syzygium–Cleistocalyx* comp., *Rhodamnia* comp., *Tristania/Meterosideros* comp. in Luly *et al.* and Sluiter	Cookson & Pike (1954b), Stover & Partridge (1973), Luly *et al.* (1980), Sluiter (1984), Martin (1973a)
Nothofagus, ancestral type (Fagaceae)	*Nothofagidites senectus* Dettmann & Playford, *N. endurus* Stover & Partridge	Dettmann *et al.* (1990), Stover & Partridge (1973)
Nothofagus subgen. *Brassospora* (Fagaceae)	*Nothofagidites* spp. (see Dettmann *et al.* 1990)	Dettmann *et al.* (1990), Cookson (1958)
Nothofagus subgen. *Fuscospora* (Fagaceae)	*Nothofagidites brachyspinulosus* (Cookson) Harris, *N. flemingii* (Couper) Potonie	Dettmann *et al.* (1990), Cookson (1958)
Nothofagus subgen. *Lophozonia* (Fagaceae)	*Nothofagidites asperus* (Cookson) Stover & Evans	Dettmann *et al.* (1990), Cookson (1958)
Nypa (Araceae)	*Spinizonocolpites prominatus* (McIntyre) Stover & Evans	Stover & Partridge (1973), Muller (1981)
Omalanthus (Euphorbiaceae)	*Omalanthus* comp.	Luly *et al.* (1980)
Phyllocladus 2-saccate type (Podocarpaceae)	*Phyllocladidites palaeogenicus* Cookson & Pike	Cookson & Pike (1954a), Pocknall (1981)
Phyllocladus 3-saccate type (Podocarpaceae)	*Trisaccites micropteris* Cookson & Pike, *Podosporites microsaccatus* Dettmann	Martin (1978), Cookson & Pike (1954a), Dettmann (1963)

Appendix 7.1. (*cont.*)

Botanical name	Fossil name	Reference(s)
Poaceae	*Graminidites media* Cookson	Martin (1978), Muller (1981)
Podocarpus (Podocarpaceae)	*Podocarpidites* spp.	Martin (1978), Dettmann (1989)
Polygalaceae	*Polycolpites esobalteus* McIntyre	Stover & Partridge (1973), Sluiter (1984)
Polygonum (Polygonaceae)	*Polygonum*, pantoporate pollen type	Martin (1982)
Polyosma (Escalloniaceae)	*Polyosma* sim.	Luly *et al.* (1980)
Portulacaceae	Portulacaceae Martin, *Lymingtonia* sp. Macphail & Truswell	Martin (1973b), Macphail & Truswell (1989)
Quintinia (Escalloniaceae)	*Quintinia psilatispora* Martin	Martin (1973a, 1978)
Restionaceae, *Hypolaena* type	*Milfordia hypolaenoides* Erdtman	Martin (1973a, 1978), Stover & Partridge (1973), Sluiter (1988)
Restionaceae, *Restio* type	*Milfordia homeopunctata* (McIntyre) Partridge	Stover & Partridge (1973), Martin (1978), Sluiter (1988)
Rhamnaceae	Rhamnaceae	Luly *et al.* (1980), Sluiter (1984), Truswell *et al.* (1987)
Rhipogonum (Liliaceae)	*Rhipogonum* comp.	Luly *et al.* (1980)
Rhizophoraceae	*Zonocostites ramonae* Germeraad *et al.*	Hekel (1972), Germeraad *et al.* (1968)
Rutaceae	*Melicope* comp. Luly *et al.*, *Acronychia* sim. Sluiter, *?Tricolporopollenites ivanhoensis* Martin, *Euodia* sim. Sluiter	Luly *et al.* (1980), Sluiter (1984), Martin (1973a; unpubl.)
Santalum type	*Santalumidites cainozoicus* Cookson & Pike	Cookson & Pike (1954b), Stover & Partridge (1973)
Sapotaceae	*Sapotaceoidaepollenites rotundus* Harris	Stover & Partridge (1973), Muller (1981)
Sparganiaceae	*Sparganiaceaepollenites* spp.	Martin (1978), Sluiter (1988)
Sphenostemon (Aquifoliaceae)	*Sphenostemon*	Sluiter (1984), Truswell *et al.* (1987)
Sterculiaceae	Sterculiaceae, *Brachychiton* sim. Sluiter	Sluiter (1984), Truswell *et al.* (1987)
Symphyonema (Proteaceae)	*Proteacidites symphyonemoides* Cookson	Cookson (1950), Stover & Partridge (1973), Sluiter (1984)
Symplocaceae	*Symplocoipollenites austellus* Stover & Partridge	Stover & Partridge (1973), Martin (1978)
Urticaceae/Moraceae	Urticaceae/Moraceae comp. Sluiter	Luly *et al.* (1980), Sluiter (1984), Truswell *et al.* (1987)
Vitaceae	Vitaceae	Luly *et al.* (1980), Sluiter (1984), Truswell *et al.* (1987)
Winteraceae	*Gephyrapollenites* spp. Stover & Partridge, *Drimys tetradites* Martin	Dettmann (1989), Martin (1973a, 1978)
?Xylomelum (Proteaceae)	*Propylipollis (Proteacidites) annularis* (Cookson) Dettmann	Cookson (1950), Dettmann (1989)

The most pertinent references have been selected. Qualification of degree of confidence in identification used by some authors is indicated thus: Comp., compares favourably with reference material; sim., bears a strong similarity to present-day type.

Appendix 7.2. *Definition of the major pollen groups used in the text*

Major pollen group	Botanical names
Spores	Pteridophytes and bryophytes
Gymnosperms	Araucariaceae
	Cupressaceae
	Dacrycarpus
	Dacrydium
	Lagarostrobos
	Microcachrys
	Phyllocladus type 1 (3-saccate)
	Phyllocladus type 2 (2-saccate)
	Podocarpus
Casuarinaceae	Casuarinaceae
Myrtaceae	*Eucalyptus*, bloodwood/ *Angophora* type
	Eucalyptus spathulata
	Myrtaceae
Nothofagus	*Nothofagus*, ancestral type
	Nothofagus subgen. *Brassospora*
	Nothofagus subgen. *Fuscospora*
	Nothofagus subgen. *Lophozonia*
Proteaceae	Banksieae
	Beauprea
	Carnarvonia
	Gevuina/Hicksbeachia
	Hakea/Grevillea
	Isopogon
	Macadamia, Helicia/Orites complex
	Symphyonema
	?*Xylomelum*
	Many unidentified species
	Many extinct species
Poaceae	Poaceae
Asteraceae	Asteraceae

For fossil names and references, see Appendix 7.1

REFERENCES

ALLEY, N. F. (1985). Latest Eocene palynofloras of the Pidinga Formation, Wilkinson No. 1 Well, Western South Australia. *South Australian Geological Survey Report Book*, 85/13.

AMBROSE, G. J., CALLEN, R. A., FLINT, R. B. & LANGE, R. T. (1979) *Eucalyptus* fruits in stratigraphic context in Australia. *Nature*, 280, 387–9.

ARCHER, M., GODTHELP, H., HAND, S. J. & MEGIRIAN, D. (1989*a*). Fossil mammals of Riversleigh, northwestern Queensland: preliminary overview of biostratigraphy, correlation and environmental change. *Australian Zoologist*, 25, 29–65.

ARCHER, M., HAND, S. & GODTHELP, H. (1989*b*). Ghosts from green gardens. Preliminary hypotheses about changes in Australia's rainforest mammals through time based on evidence from Riversleigh. *Riversleigh Notes*, 7, 4–7.

ASHTON, D. H. (1981). Tall open-forests. In *Australian Vegetation*, ed. R. H. Groves, pp. 121–51. Cambridge: Cambridge University Press.

BARKER, M. (1990). Effects of fire on the floristic composition, structure and flammability of rainforests. *Tasforests*, 2, 117–20.

BARLOW, B. A. (1981). The Australian flora: its origin and evolution. *Flora of Australia*, vol. 1, pp. 25–75. Netley, South Australia: Griffen Press.

BAUR, G. N. (1957). Nature and distribution of rainforests in New South Wales. *Australian Journal of Botany*, 5, 190–233.

BERRY, P. E., SKVARLA, J. J., PARTRIDGE, A. D. & MACPHAIL, M. K. (1990). *Fuchsia* pollen from the Tertiary of Australia. *Australian Systematic Botany*, 3, 739–44.

BINT, A. N. (1981). An Early Pliocene pollen assemblage from Lake Tay, south-western Australia and its phytogeographic implication. *Australian Journal of Botany*, 29, 277–91.

BIRKS, H. J. B. & BIRKS, H. H. (1980). *Quaternary Palaeoecology*. London: Edward Arnold.

BOLAND, D. J., BROOKER, M. I. H., CHIPPENDALE, G. M., HALL, N., HYLAND, B. P. M., JOHNSTON, R. O., KLEINIG, D. A. & TURNER, J. D. (1984). *Forest Trees of Australia*. Melbourne: CSIRO.

BOWEN, G. D. (1981). Coping with low nutrients; In *The Biology of Australian Plants*, ed. J. S. Pate & R. J. McComb, pp. 34–64. Nedlands, WA: University of Western Australia Press.

BOWLER, J. M. (1982). Aridity in the late Tertiary and Quaternary of Australia. In *Evolution of the Flora and Fauna of Arid Australia*, ed. W. R. Barker & P. J. M. Greenslade, pp. 35–45. Frewville: Peacock Publications.

BROWN, C. M. (1989). Structural and stratigraphic framework of groundwater occurrence and surface discharge in the Murray Basin, southeastern Australia. *BMR Journal of Australian Geology and Geophysics*, 11, 127–46.

BURBIDGE, N. T. (1960). The phytogeography of the Australian region. *Australian Journal of Botany*, 8, 75–212.

CALLAN, R. A. (1977). Late Cainozoic environments of part of northeastern South Australia. *Journal of the Geological Society of Australia*, 24, 151–69.

CHRISTOPHEL, D. C., HARRIS, W. K. & SYBER, A. K. (1987). The Eocene flora of the Anglesea Locality, Victoria. *Alcheringa*, 11, 303–23.

CHURCHILL, D. M. (1968). The distribution and prehistory of *Eucalyptus diversicolor* F. Muell, *E. marginata*

Dann ex Sm., and *E. calophylla* R.Br. in relation to rainfall. *Australian Journal of Botany*, 16, 125–51.

CHURCHILL, D. M. (1970). Observations on pollen harvesting by brush-tongued lorikeets. *Australian Journal of Zoology*, 18, 427–37.

COETZEE, J. A. & MULLER, J. (1984). The phytogeographic significance of some extinct Gondwana pollen types from the Tertiary of the Southwestern Cape (South Africa). *Annals of the Missouri Botanical Garden*, 75, 1088–99.

COETZEE, J. A. & PRAGLOWSKI, J. (1984). Pollen evidence for the occurrence of *Casuarina* and *Myrica* in the Tertiary of South Africa. *Grana*, 23, 23–41.

COOKSON, I. C. (1950). Fossil pollen grains of proteaceous type from Tertiary deposits in Australia. *Australian Journal of Scientific Research*, 2, 166–77.

COOKSON, I. C. (1954). The Cainozoic occurrence of *Acacia* in Australia. *Australian Journal of Botany*, 2, 52–9.

COOKSON, I. C. (1956). On some Australian Tertiary spores and pollen grains that extend the geological and geographical distribution of living genera. *Proceedings of the Royal Society of Victoria*, 69, 41–54.

COOKSON, I. C. (1958). Fossil pollen grains of *Nothofagus* from Australia. *Proceedings of the Royal Society of Victoria*, 71, 25–30.

COOKSON, I. C. & DUIGAN, S. L. (1950). Fossil Banksieae from Yallourn, Victoria, with notes on morphology and anatomy of living species. *Australian Journal of Scientific Research, Series B*, 3, 133–65.

COOKSON, I. C. & PIKE, K. M. (1953). The Tertiary occurrence and distribution of *Podocarpus* (Section *Dacrycarpus*) in Australia and Tasmania. *Australian Journal of Botany*, 1, 71–82.

COOKSON, I. C. & PIKE, K. M. (1954a). The fossil occurrence of *Phyllocladus* and two other podocarpaceous types in Australia. *Australian Journal of Botany*, 2, 60–68.

COOKSON, I. C. & PIKE, K. M. (1954b). Some dicotyledonous pollen types from Cainozoic deposits in the Australian region. *Australian Journal of Botany*, 2, 197–219.

CROME, F. H. J. & IRVINE, A. K. (1986). 'Two bob each way': the pollination and breeding system of the rainforest tree *Syzygium cormiflorum* (Myrtaceae). *Biotropica*, 18, 115–25.

DETTMANN, M. E. (1963). Upper Mesozoic microfloras from southeastern Australia. *Proceedings of the Royal Society of Victoria*, 77, 1–148.

DETTMANN, M. E. (1973). Angiospermous pollen from Albian to Turonian sediments of eastern Australia. In *Mesozoic and Cainozoic Palynology: Essays in Honour of Isabel Cookson*, ed. J. E. Glover & G. Playford. *Geological Society of Australia Special Publication* 4, 3–34.

DETTMANN, M. E. (1989). Antarctica: Cretaceous cradle of austral temperate forests? In *Origins and Evolution of the Antarctic Biota*, ed. J. A. Crame, pp. 89–105. *Geological Society of London Special Publication*, 47, 89–105.

DETTMANN, M. E. & PLAYFORD, G. (1969). Palynology of the Australian Cretaceous: a review. In *Stratigraphy and Palaeontology. Essays in Honour of Dorothy Hill*, ed. K. S. W. Campbell, pp. 174–210. Canberra: Australian National University Press.

DETTMANN, M. E., POCKNALL, D. T., ROMERO, E. J. & ZAMALOA, M. DEL C. (1990). *Nothofagidites* Erdtman ex Potonié 1960; a catalogue of species with notes on paleogeographic distribution of *Nothofagus* Bl (southern beech). *New Zealand Geological Survey Paleontological Bulletin* 69.

DUDGEON, M. J. (1983). Eocene pollen of probable proteaceous affinity from the Yaamba Basin, Central Queensland. *Memoirs of the Association of Australasian Palaeontologists*, 1, 339–62.

FERGUSON, I. K., VERDCOURT, B. & POOLE, M. M. (1977). Pollen morphology in the genera *Merremia* and *Operculina* (Convolvulaceae) and its taxonomic significance. *Kew Bulletin*, 31, 763–73.

FOSTER, C. B. (1982). Illustrations of Early Tertiary (Eocene) plant microfossils from the Yaamba Basin, Queensland. *Geological Survey of Queensland, Publication*, 381, 1–33.

FRAKES, L. A. & FRANCIS, J. E. (1990). Cretaceous palaeoclimates. In *Cretaceous Resources Events and Rhythms*, ed. R. N. Ginsberg & B. Beaudoin, pp. 273–87. Dordrecht: Kluwer Academic Publishers.

FRANCIS, J. E. (1990). Polar fossil forests. *Geology Today*, May–June, pp. 92–5.

GALLOWAY, R. W. (1986). Australian snowfields past and present. In *Flora and Fauna of Alpine Australasia, Ages and Origins*, ed. B. A. Barlow, pp. 27–36. Melbourne: CSIRO.

GERMERAAD, J. H., HOPPING, C. A. & MULLER, J. (1968). Palynology of Tertiary sediments in tropical areas. *Review of Palaeobotany and Palynology*, 6, 189–345.

GOULD, R. E. (1983). Early Australasian vegetation history: evidence from Late Palaeozoic and Mesozoic plant megafossils. In *A History of the Australasian Vegetation*, ed. J. M. B. Smith, pp. 32–43. Sydney: McGraw Hlll.

GRAHAM, A. & JARZEN, D. M. (1969). Studies in neotropical botany. 1. The Oligocene communities in Puerto Rico. *Annals of the Missouri Botanical Garden*, 56, 308–57.

GREENWOOD, D. R., CALLAN, R. A. & ALLEY, N. F. (1990). *The Correlation and Depositional Environment of Tertiary Strata Based on Macrofloras in the Southern Lake Eyre Basin*. Department of Mines and Energy, South Australia, Report, 90/15.

HARRIS, W. K. (1965). Basal Tertiary microfloras from the Princetown area, Victoria, Australia. *Palaeontographica B*, 115, 75–106.

HARRIS, W. K. (1972). New form species of pollen from

southern Australian Early Tertiary sediments. *Transactions of the Royal Society of South Australia*, **96**, 53–65.

HEKEL, H. (1972). Pollen and spore assemblages from Queensland Tertiary sediments. *Geological Survey of Queensland, Palaeontological Papers*, **30**, 1–31.

HERBERT, D. R. (1933). The relationships of the Queensland flora. *Proceedings of the Royal Society of Queensland*, **44**, 2–22.

HILL, R. S., HARWOOD, D. M. & WEBB, P. N. (1991). Last remnant of Antarctica's Cenozoic flora. *Abstract, Eighth International Symposium on Gondwana, Hobart*, p. 43.

HILL, R. S. & MACPHAIL, M. K. (1985). A fossil flora from rafted Plio-Pleistocene mudstones at Regatta Point, Tasmania. *Australian Journal of Botany*, **33**, 497–517.

HOLMES, W. B. K., HOLMES, F. M. & MARTIN, H. A. (1983). Fossil *Eucalyptus* remains from the Middle Miocene Chalk Mountain Formation, Warrumbungle Mountains, New South Wales. *Proceedings of the Linnean Society of New South Wales*, **106**, 299–310.

HOOKER, J. D. (1860). Introductory essay. In *Botany of the Antarctic Voyage of H. M. Discovery Ships 'Erebus' and 'Terror', in the years 1939–1943. III. Flora Tasmaniae*. London: Keeve.

HOS, D. (1977). *Eocene Palynology of a Sample from Golden Grove, South Australia. Department of Mines and Energy, South Australia, Report* 77/88.

HOS, D. (1978). *Eocene Palynology of SADME Wilkinson No. 1, Eastern Officer Basin. Department of Mines and Energy, South Australia, Report*, 78/149.

JEFFERSON, T. H. (1982). Fossil forests from the Lower Cretaceous of Alexander Island. Antarctica. *Palaeontology*, **25**, 681–708.

KEMP, E. M. (1972). Reworked palynomorphs from the West Ice Shelf area, East Antarctica, and their possible geological and palaeoclimatological significance. *Marine Geology*, **13**, 145–57.

KEMP, E. M. (1976). Early Tertiary pollen from Napperby, central Australia. *BMR Journal of Geology and Geophysics*. **1**, 109–14.

KEMP, E. M. & BARRETT, P. J. (1975). Antarctic Glaciation and early Tertiary vegetation. *Nature*, **258**, 507–8.

KERSHAW, A. P. (1970a). Pollen morphological variation within the Casuarinaceae. *Pollen et Spores*, **12**, 145–61.

KERSHAW, A. P. (1970b). A pollen diagram from Lake Euramoo, northeast Queensland. *New Phytologist*, **69**, 785–805.

KERSHAW, A. P. (1973). The numerical analysis of modern pollen spectra from northeast Queensland forest. In *Mesozoic and Cainozoic Palynology: Essays in Honour of Isabel Cookson*, ed. J. E. Glover & G. Playford. *Geological Society of Australia Special Publication*, **4**, 191–9.

KERSHAW, A. P. & SLUITER, I. R. (1982). Late Cainozoic pollen spectra from the Atherton Tableland, Northeastern Australia. *Australian Journal of Botany*, **30**, 279–95.

KHAN, A. M. (1974). Palynology of Neogene sediments from Papua (New Guinea). Stratigraphic boundaries. *Pollen et Spores*, **16**, 266–84.

KRIEG, G. W., CALLAN, R. A., GRAVESTOCK, D. I. & GATEHOUSE, C. G. (1990). 1: Geology. In *Natural History of the North East Deserts*, ed. M. J. Tyler, C. R. Twidale, M. Davies & C. B. Wells, pp. 1–26. Adelaide: Royal Society of South Australia.

LADD, P. G. (1979). A short pollen diagram from rainforest in highland eastern Victoria. *Australian Journal of Ecology*, **4**, 229–37.

LANGE, R. T. (1978). Carpological evidence for fossil *Eucalyptus* and other Leptospermeae (subfamily Leptospermoideae of Myrtaceae) from a Tertiary deposit in the South Australian Arid Zone. *Australian Journal of Botany*, **26**, 221–33.

LEAR, R. & TURNER, T. (1977). *Mangroves of Australia*. Brisbane: University of Queensland Press.

LUKE, R. H. & McARTHUR, A. G. (1978). *Bushfires in Australia*. Canberra: Australian Government Publishing Service.

LULY, J., SLUITER, I. R. & KERSHAW, A. P. (1980). Pollen studies of Tertiary brown coals: preliminary analyses of lithotypes within the Latrobe Valley, Victoria. *Monash University Publications in Geography*, **23**, 1–78.

MACPHAIL, M. K. (1987). Palynological analysis of BMR Manilla-1 Borehole, Murray Basin. *Bureau of Mineral Resources, Geology and Geophysics Record 1987/58*.

MACPHAIL, M. K. (1990). *Australopollis obscurus* (Harris) Krutzsch emend. Stover & Partridge. *PPAA Newsletter*, **21**, 6–7.

MACPHAIL, M. K. & TRUSWELL, E. M. (1989). Palynostratigraphy of the central west Murray Basin. *BMR Journal of Australian Geology and Geophysics*, **11**, 301–31.

MARTIN, H. A. (1973a). Palynology of some Tertiary and Pleistocene deposits, Lachlan River Valley, New South Wales. *Australian Journal of Botany Supplementary Series*, **6**, 1–57.

MARTIN, H. A. (1973b). Upper Tertiary palynology in southern New South Wales. In *Mesozoic and Cainozoic Palynology: Essays in Honour of Isabel Cookson*, ed. J. E. Glover & G. Playford. *Geological Society of Australia Special Publication*, **4**, 35–54.

MARTIN, H. A. (1974). The identification of some Tertiary pollen belonging to the family Euphorbiaceae. *Australian Journal of Botany*, **22**, 271–91.

MARTIN, H. A. (1977). The history of *Ilex* (Aquifoliaceae) with special reference to Australia: Evidence from pollen. *Australian Journal of Botany*, **25**, 655–73.

MARTIN, H. A. (1978). Evolution of the Australian flora and vegetation through the Tertiary: evidence from pollen. *Alcheringa*, **2**, 181–202.

MARTIN, H. A. (1979). Stratigraphic palynology of the Mooki Valley, New South Wales. *Journal and Proceedings of the Royal Society of New South Wales*, **112**, 71–8.

MARTIN, H. A. (1980). Stratigraphic palynology from shallow bores in the Namoi River and Gwydir River Valleys, North-Central New South Wales. *Journal and Proceedings of the Royal Society of New South Wales*, 113, 81–7.

MARTIN, H. A. (1981). Stratigraphic palynology of the Castlereagh River Valley, New South Wales. *Journal and Proceedings of the Royal Society of New South Wales*, 114, 77–84.

MARTIN, H. A. (1982). Changing Cenozoic barriers and the Australian paleobotanical record. *Annals of the Missouri Botanical Garden*, 69, 625–67.

MARTIN, H. A. (1984a). The stratigraphic palynology of the Murray Basin in New South Wales. II. The Murrumbidgee area. *Journal and Proceedings of the Royal Society of New South Wales*, 117, 35–44.

MARTIN, H. A. (1984b). The stratigraphic palynology of the Murray Basin in New South Wales. III. The Lachlan area. *Journal and Proceedings of the Royal Society of New South Wales*, 117, 45–51.

MARTIN, H. A. (1984c). On the philosophy and methods used to reconstruct Tertiary vegetation. *Proceedings of the Linnean Society of New South Wales*, 107, 521–33.

MARTIN, H. A. (1986). Tertiary stratigraphic palynology, vegetation and climate of the Murray Basin in New South Wales. *Proceedings of the Royal Society of New South Wales*, 119, 43–53.

MARTIN, H. A. (1987). The Cainozoic history of the vegetation and climate of the Lachlan River Region, New South Wales. *Proceedings of the Linnean Society of New South Wales*, 109, 214–57.

MARTIN, H. A. (1988). Stratigraphic palynology of the Lake Menindee region, northwest Murray Basin, New South Wales. *Proceedings of the Royal Society of New South Wales*, 121, 1–9.

MARTIN, H. A. (1990). The palynology of the Namba Formation in Wooltana-1 bore, Callabona Basin (Lake Frome), South Australia and its relevance to Miocene grasslands in Central Australia. *Alcheringa*, 14, 247–55.

MARTIN, H. A. (1991a). Tertiary stratigraphic palynology and palaeoclimate of the inland river systems in New South Wales. In *The Cainozoic in Australia: A Re-appraisal of the Evidence*, ed. M. A. J. Williams, P. De Dekker & A. P. Kershaw. *Special Publication of the Geological Society of Australia*, 18, 181–94.

MARTIN, H. A. (1991b). Dinoflagellate and spore pollen biostratigraphy of the SADME MC63 bore, Western Murray Basin. *Alcheringa*, 15, 107–44.

MARTIN, H. A. (1993). Middle Tertiary dinoflagellate and spore/pollen biostratigraphy and palaeoecology of the Mallee Cliffs bore, Central Murray Basin. *Alcheringa*, 17, 91–124.

MARTIN, H. A. & GADEK, P. A. (1988). Identification of *Eucalyptus spathulata* pollen and its presence in the fossil record. *Memoirs of the Association of Australasian Palaeontologists*, 5, 313–27.

MARTIN, H. A., WORRALL, L. & CHALSON, J. (1987). The first occurrence of the Paleocene *Lygistepollenites balmei* Zone in the Eastern Highlands Region, New South Wales. *Australian Journal of Earth Sciences*, 34, 359–65.

McCOY, M. (1989). Pollination of two eucalypt species by flying foxes. *Macroderma*, 5, 21–2.

McEWEN MASON, J. R. C. (1989). The palaeomagnetics and palynology of late Cainozoic cored sediments from Lake George, New South Wales, southeastern Australia. Ph.D. thesis, Monash University.

McMINN, A. (1981). A Miocene microflora from the Home Rule Kaolin deposit. *Geological Survey of New South Wales, Quarterly Notes*, 43, 1–4.

McMINN, A. & MARTIN, H. A. (1992). Late Cainozoic pollen history from Site 765, Eastern Indian Ocean. In *Proceedings of the Ocean Drilling Program Scientific Results, Leg 123*, pp. 421–7.

McNAMARA, K. J. & SCOTT, J. K. (1983). A new species of *Banksia* (Proteaceae) from the Eocene Merlinleigh Sandstone of the Kennedy Range, Western Australia. *Alcheringa*, 7, 185–93.

MILDENHALL, D. C. (1975). Palynology of the *Acacia* bearing beds in the Kornako district, Pohangina Valley, North Island, New Zealand. *New Zealand Journal of Geology and Geophysics*, 18, 209–28.

MILDENHALL, D. C. (1980). New Zealand Late Cretaceous and Cenozoic plant biogeography: a contribution. *Palaeogeography, Palaeoclimatology, Palaeoecology*, 31, 197–223.

MILDENHALL, D. C. & POCKNALL, D. T. (1989). Miocene–Pleistocene spores and pollen from Central Otago, South Island, New Zealand. *New Zealand Geological Survey Palaeontological Bulletin*, 59, 1–128.

MILNE, L. A. (1988). Palynology of the late Eocene lignitic sequence from the western margin of the Eucla Basin, Western Australia. *Memoirs of the Association of Australasian Palaeontologists*, 5, 285–310.

MORLEY, R. J. (1977). Palynology of Tertiary and Quaternary sediments in Southeast Asia. *Proceedings, Indonesian Petroleum Association, Sixth Annual Conventions*, May 1977, pp. 255–76.

MULLER, J. (1966). Montane pollen from the Tertiary of N. W. Borneo. *Blumea*, 14, 231–5.

MULLER, J. (1972). Palynological evidence for change in geomorphology, climate, and vegetation in the Mio-Pliocene of Malesia. In *The Quaternary Era in Malesia*, ed. P. Ashton & M. Ashton. *University of Hull, Department of Geography, Miscellaneous Series*, 13, 6–16.

MULLER, J. (1981). Fossil pollen records of extant angiosperms. *Botanical Review*, 47, 1–142.

NELSON, E. C. (1981). Phytogeography of southern Australia. In *Ecological Biogeography of Australia*, ed. A. Keast, pp. 733–59. The Hague: W. Junk.

OLLIER, C. D. (1986). The origin of alpine landforms in

Australasia. In *Flora and Fauna of Alpine Australasia*, ed.
B. A. Barlow, pp. 3–26. Melbourne: CSIRO.

OWEN, J. A. (1975). Palynology of some Tertiary deposits
from New South Wales. Ph.D. thesis, Australian
National University.

PARR-SMITH, G. A. (1982). Biogeography and evolution
in shrubby Australian species of *Atriplex* (Chenopodiaceae). In *Evolution of the Flora and Fauna of Arid Australia*,
ed. W. R. Barker & P. J. M. Greenslade, pp. 291–9. Frewville: Peacock Publications.

PLAYFORD, G. (1982). Neogene palynomorphs from the
Huon Peninsula, Papua New Guinea. *Palynology*, 6,
29–54.

PLAYFORD, G. & DETTMANN, M. E. (1979). Pollen
of *Dacrydium franklinii* Hook. and comparable Early Tertiary microfossils. *Pollen et Spores*, 20, 514–34.

POCKNALL, D. T. (1981). Pollen and spores from the
Rifle Butts Formation (Altonian, Lower Miocene)
Otago, New Zealand. *New Zealand Geological Survey
Report PAL 40*.

POCKNALL, D. T. (1989). Late Eocene to Early Miocene
vegetation and climatic history of New Zealand. *Journal
of the Royal Society of New Zealand*, 19, 1–18.

POCKNALL, D. T. & CROSBIE, Y. M. (1982). Taxonomic revision of some Tertiary tricolporate and tricolpate
grains from New Zealand. *New Zealand Journal of Botany*,
20, 7–15.

REITSMA, T. (1970). Pollen morphology of the Alangiaceae. *Review of Palaeobotany and Palynology*, 10,
249–332.

ROMERO, E. J. (1986). Palaeogene phytogeography and
climatology of South America. *Annals of the Missouri Botanical Garden*, 73, 449–61.

SHACKLETON, N. J. & KENNETT, P. J. (1975).
Palaeotemperature history of the Cenozoic and the
initiation of Antarctic glaciation: oxygen and carbon isotope analysis of DSDP sites 277, 279 and 181. *Initial
Reports of the Deep Sea Drilling Project*, 29, 743–55.

SLUITER, I. R. K. (1984). Palynology of Oligo-Miocene
brown coal seams, Latrobe Valley, Victoria Ph.D. Thesis,
Monash University.

SLUITER, I. R. (1992). Early Tertiary vegetation and
palaeoclimates, Lake Eyre region, northeastern South
Australia. In *The Cainozoic in Australia: A Re-appraisal of
the Evidence*, ed. M. A. J., Williams, P. De Dekker &
A. P. Kershaw. *Geological Society of Australia Special Publication*, 18, 99–118.

SLUITER, I. R. & KERSHAW, A. P. (1982). The nature
of the late Tertiary vegetation in Australia. *Alcheringa*,
6, 211–22.

SMITH, J. M. B. (1986). Origins of Australasian tropi-
calpine and alpine floras. In *Flora and Fauna of Alpine Australasia*, ed. B. A. Barlow, pp. 109–28. Melbourne:
CSIRO.

SPECHT, R. A. & WOMERSLEY, J. S. (1979).

Heathlands and related shrublands of Malesia (with
particular reference to Borneo and New Guinea). In *Ecosystems of the World 9A. Heathlands and Related Shrublands, Descriptive Analysis*, ed. R. L. Specht, pp. 321–39.
Amsterdam: Elsevier.

SPECHT, R. L., RAYSON, P. & JACKMAN, M. E.
(1958). Dark Island Heath (Ninety-Mile Plain, South
Australia). VI. Pyric succession: changes in composition,
coverage, dry weight and mineral nutrient status. *Australian Journal of Botany*, 6, 59–88.

STOVER, L. E. & PARTRIDGE, A. D. (1973). Tertiary
and Late Cretaceous spores and pollen from the Gippsland
Basin, southeastern Australia. *Proceedings of the Royal
Society of Victoria*, 85, 237–86

STOVER, L. E. & PARTRIDGE, A. D. (1982). Eocene
spore–pollen from the Werillup Formation, Western Australia. *Palynology*, 6, 69–95.

TAYLOR, G., TRUSWELL, E.M., McQUEEN, K.G.
& BROWN, M.C. (1990). Early Tertiary palaeogeography,
landform evolution and palaeoclimates of the Southern
Monaro, N. S. W. Australia. *Palaeogeography, Palaeoclimatology, Palaeoecology*, 78, 109–34.

TRUSWELL, E. M. (1982). Antarctica: the vegetation of
the past and its climatic implications. *Australian Meteorological Magazine*, 30, 169–73.

TRUSWELL, E. M. (1987). Reconaissance palynology of
selected boreholes in the Western Murray Basin, New
South Wales. *Bureau of Mineral Resources, Geology and Geophysics Record*, 1987/24.

TRUSWELL, E. M. & HARRIS, W. K. (1982). The
Cainozoic palaeobotanical record in arid Australia: fossil
evidence for the origins of an arid-adapted flora. In *Evol-
ution of the Flora and Fauna of Arid Australia*, eds. W. R.
Barker & P. J. M. Greenslade, pp. 67–76. Frewville:
Peacock Publications.

TRUSWELL, E. M., KERSHAW, A. P. &
SLUITER, I. R. (1987). The Australian southeast Asian
connection: evidence from the palaeobotanical record. In
Biogeographic Evolution of the Malay Archipelago, ed.
T. C. Whitmore, pp. 32–49. Oxford: Clarendon Press.

TRUSWELL, E. M. & MARCHANT, N. G. (1986).
Early Tertiary pollen of probable droceracean affinity from
Central Australia. *Special Papers in Palaeontology*, 35,
163–78.

TRUSWELL, E. M., SLUITER, I. R. & HARRIS, W. K.
(1985). Palynology of the Oligocene–Miocene
sequence in the Oakvale-1 corehole, western Murray
Basin, South Australia. *BMR Journal of Australian Geology
and Geophysics*, 9, 267–95.

WEBB, L. J. (1970). Eastern Australian environments in
relation to fire. In *Environments to Order?* Proceedings
of an unpublished symposium, Australian Museum, 8
August 1970, pp. 5–8.

WEBB, L. J. & TRACEY, J. G. (1972). An ecological
comparison of vegetation communities on each side of

Torres Strait. In *Bridge and Barrier: The Natural and Cultural History of Torres Strait*, ed. D. Walker, pp. 109–30. School of Pacific Studies Publication BG/3. Canberra: Australian National University.

WEBB, L. J. & TRACEY, J. G. (1981). Australian rain-forests: pattern and change. In *Ecological Biogeography of Australia*, ed. A. Keast, pp. 605–94. The Hague, W. Junk.

WEST, J. G. (1984). A revision of *Dodonaea* Miller (Sapindaceae) in Australia. *Brunonia*, 7, 1–194.

WOOD, G. R. (1986). Late Oligocene to early Miocene palynomorphs from GSQ Sandy Cape 1–3R. *Geological Survey of Queensland Publication*, 387, 1–27.

WOPFNER, H., CALLEN, R. & HARRIS, W. K. (1974). The Lower Tertiary Eyre Formation of the southwestern Great Artesian Basin. *Journal of the Geological Society of Australia*, 21, 17–52.

8 Cretaceous vegetation: the microfossil record

M. E. DETTMANN

During the past decade, the three-element invasion theory that was initially advocated by Hooker (1860) to explain the present-day Australian flora has been questioned or dismissed (Barlow, 1981; Webb *et al.*, 1986). Current concepts developed from ecological evidence indicate autochthonous differentiation from an ancient Gondwanan flora during the Late Cretaceous and early Tertiary (Webb *et al.*, 1986). However, Truswell *et al.* (1987) believed that the pollen record known to them favoured a Late Cretaceous–early Tertiary phase of floristic exchange between Australia and regions to the north, with dispersal occurring in both directions. Evidence that countered invasion from the north during the Late Cretaceous has since accrued (Dettmann & Thomson, 1987; Dettmann & Jarzen, 1988, 1990; Dettmann, 1989; Dettmann *et al.*, 1990; Jarzen & Dettmann, 1990). From patterns of pollen introductions in separate regions of the southern Gondwana assembly, it was concluded that many elements of the Australian Cretaceous flora either evolved within the Austro-Antarctic region or entered Australia using an Antarctic route (Dettmann, 1989; Dettmann & Jarzen, 1990), as had been suggested previously (Dettmann, 1981).

The earliest angiosperms in Australia were almost certainly immigrants. The pollen record emphasises a 5–10 million years (Ma) time lag between initial inceptions of angiosperms in the southern Laurasian–northern Gondwanan region (Hauterivian or earlier) than introduction in Australia (late Barremian–Aptian). This evidence argues against Australia as a cradle region of the angiosperms (Takhtajan, 1969) and provides little support for inception and diversification of earliest angiosperms on fragments of the Australian plate that rafted northwards during the Late Jurassic (Takhtajan, 1987). Migration routes taken by the early angiosperms to Australia probably involved southern Gondwana (Dettmann 1981, 1989; Truswell *et al.*, 1987); dispersal to Australia from Southeast Asia via microcontinents detached from northern Australia (Burger, 1981, 1990) has scant support from the pollen record.

The issues of introduction and Late Cretaceous differentiation of angiosperms in Australia are explored here using up-to-date pollen evidence. This evidence, combined with knowledge of palaeogeography and palaeotemperatures, has provided a basis for interpreting floristics and structure of Cretaceous plant communities represented in Australia. The spore–pollen evidence has been gathered from Cretaceous sediments that occur in 22 of Australia's Mesozoic basins (Figure 8.1). In these have accumulated successions of marine and nonmarine (fluvial and lacustrine) sediments that reflect advancement or retreat of seas resulting from progressive isolation of Australia from the southern Gondwana assembly, from oscillations in world sea levels, and

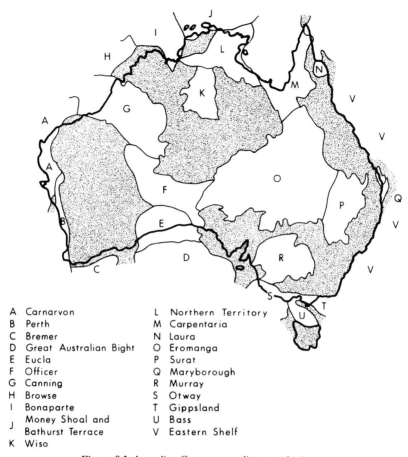

A Carnarvon
B Perth
C Bremer
D Great Australian Bight
E Eucla
F Officer
G Canning
H Browse
I Bonaparte
J Money Shoal and
Bathurst Terrace
K Wiso

L Northern Territory
M Carpentaria
N Laura
O Eromanga
P Surat
Q Maryborough
R Murray
S Otway
T Gippsland
U Bass
V Eastern Shelf

Figure 8.1 Australian Cretaceous sedimentary basins.

from alterations to drainage patterns of riverine/ lacustrine systems that fed into the seas (Dettmann *et al.*, 1992). Diverse spores and pollen represented in the sediments have been interpreted in a biostratigraphic context (Figure 8.2), and have provided a mechanism for integrating marine and nonmarine zonations (Helby *et al.*, 1987; Dettmann *et al.*, 1992). Thus the spore–pollen zones have been dated through associated planktic marine faunas and dinocyst/nannoplankton floras, and have been related to the plant macrofossil zones (Douglas, 1969, 1973) of southeastern Australia; for several of these dates there is independent radiometric evidence (Figure 8.2). While the spore–pollen record is complete for the entire Cretaceous in southeastern Australia,

it has been evaluated only for the Early to mid-Cretaceous in central, northeastern, and southwestern areas (Figure 8.2).

SPORE–POLLEN/PLANT RELATIONSHIPS

In recent years considerable advances have been made in elucidating relationships of spore–pollen taxa represented in Australian sediments. A range of cryptogam, gymnosperm and angiosperm components have been found *in situ* with macrofossils of fertile organs in Australia and overseas. Other spore–pollen taxa have been shown after rigorous scrutiny and comparison using electron and light microscopy to bear the

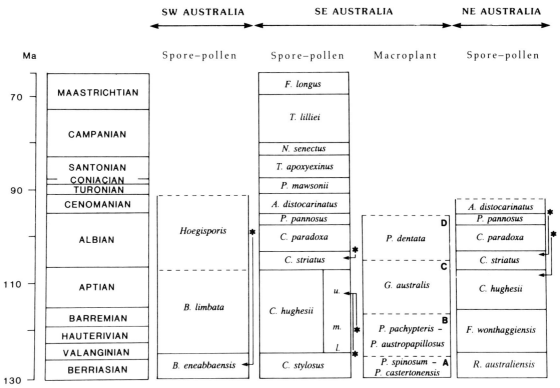

Figure 8.2 Age and relationships of spore-pollen and macroplant zones. Radiometric ages (*) are linked by arrow to corresponding spore–pollen zonal determinations, but are considered unreliable (Dettmann *et al.*, 1992). Spore–pollen zones in ascending order are as follows: SW Australia: *Biretisporites eneabbaensis* (Berriasian–early Valanginian); *Balmeiopsis limbata* (Valanginian–Aptian), and *Hoegisporis* (latest Aptian or Albian–Cenomanian). SE Australia: *Crybelosporites stylosus* (Berriasian–early Valanginian), *Cyclosporites hughesii* (Valanginian–Aptian), *Crybelosporites striatus* (latest Aptian–early Albian), *Coptospora paradoxa* (mid–late Albian), *Phimopollenites pannosus* (late Albian), *Appendicisporites distocarinatus* (latest Albian–Cenomanian), *Phyllocladidites mawsonii* (Turonian–earliest Santonian), *Tricolporites apoxyexinus* (Santonian),

Nothofagidites senectus (early Campanian), *Tubulifloriidites lilliei* (mid Campanian–early Maastrichtian), and *Forcipites longus* (latest early–late Maastrichtian or earliest Danian). NE Australia: *Ruffordiaspora australiensis* (Berriasian–early Valanginian), *Foraminisporis wonthaggiensis* (Valanginian–Barremian), *Cyclosporites hughesii* (Aptian), *Crybelosporites striatus* (latest Aptian–early Albian), *Coptospora paradoxa* (mid–late Albian), *Phimopollenites pannosus* (late Albian), *Appendicisporites distocarinatus* (latest Albian–Cenomanian). Macroplant zones for SE Australia are: *Ptilophyllum spinosum – Ptilophyllum castertonensis* (Berriasian), *Ptilophyllum pachypteris – Pachypteris austropapillosus* (Valanginian–Barremian), *Ginkgoites australis* (latest Barremian–early Albian), and *Phyllopteroides dentata* (early–late Albian).

trademarks of extant taxa. Even so, the listing of plant taxa that have entities in the Australian Cretaceous (Table 8.1; Figures 8.3–8.6) considerably understates floristic diversity, as the affinities of many Cretaceous spore–pollen taxa remain unresolved. The known ranges in Australia and on a world-wide basis of selected taxa represented in the Australian Cretaceous spore–pollen record are detailed in Figure 8.7.

VEGETATIONAL SUCCESSION

Throughout the Cretaceous much of the Australian landmass supported forests in which austral conifers (Podocarpaceae and Araucariaceae) were important elements. These two families were established in Australia during the Jurassic (Gould, 1975), and can be regarded as representing the ancestral core of present-day

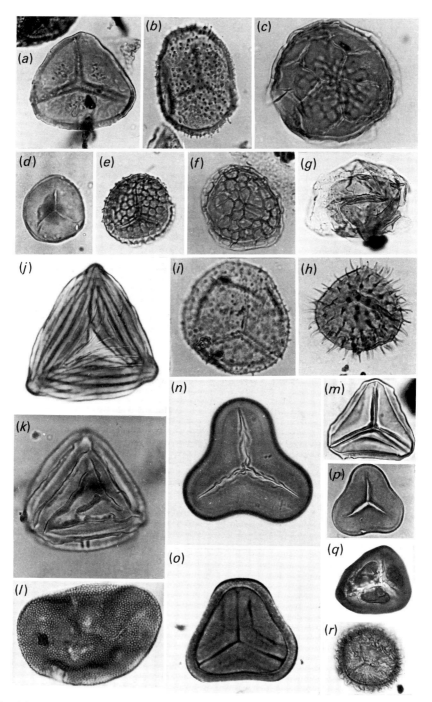

Figure 8.3 Fossil bryophytic (*a*)–(*d*), lycophytic (*e*)–(*h*) and filicean (*i*)–(*r*) spores in Cretaceous sediments of Australia. Magnifications × 500 unless specified otherwise. Asterisks below designate *in situ* association with macrofossils. (*a*) and (*b*) *Nothylas/Phaeoceros* type (*Foraminisporis dailyi* (Cookson & Dettmann) and *F. wonthaggiensis* (Cookson & Dettmann) respectively); (*c*) *Riella/Riccia* type (*Triporoletes reticulatus* (Pocock)); (*d*) *Sphagnum* type (*Stereisporites antiquasporites* (Wilson & Webster)); (*e*) and (*f*), *Lycopodium* type (*Retitriletes eminulus* (Dettmann) and *R. austroclavatidites* (Cookson), respectively); (*g*) Isoetaceae/Equisetaceae type (*Perinomonoletes* sp. of Dettmann, 1986*a*), ×750; (*h*) Lycopodiaceae/Selaginellaceae (*Ceratosporites equalis* Cookson & Dettmann); (*i*) *Osmunda/Leptopteris* type (*Baculatisporites comaumensis* (Cookson)); (*j*) and (*k*) *Anemia* type (*Plicatella* cf. *jansonii* (Pocock) and *Cicatricosisporites hughesii* Dettmann, respectively); (*l*) *Actinostachys* type (*Microfoveolatosporis* cf. *fromensis* (Cookson)); (*m*), *Gleichenia/Dicranopteris* type (*Gleicheniidites circinidites* (Cookson)); (*n*) *Adiantites lindsayoides* type (*Dictyophyllidites crenatus* Dettmann); (*o*) *Matonia* type (*Dictyophyllidites pectinataeformis* (Bolkhovitina)); (*p*) *Sphenopteris travisii* type (*Cyathidites minor* Couper); (*q*) *Lophosoria* type (*Cyatheacidites annulatus* Cookson ex Potonié), ×350; (*r*), Marsileaceae type (*Crybelosporites striatus* (Cookson & Dettmann)).

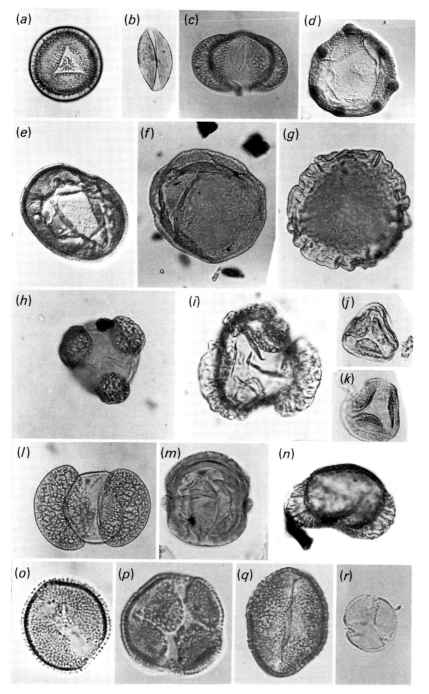

Figure 8.4 Fossil gymnospermous (*a*)–(*n*) and angiospermous (*o*)–(*r*) pollen in Cretaceous sediments of Australia. Magnifications ×500 unless specified otherwise. Asterisks below designate *in situ* association with macrofossils. (*a*) **Tomaxiella biforme* type (*Classopollis chateaunovii* Reyre); (*b*), Cycadophyta/Ginkgophyta type (*Cycadopites nitidus* (Balme)); (*c*) Pteridospermophyta type (*Alisporites similis* (Balme)); (*d*) ?Araucariaceae (*Hoegisporis uniforma* Cookson); (*e*) **Brachyphyllum irregulare* type (*Balmeiopsis limbata* (Balme)); (*f*) *Araucaria/Agathis* type (*Araucariacites australis* Cookson); (*g*) **Brachyphyllum mamillare* type (*Callialasporites dampieri* (Balme)); (*h*) *Microcachrys/Micros-* *trobos* type (*Microcachryidites antarcticus* Cookson); (*i*), *Dacrycarpus* type (*Dacrycarpites australiensis* Cookson & Pike) ×750; (*j*) and (*k*) **Trisacocladus tigrensis* type (*Trichotomosulcites subgranulatus* (Couper)); (*l*) *Podocarpus* type (*Podocarpidites ellipticus* Cookson); (*m*) *Lagarostrobos* type (*Phyllocladidites mawsonii* Cookson ex Couper) ×750; (*n*) *Dacrydium* type (*Lygistepollenites florinii* (Cookson & Pike)); (*o*), *Ascarina* type (*Clavatipollenites hughesii* Couper) ×1000; (*p*) cf. *Hedyosum* type (*Asteropollis asteroides* Hedlund & Norris) ×1000; (*q*) ?Liliaceae (*Liliacidites* cf. *kaitangataensis* Couper); (*r*) Platanaceae (*Tricolpites minutus* (Brenner)) ×1000.

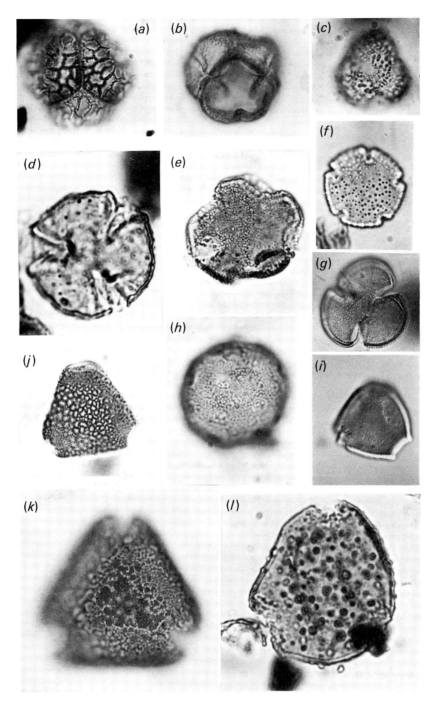

Figure 8.5 Fossil angiospermous pollen in Cretaceous sediments of Australia. Magnification ×1000. (*a*), *Belliolum/Bubbia* type (*Pseudowinterapollis wahooensis* (Stover)); (*b*), Epacridaceae type (*Ericipites scabratus* Harris); (*c*) *Ilex* type (*Ilexpollenites* cf. *anguloclavatus* McIntyre); (*d*) *Clematis* type (*Tubulifloridites lilliei* (Couper)); (*e*) *Callitriche* type (*Australopollis obscurus* (Harris)); (*f*) *Nothofagus* (ancestral) type (*Nothofagidites senectus* Dettmann & Playford); (*g*), *Gunnera* type (*Tricolpites reticulatus* Cookson ex Couper); (*h*) Trimeniaceae type ('*Polyporina*' *fragilis* Harris); (*i*) ?Proteaceae (*Triorites punctulatus* Dettmann); (*j*) *Carnarvonia* type (*Proteacidites* cf. *pseudomoides* Stover); (*k*) and (*l*), *Beauprea* type (*Beaupreaidites elegansiformis* Cookson and *B. orbiculatus* Dettmann & Jarzen, respectively).

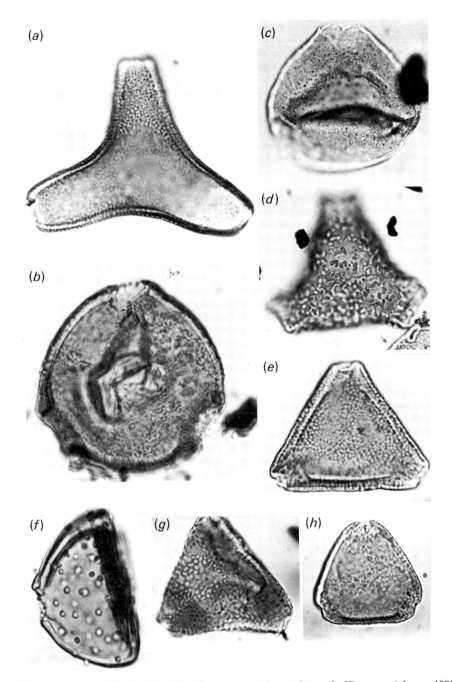

Figure 8.6 Fossil proteaceous pollen in Australian Cretaceous sediments. Magnifications ×1000. (*a*) *Adenanthos* type (*Proteacidites adenanthoides* Cookson); (*b*) *Stirlingia* type (*Proteacidites* sp. 1 of Dettmann & Jarzen, 1991); (*c*), *Persoonia* type (*Proteacidites* sp. A of Hill & Macphail, 1983); (*d*) *Grevillea* type (*Proteacidites* sp. 2 of Dettmann & Jarzen, 1991); (*e*) *Knightia* type (*Proteacidites amolosexinus* Dettmann & Playford); (*f*) *Telopea* type (*Triporopollenites ambiguus* Stover); (*g*) *Gevuina/ Hicksbeachia* type (*Propylipollis reticuloscabratus* (Harris)); (*h*) *Macadamia* type (*Propylipollis* cf. *crassimarginus* Dudgeon).

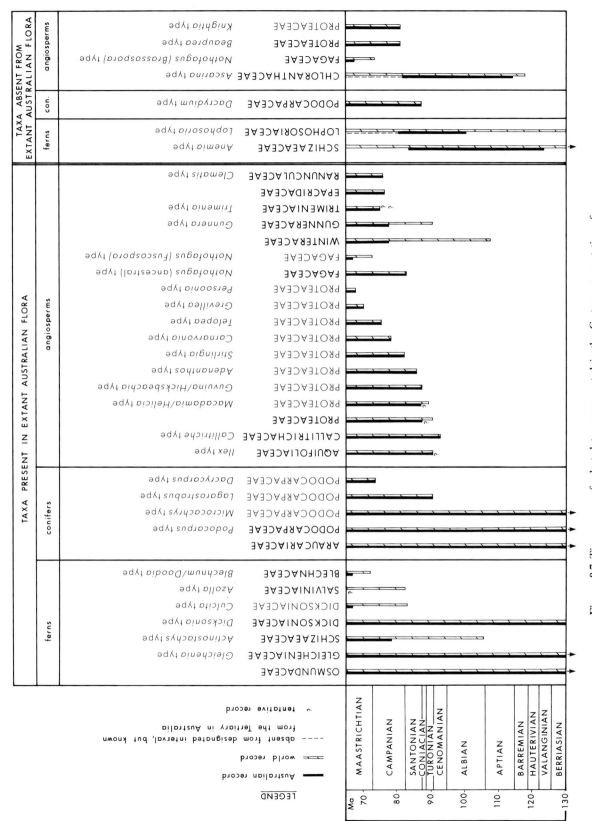

Figure 8.7 Time-ranges of selected taxa represented in the Cretaceous vegetation of Australia. Time-ranges on a world-basis are also shown for each taxon (updated from Muller, 1981). Con. conifer.

Table 8.1. *Suggested relationships of spore–pollen taxa recorded from the Australian Cretaceous (selected taxa are illustrated in Figures 8.3 to 8.6)*

Extant and/or fossil plant taxa	Spore–pollen taxa	Cretaceous range in Australia
Musci		
Sphagnaceae	*Stereisporites*	Ranges throughout Cretaceous
Anthocerotae		
Nothylas/Phaeoceros	*Foraminisporis dailyi*	Berriasian–Campanian
Nothylas	*F. wonthaggiensis*	Valanginian–Maastrichtian
?Anthocerotaceae	*F. asymmetricus*	Barremian–Campanian
	Stoverisporites microverrucatus	Mid-Albian–Turonian (N Aust.)
		Cenomanian–Campanian (SE Aust.)
	Interlobites intraverrucatus	Mid-Albian–Turonian (N Aust.)
		Cenomanian–Turonian (SE Aust.)
Hepaticae		
Riella/Riccia/	*Triporoletes*	Valanginian–Maastrichtian (N & W
ªHepaticites disciformis		Aust.)
		Barremian–Maastrichtian (SE Aust.)
?Hepaticae	*Coptospora paradoxa*	Mid–late Albian (N Aust.)
		Mid-Albian–Cenomanian (S Aust.)
Lycopodiopsida		
Lycopodiaceae		
Lycopodiella	*Camarozonosporites ambigens*	Mid-Albian–Cenomanian (N Aust.)
		Turonian–Maastrichtian (SE Aust.)
	C. amplus	Campanian–Maastrichtian (SE Aust.)
	C. ohiensis	Campanian–Maastrichtian (SE Aust.)
Lycopodium	*Retitriletes*	Ranges throughout Cretaceous
Huperzia/Phylloglossum	*Sestrosporites pseudoalveolatus*	Valanginian–Santonian
	Coronatispora	Valanginian–Albian
Lycopodiaceae/	*Ceratosporites equalis*	Ranges throughout Cretaceous
Selaginellaceae		
	Neoraistrickia truncata	Ranges throughout Cretaceous
Selaginellaceae	*Herkosporites elliottii*	Maastrichtian (SE Aust.)
	Densoisporites velatus	Ranges throughout Cretaceous
	Perotrilites	Ranges throughout Cretaceous
Isoetaceae	*Minerisporites marginatus*	Valanginian–Albian
	Paxillitriletes phyllicus	Valanginian (W Aust.)
Isoetaceae/Equisetaceae	*Perinomonoletes* sp. (Dettmann, 1986*a*)	Ranges throughout Cretaceous
Filicopsida		
Marattiaceae	*Punctatosporites walkomii*	Ranges throughout Cretaceous
Osmundaceae		
ªCacumen expansa	*Osmundacidites mollis*	Barremian–Albian
Osmunda/Leptopteris	*Baculatisporites comaumensis*	Ranges throughout Cretaceous
	Osmundacidites wellmanii	Ranges throughout Cretaceous
Schizaeaceae		
Anemia subgen. *Anemia*	*Appendicisporites* cf. *cristatus*	Cenomanian
	A. cf. *potomacensis*	Cenomanian
	A. cf. *problematicus*	Late Albian–Cenomanian
Anemia subgen.	*Plicatella distocarinata*	Mid-Albian–Turonian (N Aust.)
Coptophyllum		Late Albian–Turonian (S Aust.)
	P. cf. *gigantica*	Barremian–Aptian (N Aust.)
	P. cf. *jansonii*	Albian–Cenomanian (N Aust.)
Anemia	*Cicatricosisporites cuneiformis*	Mid-Albian–Turonian
	C. hughesii	Berriasian–Albian (W & N Aust.)
		Aptian–Turonian (SE Aust.)

Table 8.1. (*cont.*)

Extant and/or fossil plant taxa	Spore–pollen taxa	Cretaceous range in Australia
	C. pseudotripartitus	Hauterivian–Albian (W Aust.)
		Albian (SE Aust.)
	C. venustus	Mid-Albian–Cenomanian
[a]*Ruffordia goeppertii*	*Ruffordiaspora australiensis*	Valanginian–Campanian
[a]*?Ruffordia*	*R. ludbrookiae*	Valanginian–Albian
Actinostachys	*Microfoveolatosporis fromensis*	Campanian–Maastrichtian
?Schizaeaceae	*Ischyosporites*	Berriasian–Cenomanian
	Klukisporites scaberis	Berriasian–Ceonomanian
Pteridaceae		
?Pteris	*Camarozonosporites bullatus*	Santonian–Maastrichtian (SE Aust.)
	Polypodiaceoisporites elegans	Barremian–Cenomanian (N Aust.)
Gleicheniaceae		
Gleichenia/Dicranopteris	*Gleicheniidites circinidites*	Ranges throughout Cretaceous
?Gleicheniaceae	*Clavifera triplex*	Mid-Albian–Maastrichtian (N & W Aust.)
		Cenomanian–Maastrichtian (SE Aust.)
Dicksoniaceae		
Dicksonia	*Trilites tuberculiformis*	Berriasian–Maastrichtian
Culcita	*Rugulatisporites mallatus*	Maastrichtian (SE Aust.)
?Dicksoniales		
[a]*Adiantites lindsayoides*	*Dictyophyllidites crenatus*	Ranges throughout Cretaceous
[a]*Sphenopteris travisii*	*Cyathidites australis/C. minor*	Ranges throughout Cretaceous
Lophosoriaceae		
Lophosoria	*Cyatheacidites annulatus*	Albian–Cenomanian (W & N Aust.)
		Turonian–Campanian (SE Aust.)
Matoniaceae	*Dictyophyllidites pectinataeformis*	Albian (S Aust.)
	Matonisporites cooksoniae	Berriasian–Albian
Blechnaceae		
Blechnum/Doodia	*Peromonoletes densus*	Maastrichtian
Marsileaceae	*Arcellites*	Barremian–Cenomanian
	Balmeisporites	Late Aptian–Turonian
	Crybelosporites	Berriasian–Santonian
Salviniaceae		
?Azolla	*Grapnelispora evansii*	Maastrichtian (S Aust.)
Pteridospermophyta		
	Alisporites	Ranges throughout Cretaceous
	Vitreisporites pallidus	Ranges throughout Cretaceous
Cycadophyta/Ginkgophyta		
	Cycadopites nitidus	Ranges throughout Cretaceous
Coniferophyta		
Cheirolepidaceae		
[a]*Tomaxellia biforme*	*Classopollis*	Berriasian–Santonian
Araucariaceae		
[a]*Brachyphyllum mamillare*	*Callialasporites dampieri*	Berriasian–Cenomanian
[a]*Apterocladus lanceolatus*	*C. trilobatus*	Berriasian–Cenomanian
	C. dampieri	Berriasian–Cenomanian
[a]*Brachyphyllum irregulare*	*Balmeiopsis limbata*	Berriasian–Cenomanian (N & W Aust.)
Araucaria/Agathis	*Araucariacites australis*	Ranges throughout Cretaceous

Table 8.1. (*cont.*)

Extant and/or fossil plant taxa	Spore–pollen taxa	Cretaceous range in Australia
Podocarpaceae		
Dacrycarpus	*Dacrycarpites australiensis*	Campanian–Maastrichtian (S Aust.)
Dacrydium	*Lygistepollenites* spp.	Santonian–Maastrichtian (S Aust.)
Microcachrys/Microstrobos	*Microcachrydites antarcticus*	Ranges throughout Cretaceous
Lagarostrobos	*Phyllocladidites mawsonii*	Turonian–Maastrichtian (S Aust.)
	P. verrucosus	Maastrichtian (S Aust.)
Podocarpus/Dacrydium	*Podocarpidites*	Ranges throughout Cretaceous
[a]*Trisacocladus tigrensis*	*Trichotomosulcites subgranulatus*	Ranges throughout Cretaceous
Anthophyta		
Aquifoliaceae		
Ilex	*Ilexpollenites anguloclavatus*	Turonian–Maastrichtian (S Aust.)
Callitrichaceae		
Callitriche	*Australopollis obscurus*	Cenomanian–Maastrichtian
Chloranthaceae		
Ascarina	*Clavatipollenites hughesii*	Barremian/Early Aptian–Campanian (S Aust.)
		Early Albian–Turonian (N Aust.)
?*Hedyosmum*	*Asteropollis asteroides*	Early Albian
Epacridaceae	*Ericipites scabratus*	Campanian–Maastrichtian (S Aust.)
Fagaceae		
Nothofagus subgen. *Fuscospora*	*Nothofagidites brachyspinulosus*	Maastrichtian (S Aust.)
Nothofagus subgen. *Brassospora*	*N. emarcidus*	Maastrichtian
Nothofagus (ancestral)	*N. senectus, N. endurus*	Campanian–Maastrichtian (S Aust.)
Gunneraceae		
Gunnera	*Tricolpites reticulatus*	Campanian–Maastrichtian
?Platanaceae	*Tricolpites minutus*	Late Albian
Proteaceae		
Adenanthos	*Proteacidites adenanthoides*	Santonian–Maastrichtian (S Aust.)
Beauprea	*Beaupreaidites*	Campanian–Maastrichtian (S Aust.)
Carnarvonia	*Proteacidites pseudomoides*	Campanian–Maastrichtian (S Aust.)
Grevillea	*Propylipollis* sp.2 (Dettmann & Jarzen, 1991)	Maastrichtian (S Aust.)
Gevuina/Hicksbeachia	*Propylipollis reticuloscabratus*	Santonian–Maastrichtian (S Aust.)
Knightia	*Proteacidites amolosexinus*	Campanian–Maastrichtian
Macadamia/Helicia	*Propylipollis* cf. *crassimarginis*	Santonian–Maastrichtian
Persoonia	*Proteacidites* sp.A (Hill & Macphail, 1983)	Maastrichtian
Stirlingia	*Proteacidites* sp.1 (Dettmann & Jarzen, 1991)	Campanian–Maastrichtian (S Aust.)
Telopea	*Triporopollenites ambiguus*	Campanian–Maastrichtian (S Aust.)
Ranunculaceae		
cf. *Clematis*	*Tricolporites lilliei*	Campanian–Maastrichtian
Trimeniaceae		
Trimenia	*Periporopollenites fragilis*	Campanian–Maastrichtian (S Aust.)
Winteraceae		
Belliolum/Bubbia	*Pseudowinterapollis wahooensis*	Campanian–Maastrichtian
?Winteraceae	?*Afropollis* sp. (Burger, 1990)	Late Albian

[a]Fossil plant taxa.

perhumid forests (rainforests) that occur in
eastern Australia. As revealed by the spore–
pollen record, floristics and structure of the mixed
podocarp/araucarian Cretaceous forests were progressively modified prior to, and after, the
introduction of angiosperms into the Australian
region. The successive vegetational changes have
been linked to environmental disturbances associated with break-up of southern Gondwana, world
sea level changes and climatic fluctuations
(Dettmann, 1986b). The history of the Cretaceous
vegetation of Australia is discussed below, with
reference to the changes in geography, climates,
and environments of the continent.

The flora prior to angiosperm invasion (Berriasian–Barremian)

In earliest Cretaceous times Australia, Antarctica,
New Zealand, the Lord Howe Rise, the Queensland Plateau, and greater India were conjoined as
part of the Gondwana landmass, with Australia
situated between latitudes 50° and 80° S. Sea-
floor spreading initiated in the Late Jurassic on
the northwestern margin progressed southward
and eastward around the Australian continent
during the earliest Cretaceous (Figure 8.8). The
Canning, Carpentaria and Eromanga and Surat
Basins were drained by rivers that fed into the
sea on the newly formed northwest and northeast
coastlines. River systems occurred in the Perth
Basin in the southwest prior to rifting from India.
The southern rift valley between Australia and
Antarctica widened, and braided rivers flowed
south into it from the Eucla Basin. Further to the
west there were riverine systems and lakes in the
narrow depressions of the Otway, Gippsland and
Bass Basins (Figure 8.8).

The earliest Cretaceous spore–pollen record
from the *Ruffordiaspora australiensis/Crybelosporites
stylosus* Zones of eastern Australia and the
Biretisporites eneabbaensis Zone of Western Australia (Figure 8.2) reveals that araucarians and
podocarps were important elements in the
pre-angiospermous Cretaceous vegetation of
Australia. Abundant pollen of these austral conifers, including pollen of the *Araucaria/Agathis*-,

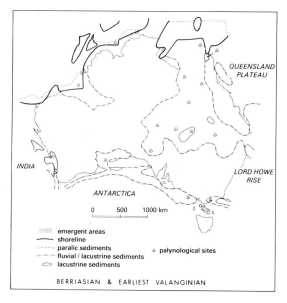

Figure 8.8 Palaeogeographical reconstruction of Australia for
the Berriasian and earliest Valanginian, showing shorelines,
riverine/lacustrine systems and palynological sites (adapted
from Dettmann *et al.*, 1992).

Podocarpus- and *Microcachrys* types (Table 8.1), are
associated with frequent pollen indicative of pteridosperms and/or cycadeoids, and cycadophytes
and/or ginkgophytes. *Classopollis* and *Corallina*,
which are indicators of cheirolepidacean conifers,
occur in low frequencies in fluvio-lacustrine sediments, but are usually common in marginal
marine sediments. Quantitative data suggest that
mixed podocarp/araucarian forests surrounded
riverine/lacustrine systems, but in coastal regions
the forests gave way to cheirolepidacean woodlands. Cryptogam spores representing ferns, fern
allies and bryophytes are both abundant and
diverse in sediments that accumulated in coastal
and inland regions. Osmundalean ferns, which
probably occurred on forest margins and in the
understorey, are indicated by abundant *Osmundac-
idites* and *Baculatisporites*; one species, *Osmundaci-
dites mollis*, is known from fertile spikes of
Cacumen expansa, which has been assigned to the
Osmundaceae (Table 8.1). Also abundant are
Cyathidites and *Dictyophyllidites* which are known
in situ from *Sphenopteris travisii* and *Adiantites lind-*

sayoides respectively; affinities of the macrofossil species may be with the Dicksoniales (Table 8.1). *Gleicheniidites, Matonisporites, Klukisporites* and *Ruffordiaspora* indicate the presence of the Gleicheniales, Matoniales, and Schizaeales (Table 8.1), present-day members of which occur in streamside to dry-zone habitats. Equisetalean and/or isoetalean reeds occupied intermittently wet sites along streams and lake margins; these are indicated in the palynofloras by microspores of *Peromonoletes* spp. and the megaspores *Minerisporites marginatus* and *Paxillitriletes phyllicus*. *Sphagnum*-type spores are represented, and are occasionally common in the southern Eromanga Basin. Meagre development of a floating aquatic fern flora is suggested by the occurrence of *Crybelosporites* (Marsileales). Also common are spores of terrestrial and epiphytic lycopods (*Lycopodium, Lycopodiella, Huperzia* and/or *Phylloglossum*, and members of the Selaginellaceae) which would have ranged from open sites to forest margins.

The floral evidence argues for some regionalism amongst overstorey and understorey communities in the earliest Cretaceous vegetation of Australia. The broad riverine area of the Eromanga and Surat Basins supported a more diverse fern/lycopod/moss flora than the narrow depressions of the Otway and Gippsland Basins in the southern rift valleys, which appear to have been more uniformly forested. In the Perth Basin diverse assemblages of ferns were represented, including several taxa that are unknown from the southeast prior to the Barremian (Dettmann, 1986a; Helby *et al.*, 1987; Dettmann *et al.*, 1992). Migration of the plants from the west may have been triggered by tectonic or volcanic disturbances that heralded rifting of India from southwestern Australia and widening of the southern rift valley.

A major rise in relative sea level during Late Valanginian–Early Barremian times led to marked changes in depositional patterns (Figure 8.9). The sea entered the Eromanga Basin, via the Carpentaria Basin, from the north, following pre-existing valleys. At the southern end of this basin there was a broad zone of rivers and lagoons, and to the east an embayment developed on the

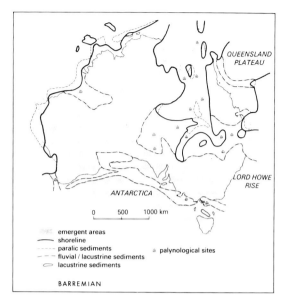

Figure 8.9 Palaeogeographical reconstruction of Australia for the Barremian, showing shorelines, riverine/lacustrine systems and palynological sites (adapted from Dettmann *et al.*, 1992).

site of the Surat Basin. A fluvial system existed in the incipient rift valleys of the Coral Sea and granites were intruded along the central Queensland coast. The Perth Basin was invaded by the sea as new ocean floor formed and India broke away from the Australian continent. Widespread faulting and extrusion of basalts accompanied this event. The rift between Australia and Antarctica slowly widened but drainage patterns changed little.

Sediments deposited during this time (lower and middle *Cyclosporites hughesii* Zone, southeast Australia; *Foraminisporis wonthaggiensis* Zone, northeast Australia; lower *Balmeiopsis limbata* Zone, Western Australia) contain palynofloras that indicate a response of the vegetation to these environmental disturbances. The increase in abundance of podocarp pollen and concomitant decline in frequencies of araucarian pollen in fluvio-lacustrine sediments signifies increased representation of the Podocarpaceae, possibly in response to higher base levels and/or increased precipitation. However, marine and marginal marine sediments in the Perth Basin also have plentiful cheirolepidacean pollen (Backhouse, 1988), suggesting that cheirolepidacean woodlands developed in coastal regions. The Perth Basin sediments also contain the earliest Australian occurrences of the fern *Anemia*, as confirmed by the spore taxa *Cicatricosisporites hughesii* and *C. pseudotripartitus* (Table 8.1).

Concepts of a uniform flora over Australia during earliest Cretaceous times must be revised. Distinct community associations recorded from the disparate depositional areas indicate a regionalised vegetation that reflects a latitudinal control as well as habitat differences related to topographical and substrate variations (Dettmann, 1986*a,b*; Dettmann *et al.*, 1992). Forests of mixed podocarp/araucarian conifers associated with pteridosperms/cycadeoids and *Ginkgo* and/or cycadophytes occurred over much of the continent, but cheirolepidacean conifers appear to have been more important in the vegetation of coastal regions and *Ginkgo*/cycadophytes in northern areas. Moreover, cryptogam communities varied across Australia, with more diverse ferns in the southwest, and greater development of lycopod communities about the river systems in the southeast and northeast.

The structure of the forests can be predicted using present-day ecophysiological parameters, combined with determinations of palaeolatitudes and palaeotemperatures (Specht *et al.*, 1992). Northern regions of Australia were situated at 45–50° S and southern areas in latitudes as high as 75–80° S. At these latitudes, energy input is sufficient to maintain forests, provided the mean daily air temperature during the growing season is above 13° C. In closed-forests situated at latitudes higher than 60° S, the tree canopies would have had low height:width ratios and there would be no understorey. For an understorey to exist in the earliest Cretaceous forests of Australia, canopy taxa would have been widely spaced and have had conical crowns of height: width ratios up to 4 : 1 (Creber & Chaloner, 1987), and the structure of the forests would have been comparable to that of present-day open forests in Australia (Specht *et al.*, 1992). There are no palaeotemperature determinations for the Australian earliest Cre-

taceous, but computer models predict a strongly seasonal climate for interior areas of southern Gondwana (Schneider *et al.*, 1985), and, on the basis of leaf size and cuticle structures of macrofossils from the Otway and Gippsland Basins, Douglas (1986) postulated humid environments under temperature regimes similar to those of present-day warm temperate–subtropical regions.

Early angiosperms in Australia (latest Barremian–Cenomanian)

Angiosperms were introduced into Australia during latest Barremian or earliest Aptian times, with magnoliid plants being the first invaders. Earliest occurring pollen are monosulcate, retitectate types referable to *Clavatipollenites*, which accommodates pollen comparable to that of *Ascarina* of the Chloranthaceae (Table 8.1). Barremian–earliest Aptian occurrences of *Clavatipollenites* are known from the Eromanga and Gippsland Basins (Dettmann, 1986*a*; Burger, 1990), and in the latter area the pollen is associated with the oldest known flower branchlet, which has been interpreted to represent a plant of prostrate herbaceous habit (Taylor & Hickey, 1990). Habitats of the early angiosperms included lakeside and riverine areas in the southern rift valley as well as the broad coastal plains adjacent to the estuary of the Eromanga Basin (Figures 8.9, 8.10).

There is general agreement that the earliest angiosperms represented in the Australian region migrated from a source in northern Gondwana or southern Laurasia (Muller, 1981; Dettmann, 1986*a*; Burger, 1990). Migration was more rapid than initially postulated (e.g. Dettmann, 1973) for there is only a brief time lag between latest Barremian–Aptian introduction in Australia and Hauterivian first appearances in the northern Gondwanan – southern Laurasian region (Brenner, 1984; Hughes & McDougall, 1987). Radiation from the source region was concurrent with early opening of the North and South Atlantic Oceans, and it has been argued that associated environmental disturbances provided the trigger for dispersal (Dettmann, 1986*b*). The route to Australia may well have been via Antarctica and

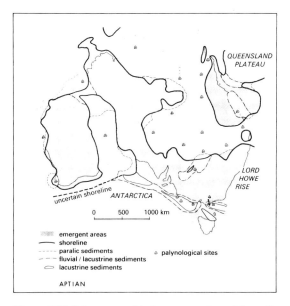

Figure 8.10 Palaeogeographical reconstruction of Australia for the Aptian, showing shorelines, riverine/lacustrine systems and palynological sites (adapted from Dettmann *et al.*, 1992).

thence into southern Australia, with the southern rift valley forming the vehicle for floral channelling (Dettmann, 1986*a*, 1989). Hypotheses for migration from East Asia to northern Australia using a series of postulated microcontinents (Burger, 1981, 1990) await substantiation from knowledge of early angiosperm history in Indonesia and Papua New Guinea.

Vegetational modifications associated with angiosperm invasion in Australia are reflected in the pollen record, which emphasises changes to understorey communities more so than to canopy associations. The mixed podocarp/araucarian forests in the southeast and in the Eromanga and Surat Basins were associated with more diverse hepatic communities than previously encountered, and several fern taxa including *Pteris* and *Anemia* were introduced into the Eromanga and Surat Basins.

With a major rise in sea level during the early Aptian, shallow seas flooded the intracratonic basins, and the Australian landmass was reduced to several large islands, the eastern ones of which retained connections with Antarctica, the Lord

Howe Rise and the Queensland Plateau (Figure 8.10). Palynofloras from sediments of the *C. hughesii* Zone of the interior basins contain high frequencies of *Podocarpus*- and *Microcachrys*-type pollen with lesser proportions of *Ginkgo*/cycadophyte, araucarian, cheirolepidacean, and pteridospermous/cycadeoid pollen. A forest vegetation in which podocarps were important canopy components is inferred. Understorey in the forests included pteridosperms/cycadeoids, together with osmundalean and dicksonialean tree ferns, and there were communities of diverse terrestrial ferns among which gleichenialean and schizaealean taxa were well represented. *Lycopodium*, which today comprises mainly terrestrial taxa, is represented by diverse and abundant spores, and spores of sphagnalean mosses, isoetaleans and/or equisetaleans are locally common. However, anthocerotalean and ricciaceous spores are poorly represented and angiospermous pollen is thus far unrecorded from the palynofloras. High base levels and flooding of the formerly broad coastal and lakeside plains may signify denudation of habitats of the early angiosperm invaders within the Eromanga Basin during the Aptian.

Podocarp forests also occurred on the western island (comprising southwestern areas of Western Australia), but knowledge of the vegetation is only sketchily known, and except for the Perth Basin (Backhouse, 1988) is based on preliminary accounts (Balme, 1957, 1964; Helby *et al.*, 1987). Palynofloras from lower Aptian sediments (upper *B. limbata* Zone) of the Perth Basin reflect a vegetation of podocarps (including substantial representation of those that shed *Microcachrys*-type pollen) and cheirolepidacean woodlands. High representation of sphagnalean spores suggests development of mosslands, whilst the diverse spore suites imply dry-zone and riparian ferns including substantial representation of the Gleicheniales, the spores of which are a common component of the palynofloras. Angiospermous pollen has not been reported.

By contrast, coeval sediments (upper *C. hughesii* Zone) in the Gippsland and Otway Basins contain *Clavatipollenites*, which confirms the presence of shrubby or herbaceous angiosperms in the southeast during the Aptian. The forests of the rift valley are interpreted to have been floristically simpler than those surrounding the Eromanga, Surat and Perth Basins, where terrestrial ferns were more diverse and the Gleicheniales were abundantly represented. Moreover, the coastal vegetation of the Perth Basin included communities of Cheirolepidaceae, which were poorly developed about the inland seas and in the southern rift valley.

Near the Aptian/Albian boundary a regressive phase produced major configurations of coastlines (Figure 8.11). The inland sea retreated from the Surat and Maryborough Basins, and in the Eromanga Basin the eastern exit was closed and the southern limit advanced southwards. The southwestern seaway extended east of the Officer–Eucla embayment to the Otway Basin, where brackish conditions occurred in low-lying areas. Volcanics were extruded in the Otway Basin, and nonmarine deposition occurred in the Bass and Gippsland Basins.

Spore–pollen evidence from the *Crybelosporites striatus* Zone (early Albian) indicates increasing

Figure 8.11 Palaeogeographical reconstruction of Australia for the latest Aptian–early Albian, showing shorelines, riverine/lacustrine systems and palynological sites (adapted from Dettmann *et al.*, 1992).

regionalism of the Australian vegetation, and more widely distributed angiosperms. Angiosperms persisted in the southeast, and after retreat of the inland sea were re-established in the Eromanga Basin and had spread to the Surat Basin (Burger, 1990). Only chloranthaceous types are known; pollen represented include *Clavatipollenites* and *Asteropollis*, the latter a trichotomosulcate form similar to pollen of *Hedyosmum* (Walker & Walker, 1984). The angiosperms are believed to have colonised the newly exposed areas, which were also invaded by aquatic, riparian, and dry-zone communities of sphagnalean mosses and ferns of the Marsileales, Gleicheniales and Schizaeales. However, lycopods, whose spores are infrequent in the palynofloras, appear not to have advanced into the shoreline communities and may have been confined to the mixed podocarp/araucarian forests on better drained sites. In the southeast portion of the Australian–Antarctic rift valley podocarp/araucarian forests persisted, but the floodplain flora expanded in response to widening of the riverine/lacustrine systems. Even so, terrestrial and aquatic ferns were considerably less diverse than those of northern regions, and the vegetation retained a substantial lycopod component as *Lycopodium*-type spores remain quantitatively important in Gippsland and Otway Basin palynofloras.

During the mid-Albian the sea in the Eromanga and Surat Basins contracted further and was fringed by brackish lagoons and estuaries. Inundation of the Eromanga Basin by the sea occurred in the early late Albian, but the sea regressed again in the latest Albian, with re-establishment of broad lagoonal and estuarine areas. To the east, coal-forming swamps occurred in depressions of the Maryborough and Styx River Basins. The Eucla embayment was of lesser extent, but the easterly-extending estuary was flanked by rivers and lakes. River and lake systems occurred in the Bass and Gippsland Basins (Figure 8.12).

Palynological evidence from the *Coptospora paradoxa* and lowermost *Phimopollenites pannosus* Zones (mid–late Albian) confirms that non-magnoliid angiosperms were introduced into the Australian vegetation by the mid-Albian and

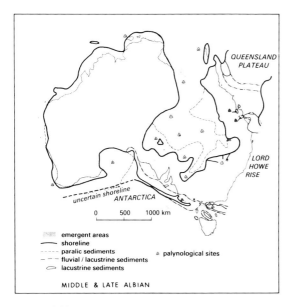

Figure 8.12 Palaeogeographical reconstruction of Australia for the mid–late Albian, showing shorelines, riverine/lacustrine systems and palynological sites (adapted from Dettmann *et al.*, 1992).

diversified rapidly during the late Albian (Burger, 1990; Dettmann *et al.*, 1992). Diversification was coincident with regression of the sea in the Eromanga and Surat Basins, and with widening of the floodplain in the southern rift valley. In addition to *Clavatipollenites* and *Asteropollis*, tricolpate and tricolporoidate forms that represent nonmagnoliid dicotyledons are represented. Precise affinities of the majority are unknown, but *Tricolpites minutus* has an *in situ* association with fossil flowers of the Platanaceae in North America (Friis *et al.*, 1988). Several of the angiosperm pollen taxa have restricted distribution, and assemblages from northern areas of Australia are more diverse than those from the southeast (Dettmann, 1973; Burger, 1990). There are few data from Western Australia, but tricolpate angiosperm pollen occurs in sediments containing the *Hoegisporis* Microflora (Balme, 1964) of Albian age.

Cryptogam spore and gymnospermous pollen fractions of Australian Albian palynofloras also express regionalism of the vegetation, particularly with respect to understorey communities

(Dettmann, 1981; Dettmann & Thomson, 1987; Dettmann *et al.*, 1992). Fern taxa with restricted distribution included *Lophosoria*, which, as indicated by *Cyatheacidites annulatus*, was confined in Australia at this time to western and northeastern areas (Dettmann 1986*b*). Several *Anemia* and gleicheniaceous spore taxa display a distribution range similar to that of *Hoegisporis*, which is thought to represent pollen of a brachyphyllous conifer. Community associations were also regionalised. Palynofloras of the intracratonic basins imply heathland and aquatic associations of diverse ferns, mosses and hepatics, whereas in the Styx and Burrum coal swamps the moss/hepatic element was poorly represented (Barnbaum, 1976). Understorey of the Otway Basin forests included dry-zone and aquatic ferns, hepatics and lycopods, but the ferns were less diverse and the gleicheniaceous types less prolific than in northern regions.

Early in the Cenomanian there was a major change in environmental patterns and Australia began to assume its present shape. Lakes and swamps developed in the Eromanga Basin after retreat of the sea. Subsidence occurred along the western margin and sea-floor spreading commenced along the southern margin, where marine conditions were established as far east as the Otway Basin. A lacustrine system persisted in the Bass Basin, and, to the east of the Gippsland Basin, the proto-Tasman Sea was initiated (Figure 8.13).

Palynofloras from the upper *Phimopollenites pannosus* and *Appendicisporites distocarinatus* Zones indicate that profound vegetational changes were associated with the dramatic changes in environment. In the southeast riparian and aquatic communities appear to have been largely replaced by dry-zone associations. Notable is the increased diversity and representation of *Anemia* and *Selaginella*-type spores. Araucarian pollen is common and canopy associates of the araucarians included podocarps that shed pollen of the *Microcachrys* and *Podocarpus* type. *Classopollis* is occasionally frequent, implying the presence of cheirolepidacean woodlands. Magnoliid and nonmagnoliid dicotyledonous angiosperms are indicated by sulcate, tricolpate, tricolporoidate and tricolporate pollen.

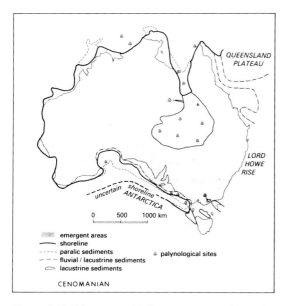

Figure 8.13 Palaeogeographical reconstruction of Australia for the Cenomanian, showing shorelines, riverine/lacustrine systems and palynological sites (adapted from Dettmann *et al.*, 1992).

Also represented are sulcate types similar to lili-aceous monocotyledonous pollen and there are rare grains of *Australopollis obscurus*, a periporate form reminiscent of pollen of *Callitriche* (Partridge & Macphail, 1990).

There are notable differences between latest Albian–Cenomanian palynofloras from the southeast and those from the Eromanga Basin. The gymnospermous fraction of Eromanga palynofloras includes a high proportion of podo-carp pollen, and the cryptogam spore assemblages are more diverse and contain greater proportions of *Gleichenia*- and *Anemia*-type spores than those from the southeast. Also present are abundant marsilealean, hepatic and equisetalean and/or iso-etalean spores. Angiosperm pollen assemblages share many features in common with those of coeval sediments from the southeast but are more diverse with respect to the tricolpates, tricolporo-idates and tricolporates, and contain ?*Afropollis*, which includes zonasulculate pollen of wintera-ceous affinity (Doyle *et al.*, 1990). Moreover, *Cal-litriche*-type pollen has not been reported from

the latest Albian–Cenomanian of the Eromanga Basin. Palynofloras similar to those of the Erom-anga Basin are known from Bathurst and Melville Islands, north of Darwin, Northern Territory, but they contain higher frequencies of *Classopollis*, *Balmeiopsis* and *Hoegisporis* (Burger, 1976). Brachyphyllous pollen and common gleichenia-ceous spores have been reported (Balme, 1964; Ingram, 1968) from the latest Albian–Cenomanian of the Perth and Eucla Basins, but few other components of the palynofloras have been detailed.

The palynological record confirms that there were regional differences in the Australian vegeta-tion during latest Albian–Cenomanian times. The Eromanga lakes and swamps were sur-rounded by mixed conifer/cycad woodlands and heathlands of ferns. Lakes and swamps supported aquatic and littoral communities of ferns, fern allies and hepatics. Angiosperms probably ranged throughout these habitats. The Cheirolepidaceae appears to have been more strongly represented in coastal vegetation than in inland areas. Com-pared to the Eromanga Basin, the Otway and Gippsland Basins had less diverse fern communi-ties and only slight development of aquatic com-munities. Cheirolepidacean conifers probably occurred in coastal sites, but the brachyphylls that shed *Hoegisporis* appear to have been poorly rep-resented in the southeast.

The forests and heathlands of the Aptian–Cenomanian grew at high latitudes, and, as implied by the palynological evidence, the forests mostly had a well-developed understorey and the canopy would therefore have had an open struc-ture. Aptian palaeotemperatures of 12 °C for the Eromanga sea (Stevens & Clayton, 1971) and 0–5 °C for the Otway Basin (Gregory *et al.*, 1989) confirm cool–cold climates, possibly with strong seasonality (Dettmann *et al.*, 1992). On the basis of present-day ecophysiological parameters, forest canopy taxa would have been mostly of micro-therm groups having notophyllous and microphyl-lous leaves in the leaf size classification of Raunkiaer (1934) and Webb (1959). Mid–late Albian palaeotemperatures signify warmer seas than for the Aptian. The highest value (16 °C) is from the southwestern Eromanga Basin, where circulation was restricted. In the northern part of the basin, recorded temperatures (12 °C) are similar to those reported from the Carnarvon Basin, Western Australia (Dettmann *et al.*, 1992). Cool to warm temperate climates with moderate to high precipitation occurred in the forested areas of the southeast and northeast, whereas in the Eromanga and Surat Basins, where there was extensive development of fern heathlands, rainfall may have been less or seasonal (Douglas, 1986; Dettmann *et al.*, 1992).

Differentiation of Gondwanan elements and evolution of Australian elements (Turonian–Maastrichtian)

By Turonian times, most of the continent was emergent. Uneven upwarping of eastern and southern areas radically altered drainage patterns. A riverine system developed in the former Eromanga Basin and drained south into the embryonic Southern Ocean. Australia and Antarctica were linked through Tasmania, which separated the embryonic Southern Ocean to the west from the advancing Tasman Sea to the east. Lacustrine deposition continued in the Bass Basin and basalts were extruded in the Gippsland Basin (Figure 8.14).

Profound vegetational changes involving understorey and overstorey communities occurred in the southeast at or near the Cenomanian/Turonian boundary. *Lagarostrobos, Dacrydium* and early Proteaceae were introduced in the Turonian (*Phyllocladidites mawsonii* Zone) followed by *Dacrycarpus* in the Santonian (*Tricolporites apoxyexinus* Zone); these are important components of present-day austral floras. *Ilex*, which today is cosmopolitan, may have originated in the region (Martin, 1977). As implied by the palynofloras, the mixed podocarp/araucarian forests had angiosperms associated in the canopy by at least Santonian times. Amongst the canopy taxa were *Macadamia* and *Gevuina/Hicksbeachia* of the Proteaceae (Dettmann & Jarzen, 1990). Angiosperm pollen assemblages display a steady increase in diversity and there was considerable turnover of taxa

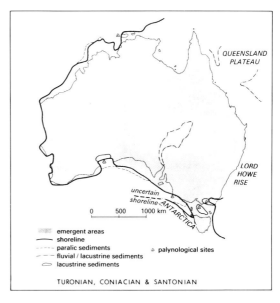

Figure 8.14 Palaeogeographical reconstruction of Australia for the Turonian, Coniacian and Santonian, showing shorelines, riverine/lacustrine systems and palynological sites (adapted from Dettmann *et al.*, 1992).

among the cryptogams. *Ascarina*-type and liliaceous-like sulcate pollen occur infrequently, but *Callitriche*-type pollen is well represented; aquatic and terrestrial herbs and shrubs are implied. Spores of hepatics and *Lycopodium* abruptly declined in diversity, whereas those of *Selaginella* were of increased diversity. The fern component included *Lophosoria*, Gleicheniaceae, Osmundaceae and Marsileaceae, but the Pteridaceae and Schizaeaceae were poorly represented. Several of the taxa evidently migrated into the region during Turonian–Santonian times after earlier appearances elsewhere in Australia or southern Gondwana. *Lophosoria* and the gleicheniaceous fern that shed *Clavifera triplex* are both known from older sediments outside the region (Dettmann, 1986*b*; Askin, 1989).

There are few published data from other areas. Sediments on Melville Island, off the coast of the Northern Territory, contain angiosperm pollen assemblages with several taxa known from the southeast, but the assemblages are more diverse (Dettmann, 1973). Proteaceous-like pollen and abundant gleicheniaceous spores occur in Santonian spore–pollen assemblages from the Eucla Basin (Ingram, 1968).

During the Campanian and Maastrichtian, sea-floor spreading continued along the southern margin, but Australia remained linked to Antarctica (Figure 8.15). Lakes occurred in the Bass Basin, and drained into the estuaries of the Gippsland and Otway Basins. The river system emptying into the Southern Ocean persisted and a northward-flowing river is believed to have developed in the Officer and Canning Basins. Igneous activity occurred along the eastern margin and may have been associated with the onset of sea-floor spreading in the southeast and of rifting in the northeast. Emergent areas existed in the northeast, linking Australia to eastern Papua New Guinea and New Caledonia (Figure 8.15).

Knowledge of the Australian Campanian–Maastrichtian vegetation is mainly from the Gippsland, Otway and Bass Basins, but there are limited data from the Perth Basin and from possible latest Cretaceous sediments from a site near the Olgas, Northern Territory, and one in the

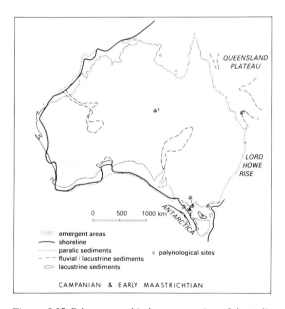

Figure 8.15 Palaeogeographical reconstruction of Australia for the Campanian and Maastrichtian showing shorelines, riverine/lacustrine systems and palynological sites (adapted from Dettmann et al., 1992).

Capricorn Basin, off the central Queensland coast (Figure 8.15). Palynological data from the *Nothofagidites senectus*, *Tubulifloriidites lilliei* and *Forcipites longus* Zones of southeastern sequences confirm that the area was clothed by forests comprising components of the older (Turonian–Santonian) vegetation together with newly introduced taxa. There were araucarians, diverse podocarps (*Lagarostrobos, Dacrydium, Dacrycarpus, Podocarpus, Microcachrys*), newly introduced *Nothofagus*, and increasingly diverse Proteaceae.

Amongst the Proteaceae, several taxa (*Knightia, Macadamia, Gevuina/Hicksbeachia, Grevillea*) were probably in the canopy; others (*Carnarvonia, Telopea, Persoonia*) in the forest understorey; and yet others (*Stirlingia, Adenanthos, Beauprea*) in sclerophyllous communities on forest fringes and/or nutrient deficient soils (Dettmann & Jarzen, 1991). Other rainforest associates included Winteraceae, *Ascarina, Ilex* and *Gunnera*. Pollen that may represent *Clematis* (Ranunculaceae), Trimeniaceae, *Callitriche* and Epacridaceae also occur in the palynofloras (Dettmann & Jarzen, 1990; Specht et al., 1992). The cryptogam flora included mosses, hepatics and diverse selaginellids. Gleicheniaceae, Osmundaceae, *Culcita* (Dicksoniaceae), *Blechnum/Doodia* (Blechnaceae), *Pteris* (Pteridaceae), *Actinostachys* (Schizaeaceae) and possibly *Azolla* (Salviniaceae) were represented in fern communities (Table 8.1). Although the vegetation was modified considerably at or near the Cretaceous/Tertiary boundary (Helby et al., 1987), a 'mass' extinction event is not evident. Many of the Cretaceous taxa range into the Tertiary, but some have broken stratigraphic ranges involving the Paleocene. This phenomenon has been interpreted as indicating short-distance retreat of the plants during the earliest Tertiary, followed by advance during latest Paleocene–Eocene times (Dettmann & Jarzen, 1991). Causative mechanisms may have involved alterations to drainage patterns near the close of the Cretaceous.

A site near the Olgas, Northern Territory, has yielded palynofloras rich in proteaceous pollen associated with less frequent pollen of podocarps and very rare *Nothofagus* (Twidale & Harris, 1977; Harris & Twidale, 1991). Also represented are pollen of *Gunnera* and microsporangia of possible *Azolla*. From another site, in the Capricorn Basin, off the central Queensland coast, Hekel (1972) reported common Casuarinaceae pollen and fern spores associated with rare *Clematis*-type pollen in assemblages that are questionably of latest Cretaceous age. Clearly the vegetation of the two northern sites was dissimilar and distinct from that in the southeast. It is uncertain whether these dissimilarities represent geographical or age differences.

During Turonian–Maastrichtian times Australia slowly drifted northwards and by the close of the Cretaceous the southeast was at latitudes of 65° S. Evidence for types of climate in the Turonian is conflicting; a single ammonite from the Otway Basin provided a palaeotemperature of 28 °C (Dorman, 1966), but associated foraminiferal faunas indicate cold waters (Taylor, 1964). Whatever the temperature regime, the palynological evidence confirms forests dominated by podocarp conifers (*Podocarpus, Dacrydium, Lagarostrobos*) in the canopy, which would have had an open structure. During Santonian–Maastrichtian times, proteaceous angiosperms (*Knightia, Macadamia, Gevuina/Hicksbeachia*) were represented in the canopy. The cool temperate conditions implied by palaeotemperature determinations (16.5–19 °C) and foraminiferal faunas (Taylor, 1964; Dorman, 1966) indicate tall open forests (up to 30 m in height) with canopy taxa (mainly podocarps and Proteaceae, but also rare *Nothofagus* (subgenus *Brassospora*)) having conical-shaped crowns and coriaceous, notophyll-sized leaves (Specht *et al.*, 1992). An open-forest structure at the high latitudes would allow some light to penetrate to a shrubby understorey of Proteaceae (*Carnarvonia, Telopea*), Winteraceae, Trimeniaceae, and *Ilex*, and a ground stratum of diverse cryptogams. Many of the taxa represented in the Late Cretaceous open forests survive today in closed forests (rainforests) in the Australasian flora and some (e.g. *Dacrydium, Dacrycarpus* of the Podocarpaceae; and *Knightia, Gevuina/Hicksbeachia, Macadamia* and *Carnarvonia* of the Proteaceae) are restricted to northeastern Australasia. Although rarely dominant in present-day per-

humid forests, canopy taxa with a Cretaceous history have slender, conical crowns that are sometimes emergent above the dominants which have spreading, dome-shaped crowns. Perhaps, as Australia drifted into lower latitudes during the Tertiary, the gaps between the canopy taxa of the Cretaceous forests were filled by invasive or newly evolved autochthonous taxa with growth habits that maximised solar absorption in mid to low latitudinal regions.

SOUTHERN MID–HIGH LATITUDINAL PHYTOGEOGRAPHICAL REGION: SOURCE AND DISPERSAL CORRIDOR OF AUSTRAL PLANTS

Spore–pollen evidence confirms that close floral relationships existed between Australia and associated high latitudinal areas of southern Gondwana throughout the Cretaceous (Dettmann, 1981, 1986*b*; Dettmann & Thomson, 1987; Dettmann *et al.*, 1992). This region, termed the Southern Gondwana Floristic Province (Brenner, 1976), was characterised by a series of mixed podocarp/araucarian forests, and was invaded by early angiosperms during latest Barremian–Aptian times from a source in northern Gondwana. In the Antarctic Peninsula–South American region, floral zonation across the latitudes was steep, with an interfingering of austral podocarp/araucarian and northern Gondwanan cheirolepidacean communities (Figure 8.16(*a*)). The geographical limits of the podocarp/araucarian forests fluctuated during opening and enlargement of the South Atlantic, Indian and Southern oceans, and with progressive opening of these oceans austral planktonic dinocyst floras achieved circumpolar distribution (Dettmann & Thomson, 1987).

The Cretaceous preangiospermous vegetation of southern Gondwana was characterised by mixed araucarian/podocarp forests. Other communities included woodlands of cheirolepidaceans, heathlands of ferns, and aquatic or semi-aquatic moss/hepatic/isoetalean/equisetalean associations. Regionalism was evident within

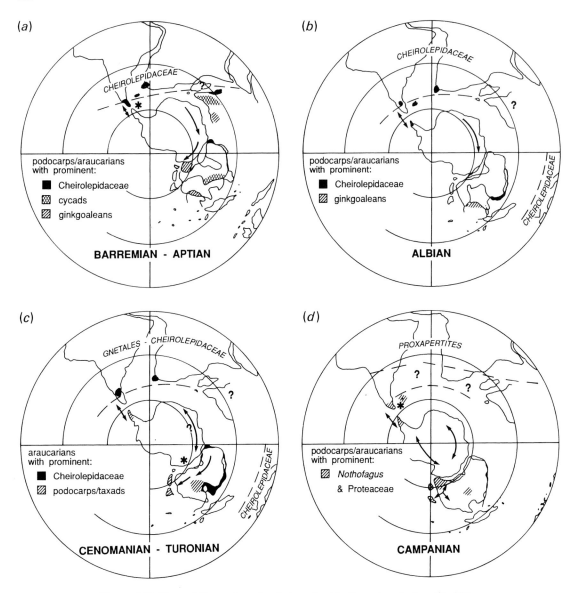

Figure 8.16 Maps of the southern hemisphere for (*a*), Barremian–Aptian, (*b*), Albian, (*c*) Cenomanian–Turonian, and (*d*), Campanian showing distribution of vegetational zones delineated in southern Gondwana. Stars denote centres of origination and/or diversification.

both overstorey and understorey associations, and the regionalism is believed to have been influenced by climatic as well as topographical and edaphic factors. Associated with the forests and forest fringes of podocarps and araucarians were ginkgos, cycads, taeniopterids, pteridosperms and cycadeoids, together with ground communities of ferns, lycopods and bryophytes. In northern areas (India, Patagonia, northern Australia), araucarians were more important in the canopy than were podocarps. Cycads were more plentiful in interior and northern coastal areas than ginkgos, which appear to have preferred more humid habitats. Woodlands of cheirolepidacean conifers intermingled with the forests in coastal regions of northern and western Australia as the Indian Ocean opened, and formed a substantive part of the vegetation in the southernmost South America–Falkland Plateau–southernmost South Africa region during early opening of the southern South Atlantic Ocean (Figure 8.16(*a*)). Cryptogam associations also varied across southern Gondwana; fern spores are more plentiful and diverse in what were coastal regions, whereas lycopods (mainly *Lycopodium*) thrived in inland areas. Spores of sphagnalean mosses and hepatics have greater abundance in depositional basins that were set in low relief terrains.

As the oceans opened, disturbances to adjacent land areas appear to have been the driving force for floral evolution and exchange (Dettmann, 1989). Many taxa migrated to Australia after origination elsewhere. Migration was in an easterly direction from other land masses in the southern Gondwanan assembly; South America, India and Antarctica appear to have been crucial migrational pathways during the earliest Cretaceous (Figure 8.16(*a*)).

Although inconclusive, the evidence for angiosperm introduction into Australia points to routes involving southern Gondwana: radiation was contemporaneous with opening of the South Atlantic and Indian Oceans. As in other regions of southern Gondwana, the first angiosperms to arrive in Australia were herbaceous or shrubby chloranthaceous taxa, which would have competed with taxa in the ground or shrubby strata at forest fringes or those of the understorey within the forests. The latest Barremian–Aptian vegetation existed under cool–cold, humid climates, and the forests had open rather than closed canopies, as light regimes at the high southern latitudes were insufficient for growth of closed forests (Creber & Chaloner, 1987; Specht *et al.*, 1992).

A second wave of angiosperms migrated to Australia during mid-Albian to Cenomanian times and involved nonmagnoliid dicotyledonous taxa whose lineages evolved in northern Gondwana near the Barremian/Aptian boundary (Brenner, 1976; Doyle *et al.*, 1982). Once again, migration appears to have been in an easterly direction, and

involved southern South America and Antarctica. India is believed to have become isolated from the southern Gondwanan assembly, and the meagre information on its mid-Cretaceous vegetation suggests that floral links with Australia had diminished. Precise affinities of many of the immigrant angiosperms are unknown, but Platanaceae and winteraceous lineages are indicated in the Australian pollen record. Introduction coincided with partial sea retreat in intracratonic basins, and with widening of the floodplain in the southern rift valley. Temperatures were warmer than for the Aptian, but precipitation may have been less or seasonal in interior regions. By the Cenomanian, Australia had commenced its slow drift north (Veevers & Eittreim, 1988), but remained in high southern latitudes, and in southern areas the forests would have had an open structure. Niches occupied by the nonmagnoliid dicotyledons included shoreline and floodplain situations, but distribution patterns of the two separate pollen associations delineated by Burger (1990) seem to imply that some of the angiosperms were established in the open forests of hinterland regions of the Eromanga Basin. Near the Cenomanian/Turonian boundary, angiosperms with pollen similar to that of *Callitriche* (Callitrichaceae) were introduced into the vegetation of the southern rift valley. *Callitriche*, which today occupies aquatic and semi-aquatic habitats, may well have had origins in northern Gondwana, as the *Callitriche*-like pollen taxon *Cretaceiporites scabratus* has been recorded from older sediments (Albian–Cenomanian) in Brazil (Herngreen, 1973).

During the Albian–Cenomanian, Australasia, Antarctica, southern South America and the Falkland Plateau retained connections and formed the southern Gondwana assembly (Figure 8.16(*b*) and (*c*)). Most regions of this assembly were situated in high southern latitudes and the mixed podocarp/araucarian forests would have had an open structure. As the South Atlantic Ocean opened, the forests advanced over the Falkland Plateau and partially replaced the cheirolepidacean communities (Figure 8.16(*b*)). Here and on the Antarctic Peninsula, the forests were associated with ginkgos/cycadophytes, pteridosperms, and diverse terrestrial ferns, several of which are unknown from Australasia (Dettmann & Thomson, 1987). Other taxa had restricted geographical ranges within the southern Gondwana region. For instance, the brachyphyll that shed *Hoegisporis* was restricted to northern and western areas of Australia where the flora was richer in angiosperm, fern and hepatic taxa than was the vegetation of south-eastern Australia. Cheirolepidacean conifers were well represented in northern coastal areas, but were only of local development about the estuary in the southeast rift valley.

Angiosperm immigrants continued to arrive in Australia from northern Gondwana during the remainder of the Cretaceous (Turonian–Maastrichtian), but many newly introduced taxa differentiated from northern Gondwanan lineages in the Australasian–Antarctic region. Northern Gondwanan origins have been demonstrated for *Gunnera* (Jarzen & Dettmann, 1990), and *Belliolium/Bubbia* of the Winteraceae may have evolved in the same region (Dettmann & Jarzen, 1990). Pollen that may represent earliest Proteaceae occur in the Cenomanian–Turonian of northern Gondwana (Muller, 1981), but subsequent Late Cretaceous diversification of the family was centred in southern high latitudes (Dettmann & Jarzen, 1991). Four of the five subfamilies of extant Proteaceae are represented in the Campanian–Maastrichtian pollen record of the Austro-Antarctic rift valley. Amongst diverse assemblages of proteaceous pollen recorded from the Otway Basin, southeastern Australia are pollen indicating *Adenanthos*, *Beauprea* and *Stirlingia* (Proteoideae), *Persoonia* (Persoonioideae), *Carnarvonia* (Carnarvonioideae) and *Grevillea*, *Telopea*, *Macadamia*, *Gevuina/Hicksbeachia* and *Knightia* (Grevilleoideae). Diversification of the family coincided with habitat changes associated with early opening of the Southern Ocean. This region may also have been the origination centre of *Ilex* (Aquifoliaceae) in the Turonian (Martin, 1977) and of lineages of the Trimeniaceae and Epacridaceae in the Campanian (Dettmann & Jarzen, 1990).

Lagarostrobos, *Dacrydium* and *Dacrycarpus* (Podocarpaceae) and *Nothofagus*, which are impor-

tant elements of temperate rainforests of the southern hemisphere, evolved in the Austro-Antarctic region during the Late Cretaceous (Dettmann *et al.*, 1990, 1992). The podocarps had successive introductions during Turonian–Santonian times, and ancestral *Nothofagus* appeared in the Campanian. Differentiation of *Nothofagus*, and appearance of its four extant subgenera has been shown to have occurred during the late Campanian-Maastrichtian in the southern South America–Antarctic Peninsula region. Diversification was concurrent with volcanic and tectonic activity under climates that were cooler than those of southern Australasia, the Late Cretaceous diversification centre of the Proteaceae. Migration of extant lineages of *Nothofagus* to Australia occurred during latest Cretaceous–early Tertiary times and routes must have involved Antarctica (Dettmann *et al.*, 1990). The same route may have been utilised by the Myrtaceae, Olacaceae, Loranthaceae and Sapindaceae. Members of these families were established on the Antarctic Peninsula during Campanian–Maastrichtian times, but are unknown from Australia prior to the Tertiary (Dettmann & Thomson, 1987; Askin, 1989; Dettmann, 1989). Antarctica probably served as a dispersal route for several other taxa, with migration in the opposite direction. Present records indicate that several Proteaceae and Epacridaceae were introduced into Australia prior to their latest Cretaceous–Tertiary arrival in western Antarctica (Askin, 1989; Dettmann, 1989).

It has been postulated (Truswell *et al.*, 1987) that several of these elements (Olacaceae, Sapindaceae, Myrtaceae) may have been introduced to northern Australia using 'stepping stones' from Southeast Asia. Whilst this route is not supported by the pollen record for the Olacaceae and Sapindaceae, it should not be dismissed for the Myrtaceae. In addition to Campanian–Maastrichtian records from the Antarctic Peninsula, pollen of the Myrtaceae occur in the Santonian of Gabon, Maastrichtian of Colombo, and the Senonian of Borneo (Muller, 1981). Migration from the north does not conflict with the pollen record, but there is only meagre evidence for stepping stones between Borneo and northern Australia in latest Cretaceous times (Audley-Charles, 1990). Knowledge of Upper Cretaceous palynofloras from northern depositional basins of Australia will be important in elucidating whether floral exchange occurred between Southeast Asia and northern Australia during the Late Cretaceous.

The pollen record firmly establishes that several important elements of the present-day vegetation were established by the close of the Cretaceous. Most were derived from Gondwanic stock; some evolved in northern Gondwana, and others in the Austro-Antarctic region. Many of the taxa are now associated in rainforests, but several (*Adenanthos*, *Stirlingia* and some epacrids) are restricted to nutrient-deficient soils in the Mediterranean climatic region of southern Australia. Thus, the pollen record implies that sclerophylly in the Australian vegetation dates to the latest Cretaceous.

The southern Gondwanan Late Cretaceous vegetation was strongly regionalised within the same latitudinal belt and across the latitudes (Figure 8.16 (*c* and *d*)). The Proteaceae appears to have been important in the central Australian vegetation, but at higher southern latitudes (60–65° S) in southern Australia both podocarps and Proteaceae were well represented in the canopy of the open forests (Specht *et al.*, 1992). Understorey to the forests included shrubby Proteaceae, Winteraceae, Trimeniaceae and *Ilex*, as well as a ground stratum of diverse ferns. Temperature regimes in southeastern Australia were warmer (mean annual temperature 16.5–22 °C) than at similar high latitudes in New Zealand (15–18 °C) where canopy taxa of the open forests were mostly podocarps, and the proteaceous understorey taxa less diverse. The Antarctic Peninsula was also situated at 60–65° S, but temperatures were even cooler (11–13 °C) than those of southern Australia and New Zealand. Podocarps and *Nothofagus* were the major canopy taxa of the western Antarctic tall open forests, and the Proteaceae was less diverse than in Australasia (Specht *et al.*, 1992). Moreover, the Antarctic forests contained myrtaceous taxa and were

associated with wetland communities of sphagnalean mosses and salviniaceous ferns (Askin, 1990); these wetland communities were poorly represented in southern Australasia. *Nothofagus* and podocarps also occurred in southern South America and on the Falkland Plateau (Dettmann & Thomson, 1987). Palynofloras from these areas are incompletely documented, but available evidence suggests substantial representation of an angiosperm element that was rare or absent in other southern Gondwanic regions (Dettmann & Thomson, 1987).

Although southern South Africa was separated from adjacent land masses of southern Gondwana by the South Atlantic Ocean after the mid-Cretaceous, latest Cretaceous–Paleocene palynofloras contain pollen of podocarps (*Podocarpus*, *Microcachrys* and *Dacrydium* types), *Gunnera*, and several proteaceous taxa, including possible *Telopea/Embothrium*. (McLachlan & Pieterse, 1978; Scholtz, 1985). Podocarps were established in the flora during Jurassic–Early Cretaceous times and *Gunnera* probably migrated to the region from the north contemporaneously with radiation to other high southern latitudinal regions (Jarzen & Dettmann, 1990). The presence of *Telopea/Embothrium*, pollen of which is also known from the Late Cretaceous of southern South America and Australia (Dettmann & Jarzen, 1991), also implies evolution of one or both taxa in northern latitudes of Gondwana.

CONCLUSIONS

From the foregoing it is evident that the southern Gondwana vegetation, including that of Australia, was floristically heterogeneous throughout the Cretaceous. Much of the area was forested, but woodlands, heathlands and aquatic communities were also represented. The high latitude forests had an open structure and podocarps and araucarians were important canopy components. Understorey communities fringing and associated with the earliest Cretaceous forests were invaded by herbaceous and shrubby angiosperms during the latest Barremian-Aptian, and by the Santonian angiosperms had entered the canopy. Australia occupied a peripheral position in the southern Gondwana assembly, and routes traversed by the earliest angiosperm invaders from northern Gondwana probably involved other landmasses in the assembly rather than Southeast Asia. The first migratory wave of magnoliid angiosperms occurred no later than the latest Barremian, and was coincident with lowered world sea levels, and early opening of the South Atlantic and Indian Oceans.

A subsequent invasion of angiosperms during mid-Albian to Cenomanian times included non-magnoliid lineages that had evolved in northern Gondwana during the Aptian. Once again, the angiosperms invaded shoreline communities after regression of the Aptian sea in intracratonic basins and widening of the floodplain of the Austro-Antarctic rift valley. Angiosperm migration from northern Gondwana continued during Turonian–Maastrichtian times and was concurrent with *in situ* evolution and differentiation of austral groups in the Australian–Antarctic region. Two loci of evolution and diversification have been identified. Areas surrounding the embryonic Southern Ocean were a diversification centre of the Proteaceae and may have been the cradle of *Ilex*. The diametrically opposite region embracing southern South America and the Antarctic Peninsula was a diversification centre of early *Nothofagus* during phases of volcanic and tectonic activity.

The pollen evidence counters Takhtajan's (1969) argument that Australia was an origination centre for early angiosperms, and supports Webb *et al.*'s (1986) thesis that Australian rainforests are remnants of a heterogeneous Gondwanan flora. Several of the proteaceous rainforest elements with a Cretaceous history are today concentrated in northeastern Australasia under temperate climates similar to those adduced for the Late Cretaceous in their likely cradle region in southern Australia. *Nothofagus* and possibly some of the podocarps differentiated in cooler climates which is to some extent reflected by their present distribution at higher latitudes and/or altitudes. Evidence for evolution of sclerophylly is also expressed in the Late Cretaceous pollen record.

The sclerophyllous taxa probably formed communities on low nutrient or waterlogged soils on forest fringes (Specht *et al.*, 1992).

Support from the Australian Research Council during tenure of an Australian Research Fellowship is gratefully acknowledged.

REFERENCES

ASKIN, R. E. (1989). Endemism and heterochroneity in the Seymour Island Campanian to Paleocene palynofloras: implications for origins and dispersal of southen floras. In *Origins and Evolution of the Antarctic Biota*, ed. J. A. Crame. *Geological Society of London Special Publication*, **47**, 107–19.

ASKIN, R. E. (1990). Cryptogam spores from the upper Campanian and Maastrichtian of Seymour Island, Antarctica. *Micropaleontology*, **36**, 141–56.

AUDLEY-CHARLES, M. G. (1990). Evolution of the southern margin of Tethys (North Australian region) from early Permian to late Cretaceous. In *Gondwana and Tethys*, ed. M. G. Audley-Charles & A. Hallam. *Geological Society of London Special Publication*, **37**, 79–100.

BACKHOUSE, J. (1988). Late Jurassic and Early Cretaceous palynology of the Perth Basin, Western Australia. *Geological Survey of Western Australia Bulletin*, **135**.

BALME, B. E. (1957). Spores and pollen grains from the Mesozoic of Western Australia. *Coal Research, CSIRO, Technical Communication*, **25**, 1–48.

BALME, B. E. (1964). The palynological record of Australian pre-Tertiary floras. In *Ancient Pacific Floras*, ed. L. M. Cranwell, pp. 49–80. Honolulu: University of Hawaii Press.

BARLOW, B. A. (1981). The Australian flora: its origin and evolution. In *Flora of Australia*, vol. 1, pp. 25–75. Canberra: Australian Government Printing Service.

BARNBAUM, D. (1976). The geology of the Burrum Syncline, Maryborough Basin, southeast Queensland. *Papers of the Department of Geology and Mineralogy, University of Queensland*, **7**, 1–45.

BRENNER, G. J. (1976). Middle Cretaceous floral provinces and early migration of angiosperms. In *Origin and Early Evolution of Angiosperms*, ed. C. B. Beck, pp. 23–47. New York: Columbia University Press.

BRENNER, G. J. (1984). Late Hauterivian angiosperm pollen from the Helez Formation, Israel. *Abstracts, Sixth International Palynological Conference, Calgary, 1984*, p. 15.

BURGER, D. (1976). Cenomanian spores and pollen grains from Bathurst Island. *Bulletin of the Bureau of Mineral Resources, Geology and Geophysics Australia*, **151**, 114–54.

BURGER, D. (1981). Observations on the earliest angiosperm development with special reference to Australia. *Proceedings of the Fourth International Palynological Conference, Lucknow (1976–1977)*, **3**, 418–28.

BURGER, D. (1990). Early Cretaceous angiosperms from Queensland, Australia. *Review of Palaeobotany and Palynology*, **65**, 153–63.

CREBER, G. T. & CHALONER, W. G. (1987). The contribution of growth rings to the reconstruction of past climates. In *Applications of Tree-ring Studies*, ed. R. G. W. Ward, pp. 37–67. Oxford: BAR International Series 333.

DETTMANN, M. E. (1973). Angiospermous pollen from Albian to Turonian sediments of eastern Australia. In *Mesozoic and Cainozoic Palynology: Essays in Honour of Isabel Cookson*, ed. J. E. Glover & G. Playford. *Geological Society of Australia Special Publication*, **4**, 3–34.

DETTMANN, M. E. (1981). The Cretaceous flora. In *Ecological Biogeography of Australia*, ed. A. Keast, pp. 357–75. The Hague: Junk.

DETTMANN, M. E. (1986a). Early Cretaceous palynoflora of subsurface strata correlative with the Koonwarra Fossil Bed, Victoria. *Memoirs of the Association of Australasian Palaeontologists*, **3**, 79–110.

DETTMANN, M. E. (1986b). Significance of the spore genus *Cyatheacidites* in tracing the origin and migration of *Lophosoria* (Filicopsida). *Special Papers in Palaeontology*, **35**, 63–94.

DETTMANN, M. E. (1989). Antarctica: Cretaceous cradle of austral temperate rainforests? In *Origins and Evolution of the Antarctic Biota*, ed. J. A. Crame. *Geological Society of London Special Publication*, **47**, 89–105.

DETTMANN, M. E. & JARZEN, D. M. (1988). Angiosperm pollen from uppermost Cretaceous Strata of Southeastern Australia and Antarctica. *Memoirs of the Association of Australasian Palaeontologists*, **3**, 217–37.

DETTMANN, M. E. & JARZEN, D. M. (1990). The Antarctic/Australian Rift Valley: Late Cretaceous cradle of northeastern Australasian relicts? *Review of Palaeobotany and Palynology*, **65**, 131–44.

DETTMANN, M. E. & JARZEN, D. M. (1991). Pollen evidence for Late Cretaceous differentiation of Proteaceae in southern polar forests. *Canadian Journal of Botany*, **69**, 901–6.

DETTMANN, M. E., MOLNAR, R. E., DOUGLAS, J. G., BURGER, D., FIELDING, C., CLIFFORD, H. T., FRANCIS, J., JELL, P., RICH, T., WADE, M., RICH, P. V., PLEDGE, N., KEMP, A. & ROZEFELDS, A. (1992). Australian Cretaceous terrestrial faunas and floras: biostratigraphic and biogeographic implications. *Cretaceous Research*, **13**, 207–62.

DETTMANN, M. E., POCKNALL, D. T., ROMERO, E. J. & ZAMALOA, M. DE C. (1990). *Nothofagidites* Erdtman ex Potonié, 1960: a catalogue species with notes on the paleogeographic distribution of *Nothofagus* Bl. (southern Beech). *New Zealand Geological Survey Paleontological Bulletin*, **60**, 79 pp.

DETTMANN, M. E. & THOMSON, M. R. A. (1987). Cretaceous palynomorphs from the James Ross Island area, Antarctica – a pilot study. *British Antarctic Survey Bulletin*, **7**, 13–59.

DORMAN, F. H. (1966). Australian Tertiary paleotemperatures. *Journal of Geology*, **74**, 49–61.

DOUGLAS, J. G. (1969). The Mesozoic floras of Victoria. Parts 1 and 2. *Geological Survey of Victoria Memoir*, **28**, 310 pp.

DOUGLAS, J. G. (1973). The Mesozoic floras of Victoria. Part 3. *Geological Survey of Victoria Memoir*, **29**, 185 pp.

DOUGLAS, J. G. (1986). The Cretaceous vegetation, and palaeoenvironments of Otway Basin sediments. In *Second South-eastern Australia Oil Exploration Symposium*, ed. R. C. Glenie, pp. 233–40. Melbourne: Petroleum Exploration Society of Australia.

DOYLE, J. A., HOTTON, C. L. & WARD, J. V. (1990). Early Cretaceous tetrads, zonasulcate pollen, and Winteraceae. I. Taxonomy, morphology, and ultrastructure. *American Journal of Botany*, **77**, 1544–57.

DOYLE, J. A., JARDINÉ, S. & DOERENKAMP, A. (1982). *Afropollis*, a new genus of early angiosperm pollen, with notes on the Cretaceous palynostratigraphy and palaeoenvironments of northern Gondwana. *Bulletin des Centres de Recherches Exploration-Production Elf-Aquitaine*, **6**, 39–117.

FRIIS, E. M., CRANE, P. R. & PEDERSEN, K. R. (1988). Reproductive structures of Cretaceous Platanaceae. *Biologiske Skrifter*, **31**, 55 pp.

GOULD, R. E. (1975). The succession of Australian pre-Tertiary megafossil floras. *Botanical Review*, **41**, 453–83.

GREGORY, R. T., DOUTHITT, C. B., DUDDY, I. R., RICH, P. V. & RICH, T. H. (1989). Oxygen isotopic composition of carbonate concretions from the Lower Cretaceous of Victoria, Australia: implications for the evolution of meteoric waters on the Australian continent in a paleopolar environment. *Earth and Planetary Science Letters*, **92**, 27–42.

HARRIS, W. K. & TWIDALE, C. R. (1991). Revised age for Ayers Rock and the Olgas. *Transactions of the Royal Society of South Australia*, **115**, 109.

HEKEL, H. (1972). Pollen and spore assemblages from Queensland Tertiary sediments. *Publications of the Geological Survey of Queensland*, **355**.

HELBY, R. J., MORGAN, R. & PARTRIDGE, A. D. (1987). A palynological zonation of the Australian Mesozoic. *Memoirs of the Association of Australasian Palaeontologists*, **4**, 1–94.

HERNGREEN, G. F. W. (1973). Palynology of Albian–Cenomanian strata of Borehole 1–QS–1–MA, State of Maranhao, Brazil. *Pollen et Spores*, **15**, 515–55.

HILL, R. S. & MACPHAIL, M. K. (1983). Reconstruction of the Oligocene vegetation at Pioneer, northeast Tasmania. *Alcheringa*, **7**, 281–9.

HOOKER, J. D. (1860). Introductory essay. In *Botany of the Antarctic Voyage of H. M. Discovery Ships 'Erebus' and 'Terror', in the Years 1839–1843*. III. *Flora Tasmaniae*. London: Lovell Reeve.

HUGHES, N. F. & McDOUGALL, A. G. (1987). Records of angiospermid pollen entry into the English

Early Cretaceous succession. *Review of Palaeobotany and Palynology*, **50**, 255–72.

INGRAM, B. S. (1968). Stratigraphical palynology of Cretaceous rocks from bores in the Eucla Basin, Western Australia. *Department of Mines of Western Australia Report for 1967*, 102–5.

JARZEN, D. M. & DETTMANN, M. E. (1990). Taxonomic revision of *Tricolpites reticulatus* Cookson ex Couper, 1953 with notes on the biogeography of *Gunnera* L. *Pollen et Spores*, **31**, 97–112.

MARTIN, H. A. (1977). The history of *Ilex* (Aquifoliaceae) with special reference to Australia: evidence from pollen. *Australian Journal of Botany*, **25**, 655–73.

McLACHLAN, I. R. & PIETERSE, E. (1978). Preliminary palynological results: Site 361, Leg 40, Deep Sea Drilling Project. In *Initial Reports of the Deep Sea Drilling Project*, ed. H. Bolli, W. B. F. Ryan, B. K. McKnight *et al.*, **40**, pp. 857–81. Washington, DC: US Government Printer.

MULLER, J. (1981). Fossil pollen records of extant angiosperms. *Botanical Review* **47**, 1–142.

PARTRIDGE, A. D. & MACPHAIL, M. K. (1990). A proposal to establish a 'Stover & Williams' style catalogue of Mesozoic and Cenozoic spore-pollen in the Australian region. *Palynological and Palaeobotanical Association of Australasia Newsletter*, **21**, 5–7.

RAUNKIAER, C. (1934). *The Life Form of Plants and Statistical Plant Geography*. Oxford: Oxford University Press.

SCHNEIDER, S. H., THOMPSON, S. L. & BARRON, E. J. (1985). Mid-Cretaceous continental surface temperatures: are high CO_2 concentrations needed to simulate above-freezing conditions? In *The Carbon Cycle and Atmospheric CO_2: Natural variations Archane to Present*, ed. E. T. Dundquist & W. S. Broecher, pp. 554–9. Washington, DC: American Geophysical Union.

SCHOLTZ, A. (1985). The palynology of the upper lacustrine sediments of the Arnot Pipe. Namaqualand. *Annals of the South African Museum*, **95**.

SPECHT, R. L., DETTMANN, M. E. & JARZEN, D. M. (1992). Community associations and structure in the Late Cretaceous vegetation of southeast Australasia and Ant-

arctica. *Palaeogeography, Palaeoclimatology, Palaeoecology*, **94**, 283–309.

STEVENS, G. R. & CLAYTON, R. N. (1971). Oxygen isotope studies on Jurassic and Cretaceous belemnites from New Zealand and their biogeographic significance. *New Zealand Journal of Geology and Geophysics*, **14**, 829–79.

TAKHTAJAN, A. (1969). *Flowering Plants: Origin and Dispersal*. Edinburgh: Oliver & Boyd.

TAKHTAJAN, A. (1987). Flowering plant origin and dispersal: the cradle of the angiosperms revisited. In *Biogeographical Evolution of the Malay Archipelago*, ed. T. C. Whitmore, pp. 26–31. Oxford: Clarendon Press.

TAYLOR, D. J. (1964). Foraminifera and the stratigraphy of the western Vicorian Cretaceous sediments. *Proceedings of the Royal Society of Victoria*, **77**, 535–603.

TAYLOR, D. W. & HICKEY, L. J. (1990). An Aptian plant with attached leaves and flowers: implications for angiosperm origin. *Science*, **247**, 702–4.

TRUSWELL, E. M., KERSHAW, A. P. & SLUITER, I. R. (1987). The Australian–southeast Asian connection: evidence from the palaeobotanical record. In *Biogeographical Evolution of the Malay Archipelago*, ed. T. C. Whitmore, pp. 32–49. Oxford: Clarendon Press.

TWIDALE, C. R. & HARRIS, W. K. (1977). The age of Ayers Rock and the Olgas, central Australia. *Transactions of the Royal Society of South Australia*, **101**, 45–50.

VEEVERS, J. J. & EITTREIM, S. L. (1988). Reconstruction of Antarctica and Australia at breakup (95 ± 5 Ma) and before rifting (160 Ma). *Australian Journal of Earth Science*, **35**, 355–62.

WALKER, J. W. & WALKER, A. G. (1984). Ultrastructure of Lower Cretaceous angiosperm pollen and the origin and early evolution of flowering plants. *Annals of the Missouri Botanical Garden*, **71**, 464–521.

WEBB, L. J. (1959). A physiognomic classification of Australian rainforests. *Journal of Ecology*, **47**, 551–70.

WEBB, L. J., TRACEY, J. G. & JESSUP, L. W. (1986). Recent evidence for autochthony of Australian tropical and subtropical rainforest floristic elements. *Telopea*, **2**, 575–89.

9 Cretaceous vegetation: the macrofossil record

J. G. DOUGLAS

Modern Australian vegetation was cradled in the enormously long span of time termed the Cretaceous Period. The Early Cretaceous floras contained the elements that had persisted throughout the Mesozoic, but by the end of the period many of these had become extinct and the angiosperms had come into prominence. Some associations would have looked very like their present-day counterparts – the araucarian and podocarp conifers that had dominated the forests from the Permian, the tree fern glades, and the moss and liverwort cover on the fallen logs and shaded outcrop. Sediments preserved in over 20 major deposition basins (see Dettmann, Chapter 8, this volume) contain a record of this vegetation. The most pertinent are those of lacustrine or fluvial origin, but marginal marine beds also supplement the story. During the Early Cretaceous more or less continuous accumulation in several basins resulted in thick sedimentary piles, and major plant assemblages have been identified and used to nominate floral zones. There were marked differences between the floras of the dynamic intercontinental troughs in the south and those of the more stable inland depressions. The length of the period, 80 million years (Ma) by some calculations, easily the greatest time span of the Phanerozoic, necessitates recourse to the recognised subdivision or 'Stage' names, and these are presented in Figure 9.1. As the Late Cretaceous progressed, the inland basins became

MILLION YEARS	PERIOD	STAGE	MACROPLANT ZONES SE Australia
65	LATE CRETACEOUS	MAASTRICHTIAN	
75		CAMPANIAN	
84		SANTONIAN	
87		CONIACIAN	
88		TURONIAN	
90		CENOMANIAN	WAARRE FLORA
100	EARLY CRETACEOUS	ALBIAN	ZONE D
115		APTIAN	ZONE C
120		BARREMIAN	
125		HAUTERIVIAN	ZONE B
130		VALANGINIAN	
135		BERRIASIAN	
145		TITHONIAN	ZONE A

Figure 9.1 Stages (subdivisions) of the Cretaceous, with macrofossil zonation applicable to southeastern Australia.

drastically reduced, and much of our knowledge is from the sediments of the southeastern intercontinental basins that have been drilled extensively for hydrocarbons.

This chapter, dealing specifically with macrofloras, seems the logical place to emphasise that the palaeobotanist almost never has a whole plant available for examination, and that those portions which are preserved often provide very limited information. Much palaeo-environmental speculation is still based not on anatomical examination

[171]

Figure 9.2 *Cladophlebis* sp. cf. *C. australis*. Fern fronds are a feature of many leaf beds of Cretaceous age and these highly weathered compressions in a mudstone slab from Whitelaw, west Gippsland, Victoria, are of Neocomian (mid-Early Cre- taceous) age. The absence of fertile pinnae makes accurate allocation to present-day family difficult and inadvisable, and a form-genus name is used, which may cover several quite unrelated taxa. The scale bar represents 1 cm.

of this or that organ, nor on fruit or flower mor- phology, nor even on leaf anatomy, but on the impressions made by the laminae in the soft sedi- ments (Figure 9.2)! Microreproductive systems (spores and pollen grains) supply a great deal of information, especially in the elucidation of strati- graphic relationships. Other forms of fossilisation such as coalification or silicification may also add substantially to our knowledge, but because it forms the vast bulk of the fossilised material, foli- age, usually the leaf, forms the basis of most palaeobotanical studies. Leaf cuticle, offering a resistance to decay similar to that of spores and pollen grains, provides anatomical evidence of

affiliation and is often identifiable as minute frag- ments in macerations from samples such as bore cores.

It should be kept in mind that, although vegeta- tion and microreproductive organs in non-marine or marginal marine beds form the basis of our knowledge of the floras, beds deposited under marine conditions loom equally large if not larger in the sedimentary sequence. The name Cre- taceous in fact comes from the latin *creta* (= chalk), a material composed principally of the comminuted remains of marine invertebrates (shells). The 'type', or standard, Cretaceous sec- tions have been nominated in European marine beds and hence the most meaningful correlation is via invertebrate fossils, with the extinct cephalopod group called ammonities very prominent in this regard. The ages of spore–pollen (Chapter 8, this volume) and macroplant zones ultimately depend on correlation with these marine fossil zones. Finally, although the microfossil and macrofossil records are allocated separate chapters in this book, they provide complementary information that must be coordinated to facilitate definition of the Cretaceous vegetation.

EARLIEST CRETACEOUS (BERRIASIAN–VALANGINIAN)

The continent should be examined, at the beginning of the Cretaceous, in the context of Gondwana, a landmass composed of Australia, Antarctica, New Zealand and India, from which South America and Africa had earlier been detached (see Wilford & Brown, Chapter 2, this volume). Deposits in the great rift valley at the Australia–Antarctica conjunction, and extensive fluviatile and lacustrine beds in the Eromanga–Surat, Maryborough and Laura Basins (Dettmann *et al.*, 1992) provide most of our knowledge of the earliest Cretaceous biota.

Flora

All of the plant groups of the present-day contribute to the floral record of the earliest Cretaceous in Australia, with the notable exception of the angiosperms. It is possible that the dicotyledons at least may have been a component of upland assemblages, even though the environment of the oldest fossils (see below) does not support this suggestion. But this was the Mesozoic, separated from the present by an enormous time gulf of over a hundred million years, and with a host of species now extinct (and not just species, even orders), or certainly extinct on the Australian continent (see Figure 9.3). These oldest Zone A (Douglas, 1969) floras in the best-researched area, the Otway Basin of southeastern Australia, are little known because of poor preservation and fortuitous discovery in the older subsurface part of the suc-

cession. The Zone A element we know most about is the cycadeoids, represented as three leaf compression taxa reminiscent of classical Jurassic (e.g. Yorkshire) floras, which are considered to have been understorey components. In central Australia silcrete layers in the Algebuckina Sandstone, Eromanga Basin (Hopgood, 1987) suggest a canopy of conifers (*Brachyphyllum*) with understorey pteridosperms, cycads, cycadeoids, and cryptogams. Terrestrial ferns, an *Isoetes*-like fossil, and two common leaves occur here (*Rienitsia variabilis* and *Taeniopteris spatulata* (= *T. daintreei* ?Pentoxylales)), which have more affinities with the later (Zone B) floras of the southeast, but any correlation should take into account paucity of information on these earliest floras. Collections from the Surat Basin have not been studied in detail (Cantrill & Webb, 1987) but indicate a conifer and *Ginkgo* (*Ginkgoites digitata*) canopy over ferns including *Phyllopteroides*, an *Equisetum*-like sphenopsid and pteridosperms. Rozefelds (1988) reported the corystosperm *Pachypteris* (see below) and cryptogams from the Laura Basin, North Queensland. There have not been any detailed studies on the Cretaceous flora of Western Australia, but McLoughlin & Guppy (1993) commented on plant remains from the Canning, Carnarvon and Perth Basins.

NEOCOMIAN–BARREMIAN

Ten million years later, steadily rising sea levels resulted in submergence of the Eromanga Basin. In the immediate south, a lowland belt continued to receive marginal marine, lacustrine and fluvial deposits. High energy fluvial deposition continued in the southeast, and deposits in the Gippsland and Otway Basins provide a first good look at the macroplants.

Flora

Zone B floras have been described from bore cores obtained from several key sections, but the most accessible are the Boola Boola Forest localities in the oldest outcrop beds of the Strzelecki Group (Gippsland Basin). In the Boola Flora,

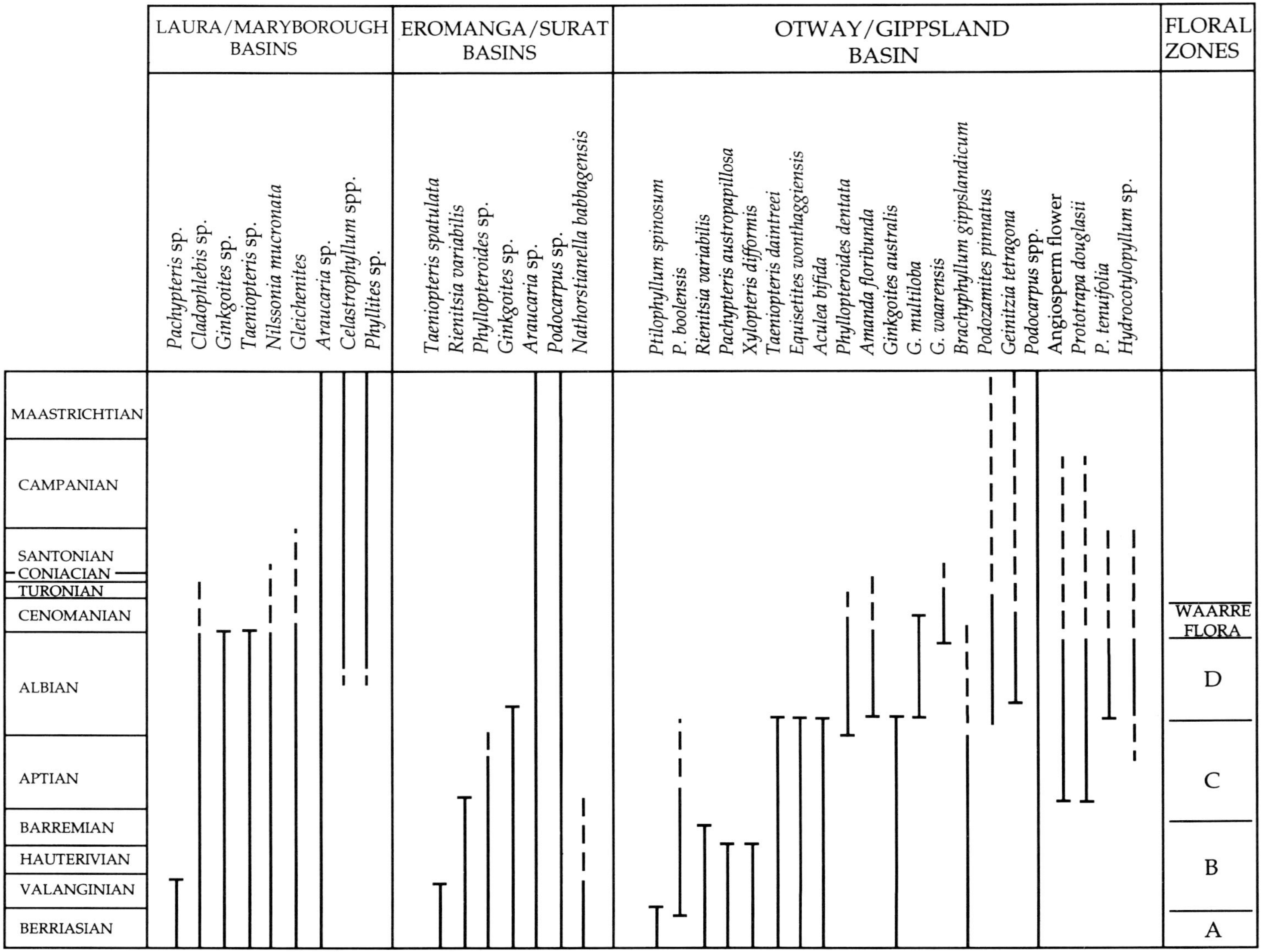

Figure 9.3 Relationship of selected Cretaceous macrofossil taxa (principally foliage) in the more intensively collected depositional basins in Australia. Dashed lines denote ranges suspected but unverified.

podocarps formed the canopy of the climax vegetation. Extremely digitate ginkgos may have formed a second storey, with well-preserved leaves of understorey pteridosperms and cycadeoids providing some information on these typical elements of the earlier Cretaceous assemblages (Figures 9.4, 9.5, 9.6(*a*)).

The pteridosperms have been loosely defined as 'gymnospermous plants, with leaves, pollen and seed organs built on a pinnate plan and reproductive organs not aggregated into cones or flowers' (Townrow, 1965). The relationship of several orders, for example Peltaspermales, Corystospermales and Caytoniales, is not clearly defined (Harland *et al.*, 1967). *Pachypteris austropapillosa* (Corystospermales), found as fern-like foliage with very thick cuticle and pinnae, is very similar to several other *Pachypteris* spp. from both the northern and southern hemispheres of Jurassic and earlier Cretaceous age. These leaves have cuticles so resistant and strong that they can often be peeled or easily released from their siltstone matrix. They belong to an otherwise unknown plant preserved in lenses of mud that settled in pools amid thunderous cascades and torrents sweeping from uplands fringing the Australian inland.

Xylopteris, another corystosperm with very thin leaves, forked and phyllode-like (e.g. *X. difformis*), appears to have been a very small herb. *?Rienitsia variabilis*, of uncertain affiliation (pteridosperm or cycad), is found in larger siltstone bands and seems to be derived from understoreys that flourished around pond margins. *Ptilophyllum boolensis* (Cycadeoidales), with beautiful fine pinnate leaves, has been interpreted as a semi-prostrate ground cover, or climber. For some reason cycadeoids are not preserved as silicified palm or cycad-like trunks as in many overseas localities, and the only indication of their presence is stem fragments bearing two or three pinnae, or disseminated cuticular debris. This is readily identifiable because of the syndetocheilic* nature

of the stomata, which are massed in interveinal rows on the underside of the pinnae, with the upper side also very readily recognisable because of the highly sinuous outlines of the cell walls. Apart from a few rare occurrences in parts of the section of disputed age, the order has not been found in younger assemblages in southeastern Australia. Leaf fragments (?*Ctenis coronata*) provide one of the very few and tantalising glimpses of the Cycadales in the Cretaceous of southeastern Australia, where the order is not otherwise sighted until the early Tertiary (see Christophel, Chapter 11; Carpenter *et al.*, Chapter 12, this volume).

Ginkgoites australis, a key zone fossil in younger beds of the southeast of the continent, appears for the first time in the fossil record, with *Taeniopteris* and *Phyllopteroides*, in the Surat Basin in Queensland. Cantrill & Webb (1987) regarded the Stanwell Coal Measures flora (Queensland), which also contains these two latter genera, as equivalent in age to the southeastern Zone B assemblage.

McLoughlin & Guppy (1993), suggested that the absence of ginkgophyte and certain pteridosperm index species from Western Australian assemblages indicated that the closest similarity of these was to the Victorian Neocomian–Barremian assemblages.

Small dinosaurs formed part of the biota in the Gippsland Basin at this time, notably *Allosaurus* sp. (Molnar *et al.*, 1981) and *Fulgurotherium australe* (Rich & Rich, 1989).

APTIAN

The sea transgressed still further until much of the continent was submerged: the Eromanga–Surat southern shoreline remained many hundreds of kilometres north of the Austro-Antarctic Trough, where fluvial and lacustrine systems drained the Eucla Basin and Great Australian Bight in the western portion and the Otway, Bass and Gippsland Basins in the east. Volcanism is one of many factors that may have made major contributions to floral change. The earlier of two major volcanic episodes proposed by Gleadow & Duddy (1981) at 126 Ma pre-dated by a very few

---

* The presence of a pair of subsidiary cells opposite the guard cells, the whole group appearing to be formed by the division of a single cell, enabling in almost all cases immediate distinction from other Mesozoic gymnospermophyta.

Figure 9.4 The Neocomian forest of Boola Boola, Victoria. Conifer (principally podocarp) stands dominate the hills surrounding a tumultuous stream carrying a heavy load of sediment from the neighbouring uplands. A *Ginkgo* grove of form akin to that of the present-day *Acacia melanoxylon* stands above the falls to the left. Tree ferns (*Cladophlebis*) are in the foreground. *Phyllopteroides* (?Osmundaceae), *Taeniopteris* (pteridosperm) and small cycads provide ground cover near the water. The cycadeoidean *Ptilophyllum boolensis* clambers around the rocks, with numerous other ferns on mudbanks. A leaping fish is the only indication of the extensive fauna. (From Douglas, 1983.)

Figure 9.5 The lowlands at Moonlight Head, Victoria. Twenty million years after the cascades at the margin of the Gippsland Basin piled up the boulders and rock debris that now comprises the rock unit called the Tyers Conglomerate at Boola Boola (Figure 9.4), the Early Cretaceous fluvio-lacustrine regime was nearing its end. The floodplain flora here again has tree ferns (*Cladophlebis*) with other ferns and several liverworts. *Phyllopteroides dentata* (?Osmundaceae) is prominent in several situations as a small bush with semi-weeping habit. Semi-aquatic ferns (*Amanda, Aculea, Adiantites* etc.) flourish in partly submerged banks, and the angiosperm *Prototrapa* trails in backwater pools. The conifers in the right background are *Agathis* like, framed by *Ginkgo* leaves. Dinosaurs were present somewhere in the environs. (From Douglas, 1983.)

(a)

(b)

Figure 9.6 (a) A large mudstone specimen from Boola Boola Forest, central Victoria, displays two types of identifiable leaf, an unidentified seed, a fragment of fern pinnule (bottom left) and nondescript stem and leaf debris. The pinnate leaf (across the top) is derived from a conifer (perhaps an undescribed podocarp) but is placed in the form-genus *Elatocladus* for convenience. The second leaf featured is *Taeniopteris daintreei*, allocated to the long extinct pteridosperms. A long, narrow specimen on the left contrasts with a much broader specimen on the right, which incidentally shows leaf apex form and incision in margin due perhaps to insect action. The age is Barremian (Early Cretaceous). (b) Another mudstone specimen, from the Bellarine Peninsula, Victoria, although still of Early Cretaceous age is considerably younger (probably Albian). This features leafy branches of the conifer *Bellarinea barklyi*. Scale bar represents 1 cm.

million years the disappearance of the Zone B assemblage in the late Barremian.

Flora

The Zone C flora is the best collected and documented Cretaceous assemblage in the continent (e.g. Figure 9.7). *Ginkgoites australis*, a leaf similar to *Ginkgo biloba*, but more digitate, occurs over much of the outcrop assigned to this zone, and its cuticular debris is readily distinguishable in bore core macerations, for example in the Wonthaggi coalfield, where there was an intensive search for bituminous coal at the turn of the century. But the most common fossil, a plant that thrived in a range of environments, is a spatulate Pteridosperm (Pentoxylales) leaf, up to 200 mm in length, *Taeniopteris daintreei*. Like most fossil leaves this is most commonly found in carbonaceous mudstones, or fine sandstone. It is very obvious in the pink-white highly weathered siltstone road cuttings of the elevated areas of the Strzelecki Ranges in the Gippsland Basin, where it is often the only species fossilised. Sometimes it is accompanied by masses of seeds. The pollen-bearing organ ascribed to this plant, *Sahnia laxiphora*, has been recognised from two localities, among which the Koonwarra Fish Beds is easily the best known. This Zone C locality, which comprises fine lamellae of siltstone deposited during

Figure 9.7 Identification of fossil taxa using morphological criteria only may lead (and has led) to incorrect diagnoses. In the case of fossil vegetation, anatomical evidence (obtained in several ways depending on the form and state of preservation) usually results in much more satisfactory understanding of allegiance. This figure attempts to convey the value of maceration and chemical clearing processes for facilitating examination of compression fossils either already visible or buried undetected in the rock.

(a) Leaf tip of an unnamed conifer (perhaps with similar biseriate leaf arrangement to the conifer illustrated in Figure 9.6(a)). Illustrated are the adhering upper and lower cuticles. From the Merriman Bore 1 South Gippsland, Victoria. The age is Early Cretaceous (Aptian). Scale bar represents 250 μm. (b) Unidentified sac, Moonlight Head, Victoria. The age is Early Cretaceous (Albian). Scale bar represents 250 μm. (c) It has been noted in the text that knowledge of the vegetation of the Late Cretaceous in Australia is largely based on leaves derived from bore core obtained from a sedimentary

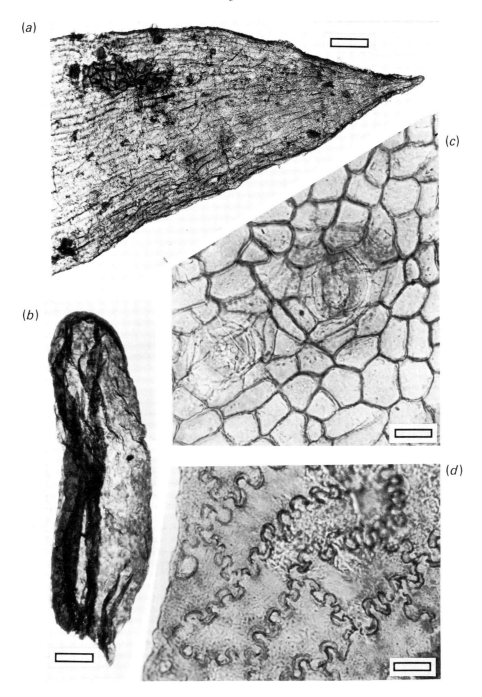

(a)

(c)

(b)

(d)

sequence in western Victoria that does not outcrop. Illustrated are stomata from a leaf macerated from core at 1500 m in Port Campbell Bore 4. These relatively large stomata are sunken in pits. Nature, size, arrangement etc. of stomata are key criteria for recognition and establishment of affiliation of such material. The age is (Late Cretaceous) Cenomanian. Scale bar represents 20 μm. (d) Characteristic 'jigsaw'-like cell wall thickening on the upper surface cuticle of a *Ptilophyllum* pinna, a representative of the extinct cycadeoides. From outcrop in the tidal zone, Sunnyside Road Beach, Mornington, Victoria. The age is Early Cretaceous (Barremian). Scale bar represents 20 μm.

episodic flooding of a backswamp (Douglas & Williams, 1982), is the subject of two major publications (Waldman, 1971; Jell & Roberts, 1986), and is one of only two fossil localities included in the Register of the National Estate in Victoria. Waldman described five fish species including two new genera. Jell & Duncan (1986) made a major contribution to entomology with a description of more than 70 insects, including 13 new genera. There are several other elements of the biota e.g. Crustacea, Arachnida, Clitellata, Phylactolaemata and Bivalvia, and the locality provides a notable addition to our knowledge of Cretaceous life. The flora is unfortunately represented only by impressions or compressions too highly weathered to yield anatomical information. Although *Taeniopteris daintreei* and *G. australis* are key fossils for Zone C and therefore for stratigraphic purposes, the main scientific interest in the flora is in a leafy specimen (Figure 9.8(*a*) and (*b*)) containing what is regarded by Taylor & Hickey (1990) to be the world's oldest flower, affiliated to the Magnoliidae, and referred to a plant of prostrate herbaceous, possibly aquatic habit.

Seeds tentatively allocated to the angiosperms (Douglas, 1963) are an enigmatic and important part of the nearly 60 plant taxa recorded from Koonwarra and the hundred or so recognised in Zone C. The unusual (but not unique in the eastern Australian Early Cretaceous) deposit of nearly 8 m of finely laminated siltstone, isolated between massive sandstone with no sign of similar deposit in hundreds of metres of more or less continuously exposed section, indicates a microenviroment within the prevailing ecosystem. The sediment and fossil content indicates quiet pools with aquatic dicotyledons, and ferns, flanked by mudflats with *Equisetum*, liverworts, fringing ferns and lycophytes. Near-stream pentoxylaleans (*Taeniopteris*), more xerophytic ferns, including osmundaceous types, and scrubby conifers appear to have prevailed in the lowland areas, with tall araucarians forming groves in the gullies. Other *Araucaria* species, podocarps and *Ginkgo* dominated the slopes and the hills, with mosses greening fallen trunks and damp rocky outcrops in shaded situations.

Elsewhere the best-documented flora is a meagre assemblage from the Maryborough Basin, eastern Queensland (Cantrill & Webb, 1987). The forest canopy here appears to have been formed by araucarian conifers and *Ginkgo*, with an understorey of elements not dissimilar to those of Zone C in the southeast, and indeed Cantrill & Webb extended this zonation up the east coast on the basis of this similarity. Far away to the northwest, at Bauhinia Downs in the Northern Territory Basin, cycads, cycadeoids, pteridosperms, conifers and ferns, as illustrated by White (1986), may be Aptian in age. Frakes & Francis (1988) described silicified wood in the western part of the Eromanga Basin (South Australia).

ALBIAN

The inland sea retreated from the Surat and Maryborough Basins leaving large areas of fringing swamps and lagoons. There was transgression over much of the Murray Basin and the sea in the Eucla Basin extended over the low-lying margins of the Otway Basin. The second volcanic episode (106 Ma) of Gleadow & Duddy (1981, see above) in the Otway Basin may have given impetus to the massive floral change marked by the incoming of the Zone D assemblages.

In the Otway Basin the Zone C/Zone D boundary features a major vegetational change, with the disappearance of several plant groups (Figure 9.2). Fossil assemblages are dominated by the osmundaceous fern *Phyllopteroides dentata* (often fossilised in company with its spore-bearing organ *Cacumen*) and a new suite of araucarian and podocarp conifers (Figure 9.6(*b*)). Cantrill (1989, 1991) described several Araucariaceae, comprising seven foliage, one wood, and nine fertile organ taxa, three Podocarpaceae, with one foliage and two wood taxa, and three Taxodiaceae, with one foliage, one root and one wood taxa. *Geinitzia tetragona* (Taxodiaceae), associated with mycorrhizal rootlets in a palaeosol, is considered to indicate nutrient deficiency (Cantrill & Douglas, 1988). This 'Dentata' Flora vegetation prevailed over a much more subdued relief with floodplain prominent. *Ginkgoites australis* has not been recorded,

(a)

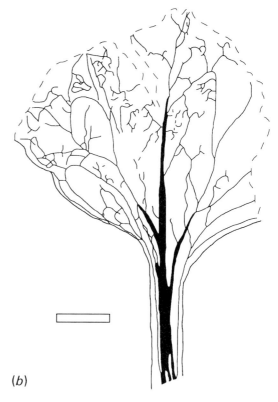

(b)

(c)

Figure 9.8 (a) 'The world's oldest flower', from the Koon-warra fish beds, Victoria (Aptian). These 'very small leaves and attached female inflorescences . . . are the oldest unequivocal angiosperm reproductive structures' (Taylor & Hickey 1990). They were previously described and figured by Drinnan & Chambers (1986) as 'Marsileales? indet.', and there was a suggestion that the 'female inflorescences' represented a possible sporocarp, not conclusively attached to the leaves. The identification of Taylor & Hickey, with discussion of possible present-day affiliates and the form of the earliest angiosperms, received a great deal of publicity in the daily press. Scale bar represents 2 mm. (b) Drawing of the leaf shown in (a). Modified from Taylor & Hickey (1990). Scale bar represents 2 mm. (c) Angiosperm leaf referred to *Hydrocotylophyllum lusitanicum*. Some of the best material referred to very early angiosperms has been obtained from bore core in otherwise inaccessible sediment. This tiny leaf, regarded as a possible aquatic plant, was noted when a cylinder of rock from the coring device was split after carbonaceous traces were evident. The edge of the core cylinder is on the right. From the Yangery Bore 1, 1317 m, near Warrnambool, Victoria. The age is Early Cretaceous (Albian). Scale bar represents 1 cm.

being replaced by less digitate smaller-leaved species (Table 9.1). Terrestrial and aquatic ferns and hepatics prevailed in damp environments. Two small ferns with very long, narrow pinnae, *Alamatus bifarius* and *Amanda floribunda*, formed part of the community around the mud banks. *Alamatus* bore oval sori in rows on each side of a main vein. *Amanda* had unusual fertile spikes with sori in elongate clusters on modified pinnae. The angiosperms, although still rare at most outcrop localities, are represented by crenate-leaved species, including *Hydrocotylophyllum lusitanicum*, (Figure 9.8(*c*); regarded as aquatics), the first lanceolate leaf (Medwell, 1954*b*) and cuticular debris in bore core.

Walkom (1919) described angiosperm leaves (?*Celastrophyllum* and *Phyllites* sp.) from the Styx River Coal Measures in Queensland, but *Taeniopteris*, an indicator of Aptian or earlier deposits in the southeast, survives in this assemblage.

The Burrum assemblages of the Maryborough Basin in Queensland, which is slightly older than the Styx River assemblage, contains equisetaleans and *Ginkgoites australis*, cycads and Araucariaceae similar to those of the southeastern Zone D 'Dentata' Flora. Rozefelds (1986) reported the cycad *Nilssonia mucronatum*. A controversy has arisen out of the implications of palaeomagnetic measurements used to plot a wander path for the South Pole. Data from late Albian determinations (100 Ma), for example, place Otway Basin localities in the latitude range 70–85° S. This is within the present-day polar icecap but, more to the point, in view of the evidence that this condition is ephemeral, is a situation with several months of polar night. Therefore, if the palaeomagnetic determinations are correct, the forests of this part of the Cretaceous at least must have existed under a regime totally foreign to our experience at the present time, and the concept raises many questions concerning photosynthesis, transpiration, reproduction, etc., in plants, and a multitude of similar questions with the fauna. Axelrod (1984) advocated that if temperature conditions in particular were suitable, a forest of deciduous taxa might well have flourished in such an environment. Douglas & Williams (1982) found no mor-

Table 9.1. *Lamina area of selected Early Cretaceous leaves from southeastern Australia*

Taxon	Average lamina area (cm²)
Ginkgoites australis (Koonwarra Fish Beds)	17
Ginkgoites australis (various localities)	11
Ginkgoites waarrensis (Port Campbell Bore 1)	9
Taeniopteris daintreei (various localities)	25
Ptilophyllum boolensis (various localities)	0.04
Phyllopteroides dentata (various localities)	6
Podozamites sp. (various localities)	0.6
Hydrocotyllophyllum sp.? (Otway Basin bore core)	1.5

All taxa with the exception of *Podozamites* sp. are discussed in the text. The *P. boolensis* measurement pertains to individual pinnae on a leaf of average length 6cm.
Modified from Douglas, 1969.

phological features in the plants indicating adaptation to such a specialised environment and could not accept growth, reproduction and the apparently normal interaction of the flora and the other elements of the biota in a forest-of-the-night context. They advocated a reduction in angle of the earth's obliquity (tilt) as a mechanism that would reduce the polar night duration.

The role of the leaf in palaeobotany was outlined earlier. Details of leaf size and form in the Early Cretaceous of southeastern Australia are included in Douglas (1969). Two features were emphasised.

1. There is a preponderance of species with small leaves.
2. Many species bear well-developed papillae on leaf and stem.

These observations have been used for palaeoclimate speculation, and extended to explain floral change within the period. For example, the replacement of the cycadeoids of Zone A bearing leaves small in comparison with those of classic Mesozoic floras (cf. Yorkshire Jurassic) by plants with tiny leaves in Zone C has been interpreted as the result of the deterioration of a subtropical regime. Leaf size of some common species is shown in Table 9.1.

LATE CRETACEOUS CENOMANIAN

At the end of the Albian a major change in environmental pattern saw the retreat of the sea from much of the Eromanga Basin and associated basins, and the cessation of deposition in the Otway and Gippsland Basins, which had proceeded without major interruption from the beginning of the Cretaceous. The invasion of the sea in the southeast was the commencement of cycles of transgression and regression (Bock & Glenie, 1965) that continued there throughout the Tertiary, and the quartz sandstone that replaced the felspathic sandstone and mudstone of the Early Cretaceous gives evidence of another major floral change.

Flora

Unfortunately dispersed cuticle supplemented by fortuitously obtained vegetable organs, mainly leaves (plus of course the microreproductive elements, see Figure 9.9) is the only floral record obtainable from this extensive Late Cretaceous section because it is covered by overlying Tertiary sediments and basalts. Over the last 40 years, however, drilling for hydrocarbon and groundwater resources has provided a great number of bore cores containing information on the biota. The drastic change in environment devastated the existing plant communities, particularly the riparian. Fern communities changed (*Phyllopteroides* disappears from the fossil record), and araucarian dominance of the canopy is evidenced by plentiful foliage and carpets of resin. Small leaves of *Ginkgoites waarrensis* show that the Ginkgoales still survived, and a recognisable but as yet still minor angiosperm component in the dispersed cuticles probably represents ground cover. In the Eromanga Basin, Queensland, the Winton Formation contains conifers including:

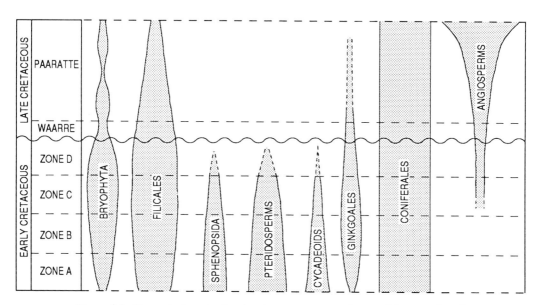

Figure 9.9. Change in the role of major plant groups as constituents of the vegetation of the Cretaceous of southeastern Australia. Dashed lines indicate suggested presence as relict or precursor species, although not found in the fossil record. (Modified from Douglas, 1969).

Austrosequoia wintonensis (Peters & Christophel, 1978), a cone considered to belong to the Taxodiaceae; a cycad; *?Lygodium*; sphenopsids (the youngest on the continent); and the only Late Cretaceous angiosperm leaves documented from outcrop in Australia. Regional differences in the vegetation are illustrated in the spore–pollen record. Dinosaurs (Thulborne & Wade, 1984), other reptiles, vertebrates and insects are also recorded from the Winton Formation.

TURONIAN–SANTONIAN

The sea had retreated from the Eromanga Basin and the residual swamps drained to the newly opened depressions to the south. The cycle in this marginal area was transgressive, with the sea inundating all but the uplifted parts.

Flora

There has been nothing described from these or any other beds in the way of macrofossils: cuticular fragments of conifer leaves and occasional angiosperms in the Otway Basin give little idea of understorey components, and our knowledge depends on the microfloras (Dettmann, Chapter 8, this volume).

CAMPANIAN–MAASTRICHTIAN

Australia retained its link with Antarctica through Tasmania and there was a regressive regime with nonmarine deposition in the Otway and Gippsland Basins.

Flora

Angiosperms become prominent in the cuticular debris, but there have been no macroplants described from the youngest Cretaceous sediments.

CRETACEOUS/TERTIARY BOUNDARY EVENT

There are no well-exposed Cretaceous/Tertiary boundary sections in Australia, and hence no evidence, in the form of an iridium layer or otherwise, of a catastrophic event. Bore core macerations and microfloras (Dettmann, Chapter 8, this volume) indicate that many of the angiosperm families that achieved prominence as the Tertiary proceeded were present in the latest Cretaceous. Although there has been no systematic examination of core from the 40 or so bores that may provide information on Cretaceous/Tertiary boundary events in the Otway Basin, preliminary observations reveal no floral changes anywhere as marked as that between Zones C and D or Zone D and the Waarre (Cenomanian) Flora.

RESUMÉ OF AUSTRALIAN CRETACEOUS MACROPLANTS RECORDED

The following is intended to give an appreciation of what we know of Australian Cretaceous plants. It is not feasible, and certainly would be misleading, to list every collection, some from sediments dubiously Cretaceous, and often named by collectors with little knowledge of the ancient floras.

Algae

No algae have been described from the nonmarine Cretaceous sediments that have yielded the bulk of the macroplants. Douglas (1981) suggested that markings and compressions in Albian beds of the Otway Basin might originate from algae.

Fungi

Florin (1952) allocated fungi found on a conifer leaf cuticle to the Pyrenomycetes. Douglas (1973) described and figured four Early Cretaceous fungi, and placed two Late Cretaceous epiphyllous fungi in the Microthyriales.

Hepatophyta

Liverworts have been collected from many eastern Australian Cretaceous localities but are well documented only from the southeast, where they sometimes occur massed along bedding planes that contain few other plants. Drinnan & Chambers

(1986) described *?Jungermannia*, *?Plagiochila* (Jungermanniales), *Riccardia koonwarriensis* (Metzgeriales), *Dendroceros victoriensis* (Anthocerotales), *Hepaticites undulatus*, *Hepaticites* sp., and *Thallites* sp. (Incertae Sedis) from the Koonwarra Fish Beds (Aptian).

Douglas (1973) described fertile structures and sporomorphs from *Hepaticites discoides* from Tullich Bore 1, 470 m (Otway Basin).

Bryophyta

Mosses, although probably fossilised at many localites, have very rarely been recognised, often because the nature of the leafy shoot results in only a faint impression or fragile compression, or because they are obscured by larger vegetative organs. Drinnan & Chambers (1986) described two sporophyte and two gametophyte types from the Koonwarra Fish Beds.

Lycophyta

Lycopodiales

The sporomorph story (Dettmann, Chapter 8, this volume) indicates that this group is strangely underrepresented in macrofossil collections. Seward (1904) described *Lycopodites victoriae*, a small ?semiprostrate plant from the Gippsland Basin Neocomian.

Isoetales

Plants regarded as isoetalean have been described from the Koonwarra Fish Beds (Aptian) and Mount Babbage (Eromanga Basin) South Australia (?Berriasian). Unlike other cryptogams they occur only as rare specimens at each locality.

Sphenophyta

Along with the liverworts and ferns this cryptogam group may occur as masses of impressions and compressions, in this case usually as broken stems, to the exclusion of other plants. In southeastern Australia the sphenopsids are commonest in Gippsland Basin beds of Aptian age. They are not rare in other Zone C and some Zone B assemblages, but have not been found in Zone D. In Queensland they have been recorded from the older sequence (Berriasian) in the Surat Basin and the much younger Albian beds of the Burrum Formation, Maryborough Basin.

Pteridophyta

The fossil record of the ferns in Australia reached an apogee in the Early Cretaceous, and it is a rare assemblage that lacks some representative. Sterile and fertile foliage has been tentatively referred to a wide variety of families, although a great deal of the most common sterile foliage is included in form genera such as '*Cladophlebis*', or '*Sphenopteris*'. Silicified Osmundaceae stems of Neocomian or Aptian age from the Gippsland Basin were described by Medwell (1954a). Fertile foliage has yielded spores in more than a dozen cases, and several of these are attributed to families (see Table 8.1, pp. 151–3). The affiliation of many other specimens that have not yielded microspores has also been discussed. The morphology of *Gleichenites nanopinnatus* strongly suggests affinity with the Gleicheniaceae. *Amanda floribunda* appears to have much in common with *Anemia* (subgenus unspecified, Schizaeaceae), and *Alamatus bifarius* is reminiscent of the Parkeriaceae (J. Lovis, personal communication). Foliage found in association with megaspores has tentatively been assigned to aquatic ferns (Marsiliaceae).

Pteridospermophyta

The apparent extinction of the 'Pteridosperms' (Figure 9.9) at the Aptian/Albian boundary in southeastern Australia, and later in northern assemblages is a landmark in floral history. *Taeniopteris*, for example, is an ubiquitous element of fossil assemblages for 30 million years preceding the Albian, often to the exclusion of all other taxa, and its demise surely had a vast impact on landscapes.

Cycadeoidales

In the best-documented section (Otway and Gippsland Basins) this order of extinct Cycadopsida shows a marked diminution in importance as the Cretaceous progresses. The youngest Cycadeoidales recorded are from the Burrum Formation (Albian) of the Maryborough Basin, Queensland.

Ginkgoales

Foliage, and in some cases fertile organs (Drinnan & Chambers, 1986), very similar anatomically to *Ginkgo biloba* is found throughout the Early Cretaceous in Australia, and *Ginkgoites waarrensis* survived into the early part of the Late Cretaceous. In view of the survival of *Ginkgo* to the present day in the northern hemisphere, its extinction (compare also *Equisetum*) in Australia is an enigma.

Coniferales

Sporomorphs indicate that some ancient families, now extinct (e.g. Cheirolepidaceae) were present in the very earliest part of the Cretaceous, and that the Taxodiaceae played a part in local assemblages, but essentially the conifer story, best illustrated by foliage remains, is that of the Araucariaceae and Podocarpaceae. These families supplied the large trees of the period and continued to dominate the forest scene for tens of millions of years throughout the Tertiary. Silicified stems and small trunks are not un-common in Australia and have been described from the Albian deposits of the south (Cantrill, 1989).

Angiosperms

The appearance of the angiosperms in the Early Cretaceous is a landmark in palaeobotany. There has already been some discussion at appropriate places, but the following represents angiosperm and putative angiosperm identification. All these are regarded as dicotyledons. Despite intensive searching in pertinent outcrop localities in south-eastern Australia, no monocotyledon fossils have been recognised, and the components of the earli-est grasslands are unknown.

Aptian, Strzelecki Group *Prototrapa douglasii* Vas-siljev (1967). Gemmills Hill, Koonwarra, Victoria.

Prototrapa praepomelii Vassiljev (1967). Gemmills Hill, Koonwarra, Victoria.

Leaves with flower. Taylor & Hickey (1990). Koonwarra, Victoria.

?Angiosperm sp. 'b'. Douglas (1969). Koonwarra, Victoria.

Albian, Otway Group *Prototrapa tenuifolia* Vassil-jev (1967). (Syn. *Lappacarpus aristata* Douglas 1969.) Yangery Bore 1, Victoria.

Hydrocotylophyllum lusitanicum Teixeira. Douglas (1965). Yangery Bore 1, Victoria.

?Angiosperm cuticle type 'a'. Douglas (1969). Yangery Bore 1, Victoria.

Angiosperm sp. 'a'. Medwell (1954b). Killara Bluff, Victoria.

?Albian Styx River Coal Measures *?Celastrophyl-lum* cf. *hunteri* Ward. Styx Bore 3, Queensland.

?Celastrophyllum sp. Styx Bore 1, Queensland.

Phyllites sp. Walkom (1919). Styx Bore 3, Queensland.

Cenomanian, Waarre Formation Small crenulate leaves with cuticle. Unpublished, Fergusons Hill Bore 1, Victoria.

?Cenomanian, Winton Formation Lanceolate leaves, Douglas (1977). Rosebrook, Longreach; Portland Bore 3, McKonkeys Creek, Queensland.

CONCLUSION

Our knowledge of Australian Cretaceous macro-floras is based largely on the wonderful fossil record in the Gippsland and Otway Basins. This is supplemented by some excellent but largely unstudied material from the Queensland sea-board, and some central Australia and Northern Territory sites. We have little or no information from Tasmania, New South Wales and Western Australia, although the extent of marginal marine and continental deposits in the latter State make it very likely that after more precise dating of the mid-Mesozoic sections, vital information will be forthcoming. When the spore–pollen information is used in conjunction with the macroflora, as it always must, regional differences are apparent, the extent of which may never be fully known, but which can be assessed only after the Western Australian scene is clearer and the northern and eastern Queensland localities are studied by modern methods. The magnificent sequence in the Otway Basin, however, shows very clearly that over the immense time span of the Cretaceous there were several drastic, if not catastrophic, changes that locally saw the demise of the floral status quo, and the establishment of new communities.

I acknowledge the cooperation of Dr M. E. Dettmann in facilitating the coordination of chapters 8 and 9 on Cretaceous vegetation. I also thank my former employers of a lifetime, the Geological Survey of Victoria (Australia) for support in the earlier stages of compilation.

REFERENCES

AXELROD, D. I. (1984). An interpretation of Cretaceous and Tertiary biota in polar regions. *Palaeogeography, Palaeoclimatology, Palaeoecology*, **45**, 105–47.

BOCK, P. E. & GLENIE, R. C. (1965). Late Cretaceous and Tertiary depositional cycles in southwestern Victoria. *Proceedings of the Royal Society of Victoria*, **79**, 153–63.

CANTRILL, D. J. (1989). An Albian coniferous flora from the Otway Basin, Victoria, Australia. Ph.D. thesis, University of Melbourne.

CANTRILL, D. J. (1991). Broad leaved coniferous foliage from the Lower Cretaceous Otway Group, southeastern Australia. *Alcheringa*, **15**, 177–90.

CANTRILL, D. J. & DOUGLAS, J. G. (1988). Mycorrhizal conifer roots from the Lower Cretaceous of the Otway Basin, Victoria. *Australian Journal of Botany*, **36**, 257–72.

CANTRILL, D. J. & WEBB, J. A. (1987). A reappraisal of *Phyllopteroides* Medwell (Osmundaceae) and its stratigraphic significance in the Lower Cretaceous of eastern Australia. *Alcheringa*, **11**, 59–85.

DETTMANN, M. E., MOLNAR, R. E., DOUGLAS, J. G., BURGER, D., FIELDING, C., CLIFFORD, H. T., FRANCIS, J., JELL, P., RICH, J., WADE M., RICH, P. V., PLEDGE, N., KEMP, A. & ROZEFELDS, A. (1992). Australian Cretaceous terrestrial faunas and floras: biostratigraphic and biogeographic implications. *Cretaceous Research*, **13**, 207–62.

DOUGLAS, J. G. (1963). Nut-like impressions attributed to aquatic dicotyledons from Victorian Mesozoic sediments. *Proceedings of the Royal Society of Victoria*, **76**, 23–28.

DOUGLAS, J. G. (1965). A Mesozoic dicotyledonous leaf

from the Yangery No. 1 bore Koroit, Victoria. *Mining and Geological Journal*, **6**, 64–7.

DOUGLAS, J. G. (1969). The Mesozoic floras of Victoria. Parts 1 and 2. *Memoirs of the Geological Survey of Victoria*, **28**, 1–310.

DOUGLAS, J. G. (1973). The Mesozoic floras of Victoria. Part 3. *Memoirs of the Geological Survey of Victoria*, **29**, 1–185.

DOUGLAS, J. G. (1977). Cretaceous plant fossils from the Geological Survey of Queensland. Geological Survey of Victoria Unpublished Report 1977/44.

DOUGLAS, J. G. (1981). Fossil algae, Victoria, Australia. *Palaeobotanist*, **28–29**, 8–14.

DOUGLAS, J. G. (1983). *What Fossil Plant is That?* Melbourne: Field Naturalists Club of Victoria.

DOUGLAS, J. G. & WILLIAMS, G. E. (1982). Southern polar forests: the Early Cretaceous floras of Victoria and their palaeoclimatic significance. *Palaeogeography, Palaeoclimatology, Palaeoecology*, **39**, 171–85.

DRINNAN, A. N. & CHAMBERS, T. C. (1986). Flora of the Lower Cretaceous Koonwarra fossil bed (Korumburra Group) South Gippsland, Victoria. *Memoirs of the Association of Australasian Palaeontologists*, **3**, 1–72.

FLORIN, R. (1952). On two conifers from the Jurassic of south-eastern Australia. *Palaeobotanist*, **1**, 177–82.

FRAKES, L. J. & FRANCIS, J. E. (1988). A guide to Phanerozoic cold polar climates from high latitude ice rafting in the Cretaceous. *Nature*, **333**, 547–49.

GLEADOW, A. J. W. & DUDDY, I. R. (1981). Early Cretaceous volcanism and the early breakup history of south-eastern Australia: evidence from fission track dating of volcaniclastic sediments. In *Gondwana Five*, ed. M. M. Creswell & P. Vella, pp. 295–300. Rotterdam: A. A. Balkema.

HARLAND, W. B. *et al.* (1967). *The Fossil Record*. London: The Geological Society.

HOPGOOD, L. S. (1987). The taxonomic and palaeoclimatic interpretation of Late Mesozoic fossil floras from the southwestern Eromanga Basin. Honours thesis, University of Adelaide.

JELL, P. A. & DUNCAN, P. N. (1986). Invertebrates, mainly insects, from the freshwater, Lower Cretaceous, Koonwarra Fossil Bed (Korumburra Group), South Gippsland, Victoria. *Memoirs of the Association of Australasian Palaeontologists*, **3**, 111–205.

JELL, P. A. & ROBERTS, J. (eds.) (1986) Plants and invertebrates from the Lower Cretaceous Koonwarra Fossil Bed, South Gippsland, Victoria. *Memoirs of the Association of Australasian Palaeontologists*, **3**, 1–205.

McLOUGHLIN, S. B. & GUPPY, L. (1993). Western Australia's Cretaceous floras. *Fossil Collector*, **39**, 11–21.

MEDWELL, L. M. (1954a). A review and revision of the flora of the Victorian Lower Jurassic. *Proceedings of the Royal Society of Victoria*, **65**, 63–111.

MEDWELL, L. M. (1954b). Fossil plants from Killara, near Casterton, Victoria. *Proceedings of the Royal Society of Victoria*, **66**, 17–23.

MOLNAR, R. E., FLANNERY, T. F. & RICH, T. H. V.

(1981). An allosaurid theropod dinosaur from the early Cretaceous of Victoria, Australia. *Alcheringa*, **5**, 141–6.

PETERS, M. D. & CHRISTOPHEL, D. C. (1978). *Austrosequoia wintonensis*, a new taxodiaceous cone from Queensland, Australia. *Canadian Journal of Botany*, **56**, 3119–28.

RICH, T. H. & RICH, P. V. (1989). Polar dinosaurs and biotas of the Early Cretaceous of south eastern Australia. *National Geographic Research*, **5**, 15–53.

ROZEFELDS, A. C. (1986). Type, figured and mentioned fossil plants in the Queensland Museum. *Memoirs of the Queensland Museum*, **22**, 141–53.

ROZEFELDS, A. C. (1988). Lepidoptera mines in *Pachypteris* leaves (Corystospermaceae: Pteridospermaphyta) from the Upper Jurassic–Lower Cretaceous Battle Camp Formation, North Queensland. *Proceedings of the Royal Society of Queensland*, **99**, 77–81.

SEWARD, A. C. (1904). On a collection of Jurassic plants from Victoria. *Records of the Geological Survey of Victoria*, **1**, part 3, 155–211.

TAYLOR, D. W. & HICKEY, L. J. (1990). An Aptian plant with attached leaves and flowers: implications for angiosperm origin. *Science*, **247**, 702–4.

THULBORNE, R. A. & WADE, N. (1984). Dinosaur trackways in the Winton Formation (mid-Cretaceous) of Queensland. *Memoirs of the Queensland Museum*, **21**, 413–577.

TOWNROW, J. A. (1965). A new member of the Corystospermaceae Thomas. *Annals of Botany*, **29**, 495–511.

VASSILJEV, V. N. (1967). New species of Trapaceae. *Palaeontological Journal, Academy Nauk USSR*, **2**, 107–12.

WALDMAN, M. (1971). Fish from the freshwater Lower Cretaceous of Victoria, Australia, with comments on the palaeoenvironment. *Special Papers in Palaeontology*, **9**, 1–124.

WALKOM, A. B. (1919). Mesozoic Floras of Queensland. Parts 3 and 4. The flora of the Burrum and Styx River Series. *Publication of the Geological Survey of Queensland*, **263**, 1–76.

WHITE, M. E. (1986). *The Greening of Gondwana*. Balgowah, NSW: Reed.

10 Early Tertiary vegetation: evidence from spores and pollen

M. K. MACPHAIL, N. F. ALLEY, E. M. TRUSWELL & I. R. K. SLUITER

This chapter reviews palynological evidence for the nature of the early Tertiary flora and vegetation of Australia. Because of difficulties in distinguishing between Late Oligocene and Early Miocene palynofloras, the interval of time covered is Paleocene to late Early Miocene, 65 to *ca* 18.5 million years (Ma) based on the geochronological time scale of Harland *et al.* (1990).

The period is critical in tracing the origins and rise of the modern Australasian vegetation from an early, diverse angiosperm flora, such as that sampled in rift valley sequences along the southern margin of Australia during the Maastrichtian. Whether this flora was representative of inland regions or the northwest margins is debatable (see Twidale & Harris, 1977; Harris & Twidale, 1991), but it is clear that during the Danian a floristically more simple vegetation dominated by conifers and ferns prevailed in coastal/lowland southern Australia. In most general terms, the subsequent history of the early Tertiary vegetation is the rise to prominence of floristically complex nonseasonal mesothermal–megathermal forest types and their time-transgressive replacement by more open or seasonal mesothermal–microthermal types during the Miocene. The same period saw the final separation of Australia from Antarctica, its northward drift through some 20 degrees of latitude and an irregular but overall decline in global high latitude sea surface temperatures of *ca* 13 °C from an Early Eocene maximum.

The last major reviews incorporating evidence for the early Tertiary vegetation (Barlow, 1981; Lange, 1982) concentrated upon individual elements within the flora, utilising cytogenetic, cladistic and other phylogenetic studies to augment but also to overcome deficiencies in the fossil database. Since these reviews, the substantial increase in the volume of published and unpublished information allows plant fossils to be used as primary evidence for the history of the early Tertiary flora and vegetation. Not surprisingly, this vegetation is found to have been as heterogeneous in space and labile in time as that of the Quaternary. Accordingly, this chapter concentrates on the fossil record per se, focussing on the sites and on the spore and pollen sequences that are available and what these imply, rather than adopting the broader approach of earlier reviews.

Perhaps the major weakness of our approach is that only rarely can fossil pollen and spores be identified with extant taxa beyond the level of genus and usually not below that of tribe or family. At this level of taxonomic resolution a large number of species, and wide ecological adaptation and environmental tolerance are represented normally by one fossil species. For this reason, the fossil species and relative abundance data upon which the reconstructions are based have been made as explicit as possible. Even so, only a selection can be listed: taxa that have not been cited may prove to be of greater ecological or phytogeographical significance as knowledge of the evolutionary relationships between the fossil and modern pollen floras improves.

Due to the complexity and unfamiliarity of the fossil names, all palaeofloras are discussed in terms of living analogues. This is done with considerable misgivings. To counter the unavoidable impressions of modernity: (1) where necessary, fossil species names are included in parentheses; (2) the review of the fossil data is prefaced by an analysis of the strengths and weaknesses of Tertiary pollen analysis per se; and (3) the interpreted taxonomic affinities are given in Appendix 10.1.

Macrofossil data that complement the palynological evidence at a number of sites in southern Australia are discussed by Christophel (Chapter 11), Carpenter et al. (Chapter 12) and Hill (Chapter 16, this volume). That the two lines of evidence are often difficult to reconcile (see Lange, 1982; Hill & Macphail, 1983; Macphail et al., 1991) merely restates the observations that (1) different areas of vegetation are usually represented and (2) good preservation of foliage, flowers or fruits is no guarantee that miospores of that taxon will also be preserved or able to be identified in the fossil record.

In common with other reviews of Tertiary floras and vegetation, the discussion is limited by the reliability of the chronostratigraphic framework against which evolutionary events are measured. The biassed distribution and uneven sampling of geological sections are as much a fact of life in Australia as elsewhere. For example, there are no reliably dated Tertiary palynofloras older than Late Oligocene from Queensland: three Tertiary palynofloras are known from the Northern Territory and only one younger than Eocene from northwest Western Australia.

What is certain is that few, if any, early Tertiary plant communities have any close analogue in the modern Australasian vegetation (see also Carpenter et al., Chapter 12, this volume), although a broad resemblance to the lowland-montane rainforests of New Caledonia is becoming increasingly probable. Nor are reconstructions based on any particular area likely to be applicable across Australia as a whole, except in excessively general

terms. Accordingly the floras and vegetation are discussed in chronological order in terms of four major geographical sectors in Australia: (1) Southeast Australia, comprising central-southern New South Wales, Victoria and Tasmania and including the major reference areas for the Australian Tertiary, the Gippsland and Murray Basins; (2) Northeast Australia, comprising northern New South Wales and Queensland; (3) the central regions of South Australia and the Northern Territory; and (4) Western Australia.

It is recognised that none is a natural biological region, but the lack of suitable deposits pre-empts a more selective approach, e.g. focussing outwards from areas of high endemism and/or taxonomic diversity such as northeastern Queensland, western Tasmania or southwestern Western Australia. Positions of basins and sites mentioned in the text are shown in Figure 10.1. Some overlap in discussion is unavoidable, given that several basins extend across State boundaries and the overwhelming majority of fossil taxa are shared between basins.

HISTORICAL DEVELOPMENTS

Although Tertiary plant fossils were 'tabled' before learned societies in the early nineteenth century, the systematic study of Tertiary pollen and spore assemblages, or palynofloras in Australia is entirely a twentieth century phenomenon. It was pioneered between 1945 and 1965 by Dr Isabel Cookson and arose from research on plant macrofossils and Sub-Antarctic floras (see Baker, 1973). Some 70 Tertiary pollen and spore species were described by Cookson and associates and used as 'tools' to date Tertiary deposits in areas as widely separated as the Northern Tablelands of New South Wales, Southwest Tasmania and Western Australia. These species were added to by Harris (1965a), leading to a working fossil base of ca 100 taxa by the late 1960s.

By this time sufficient stratigraphic boreholes had been drilled to allow broad-scale palynostratigraphies to be erected for the Otway Basin (Harris, 1971) and for Oligo-Miocene sections in eastern Queensland (Hekel, 1972). However, it

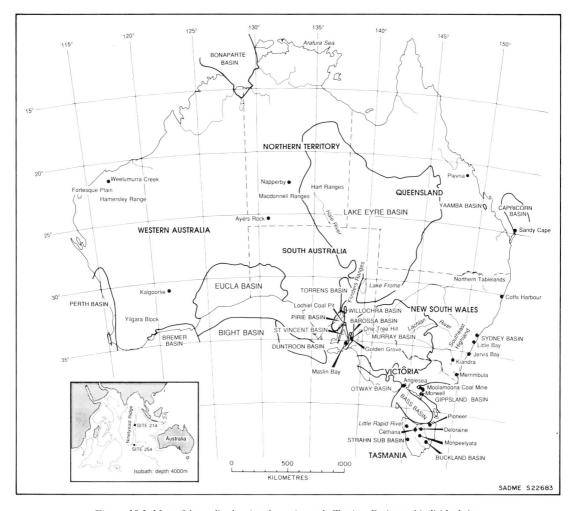

Figure 10.1 Map of Australia showing the major early Tertiary Basins and individual sites mentioned in the text.

was not until major oil fields in the offshore Gippsland Basin in the eastern Bass Strait were developed that a detailed palynological zonation scheme covering the Late Cretaceous and Tertiary was developed (Stover & Evans, 1973; Stover & Partridge, 1973; Partridge, 1976). Here, continuous subsidence has resulted in a better sedimentary record of the Late Cretaceous–Tertiary than in most other Australian basins and, deservedly, the Stover & Partridge zonation has become the standard schema against which palynosequences are dated and correlated in southern Aus-

tralia (including Tasmania and southwest Western Australia) and as far north as northern New South Wales and the Northern Territory. The zonation is supported by a well-developed geophysical and geological database.

A predictable outcome has been an upsurge in pollen-based reviews of the Australian Tertiary vegetation, complementing earlier reconstructions based on macrofossil and plant biogeographical evidence, e.g. reviews by Crocker & Wood (1947), Burbidge (1960), Melville (1975) and Smith (1975). These are divided unevenly between general overviews such as those by Martin (1978), Barlow (1981), Walker & Singh (1981), Lange (1982) and Kershaw (1988) and studies that have concentrated on particular aspects or controls, e.g. mangroves (Churchill, 1973), Euphorbiaceae (Martin, 1974), Aquifoliaceae (Martin, 1977), continental movements (Kemp, 1978; Truswell, 1981), fire (Kemp, 1981), brown coal floras (Luly *et al.*, 1980; Sluiter & Kershaw, 1982; Holdgate & Sluiter, 1991), Proteaceae (A. R. H. Martin, 1982), geographical barriers (H. A. Martin, 1982), semiarid and arid zone regions (Bint, 1981; Truswell & Harris, 1982; Sluiter, 1991), glaciation and the evolution of subalpine and alpine floras (Macphail & Hill, 1983; Macphail, 1986, Macphail *et al.*, 1991), Southeast Asian affinities of the flora (Truswell *et al.*, 1987), eucalypts (Martin & Gadek, 1988), salinity (Martin, 1989*a*) and mallee floras and environment (Martin, 1989*b*).

With improved chronological control, it is now evident that most fossil spore and pollen species have diachronous distributions across Australia. It is equally clear from the new concepts of sequence stratigraphy that, in coastal basins, eustacy has been critical in determining the sedimentary environment in which spores and pollen can be preserved and also in the positioning of zone boundaries within, and the extent of geological time represented by, particular sedimentary units. As a consequence, a number of studies are in progress to improve palynological dating of Tertiary sediments within and outside the Bass Strait region: it is to be expected that many of the interpretations in this chapter will be modified in the light of this research and the growing body of macrofossil evidence and ecophysical data on the response and competitive abilities of rainforest canopy species (e.g. Read, 1990).

ASSUMPTIONS

Two caveats are applicable to all pollen-based reconstructions of past floras and vegetation. Firstly, fossil pollen and spores are direct (albeit partial) evidence of past *floras* only with some indication of *abundance*. Secondly, in the absence of

macrofossil evidence such as stems, plant community *structure* is deduced by analogy when a fossil community appears to agree with a modern one. All vegetation reconstructions therefore are interpretations, limited not only by the usual complexities of spore and pollen production, dissemination and preservation but also by local phenomena including the geometry of the depositional environment. Unlike in the northern hemisphere (see e.g. Farley, 1990), no objective study has attempted to correlate pollen transport with specific Tertiary depositional environments in Australia. Nevertheless, as for the late Quaternary (see Birks & Birks, 1980, p.195), some six questions still can be validly asked of early Tertiary pollen data: (1) what taxa were present?; (2) what was the relative abundance of each?; (3) what plant communities were present?; (4) what space did each community occupy?; (5) what time did each community occupy?; and (6) what were the other factors operating in the ecosystem in that time and space? Not surprisingly there are few certain answers at present.

Taxonomic affinities

In common with all fossils, the further back in time, the less confidently can a spore or pollen type be related to a modern taxon except where the degree of morphological specialisation is exceptional (see Truswell & Marchant, 1986). For example, many rainforest families, including Annonaceae, Baueraceae, Cunoniaceae, Elaeocarpaceae, Eucryphiaceae, Eupomatiaceae, Meliaceac, Menispermaceae, Monimiaceae, Moraceae, Pittosporaceae, Rutaceae and Sterculiaceae are presumed to be represented amongst the large number of undescribed colpate (*Dicolpopollis, Liliacidites, Tricolpites*), colporate (*Rhoipites, Tricolporites, Tricolporopollenites, Tetracolporites*) and porate (*Tetrapollis, Triorites, Triporopollenites*) genera.

However, Australia is fortunate in that some important canopy and subcanopy species do produce distinctive pollen types, e.g. Araucariaceae, most genera of Podocarpaceae, *Acacia* (Mimosaceae), Casuarinaceae, Cupanieae (Sapindaceae), *Eucalyptus* species belonging to the 'bloodwood' group (Myrtaceae), *Nothofagus* (Fagaceae), *Quintinia* (Grossulariaceae), most genera of Proteaceae, and Sapotaceae. The same is true for some minor taxa that can be important in tracing plant migration routes after the break-up of Gondwana, e.g. *Fuchsia* (Onagraceae) and *Gunnera* (Gunneraceae) (Jarzen & Dettmann, 1989; Berry *et al.*, 1990).

Although the affinity of most widespread pollen types is known in broad terms, one unresolved problem concerns the common inaperturate granulate–tuberculate pollen types *Dilwynites granulatus* and *D. tuberculatus*. The identification of these as fossil members of the Lauraceae (tribe Cinnamomeae) by Harris (1965*a*) and Askin (1990) is rejected because Australasian species such as *Cinnamomum virens* have pollen with pointed spinules with cushion-like bases (see Raj & van der Werff, 1988), as do the tribe Perseeae (e.g. *Beilschmiedia, Cryptocarya, Endiandra* and *Litsea*). Although the alternative proposal, e.g. by Hill & Macphail (1983), that *Dilwynites* is a fossil member of the Araucariaceae is followed in this chapter, we specify whether it is this fossil genus or *Araucariacites* (see Appendix 10.1) that forms the basis for identifying the presence of Araucariaceae in the palaeovegetation. Modern samples confirm that Lauraceae pollen will not be found as fossils except under unusual conditions (Macphail, 1980).

As noted by Truswell & Harris (1982), to expect close matches between early Tertiary fossils and their descendants is to deny up to 65 Ma of evolution. The same is true of geographical variation, and subtle differences are observed in the pollen morphologies of many widely distributed pollen types, in particular *Nothofagus* and Proteaceae. Broad species definitions or amalgamated species (e.g. Stover & Evans, 1973) may conceal these differences.

Pollen production, dispersal and deposition

Tertiary spores and pollen are sedimentary particles and subject to the same sorting processes during transportation and deposition as modern palynomorphs. As such, present-day relationships between topographical relief, size of the depositional basin and the area and type of plant community represented by plant microfossils (Jacobson & Bradshaw, 1981; Solomon & Silkwood, 1986) are applicable to Tertiary situations. Assumptions that Tertiary species had the same pollen production and dispersal characteristics, habit(s) and ecological preferences are less certain. For example, there is a growing body of evidence (e.g. Suc & Drivaliari, 1991) that water has been the predominant transporting agency in coastal depositional environments. This includes bisaccate coniferous pollen types generally considered to be favoured by aerial transport.

Differences between the Cenozoic floras of Australia and temperate regions of the northern hemisphere are that: (1) in Australia, pollen transport, at least for major tree taxa, is relatively reduced due to the prevalence of insect and bird pollination; and (2) many wind-pollinated taxa are underrepresented in the fossil pollen record. The latter include shrub conifers such as *Microcachrys (Microcachrydites antarcticus)* and *Microstrobos (Podosporites erugatus)* and all rushes and sedges (*Cyperaceaepollis, Milfordia*).

By analogy with the modern Australian, New Guinean, New Zealand and Indo-Malaysian floras, the overwhelming majority of Tertiary plants will have been severely underrepresented, i.e. seldom represented by pollen or spores unless the source plants were present at, or upstream from, the site. This group includes most of the taxa now prominent in the canopy and understorey of Australasian warm temperate, subtropical and tropical rainforests and within sclerophyllous forests, woodlands, scrub and heath, although exceptions occur in many families, e.g. *Vesselowskya, Weinmannia* (Cunoniaceae), *Amperea* (Euphorbiaceae) and *Dodonaea* (Sapindaceae) (cf. Kershaw, 1988).

Rainforest taxa interpreted as being underrepresented include (fossil genera in parentheses): Cunoniaceae (*Concolpites leptos*), Elaeocarpaceae, Euphorbiaceae (*Malvacipollis, Polyorificites oblatus, Tricolporopollenites endobalteus*), Malvaceae (*Malvacearumpollis*), Mimosaceae (*Acaciapollenites, Polyadopollenites*), Myrtaceae (*Myrtaceidites eugenioides, M. parvus-mesonesus, M. verrucosus*), Proteaceae (*Beaupreaidites, Cranwellipollis, Propylipollis, Proteacidites*) and Rubiaceae (*Canthiumidites*). Underrepresented sclerophyllous taxa include all Fabaceae, most Epacridaceae (*Ericipites*), most Myrtaceae and Proteaceae including *Banksia/Dryandra, Grevillea/Hakea, Isopogon, Orites, Persoonia, Telopea* and most *Xylomelum* spp. Virtually all herbs fall into the same category, e.g. Scrophulariaceae (*Tricolpites trioblatus*), Liliaceae–Iridaceae (*Liliacidites*), Gentianaceae (*Striasyncolpites laxus*) and Onagraceae (*Corsinipollenites*).

Conversely a comparison of Cenozoic palynosequences indicates that the small class of well-represented taxa will have included some rainforest canopy and major subcanopy species. The most prominent examples are *Nothofagus* (*Brassospora*) spp. (*Nothofagidites emarcidus-heterus*). The type is overrepresented by pollen within the Latrobe Depression (Sluiter & Kershaw, 1982) and almost certainly at all Tertiary sites where the local flora was dominated by taxa producing and/or dispersing pollen in low amounts.

Examples of well-represented gymnosperms (many undoubtedly forming an overstorey in Tertiary rainforests) are: Araucariaceae (*Araucariacites australis, Dilwynites* spp.), Cupressaceae/Taxodiaceae, and Podocarpaceae such as *Dacrycarpus* (*Dacrycarpites australiensis*), *Dacrydium* Group B spp. (*Lygistepollenites florinii*), *Lagarostrobos*-type (*Phyllocladidites mawsonii*) and *Dacrydium/Podocarpus* (*Podocarpidites*).

Apart from *Nothofagus* (*Brassospora*), rainforest angiosperms that appear to have dispersed pollen widely are *Ascarina* (*Clavatipollenites glarius, C. australis*: Chloranthaceae), *Quintinia* (*Quintiniapollis*: Grossulariaceae), *Gymnostoma* (*Haloragacidites harrisii*: Casuarinaceae), Cunoniaceae (dicolpate, tricolpate species), Sapindaceae tribe Cupanieae (*Cupanieidites* spp.) and Sapotaceae (*Sapotaceoidaepollenites latizonatus, S. rotundus* and related undescribed species). Most are likely to have been subcanopy trees or shrubs.

Well-represented taxa of sclerophyllous shrubs and herbs include *Casuarinaceae* (included within *Haloragacidites cainozoicus, H. harrisii*), *Dodonaea* (*Dodonaea sphaerica, Nuxpollenites*), Asteraceae (*Tubulifloridites* spp.), halophytes such as Chenopodiaceae-Amaranthaceae (*Chenopodipollis chenopodiaceoides*) and reeds (*Aglaoreidia qualumis, Sparganiaceaepollenites*). Fern spores, in particular those of the tree fern families Cyatheaceae (*Cyathidites australis, C. paleospora* and related types) and Dicksoniaceae (*Matonisporites ornamentalis*), appear to be overrepresented in marine and fluvial sediments. *Eucalyptus* (*Myrtaceidites eucalyptoides*) pollen are widely dispersed in modest numbers but their prominence in recent sediments is likely to reflect widespread *Eucalyptus* sclerophyllous communities. A major anomaly are the Proteaceae, which now consist almost wholly of severely underrepresented species but which are a consistent component of Tertiary palynofloras.

Thus it is reasonable to expect that Tertiary palynofloras recovered from sediments accumulating in larger (> 200 m diameter) lakes, estuaries, or on the continental shelf will be biassed towards well-represented sources in the regional vegetation (including types readily transported by water), whereas palynofloras recovered from coals, lignites and other backswamp deposits will reflect a more local flora and include occasionally very high values of underrepresented taxa. This is strongly supported by a statistical analysis of brown coal lithofacies (Sluiter & Kershaw, 1982) that demonstrates that *Nothofagus* (*Brassospora*), Cunoniaceae and *Quintinia* are in fact better represented in open-water facies whereas Cyperaceae and Restionaceae are better represented in peat-swamp facies. The change in depositional environments in the Gippsland Basin from fluvio-deltaic to open marine is correlated with a marked decrease in spore–pollen abundance: the proportions of gymnosperms and ferns increases relative to angiosperms (A. D. Partridge & M. K. Macphail, unpublished data).

Reconstruction of the palaeovegetation and palaeoclimates

Whilst there is no *prima facie* case against equating pollen dominance with physical or numerical prominence of the source plant(s) in the surrounding vegetation, this remains a questionable assumption in the Tertiary. For example macrofossil data from two Oligo-Miocene sites in northern Tasmania demonstrate that *Lophozonia*, not *Brassospora*, spp. dominated local *Nothofagus* communities in spite of pollen data indicating the reverse situation (Hill & Macphail, 1983; Macphail *et al.*, 1991). More generally, as noted above, Tertiary palynofloras in Australia probably display a bias towards locally produced or water-transported spores and pollen, whilst some major canopy and subcanopy rainforest species have gone unrecorded by pollen and therefore will be 'blind-spots' in our reconstructions, e.g. Lauraceae, macrofossil remains of which are extremely common in the early Tertiary deposits in southern Australia (Carpenter *et al.*, Chapter 12, this volume). Different processing techniques may have biassed the relative abundance data. An extreme example of this is the near total loss of pollen types less than *ca* 10 μm in maximum diameter through ultra-sieving and equivalent techniques such as 'short-spinning', e.g. Cunoniaceae, Baueraceae, Elaeocarpaceae, Eucryphiaceae and many Myrtaceae including the important forest trees *Eugenia* and *Syzygium* (Macphail, 1991).

Since most plant families and fossil pollen taxa include a wide range of life forms, interpretations of structure in Australian Tertiary plant communities are the least certain of all. It is also important to note that some tall forest tree species, including *Nothofagus* and many Podocarpaceae, are capable of surviving as shrubs where adverse conditions exist: these and others occur within a number of structurally distinct associations, many of which have shown repeated synthesis and breakdown during Quaternary glacial/interglacial cycles (Walker & Flenley, 1979; Macphail & Hill, 1983).

The omission or inclusion of conifer and broad-leaved angiosperm species makes a considerable difference to the physiognomy and presumably the long-term dynamics of *Nothofagus* rainforest in New Guinea, New Zealand and South America. This calls into question the wisdom of categorising early Tertiary plant communities in terms of, for example, the detailed floristic and physiognomic classifications

developed for modern Australian rainforest by Webb and associates (Webb, 1959, 1968; Webb *et al.*, 1972, 1984) or the comprehensive structural classification developed by Specht (1970) based on life form, height and projective foliage cover of the tallest stratum. Nor has there been consistency in the use of terms such as tropical, subtropical and temperate, all of which have geographical as well as thermal connotations.

In recognition of this, Nix (1982) proposed an ecological classification that relates plant growth response and therefore competitive ability to the major environmental variables of light, air temperature and precipitation. Three thermal response groups are recognised (mean air temperature in parentheses): megatherm (>24 °C), mesotherm (>14 °C, <20 °C) and microtherm (<12 °C), linked by intermediate categories. Where moisture availability and light regimes are not limiting during one or more seasons, these correspond to the tropical, subtropical–warm temperate and cool–cold temperate categories, respectively.

The value of the Nix classification, adopted in this chapter, is that it allows tropical plant genera that are distributed primarily at mid (mesothermal) and high (microthermal) elevations on tropical mountains to be distinguished from those (megathermal) elements in the adjacent lowlands. It is emphasised that, although trends in air temperature are dictated by sea surface temperatures, the mean values will differ. For example, pollen data suggest megathermal species were present during the Early Eocene in southeast Australia (implying mean air temperatures >24 °C at palaeolatitude of *ca* 65° S), although sea surface temperatures 15 to 30 degrees of latitude to the north were at or below 18 °C at the same time (see Feary *et al.*, 1991).

Whilst much weight usually is placed upon taxa with modern megatherm–mesotherm affinities during reconstruction of the early Tertiary vegetation, the history of two genera now largely restricted to the tropics suggests that ecotypes had become adapted to seasonal/microthermal conditions during the Neogene. These are *Lophosoria* (*Cyatheacidites annulatus*: Lophosoriaceae), a fern confined mostly to tropical highlands in central South America but linked to cooling Oligo-Miocene climates in Australia and present on the Sub-Antarctic Kerguelen Islands during the Miocene (Cookson, 1947), and *Beauprea* (*Beaupreaidites elegansiformis*), a shrub now confined to New Caledonia but part of a cool–cold climate flora in western Tasmania during the Late Pliocene (M. K. Macphail, unpublished data). Accordingly, few detailed palaeoclimatic reconstructions can or should be made, based on the (assumed) ecological preferences of the (interpreted) canopy dominants.

DATING AND CORRELATION OF SEQUENCES

Geographical bias

Much of the information in this chapter comes from petroleum exploration wells, hydrogeological boreholes and commercially exploited lignite deposits in southeastern Australia, notably the Gippsland, Bass and Otway Basins in the Bass Strait, the Murray Basin straddling western New South Wales, Victoria and southeastern South Australia, and the Eucla and Bremer Basins within the Great Australian Bight (Figure 10.1). In the Bassian region, relatively continuous sedimentation has resulted in the accumulation of up to 4000 m of organic and clastic sediments during the early Tertiary but, with few exceptions, wide sample spacing does not allow the history of particular vegetation types to be traced in detail. Much of the recent data are proprietary or, if not, remain unpublished: estimates of relative pollen abundance are usually subjective. Conversely organic deposits are widespread onshore but are difficult to date with precision and usually represent very short periods of geological time.

Early Tertiary sediments exist in basins in northern Australia but remain unanalysed or, with few exceptions, the pollen data are unavailable. Records of probable Casuarinaceae (*Haloragacidites* sp.) and 'ancestral' *Nothofagus* (*Nothofagidites* sp. cf. *N. senectus*) in Late Cretaceous sediments from the Arafura Sea and Papua New Guinea

(R. Helby & M. K. Macphail, unpublished data) highlight the importance of this region in reconstructing patterns of plant migration within Australia during the early Tertiary.

Zone concepts

Zonation schema used to date and correlate Australian Tertiary palynosequences have been developed to meet the demands of industry. Not surprisingly, most zone boundaries are empirical, based upon a combination of first and last appearances of selected species (Concurrent Range zones), less frequently using quantitative criteria or associations between selected species (Oppel zones). Although the criteria tend to change over time, leading to the current situation where some *published* time-distributions are now unreliable indications of geological age, the zones per se have proved to be enduring (Figure 10.2).

Few palynological zone boundaries coincide with chronostratigraphic boundaries (see below), although in subsiding basins open to the sea, most appear to correlate with condensed sections – thin marine units that have accumulated very slowly during periods of maximum relative sea level rise and maximum transgression of the shoreline (see Haq *et al.*, 1987). Since factors influencing patterns of sedimentation within a basin (eustacy, subsidence, sediment supply and physiography) are variable over time, there is no reason to suppose that the sedimentary record is complete for any particular zone.

An example is the *Forcipites* (*Tricolpites*) *longus* Zone (see Dettmann, Chapter 8, this volume) which is considered by Helby *et al.* (1987) to be latest Early to Late Maastrichtian to possibly basal Danian. The geological reason is that global relative sea level was rising (transgressive system tract) during the period, resulting in a major condensed section within the basal Danian as defined by Haq *et al.* (1987). Due to very low depositional rates, it is probable that any palynofloras preserved in this condensed section will include recycled Late Cretaceous spores and pollen, blurring both the dating and the palaeobotanical record.

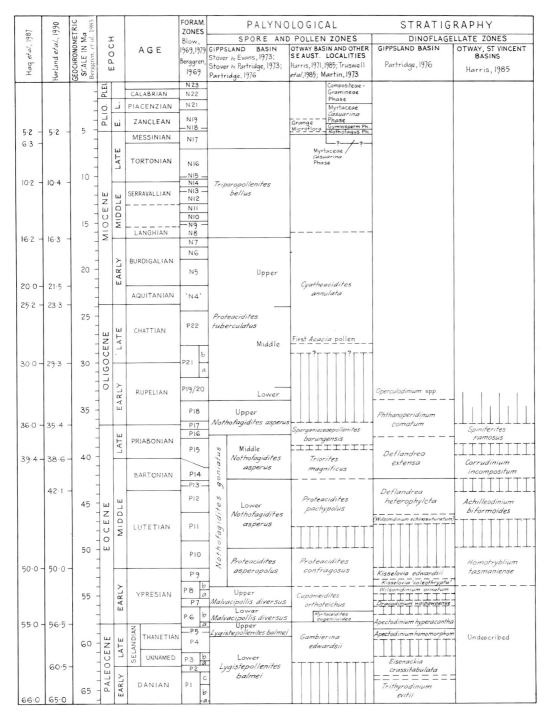

Figure 10.2 Correlation of palynological zones in southeast Australia against the global geochronometric scales of Berggren *et al.* (1985), Haq *et al.* (1987) and Harland *et al.* (1989). (Adapted from Macphail & Truswell, 1989.)

Geochronology

The Gippsland Basin zonation of Stover & Partridge (1973), subsequent proprietary revisions (A. D. Partridge & M. K. Macphail, unpublished data) and earlier regional schema of Harris (1971; Otway Basin) and Hekel (1972; Capricorn Basin) have been calibrated against the international time scale using planktonic foraminifera, marine dinoflagellates and, most recently, sequence stratigraphy (Taylor, 1971; Partridge, 1976; Harris, 1985; McGowran, 1989; Rahmanian *et al.*, 1990). These correlations are subject to the usual constant revisions. The geological time scale of Harland *et al.* (1990) is followed here, but the equivalent period proposed by Haq *et al.* (1987) is included in parentheses at the top of each section. Where marine faunas and floras are absent, e.g. the Yaamba Basin (Foster, 1982) or restricted in scope, e.g. the Murray Basin (Truswell *et al.*, 1985; Macphail & Truswell, 1989), the palynofloras are referred to zones or units defined in adjacent offshore basins. Widespread Tertiary volcanism has allowed some palynofloras in eastern Australia to be independently dated using K–Ar, although in most situations these provide only a maximum or minimum age limit, e.g. Owen (1988), Taylor *et al.* (1990), Pickett *et al.* (1990), Macphail *et al.* (1991). This approach is not usually possible in central and western Australia and here controls are restricted to broad ages determined by palaeomagnetism and pedological evidence in the form of laterites and silcretes (McGowran, 1979; Truswell & Plane, 1983; Truswell & Marchant, 1986).

ENVIRONMENTAL FORCING FACTORS

Climate is widely accepted as being the primary determinant of vegetation on the geographical scale or over periods of time exceeding several millennia. Except for plant communities subject to strong edaphic limitations, e.g. in coastal wetlands, or dependent on repeated tectonic disturbance, e.g. some *Nothofagus* rainforest communities, any sustained long-term shift in community dominance is likely to reflect regional changes in climate. For reasons that include the spatial heterogeneity of most forests and the different responses and resiliences of canopy species, the ecological effects of longer-term climatic change are likely to be time transgressive but systematic within a region. Over shorter periods, the role of climate is more ambiguous (see Walker & Flenley, 1979) and that of other factors such as soil fertility and fire more pronounced at the local level. Often these forces are associated via complex feedback relationships also operating on various scales (see Jackson, 1968*a*). The point is important, given that the time represented by a Tertiary palynoflora may be very short, both in absolute terms and relative to the time interval separating adjacent palynofloras in the typical, sparsely sampled geological section. As such, there is no guarantee that a particular flora reflects the climatically determined 'climax' formation of a region.

Climatic change in the Australian region

Changes in the geographical distribution of continents throughout the Cenozoic are a major factor in the evolution of oceanic circulation patterns and hence of climatic change, although other forcing factors such as atmospheric CO_2 concentration related to plate tectonics may be more important in the longer term (Barron, 1985; Rhea *et al.*, 1990; Barron & Peterson, 1991). Similarly, the development of significant topographical relief may exercise considerable control on local and regional climatic patterns. Events directly shaping climates and hence the evolution of the Australian Tertiary vegetation are:

1. The separation of Australia from Antarctica allowing the development of a strong Circumantarctic Current, an event which isolated Antarctica from warm oceanic gyres of equatorial origin, and thus created conditions conducive to the development of a major icecap. The associated steepening of latitudinal temperature gradients has produced the modern patterns of atmospheric circulation.

2. Uplift of the Eastern Highlands of Australia,

developing a significant regional relief by the beginning of the Tertiary. For example there may have been a regional relief of as much as 800 m by the Late Paleocene in southern South Wales (Taylor *et al.*, 1990), whereas mountains in Tasmania are thought to have achieved their present relief by Late Cretaceous times (Williams, 1989). The landscape in inland central and western Australia was more or less subdued with low river gradients: uplands were located in the same areas as now (see Harris & Twidale, 1991).

3. The development of increasingly seasonal rainfall patterns as Australia drifted into middle-low latitudes (Truswell & Harris, 1982) and the atmospheric circulation assumed its modern configuration.

The last factor is of particular importance because plant distributions are controlled not only by mean values but also by the seasonal distribution of rainfall and temperatures. For example Jackson (1968*b*) has argued that the important canopy species in microthermal rainforest, *Nothofagus cunninghamii*, requires at least 500 mm rainfall per summer month. The importance of the seasonal distribution of precipitation on the distribution of the major rainforest types is discussed by Webb (1968) and Nix (1982). Summer mean monthly temperatures above 9.5–10 °C are usually seen as the minimum required to support tree-dominated vegetation types but there is no reason to suppose that seasonal values fell below this minimum during the early Tertiary, except possibly on Tasmanian mountains. Less easily evaluated is the impact of extreme climatic events. Examples include prolonged drought and intense frosts. Hill (1982) has discussed the impact of drought on *Nothofagus* rainforest in western Tasmania and Woodroffe & Grinrod (1991) the impact of prolonged cold events on mangrove distributions.

Variation in the oxygen isotope composition of foraminiferal tests has allowed palaeotemperature curves to be constructed for shelf waters in southern and northeastern Australia (Kamp *et al.*, 1990; Feary *et al.*, 1991). These curves conform with global trends during the early Cenozoic, e.g. the earliest Eocene temperature maximum is followed by decreasing temperatures, but subsequently show marked departures from the high latitude cooling trend. This variation is related to the palaeogeography. For example, warming since the Middle Oligocene in northeast Australia reflects the transition from a mid-latitude situation in a world with little climatic zonation, to a low latitude situation in a world with pronounced latitudinal temperature gradients (Table 10.1).

Although it is widely accepted that global cooling during the Tertiary is associated with the development of seasonally dry climates, particularly in inland and northern Australia, these effects are difficult to demonstrate through the palynological record. Other lines of evidence provide a more direct record of seasonality, e.g. variation in tree ring width allows reliable estimates to be made of the nature of seasonal variations in temperature in southeast Australia (Taylor *et al.*, 1990).

Eustacy

The southern margin and presumably coastal regions elsewhere around Australia have been subjected to marine transgression on a number of occasions during the early Tertiary. These appear to have been synchronous over some tens of degrees of longitude (McGowran, 1989), despite the overprinting effects of local and regional tectonism on global variations in sea level. Although the vegetation most directly affected was coastal, a salt-water influence can be detected up to several hundred kilometres away from palaeocoastline in the Murray and Eucla Basins during the Oligo-Miocene. Whether the impact of marine flooding would be different in basins directly opening on to the sea (e.g. Gippsland, Otway, Eucla and Bremer Basins) than in effectively barred basins (e.g. Bass and Murray Basins) is unknown but in both situations, the area of coastal plain available for colonisation by plants would have been considerably reduced during the transgressive and highstand phases: plants will have been the main factor stabilising siliciclastic landforms such as braidplains

Table 10.1. *Palaeolatitude and temperature for northeastern Australia during the Paleogene*

Age	NE Australia palaeolatitude	Estimated NE Australia surface-water temperature (°C)	Comments
Early Paleocene	37–52° S	12–14 to 14–16	Warming episode following cooling in the latest Cretaceous
Early Late Paleocene	36–51° S	14–16 to 12.5–15	Slight cooling
Mid-Late Paleocene to Early Eocene	35–50° S	12.5–15 to 16–18	Relatively rapid warming to temperature maximum in high latitudes; less pronounced cooling in low latitudes
Mid-Early Eocene to mid-Middle Eocene	49–34° S	16–18 to 13–15.5	Major cooling episode from temperature maximum in high latitudes; little variation in low latitudes
Mid-Middle Eocene	48–33° S	13–15.5 to 14.5–17	Brief warming episode interrupting overall cooling trend
Mid-Middle Eocene to latest Middle Eocene	46–31° S	14.5–17 to 12.5–15	Continuation of Eocene cooling trend (but less pronounced in Pacific compared to Atlantic)
Late Eocene	44–29° S	12.5–15	Brief warming episode in high latitudes; slight cooling in low latitudes
Earliest Oligocene	43–28° S	12.5–15 to 10–13	Glaciation in Antarctica; dramatic cooling event in high latitudes and South Atlantic mid-latitudes; moderate cooling in Pacific mid-latitudes; approximately constant temperatures at equator
Early Oligocene	42–27° S	10–13 to 9.5–12.5	Gradual temperature decrease in mid-high latitudes to glacial maximum; little change in low latitudes
Early Late Oligocene	39–24° S	9.5–12.5 to 11–14.5	(Poor data) worldwide warming; reduced glaciation
Late Late Oligocene	38–23° S	11–14.5	(Poor data) slight warming in low latitudes; slight cooling in high latitudes; little change in mid-latitudes
Earliest Early Miocene	37–22° S	11–14.5 to 14.5–17.5	Little residual Antarctic ice; warming from Late Oligocene minimum; substantial temperature increase in southwest Pacific mid-latitudes
Early Miocene	34–19° S	14.5–17.5 to 17–19.5	Gradual warming in all latitudes to Miocene climatic optimum at 16 Ma; negligible Antarctic ice; major temperature increase in southwest Pacific mid-latitudes
Early and mid-Middle Miocene	32–17° S	17–19.5 to 17–20.5	Rapid fall in high latitude temperatures; establishment of East Antarctic icecap; slight warming in low and mid-latitudes

From Feary *et al.*, 1991.

and barrier sand dunes. Knighton *et al.* (1991) have provided graphic evidence of the vulnerability of freshwater ecosystems on low-lying coastal plains to salt-water intrusions.

Changes in geomorphic environments in the Gippsland Basin, resulting from the progressive encroachment of the Tasman Sea, are illustrated in Figure 10.3. Here the marine invasion was both

(a) Maastrichtian – Early Paleocene
(*T. longus*–Lower *L. balmei*)

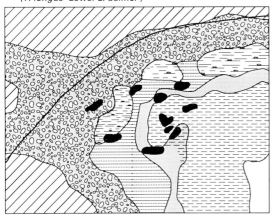

(b) Mid–Late Paleocene
(Upper *L. balmei*–Lower *M. diversus*)

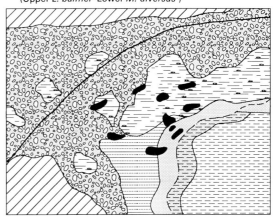

(c) Early Eocene
(Upper *M. diversus*–*P. asperopolus*)

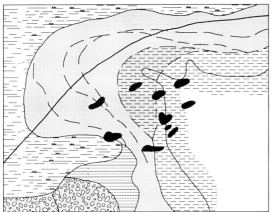

(d) Mid – Late Eocene
(Lower – Middle *N. asperus*)

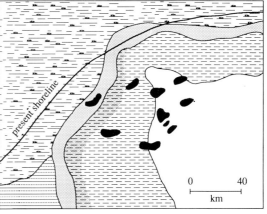

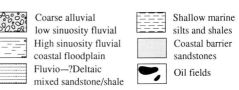

Figure 10.3 Changes in depositional environments within the Gippsland Basin during the Maastrichtian (*T. longus* Zone) to the Late Eocene (Middle *N. asperus* Zone). The progressive encroachment of the Tasman Sea into the basin is clearly shown by changes in the position of the palaeoshoreline relative to the present coastline. The locations of major oil fields, which are the source of much of the pollen data, are shown as black shapes. Blank areas represent deep marine environments. (Adapted from Boddard *et al.* 1986.)

gradual and intermittent, with barrier sands and associated lignite-rich paludal sediments accumulating during the flooding and highstand phases: the last major barrier system, associated with the peak of the Middle Miocene transgression, occurs onshore some 100 km to the west of the earliest

(Late Cretaceous) barrier (see Thompson, 1986). Not surprisingly, depositional environments preserving early Tertiary spores and pollen range from open marine carbonates, barrier sands and a range of paralic environments to freshwater swamps isolated from any marine influence. The effects of sea level fluctuations on inland climates are difficult to evaluate with any precision, although there is general agreement that marine highstands are likely to be correlated with higher temperatures and increased humidity. Periodic disruption of coastal ecosystems may have provided evolutionary opportunities for some taxa.

Soils

The low fertility of Australian soils is usually explained in terms of prolonged leaching of nutrients during the Tertiary. This may not have limited the development of Tertiary rainforest because nutrients can be efficiently recycled via organic litter accumulating under undisturbed rainforest (see Kirkpatrick, 1977; Bowman *et al.*, 1986). Nevertheless low soil fertility is widely associated with the evolution of sclerophylly in the Australian vegetation (see Barlow, 1981). Tertiary volcanism has produced small areas of highly fertile soils in the Eastern Highlands and northwestern Australia but the long-term impact, if any, of this on the Tertiary flora is not obvious in the pollen record.

Fire

A considerable body of evidence exists on the role of fire in modern and late Quaternary Australian ecosystems, in particular (1) in the maintenance of 'fire-requiring and promoting' (Jackson, 1968*b*) sclerophyllous species such as *Acacia* and *Eucalyptus* at the expense of fire-susceptible rainforest angiosperms and gymnosperms, and (2) in lowering soil fertility. The role of fire in the pre-Quaternary vegetation is less clear (Kemp, 1981) but there is a growing body of evidence for localised wildfires since at least Early Cretaceous times (e.g. Macphail & Truswell, 1989), possibly analogous to 'spot fires' in drought-affected rainforest (see Mount, 1979). Charcoal particle data (Martin, 1987) imply that fire frequencies during the Late Eocene–Middle Miocene were low in the Lachlan River region of central New South Wales.

RECONSTRUCTION OF THE FLORA AND VEGETATION

Paleocene 65–53 Ma (66.5–54 Ma)

Palaeomagnetic data indicate that the southern margin of Australia was *ca* 70° S, i.e. approximately at the present-day position of the Antarctic coastline (Veevers *et al.*, 1991). The northern margin (incorporating much of southern Papua New Guinea) was at *ca* 40° S, separated by a largely vacant ocean from Southeast Asia, then at or north of the equator (Burrett *et al.*, 1991). The period was characterised by warming climates – a trend culminating in the Early Eocene (Kennett, 1982; Feary *et al.*, 1991). Low light intensities during winter months may have been of equal biological significance. Rainfall is less easily estimated but clay mineral associations near the Paleocene/Eocene boundary in oceanic sediments in the Atlantic and adjoining Southern Oceans imply enhanced but seasonal precipitation at high latitudes and in continental hinterlands (Robert & Chamley, 1991). Taylor *et al.* (1990) proposed that climates in the Southeastern Highlands of New South Wales were moist with seasonal cooling emphasised by high local relief. Feary *et al.* (1991) estimated that sea surface temperatures in northeast Australia were 12–16 °C.

Southeast Australia (*Lygistepollenites balmei* Zone)

The possibility that some *Forcipites longus* Zone palynofloras are earliest Danian is considered too remote to justify a discussion of them in this chapter. There is no doubt that a number of pollen species were common to both the Maastrichtian and Paleocene floras, e.g. *Lygistepollenites balmei*, 'ancestral' *Nothofagus* spp. and the *Normapolles*-like taxon *Gambierina*. Another pollen type, an undescribed species of *Integricorpus*, appears to be related to the important *Triprojectacites* pollen group of the northern hemisphere.

Reliably dated Paleocene palynofloras extend from northwest Tasmania (Forsyth, 1989) northwards through the Bass (Lower Eastern View

Group), Otway (Pebble Point Formation) and Gippsland (Latrobe Group in part) Basins, South-eastern Tablelands (Taylor *et al.*, 1990) into the Sydney Basin (C. J. Jenkins, R. Helby & A. D. Partridge, unpublished data) and, less certainly, on to the Northern Tablelands of New South Wales (Martin *et al.*, 1987). All are dominated by gymnosperms (chiefly fossil members of the Araucariaceae and Podocarpaceae), ferns and fern allies. Variable pollen morphologies suggest that each fossil taxon represents a number of plant genera/species. Many of the conifer lineages still survive in Australasia: *Agathis, Araucaria, Dacrycarpus, Dacrydium* Group B species, *Halocarpus bidwilli* type, *Lagarostrobos franklinii, Microcachrys tetragona, Phyllocladus* and *Podocarpus* spp. Several are extinct, e.g. *Lygistepollenites balmei* and *Podosporites microsaccatus*.

Angiosperm representation is variable, consisting mainly of Proteaceae species (characterised by small scabrate pollen types), *Periporopollenites polyoratus* (?Caryophyllaceae) and *Australopollis obscurus*, identified by Macphail & Partridge (1991) as a fossil species of the aquatic herb family Callitrichaceae. A palynoflora recovered from the continental slope off Jervis Bay (C. J. Jenkins, R. Helby & A. D. Partridge, unpublished data) implies that Casuarinaceae had become prominent on the New South Wales south coast during the Danian (*E. crassitabulata* Zone). There is no evidence of *Lygistepollenites balmei* in the coastal vegetation at this latitude. Macrofossils (Hill, 1991) indicate that *Eucryphia* was part of a Late Paleocene flora on the Southeastern Highlands of New South Wales.

Objective techniques (Sluiter & Macphail, 1985) have allowed trends in the vegetation of the Gippsland Basin to be traced in broad detail during the Paleocene. These include: (1) an irregular decline in *Lagarostrobos* and *Dacrydium/Podocarpus* during the Danian; (2) a transient rise in Araucariaceae, *Lygistepollenites balmei* and Callitrichaceae; and (3) during the latest Thanetian (Upper *L. balmei* Zone), a variable dominance by ferns such as Cyatheaceae and Gleicheniaceae and angiosperms, in particular *Nothofagidites endurus*. Two probable plant associations have been identified.

Neither is likely to have been important in the vegetation, since the species concerned are uncommon: *Gambierina–Proteacidites–Tricolpites–Basopollis-Herkosporites* and *Lygistepollenites florinii–Haloragacidites harrisii–Dilwynites granulatus*.

The number of rare angiosperm species with modern mesotherm–megathermal affinities increase significantly during the latest Thanetian, e.g. *Anacolosa (Anacolosidites acutullus), Beauprea (Beaupreaidites elegansiformis, B. verrucosus)* and Polygalaceae (*Polycolporopollenites esobalteus*). Several Proteaceae can be confidently related to modern genera, e.g. *Gevuina/Hicksbeachia (Proteacidites reticuloscabratus)*, Musgraveinae (*Banksieaeidites arcuatus*), *Xylomelum occidentale*-type (*Proteacidites annularis*) and species related to *Agastachys* and *Telopea (Proteacidites obscurus, Triporopollenites ambiguus*). Equally characteristic of the period is the re-emergence of Proteaceae with highly ornamented pollen (e.g. *Proteacidites grandis* and *P. incurvatus*). One species of *Beauprea (Beaupreaidites orbiculatus), Gambierina*, 'ancestral' *Nothofagus (Nothofagidites endurus)* and *Lygistepollenites balmei* were extinct by earliest Eocene times, although the undescribed *Integricorpus* sp. (see above) persisted into the Early Eocene.

The Danian–Early Thanetian coniferous vegetation has few modern parallels, except possibly microthermal/mesothermal podocarp-dominated swamp forests, whereas the latest Thanetian vegetation may be analogous to *Nothofagus*–conifer–broad-leaved mesothermal rainforest, now widely distributed in New Zealand and Papua New Guinea/New Caledonia. Since many of the angiosperm taxa occur in latest Maastrichtian sediments in Antarctica (Askin, 1990) and some, e.g. *Anacolosa* and *Beauprea (Beaupreaidites verrucosus)*, in Late Cretaceous floras in central and northern Australia (Harris & Twidale, 1991; A. D. Partridge, personal communication), part of the floristic evolution is due to migration. In contrast, all gymnosperms were part of the Late Cretaceous flora, although in lesser abundance than in the Paleocene. The cause(s) of the apparently massive impoverishment of the angiosperm flora at the Cretaceous–Tertiary boundary and (?related) major expansion of gymnosperms is beyond the scope of this chapter but all subsequent trends are consistent with year-round high humidity and increasingly warm temperatures during the period.

Northeast Australia

Carbonate platform build-up shows that the northern part of northeast Australia had sea surface temperatures now characteristic of subtropical regions, whereas southern areas fluctuated about the temperate/subtropical boundary (Figure 10.4).

Harris (1965*b*) has recorded a possible Upper *L. balmei* Zone equivalent assemblage from Brisbane. The flora included *Anacolosa* (*Anacolosidites acutullus*), *Xylomelum occidentale* type and, if the age determination is correct, the earliest record to date of Tiliaceae in eastern Australia. *Lagarostrobos* was part of the regional flora here and also on the Northern Tablelands of New South Wales (see Martin *et al.*, 1987). Apart from rare specimens in Maastrichtian sediments in the Bonaparte Basin (A. D. Partridge, personal commication), there is no evidence that *Lygistepollenites balmei* was present in northern Australia.

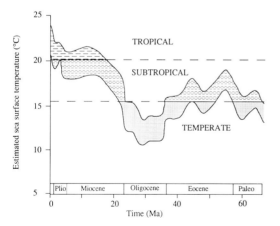

Figure 10.4 Variations in surface water temperatures in the nearshore northeast Australian region during the Cenozoic. The upper and lower curves approximate to conditions prevailing at the northern and southern limits of the region, respectively. Horizontal lines show the modern limits for the deposition of temperate/subtropical/tropical carbonates. (Adapted from Feary *et al.*, 1991.)

Central regions

The Early Paleocene was largely a period of erosion or non-deposition. Late Paleocene palynofloras dominated by gymnosperms, Proteaceae and cryptogams occur in the western Otway Basin (Pebble Point Formation and lower Pember Mudstone) and Murray Basin (lowermost Renmark Group), but have not been recorded in the Duntroon and Great Australian Bight Basins. Basins within the continental interior (e.g. Lake Eyre Basin) are infilled with thin sequences of fluviolacustrine, carbonaceous and arenaceous sediments containing numerous hiatuses. This, and subsequent strong weathering, has resulted in a generally patchy preservation of palynomorphs.

On present indications the coastal/lowland vegetation closely resembled that in southeast Australia, except that *Nothofagus* and *Lygistepollenites balmei* remained relatively uncommon and *Microcachrys* was abundant locally. Pollen dominance is variable but prominent taxa (maximum values in parentheses) include: *Dacrydium/Podocarpus* (20%), *Lagarostrobos* (14%), *Microcachrys* (12%), *Dacrydium* Group B (7%), Araucariaceae (9%), Proteaceae (12%), Callitrichaceae (4%), ?Caryophyllaceae (2%), and trilete fern spores (*Cyathidites* spp: 11%). The Proteaceae included *Gevuina/Hicksbeachia* and *Xylomelum occidentale* type. *Nothofagus* spp. (< 7%) and Gleicheniaceae (16%) increase during the latest Thanetian. *Anacolosa*, Casuarinaceae, Euphorbiaceae and Myrtaceae were minor elements within what were almost certainly conifer-dominated mesothermal rainforest types.

Inland, the only reliably dated (latest) Paleocene sediments come from the lower part of the Eyre Formation, Lake Eyre Basin (Wopfner *et al.*, 1974; Sluiter, 1991; N. F. Alley, unpublished data). Palynofloras preserved in these sediments differ from apparently contemporaneous coastal/lowland assemblages in several respects: (1) within the limits of dating, angiosperm dominance occurs earlier, although at no time was *Nothofagus* common; (2) the gymnosperm component includes irregularly high percentages of Cupressaceae and/or Taxodiaceae; and (3) the flora contained a number of species which are not recorded in the Bassian region until the Early–Middle Eocene. Examples are *Ascarina*, Tiliaceae, *Proteacidites crassus*, *P. latrobensis*, *P. leightonii* and *P. reticulatus sensu lato*. Overall, the diversity of gymnosperms and angiosperms is high. Pteridophytes are well represented, in particular tree ferns (Cyatheaceae, Dicksoniaceae; up to 14%) and *Cyathidites splendens* (up to 15%). Rare species include *Anacolosa* (*Anacolosidites acutullus*), Cupanieae, Droseraceae, Euphorbiaceae, *Gambierina*, *Gillbeea* type, *Gunnera*, *Ilex*, Trimeniaceae and Winteraceae. Pollen of Sparganiaceae and/or Typhaceae occur in a few samples from the north of the area.

Quantitative analyses demonstrate that pollen dominance fluctuated irregularly between Cunoniaceae (16–80%) and gymnosperms (chiefly Cupressaceae/Taxodiaceae, *Dacrydium/Podocarpus* and *Lagarostrobos*) associated with relatively abundant Callitrichaceae (6–11%). Other gymnosperms (including *Lygistepollenites balmei*), Casuarinaceae and Myrtaceae are usually rare to uncommon (<2–8%). Proteaceae may form up to 12% of individual palynofloras, the most frequent species being *Proteacidites fromensis* (<5%).

The source vegetation is interpreted as a shifting mosaic of mesotherm rainforest types (including *Lagarostrobos* swamp forest) in which local dominance was determined by variations in water table level. If correct, regional trends may be useful in reconstructing rainfall patterns in central Australia during the latest Paleocene. Of as yet unknown significance is the presence in the Lake Eyre Basin of *Deflandrea obliquipes*, a dinoflagellate that occurs in marginal marine sediments in the Gippsland and Otway Basins.

Western Australia

No information is available for this region. The Kings Park Formation of the Perth Basin, dated as Late Paleocene–Early Eocene by Quilty (1974) is now considered to be Middle Eocene (Quilty, 1978; Marshall & Partridge, 1988).

A more reliably dated Paleocene pollen sequence has been recorded from Ninetyeast Ridge (Deep Sea Drilling Project Site 214) located at 11° 20′ S, 88° 43′ E in the Indian Ocean

(Harris, 1974; Kemp & Harris, 1977). In spite of its remote location, this site is relevant to the flora of northwest Australia in two ways:

1. The Ninetyeast Ridge palynofloras more closely resemble the coeval gymnosperm-dominated assemblages of the Australian region than those of Africa to the west or Southeast Asia to the north. Examples are *Dacrydium* Group B species (up to 29%), *Phyllocladus* (up to 23%), *Podocarpus* (up to 10%) and *Microcachrys* (up to 1%), although Araucariaceae are absent.
2. The lignitic nature of the sediments means that the overwhelming majority of these pollen will have come from plants established on the Ridge. This is good evidence that their seed was able to be dispersed effectively over salt water.

These observations suggest that the Paleocene flora of northwest Australia almost certainly would have shared a large number of taxa with southern Australia (and probably also Antarctica).

Relatively few of the more common angiosperms can be referred to extant taxa: Arecaceae (*Arecipites*; up to 65%), *Ascarina* (up to 17%) and *Gunnera* (up to 3%). Rare taxa include a possible Asteraceae (Tubuliflorae), Callitrichaceae, Casuarinaceae, Didymelaceae and *Gillbeea* type (*Concolpites* sp. cf. *C. leptos*). Differences between the Ninetyeast Ridge floras and those of Australia include the absence of Cupanieae and ?Myrtaceae (all present on Ninetyeast Ridge by the Early Miocene).

Early Eocene 50–56.5 Ma (49–54 Ma)

In spite of the continuing very high latitude position adjacent to Antarctica, climates in Australia as a whole were warmer and, less certainly, wetter than at any other time during the Cenozoic. Global sea levels were high but fluctuating (Haq *et al.*, 1987), with coastal basins along the southern margin subject to marine transgression on at least five separate occasions in *ca* four million years (Harris, 1985). Unique assemblages of dinoflagellates allow palynofloras deposited during each transgression to be dated and correlated with considerable precision.

In southern Australia, the Upper *L. balmei*/ Lower *M. diversus* Zone boundary corresponds to a major flooding surface (correlated with the 53.5 Ma condensed section of Haq *et al.*, 1987). This event, the *Apectodinium* (*Wetzeliella*) *hyperacanthum* transgression (Partridge, 1976) appears to have influenced the flora and vegetation occupying low-lying coastal areas to a greater degree than any previous or subsequent transgression during the Cenozoic. How rapid the shift in plant community dominance was is difficult to estimate but 0.2 Ma (based on dinocysts) would appear to be well within the limits for the development of mesothermal–megathermal rainforests during the late Quaternary (see Hope, Chapter 15, this volume).

Southeast Australia (Lower *Malvacipollis diversus* to *Proteacidites asperopolus* Zones)

Early Eocene fluvio-deltaic and marginal marine sediments containing diverse palynofloras are widespread in the three Bass Strait basins and onshore extensions in Victoria (e.g. Dartmoor, Dilwyn, Upper Eastern View and Werribee Formations) and northwest Tasmania (Devonport-Port Sorrell, Wesley Vale and Sassafras Sub-basins). Contemporaneous strata outcrop in sea cliffs within Macquarie Harbour in western Tasmania (Strahan Sub-basin): rotational slipping associated with opening of the Southern Ocean has left marginal marine palynofloras of late Early Eocene age under water depths of *ca* 3500–4000 m on the adjacent abyssal plain (M. K. Macphail, unpublished data; N. F. Exon, E. M. Truswell & W. K. Harris, unpublished data).

Pollen dominance is extremely variable within the region and over time but locally prominent taxa include Araucariaceae (*Araucariacites* and *Dilwynites*), Casuarinaceae (almost certainly *Gymnostoma*), Euphorbiaceae (in particular *Malvacipollis*), Proteaceae (including *Gevuina*/ *Hicksbeachia*, Musgraveinae, *Telopea* type and *Xylomelum occidentale* type) and ferns (including

Adiantaceae, Cyatheaceae, Gleicheniaceae and Polypodiaceae). The diversity and relative abundance of unidentified angiosperms represented by triporate (*Basopollis*, *Triporopollenites*), tricolpate (*Tricolpites*) and tricolporate (*Rhoipites*, *Tricolporites*) pollen is irregularly high. Values of *Nothofagus* pollen, including *Brassospora* spp., seldom rise above 5% in coastal-marine sediments. Amongst the less common elements, there is a distinct upsurge in the frequency of occurrence of trees, shrubs and ferns with megathermal/mesothermal affinities. Some, like *Anacolosidites acutullus*, Cupanieae and Polygalaceae were very rare elements in the Paleocene flora: others, such as *Anacolosidites luteoides*, *Santalum* type, Sapotaceae, Tiliaceae and *Lygodium* (*Crassiretitriletes vanraadshoovenii*), are not recorded before the Eocene in the Gippsland and Bass Basins and ?Tasmania.

The data are emphatic that very diverse associations of angiosperms (largely excluding *Nothofagus*) had replaced conifers and 'ancestral' *Nothofagus* as the dominant rainforest communities in the coastal/lowland regions. Ferns remained prominent, although Cyatheaceae and epiphytic types had largely replaced Gleicheniaceae. The relative spore and pollen proportions are not dissimilar to that produced by modern tropical evergreen (nonseasonal, megathermal) rainforest containing emergent Araucariaceae and a very diverse subcanopy stratum of shrubs, lianes and cryptogams (compare Kershaw, 1988). High values of species believed to be underpresented strongly imply that much of the spore–pollen was derived from vegetation lining local waterways.

Both the scale and direction of ecological change link this vegetation to tide-dominated deltaic environments in which temperatures and year-round humidity were high. Direct evidence for this is the presence of tropical mangroves at least as far south as the Strahan Sub-basin on the west coast of Tasmania (see Cookson & Eisenack, 1967). Mangroves that were widespread in the Bassian region are *Nypa* palms, (*Spinizonocolpites prominatus*), and the ferns *Acrostichum aureum*/ *speciosum* (included within *Cyathidites splendens*). *Sonneratia*-type (*Florschuetzia levipoli*) pollen occurs very rarely in Late Eocene sediments in the Gippsland Basin, and it is just possible that a fourth mangrove (*Brownlowia*) is represented by high values of Tiliaceae pollen within a channel cut into marine sediments at Regatta Point, Strahan Sub-basin (M. K. Macphail, unpublished data). *Eucryphia* macrofossils are preserved in these marine sediments (Hill, 1991). The absence of Callitrichaceae (*Australopollis obscurus*) and *Lygistepollenites balmei* is consistent with tide-dominated environments: both taxa survived in Early Eocene times in the Otway Basin and *Australopollis* in inland freshwater lakes until the Miocene.

Apart from mangroves, few Early Eocene species can be associated with a specific habitat, although, by analogy with modern siliciclastic shorelines, it is reasonable to suggest that water table levels as well as exposure to salt were major controls on plant distributions. Periodic destabilisation of coastal ecosystems may have provided evolutionary opportunities for some species. One probable example, suggested by a progressive increase in species with distinctively sculptured pollen, are the Proteaceae (times of first appearance in parentheses): *Proteacidites leightonii*, *P. ornatus* and *P. tuberculiformis* (Middle *M. diversus* Zone); *P. nasus*, *P. pachypolus*, (Upper *M. diversus* Zone) and *P. asperopolus*, *P. recavus*, *P. rugulatus* (*P. asperopolus* Zone). One species, *Proteacidites pachypolus*, appears to have been restricted to marginal marine sediments during the earliest Eocene *Apectodinium* (*Wetzeliella*) *hyperacanthum* transgression, apart from one specimen recorded in latest Paleocene–earliest Eocene sediments in the Lake Eyre Basin (see below).

Other distinctive taxa appearing in the Gippsland pollen record during the same period include: Alangiaceae (*Alangium villosum* type), *Gillbeea* type, Droseraceae, *Rhodomyrtus* type (*Myrtaceidites verrucosus*) and a pollen resembling the northern hemisphere *Normapolles* group (*Myrtaceoipollenites australis*). Evolution, possibly centred in the Bassian region, is suggested to be responsible for the appearance of *Nothofagidites goniatus* (*Nothofagus* subgenus *Brassospora*) and *Myrtaceidites tenuis* (Myrtaceae). It is possible the latter is a primitive eucalyptoid species: the type

became widespread and locally common in the coastal vegetation during the late Early Eocene. Some taxa now typical of low, open vegetation types such as heath and sedgeland are very rare but persistent in the pollen record, e.g. Ericaceae/Epacridaceae and Restionaceae, but there is no evidence of Sparganiaceae/Typhaceae reed swamp communities (present in central Australia during this period; see below).

The occurrence of Early Eocene sediments in inland areas in southeast Australia (and elsewhere) is difficult to verify because the zone index species tend to be restricted to coastal sites or have extended ranges outside the Bassian region.

One probable exception is a lignite in the non-marine Buckland Basin in eastern Tasmania. This preserves a rich macrofossil flora including extinct species of *Athrotaxis* (Taxodiaceae), *Microstrobos* and *Podocarpus* (Townrow, 1965a,b). The associated palynoflora (M. K. Macphail, unpublished data) differs from coastal assemblages in two respects. The pollen dominants include *Nothofagus* (*Brassosporal*) spp. (34–37%), and megathermal angiosperm types are absent.

Consistent with the macrofossil data, gymnosperm pollen comprise *ca* 50% of the assemblages – chiefly *Dacrydium/Podocarpus* (26–28%) but also including significant amounts of Araucariaceae (4–6%), *Lagarostrobos* (up to 6%), *Dacrydium* Group B spp. (2–3%). *Ephedra* (*Ephedripites*: Ephedraceae) pollen is recorded but not that of *Microstrobos*. Angiosperms represented by frequent (3–7%) amounts of pollen are: *Nothofagus* (*Nothofagus*), Myrtaceae, ?Caryophyllaceae (*Periporopollenites polyoratus*), *Proteacidites crassus* and *P. grandis*. Rare species include *Ascarina*, Musgraveinae, Cunoniaceae (dicolpate and tricolpate), Elaeocarpaceae, Euphorbiaceae and a distinctive but undescribed tetracolporate species that ranges typically no higher than the late Early Eocene, *P. asperopolus* Zone in the Gippsland Basin.

If correctly dated, the site emphasises that mesothermal rainforests including abundant *Nothofagus* remained present in southern southeast Australia during the Early Eocene but had become confined to the hinterland. The species were the more 'evolved' *Brassospora* types, although 'ancestral' characteristics such as a deciduous habit may have been retained (see Hill & Read, 1991).

This is almost certainly the case for some *Fuscospora* spp. on the basis of the abundance of *Nothofagidites brachyspinulosus* pollen associated with a deciduous *Nothofagus* leaf in late Early Eocene sediments at Deloraine, northern Tasmania (Hill, Chapter 16, this volume). The same site confirms that mesothermal *Nothofagus (Brassospora)* communities were part of the inland vegetation in northern Tasmania, although, because of the location of the site, it is uncertain whether these grew at low elevations or upslope on the Central Plateau. The local flora included significant numbers of ?Meliaceae (pentaporate pollen types), Sapotaceae and *Proteacidites pachypolus*. Minor elements included Cupanieae and *Santalum* type.

Northeast Australia

Carbonate platforms in northeast Australia (palaeolatitude 35°–50° S) continued to build up, reflecting sea surface temperatures (16–18 °C) that were warmer than at any other time during the early Tertiary prior to the Early Miocene (Feary *et al.*, 1991).

No reliably dated palynoflora exists, although a number of the assemblages recorded at, and north of, Brisbane by Harris (1965*b*) and (Unit 1, Capricorn Basin) by Hekel (1972) are not dissimilar to earliest Eocene assemblages from the Gippsland Basin.

Otherwise, the only indication of Early Eocene floras in the region comes from a grab sample of sea-floor sediments recovered under water depths of 2100–2500 m from the continental slope, *ca* 50 km east of Coffs Harbour on the north coast of New South Wales. Dinoflagellates demonstrate that the sparse spore–pollen assemblage is a correlative of the late Early Eocene, Upper *M. diversus* Zone (M. K. Macphail & R. Helby, unpublished data). As is usually the case with deep-sea sediments, the palynoflora is dominated by water-transported spores, chiefly *Pteris* spp. (*Polypodiaceoisporites retirugatus* and related types) and *Lygodium*. The only frequent pollen type present is Casuarinaceae, although trace records provide hints that the source vegetation also contained Cupanieae, Euphorbiaceae, *Nothofagus (Brassospora, Nothofagus)*, Tiliaceae, Trimeniaceae and a diverse range of Proteaceae including *Proteacidites pachypolus*. Little can be deduced regarding this vegetation other than that the diversity of *Pteris* spp. and rarity of *Nothofagus* is consistent with megathermal–mesothermal rainforest.

Central regions

In spite of the widespread distribution of the Early Eocene Dartmoor–Dilwyn Formations in the western Otway Basin (Laing *et al.*, 1989) and marine correlatives in the Great Australian Bight, few data are available from this sector. Nevertheless it is reasonable to assume that warm, humid conditions would have allowed mesothermal–megathermal rainforests to dominate the coastal/lowland regions. What data are available (N. F. Alley, unpublished) show that Araucariaceae (chiefly *Dilwynites*), *Dacrydium* Group B, *Dacrydium/Podocarpus* and *Microcachrys* were prominent and *Nothofagus* spp. uncommon to rare. Trends within the angiosperms appear to parallel those in southeast Australia, both in the relative abundance of Casuarinaceae, Euphorbiaceae, Myrtaceae and in the increase of Proteaceae with highly ornamented pollen. Minor taxa include *Anacolosa (Anacolosidites acutullus, A. luteoides)*, *Beauprea*, Callitrichaceae, Cupanieae and Tiliaceae.

Large freshwater lakes remained a feature of the inland landscape. All sites studied appear to have been surrounded by mesothermal (some ?megathermal) rainforest communities but the local vegetation also included taxa now characteristic of low/open communities, e.g. Cyperaceae, Poaceae, Restionaceae and Sparganiaceae/Typhaceae. As for the Paleocene, the only detailed palynological records come from the Lake Eyre Basin (Sluiter, 1991; N. F. Alley, unpublished data).

Here, Early Eocene palynofloras are distinguished by (1) the absence or rarity of typically Paleocene taxa such as *Australopollis obscurus*,

Gambierina, Camarozonosporites bullatus, Lygistepol-lenites balmei and *Proteacidites fromensis* and (2) the consistent presence of Cupanieae, Didymelaceae and Tiliaceae. Apart from irregularly abundant *Dilwynites* (up to 17%), values for gymnosperms usually fall within the range recorded during the Paleocene. Other genera represented are *Dacryd-ium/Podocarpus, Dacrydium* Group B, *Halocarpus bidwillii* type, *Lagarostrobos* and *Microcachrys*. Cryptogams are locally common, e.g. Cyatheaceae (3–12%), Gleicheniaceae (1–8%) and *Lygodium* (up to 4%).

The angiosperm component is dominated by Cunoniaceae (mainly tricolpate types) and Myrta-ceae, although, as with other common types, values are extremely variable (1–19%, 1–46%, respectively). Casuarinaceae are uncommon and *Nothofagus* is rare, the only subgenera present being *Brassospora* (< 2%) and *Fuscospora* (<1%). Other rare types able to be related to extant genera include *Anacolosa* spp., Arecaceae, *Ascarina*, Cupanieae, Didymelaceae, *Eucalyptus* (possibly the earliest record to date of this important sclero-phyllous genus), Elaeocarpaceae, Euphorbiaceae, *Gevuina/Hicksbeachia, Gillbeea* type, *Ilex*, Loran-thaceae, Polygalaceae, Tiliaceae, Winteraceae and *Xylomelum occidentale* type. The Proteaceae included some species that had already died out in southeast Australia (*Proteacidites angulatus, Beaupreaidites orbiculatus*) and others that did not become established in that region before the Middle–Late Eocene (*P. confragosus* and *P. reticu-latus*). One specimen of *Proteacidites pachypolus* occurs in a palynoflora characterised by frequent *P. fromensis* and *Deflandrea obliquipes*.

Pollen data (N. F. Alley, unpublished) from the Lake Torrens and Lake Frome Basins confirm that many of the above species were in fact part of the regional flora of inland central Australia but show that community dominance was more a function of the local conditions than of regional climates. Evidence of this is provided by the very high values of Casuarinaceae (43–82%), occasional significant levels of Callitrichaceae (2–7%) and virtual absence of Cunoniaceae.

Western Australia

No reliably dated palynoflora exist, although the Kings Park Formation palynoflora (see above) is not dissimilar to earliest Eocene assemblages from southeast Australia in terms of the dominance by Proteaceae and cryptogams.

Middle–Late Eocene 35.4–50 Ma (36–49 Ma)

From the oxygen isotope record, ocean deep waters had begun to cool between 51 and 49 Ma (Shackleton, 1986), although it is not until about the Eocene–Oligocene boundary that surface waters underwent profound cooling (see Wei & Wise, 1990; Wei, 1991). Over the same period, Australia became more conspicuously separated, in the physical if not the biological sense, from Antarctica, whilst its northern margin faced on to a narrowing, but still wide, ocean with scattered continental fragments.

McGowran (1989) has argued that four, essen-tially synchronous, marine transgressions along the southern margin represent significant rever-sals of the global cooling trend between the down-wards steps in temperature at the Early/Middle Eocene and Eocene/Oligocene boundaries. This is supported by (warm temperate–subtropical) shelf water temperatures in Southern Australia during the Late Eocene (Kamp *et al.*, 1990). Feary *et al.* (1991) estimate that sea surface temperatures in northeast Australia had decreased from the Early Eocene maximum of 16–18 °C to 12.5–15 °C, less pronounced than the fall at high lati-tudes. The overall cooling trend was interrupted by brief warming episodes during the mid-Middle Eocene (see Table 10.1) but there was little vari-ation in palaeotemperatures at low latitudes.

Middle–Late Eocene sediments are widely distributed in buried channels cut into the Yilgarn Block of southern West Australia but elsewhere tend to be confined to offshore and coastal basins and adjacent uplands. Most of these accumulated in fluvio-deltaic environments, but a marine influence becomes increasing apparent during the Middle–Late Eocene in the offshore basins.

Southeast Australia (Lower and Middle *Nothofagidites asperus* Zones)

Palynofloras of Middle–Late Eocene age are widespread offshore in the Gippsland Basin (Upper Latrobe Group), Bass Basin (Upper Eastern View Group) and Otway Basin (Upper Dilwyn Formation). Onshore correlatives extend from northern Tasmania and coastal Victoria (e.g. Anglesea Sands, Brown Creek Clays, Johanna River Sands) at least as far north as the Northern Tablelands of New South Wales and into inland southwest New South Wales. Many are difficult to assign to a particular zone because of the rarity of the Middle *N. asperus* Zone index species *Triorites magnificus* and apparent diachronism in the range of accessory species away from the Gippsland Basin (see Macphail & Truswell, 1989).

In the Bass and Gippsland Basins, the boundary between the Early and Middle Eocene coincides with the emergence of *Nothofagidites emarcidus-heterus* as the major palynological dominant (av. 50–60%). Similar values are reached in the Murray Basin. Unlike the Paleocene/Early Eocene boundary, this floristic 'event' does not correspond to any *major* tectonic–eustatic event in the schema of Haq *et al.* (1987). Accordingly, the apparent rapidity and magnitude of the increase is best explained by climates within the Bassian region becoming optimal for this group of *Brassospora* spp. The other consistently common pollen taxon is Casuarinaceae (up to 30%). Macrofossils demonstrate that their source was *Gymnostoma* (Christophel, 1980; Christophel *et al.*, 1987). Gymnosperm representation shows a slight increase over late Early Eocene values (<30%): all species and the overwhelming majority of angiosperms were part of the Early Eocene flora in the region.

Isolated palynofloras may be dominated by taxa that are usually uncommon, e.g. *Nothofagidites deminutus* and *N. goniatus*, less often *N. brachyspinulosus* and *N. flemingii* in northwest Tasmania. What few quantitative data are available give the impression of significant short-term variation but overall long-term stability in the regional vegetation. The former may reflect local disturbances,

possibly necessary to the long-term survival of *Nothofagus* dominance (see Read *et al.*, 1990), or that many species had 'patchy' distributions, analogous to many living megathermal–mesothermal rainforest trees (see Webb *et al.*, 1984), or merely differences in the depositional environment.

A significant number of new pollen taxa continued to appear in the region, in particular in the Murray Basin. These include a *Brassospora* sp. (*Nothofagidites falcatus*), *Anacolosa* (*Anacolosidites sectus*), *Epacris virgata* type (*Paripollis orchesis*), and at least 12 species of Proteaceae. Some have strong sclerophyll affinities, e.g. *Banksia serrata* type (*Banksieaeidites* sp. of Macphail & Truswell, 1989), *Dodonaea triquetra* type and *Isopogon*. Others are more likely to be small rainforest trees e.g. *Quintinia* (*Quintiniapollis psilatispora*), or epiphytic species of Loranthaceae (*Cranwellia* spp., *Gothanipollis* spp., *Tricolpites simatus*, *T. thomasii*). Rare but phytogeographically interesting taxa include a species of Sterculiaceae now endemic to Norfolk Island (*Ungeria floribunda*: *Reevesiapollis reticulatus*); an Apocynaceae that is mostly restricted to Indo-Malesia and the western Pacific (*Alyxia*: *Psilodiporites* sp. of Macphail & Truswell, 1989); and a group of wet forest lianes that in Australia are now confined to the humid east coast and adjoining tablelands (*Cissus*: Vitaceae) Some plants present in the Murray and onshore Otway Basins had not extended eastwards into the Gippsland and Bass Basins or Tasmania at this time. Most of these have strong megatherm–mesotherm affinities, e.g. Malpighiaceae (*Perisyncolporites pokornyi*), Caesalpinoidae (*Margocolporites vanwijhei*), *Canthium* (*Rubipollis oblatus*: Rubiaceae) and *Pteris* (*Asseretospora*, *Polypodiaceoisporites retirugatus*). One is a possible ruderal (*Epilobium*: Onagraceae); another is widely distributed in the Pacific region, including alpine habitats – *Coprosma* type (*Palaeocoprosmadites*: Rubiaceae). An unidentified angiosperm, *Psilastephanocolporites micus*, is confined to brown coal facies in the Gippsland Basin but occurs in relatively inorganic floodplain sediments elsewhere. Extinctions during the period are few and these are mostly local. Examples occurring in the Bass and Gippsland Basins at, or shortly after, the Early/Middle

Eocene boundary include *Spinizonocolpites prominatus*, Tiliaceae, *Proteacidites ornatus* and *Myrtaceidites tenuis*, with *Proteacidites asperopolus* at the Middle/Late Eocene boundary.

In broad terms, the Middle–Late Eocene vegetation appears to have been a mosaic of *Nothofagus*-dominated rainforest associations, mesothermal rather than megathermal in character and certainly more diverse than any Australian microthermal *Nothofagus* community. A podocarp overstorey almost certainly was present. Given that *Nothofagus* spp. have poorly dispersed seed, the rapidity with which *Brassospora* spp. expanded within the coastal/lowland regions is more consistent with seed being carried down river from upland *Nothofagus* forests than by outward expansion of whatever populations survived at low elevations during the Early Eocene. This development almost certainly is due to global cooling given that some megathermal taxa such as *Nypa* became extinct and others such as Cupanieae became increasingly rare during the Late Eocene (compare Read *et al.*, 1990).

What also is highly probable is that, once established in the coastal/lowland vegetation, the gregarious behaviour of *Nothofagus* spp. allowed them competitively to exclude most other palynologically well-represented canopy angiosperm species until at least the Late Oligocene. In the Gippsland Basin, the long-term maintenance of *Nothofagus* communities is associated with the development of an extensive floodplain (Figure 10.3) and moderately stable sea levels. Since few if any living *Nothofagus* spp. can tolerate permanently high water table levels, it is deduced that sizeable areas of freely draining soils were present, e.g. on the narrow zone of barrier sands or on levee banks within the floodplain. What plant communities occupied permanently wet areas is uncertain because sedges are rare and high pollen values of swamp podocarps such as *Dacrycarpus* and *Lagarostrobos* occur only sporadically. Reeds such as *Aglaoreidia qualumis* (Sparganiaceae/Typhaceae) are first recorded in the Late Eocene in the Gippsland Basin but were present in the Murray Basin before this period. Araucariaceae (*Dilwynites*) dominate an ?early

Middle Eocene palynoflora on the Southern Highlands of New South Wales (Truswell & Owen, 1988). The flora shares many species with the Gippsland Basin, e.g. *Anacolosa*, *Beauprea* and *Santalum* type and includes a palm (subfamily ?Lepidocaryoidae) and the youngest records to date of Tiliaceae and *Myrtaceidites tenuis* in southeast Australia. A number of the tricolporate types at this site have yet to be found elsewhere. In spite of the relatively high elevation, the impression given is of a mesothermal rainforest in which *Nothofagus* was uncompetitive relative to other rainforest trees.

A Middle–Late Eocene sequence from the Northern Tablelands of New South Wales (McMinn, 1989) records the replacement of a *Nothofagus* (*Brassospora*)–Casuarinaceae-Podocarpaceae association by *Lagarostrobos*. In this instance, the associated increase in Callitrichaceae (*Australopollis obscurus*) links the development with high water table levels. ?Contemporaneous palynofloras from the same highlands include Tiliaceae and *Proteacidites pachypolus* (McMinn, 1989).

Northeast Australia

Diverse palynofloras of the Middle–Late Eocene occur in freshwater sediments in the Yaamba Basin (23° S) on the central coast of Queensland (Foster, 1982; Dudgeon, 1983). *Nothofagus* spp. present include *Brassospora*, *Fuscospora* and *Lophozonia*, but overall their pollen is infrequent relative to Casuarinaceae, Araucariaceae (*Araucariacites* and *Dilwynites*) and, less often, Myrtaceae. The minor elements include *Alangium villosum* type, *Anacolosa* (*Anacolosidites sectus*), Loranthaceae (*Tricolpites simatus*, *T. thomasii*), *Santalum* type, Sapotaceae as well as many Proteaceae found in the Bassian flora (e.g. *Proteacidites biporus*, *P. crassus*, *P. pachypolus*, *P. stipplatus*). The source vegetation is presumed to have been mesothermal–megathermal rainforest.

The Yaamba flora included a significant number of taxa that are found in contemporaneous floras in the Murray Basin but which are absent (or very rare) in the Gippsland Basin:

Alyxia, Arecaceae (*Dicolpopollis* sp.), Caesalpino-idae, *Canthium*, Convolvulaceae (*Perfotricolpites* sp. cf. *P. digitatus*), *Grevillea/Hakea* and unidentified tricolporates such as *Simpsonipollis* and *Tricolporites valvatus*. Assuming that Foster's (1982) dating of Unit 2 palynofloras in the Capricorn Basin is correct, then Late Eocene communities on Queenland's central coast included the mangroves *Nypa* and Rhizophoraceae (*Zonocostites*).

Central regions

Widespread sedimentation in the offshore/coastal basins recommenced during the Middle Eocene. Deposits of this age are mainly lignitic, accumulating within fluvio-lacustrine environments in the western Otway Basin (Dilwyn Formation), the Murray Basin (Olney Formation) and in the St Vincent Basin (North Maslin Sands). In the Eucla Basin to the west, carbonaceous paralic to estuarine clays (Pidinga Formation) accumulated in palaeochannels draining the western interior of South Australia. Middle Eocene correlatives occur inland in the Willochra Basin (Langwarren Formation), Lake Torrens Basin (Cotabena Formation), Lake Eyre Basin (Upper Eyre Formation) and in southern Northern Territory, e.g. the Hale River Basin (Ulgnamba Lignite) north of Alice Springs.

Data from the offshore basins are sparse but are adequate to demonstrate an increasing representation of *Nothofagus* (*Brassospora*) in rainforests on the coastal/lowland regions (see Harris, 1971). The same general vegetation is recorded in the western Murray Basin: *Nothofagus* (*Brassospora*), Casuarinaceae and gymnosperms dominate Middle Eocene palynofloras in the St Vincent Basin (N. F. Alley, unpublished data).

Macrofossils and quantitative analysis of pollen preserved in the North Maslin Sands confirm the mesotherm affinities of this rainforest as well as providing detailed information on its floristic make-up (McGowran *et al.*, 1970; Blackburn, 1981; Alley, 1987; Christophel, Chapter 11, this volume). For example, the Casuarinaceae pollen type represents *Gymnostoma*, and some undescribed *Rhoipites* spp. represent extinct species of

Araliaceae. *Nothofagus* remains were not found but there is no doubt that *Brassospora* spp. were present within the pollen source area.

Palynofloras recovered from the North Maslin Sands contain variable (3–23%) amounts of *Nothofagus* pollen, chiefly subgenus *Brassospora* but also including low numbers (< 3%) of *Fuscospora* and *Lophozonia*. Gymnosperm pollen values are variable but indicate that *Araucariacites* had replaced *Dilwynites* as the dominant araucarian type. Because of differences in the depositional environment, it is unclear whether the decrease in *Dilwynites* (21% to < 1%) is of local or regional significance. Values for other conifers are: *Dacrydium/Podocarpus* (2–12%), *Dacrydium* Group B (1–12%), Araucariaceae (*Araucariacites*: up to 6%) and *Lagarostrobos* (up to 3%).

Otherwise pollen dominance is shared between Casuarinaceae (4–8%, with occasional values up to 22%), Myrtaceae (up to 5%) and Proteaceae (5–12%). Spore values vary considerably, with Cyatheaceae (1–7%) being the most consistently frequent type. Isolated samples yielding up to 21% Gleicheniaceae and 52% Polypodiaceae have been found, e.g. at One Tree Hill. As is usually the case, a much richer regional flora is indicated by the minor species. This included *Anacolosa* (*Anacolosidites acutullus, A. luteoides, A. sectus*), *Ascarina*, *Beauprea*, Cunoniaceae, Cupanieae, Cupressaceae/Taxodiaceae, Ericales, *Eucalyptus*, *Halocarpus bidwillii* type, *Ilex*, Loranthaceae, Poaceae, Restionaceae, *Santalum* type, Sapotaceae and numerous species of Proteaceae including *Proteacidites pachypolus*, *Agastachys* type, *Gevuina/Hicksbeachia* and *Xylomelum occidentale* type. Some samples from the North Maslin Sand include Callitrichaceae pollen.

Probable Middle Eocene palynosequences have been recorded from the Lake Eyre and Lake Torrens Basins (Sluiter, 1991; N. F. Alley unpublished data) and from boreholes drilled in areas to the northwest of Alice Springs (Kemp, 1976; Truswell & Marchant, 1986). In all instances freshwater dinocysts and sedge and reed pollen confirm that the depositional environments were swamps or shallow lakes. As such, variations in the relative pollen abundance of many taxa may reflect no more than the proximity of, or local developments in, the shoreline vegetation.

The Lake Eyre/Torrens palynosequences are the more diverse. In contrast to the Early Eocene, *Nothofagus* spp. (11–35%) are dominant in palynofloras of the Lake Eyre and Lake Torrens Basins: *Brassospora* (up to 35%), *Fuscospora* (<5%) and *Lophozonia* (<1%). Gymnosperm representation differs little from that found during the Paleocene–Early Eocene except that *Dilwynites* and *Lagarostrobos* values are lower. Cyperaceae (1–5%), Restionaceae (up to 9%) and Sparganiaceae/Typhaceae (<3%) are distinctly more common: Poaceae are rare (<1%) throughout. Cryptogams were locally important, e.g. Cyatheaceae (up to 16%) and Gleicheniaceae (up to 12%) in the Lake Eyre Basin.

Geographical distributions of Casuarinaceae, Cunoniaceae and Myrtaceae follow the pattern found in the Early Eocene, i.e. Casuarinaceae are most abundant (up to 79%) in the Lake Torrens Basin, whereas Cunoniaceae are frequent–common (5–17%) only in the Lake Eyre Basin. Myrtaceae also tend to be more common here (16% versus 10%). Up to 2% of the taxon is the *Eucalyptus* type. Proteaceae producing highly ornamented pollen are mostly infrequent (1–5%), although *Proteacidites confragosus* may dominate the occasional sample. Contributing to the impression that many species had a patchy distribution within the regional vegetation are samples containing anomalously large numbers of Ericales (23%), *Nothofagidites falcatus* (19%), Sapotaceae (8%) and Elaeocarpaceae (5%) pollen.

Rare, but widely distributed taxa, include *Anacolosa*, *Beauprea*, Cupanieae, Didymelaceae, *Dodonaea triquetra* type, Euphorbiaceae, Gyrostemonaceae, *Isopogon* (*Proteacidites truncatus*), Loranthaceae (*Gothanipollis bassensis*, *Tricolpites thomasii*), and *Xylomelum occidentale* type.

Similar variation exists in two probable Middle–Late Eocene palynosequences recovered from lignitic sediments at Napperby and the Hale River Basin north of Alice Springs (Kemp, 1976; Truswell & Marchant, 1986). Pollen dominance is shared between *Nothofagus* (*Brassospora*) and Casuarinaceae (up to 50% each at Nap-

perby): Myrtaceae, Cunoniaceae and gymnosperms are uncommon to rare elements. Values of *Aglaoreidia qualumis* pollen (up to 17% at Napperby) confirm that local hydrological conditions had permitted the development of extensive reed swamps of Sparganiaceae or Typhaceae.

A strong facies control exists on the number of rare taxa recorded in any sample but overall the regional diversity is similar to areas to the south. Types that can be related to extant taxa include *Alyxia*, Araucariaceae, *Beauprea*, Cupanieae, Cyperaceae, *Dacrycarpus*, *Dacrydium* Group B species, *Dodonaea triquetra* type, Ericales, Euphorbiaceae, *Gunnera*, Gyrostemonaceae, *Halocarpus bidwillii* type, *Ilex*, Liliaceae, Loranthaceae (*Cranwellia*), *Micrantheum*, Musgraveinae, Polygalaceae, Restionaceae, *Santalum* type, Sapotaceae, *Telopea* type, Trimeniaceae, *Ungeria* and *Xylomelum occidentale* type. Amongst the extinct taxa are *Proteacidites pachypolus* and, in the Hale River Basin, a species of Droseraceae (*Fischeripollis halensis*) whose affinities lie with the Venus fly-trap (*Dionaea*) now confined to the southeast United States.

All sites confirm that *Nothofagus* (*Brassospora*) spp., including *Nothofagidites falcatus*, had become widely established within the interior of the continent. However, their topographical distribution is far from clear. For example the taxon could have been part of a gallery rainforest or largely restricted to upland areas such as the Flinders, Harts and MacDonnell Ranges. It is equally difficult to interpret the patchy distribution of Cunoniaceae, Myrtaceae, Casuarinaceae and gymnosperms, except to note that the sporadic prominence of Araucariaceae could be an early indication of mildly seasonal climates. Taxa with well-defined megatherm affinities are absent.

By the Late Eocene, extensive lignite deposits had accumulated along the coast in the Murray, St Vincent, Pirie and Eucla Basins and inland in the Torrens Basin. These preserve exceptionally rich palynofloras (N. A. Alley, unpublished data), but a very local origin of much of the spore–pollen has tended to mask any regional patterns in the vegetation.

One trend that seems to have been widespread

in southern central Australia is a marked increase in the number of sites at which *Nothofagus* (*Brassospora*) spp. (*ca* 50%) and Casuarinaceae (20–40%) were the major pollen dominants. Gymnosperms appear to decrease in importance, although *Dacrydium* Group B species or *Lagarostrobos* pollen can dominate occasional palynofloras: one sample from the St Vincents Basin yielded *ca* 30% *Lagarostrobos*. Contributing to the impression of increasingly simple rainforest types are decreases in the overall representation of Proteaceae (< 5%), Myrtaceae (< 4%) and Cunoniaceae (mostly absent). Minor taxa such as *Anacolosa*, *Beauprea*, Cupanieae, *Santalum* type and Sapotaceae become harder to find. With other taxa, notably the Proteaceae, extinctions appear to be matched by the appearance of new species, e.g. *Proteacidites rectomarginis*. Sedge and reed pollen tend to become more sporadic, with 'high' values (< 6%) being recorded only in the western and inland basins.

Western Australia

Palynofloras of Middle–Late Eocene age are restricted to the southern and southwestern margins and adjoining lowlands. These include the Bremer and western Eucla Basins (see Churchill, 1973; Hos, 1975; Stover & Partridge, 1982; Milne, 1988) and palaeochannels incised into the Yilgarn Block (Balme & Churchill, 1959; I. R. K. Sluiter, unpublished data). In most cases palynological dominance is shared between Casuarinaceae, *Nothofagus* (*Brassospora*) and Proteaceae. Assemblages recovered from a probable palaeochannel on Glen Ferrie Station, on basement east of the Carnarvon Basin, are dominated by *Nothofagus*, Casuarinaceae and Proteaceae (Truswell, 1987*a*), and have a number of taxa in common with the Werillup Formation of late Middle – Late Eocene age (see below). Included are *Phormium* (*Luminidites phormioides*: Agavaceae), *Proteacidites cumulus* and *Ascarina*. A short sequence from a site northwest of Kalgoorlie included samples with abundant Euphorbiaceae (*Malvacipollis* spp.) and palm pollen (I. R. K. Sluiter, unpublished data).

Stover & Partridge (1982) noted that more than 85% of the fossil species in the late Middle–Late Eocene Werillup Formation in the Bremer Basin also occur in southeast Australia. Shared taxa include *Cyathidites splendens*, *Culcita* (*Rugulatisporites mallatus*), *Dacrydium* Group B species, *Lagarostrobos*, *Microcachrys*, *Ascarina*, *Anacolosa* (*Anacolosidites luteoides*, *A. sectus*), *Beauprea* (*Beaupreadites verrucosus*), Cupanieae, Euphorbiaceae, *Ilex*, Loranthaceae (*Gothanipollis bassensis*), *Eugenia* and other Myrtaceae, *Nothofagus* (*Fuscospora*, *Nothofagus*), Polygalaceae, *Santalum* type, Sapotaceae, Trimeniaceae, *Ungeria* and numerous Proteaceae (e.g. *Proteacidites pachypolus*, *Agastachys* type, *Banksia/Dryandra* and *Telopea* type). Taxa not recorded in southeast Australia are mostly Proteaceae, e.g. *Proteacidites bremerensis*, whilst others such as *Dysoxylum*-type (*Tetracolporites palynius*: Meliaceae) and *Phormium* are distinctly rare in the Bassian flora. *Nothofagus* (*Lophozonia*) is absent.

Essentially the same fossil palynoflora is recorded in the Middle–Late Eocene sediments in the Eucla Basin, although pollen dominance varies with location. For example Cunoniaceae (3–20%) are common in palaeochannel deposits on the northeastern margin (Kumunga, Loongana) of the Eucla Basin (I. R. K. Sluiter, unpublished data). *Nothofagus* (chiefly *Brassospora*) spp. dominate Late Eocene palynofloras recovered from mostly nonmarine sediments in the Zanthus-6 borehole at the hilly western margin of the Eucla Basin (Milne, 1988). As in central Australia, the flora retained species that had become extinct in the southeast, e.g. *Phyllocladidites reticulosaccatus*: others did not extend into the southeast until the Oligocene, e.g. *Acaciapollenites* sp. cf. *A. myriosporites* (= ? *A. miocenicus*) and *Psilastephanocolporites micus*. The former is the earliest record to date of Mimosaceae pollen in Australia. Low to trace numbers of *Ascarina*, Cupanieae, *Santalum* type and Sapotaceae, confirm the (?distant) presence of mesothermal rainforest, whilst *Eugenia* (6%) was either abundant in the regional vegetation or, along with Trimeniaceae (up to 6%), *Xylomelum occidentale*-type (up to 8%), sedges and reeds, was part of a local peat-bog community.

Two developments recorded in the Zanthus-6

palynosequence have implications beyond the Eucla Basin. Firstly, there is the extraordinarily high number of distinctive Proteaceae pollen types (minimum 46 species). It is difficult to avoid the conclusion that by latest Eocene times, southwest Australia had already become a major centre of diversification for this family. Secondly, the decrease in *Nothofagus* (*Brassospora*) and gymnosperm pollen (associated with increasing Casuarinaceae and Restionaceae) is strongly correlated with an increasing marine influence (Figure 10.5). This influence is unlikely to have affected *Nothofagus* (*Brassospora*) communities if these were wholly confined to the uplands on the margin of the Yilgarn Block, indicating that some *Brassospora* populations had been present on the coastal plain.

A diverse Proteaceae–cryptogam flora including infrequent *Nothofagus* (subgenus not specified) occurs in probable Middle Eocene sediments

from the Kings Park Formation, Perth Basin. Churchill (1973) has recorded *Nypa* pollen in these sediments.

Latest Eocene–earliest Oligocene *ca* 37–34 Ma (36.5–35 Ma)

During the Late Eocene the rate of sea-floor spreading between Antarctica and Australia increased markedly, strengthening the Circumantarctic Current and hence the thermal isolation of Antarctica. At this time Tasmania became a peninsula jutting into the widening Southern Ocean and connected to the rest of Australia via a relatively narrow land-bridge formed by the Bassian Rise. In the diminishing ocean reaches to the north of Australia, various terranes that had been rifted off Gondwana during the Late Palaeozoic–Cretaceous were being swept up

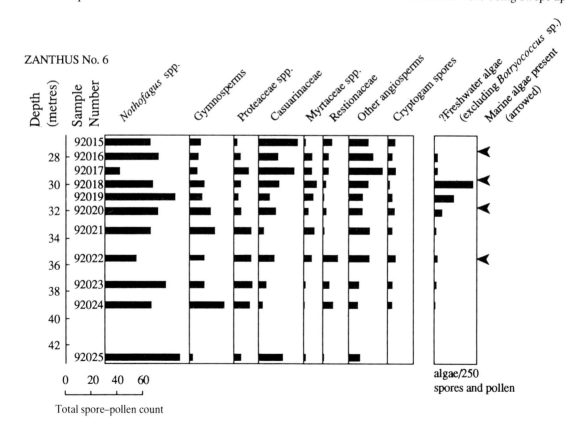

Figure 10.5 Variation in the relative pollen abundance of major tree, shrub and herb families during a period of increasing marine influence in the Late Eocene, western Eucla Basin. (Adapted from Milne, 1988.)

and accreted to the Asian continental margin, New Guinea, or embedded within the developing Malaysian archipelago (Burrett et al., 1991). These migrating continental fragments may have provided dryland 'stepping stones' for faunal and floral dispersal prior to the Middle Miocene collision between northwest Australia/New Guinea and Southeast Asia.

The earliest Oligocene is marked by one of the most important episodes of cooling of surface waters in the Southern Ocean (Wei, 1991) after a period of pronounced local volcanic activity in the South Pacific (Kennett et al., 1985). Glacial climates were already established in East Antarctica, with ice reaching sea level by ca 35.9 Ma (Wei, 1991). The decline in the temperature of southern Australian shelf waters at approximately the same time is estimated to have been as much as 7 deg.C, although there was a rapid recovery of water temperatures in shallow broad embayments during Early Oligocene (Kamp et al., 1990). The cooling event has been associated with local glacier development in northern Tasmania (Macphail et al., 1991).

In northeast Australia, seasurface temperatures fell below the minimum required for carbonate build-up (Figure 10.4) in spite of the relative low palaeolatitudes (28°–43° S).

Southeast Australia (Upper *Nothofagidites asperus* Zone)

Latest Eocene–earliest Oligocene sediments are thin and patchily distributed in southeast Australia due to falling relative sea levels and, in the Gippsland Basin, widespread erosion resulting from tectonic uplift. Marine sediments are often bioturbated and what spores and pollen are preserved are dominated by long distance/water-transported types. As a result, much of what is known of the vegetation comes from localised brown coal deposits in the onshore Gippsland Basin (Stover & Partridge, 1973) and two boreholes in the central Murray Basin (Macphail & Truswell, 1989). There is no reason to suppose either setting preserves even a moderately complete record of the southeast Australian flora.

In both basins, the palynofloras remain domi-

nated by *Nothofagus* (*Brassospora*) spp. (av. 50–60%) and Casuarinaceae (< 20%) but are transitional in other respects, notably the low diversity of angiosperms in the Gippsland Basin and the persistence of Eocene species in the flora of the Murray Basin.

The low diversity of angiosperms in the onshore Gippsland Basin almost certainly reflects the narrow range of plants able to grow in freshwater environments, then characterising the Latrobe Valley Depression, and swamp taxa are increasingly prominent and occasionally dominant in the fossil pollen record, e.g. Sparganiaceae/Typhaceae and *Lagarostrobos* (up to 80%). These are often associated with significant values of *Proteacidites rectomarginis*, *P. stipplatus* and, less consistently, *Nothofagus* (*Fuscospora*), suggesting that the latter were part of the peat-swamp vegetation. *Proteacidites pachypolus* becomes extinct in the Gippsland Basin during this zone.

Palynofloras recovered from the Bass and offshore Gippsland Basin were deposited in deep water up to 150 km away from the palaeoshoreline. Yields are low and demonstrate little more than that gymnosperms and *Nothofagus* spp. remained part of the regional vegetation. Few new species are recorded in the period. Surprisingly, *Camptostemon* (Bombacaceae), a mesothermal–megathermal taxon now confined to northern Australia and Malesia and present in the Otway Basin during the Late Eocene, appears to have been established in the Bass Basin flora only during the latest Eocene–earliest Oligocene (A. D. Partridge & M. K. Macphail, unpublished data). Several herbs and unidentified cryptogams that first appear in the pollen record near the top of the Middle *N. asperus* Zone become persistent elements in the Upper *N. asperus* Zone palynofloras, e.g. Geraniaceae (*Tricolpites geranioides*), a lycopod (*Verrucosisporites cristatus*) and several undescribed Polypodiaceae spp.

A wide range of depositional environments remained present in the Murray Basin and this is reflected in the range of plant communities recorded by spores and pollen. Nevertheless warm shelf waters in, or adjoining, the Murray Basin are almost certainly the reason for the retention there of several thermophilous species after these had become extinct in the Bass and Gippsland Basins, e.g. *Anacolosa* (*Anacolosidites sectus*), *Beauprea* (*Beaupreaidites verrucosus*), *Proteacidites crassus*, *P. pachypolus* and *P. reticulatus*. This interpretation is supported by the appearance of a Mimosaceae pollen related to the tropical African fossil species *Polyadopollenites granulosus*. Quantitative data indicate that mesothermal rainforest stands were dominated by *Nothofagus* and a range of podocarps.

Developments common to the region (including northern Tasmania) include (1) the apparently synchronous extinction of a *Sphagnum* sp., *Stereisporites* (*Tripunctisporis*) sp., present in the flora since the Late Maastrichtian, (2) the appearance of *Granodiporites nebulosus*, a pollen type now produced by only the southern South American waratah *Embothrium coccineum* (Proteaceae), and (3) a transient expansion of *Proteacidites rectomarginis* and *Nothofagus* (*Fuscospora*). *Fuchsia* (Onagraceae), now absent from Australia, was present in both the Gippsland and Murray Basins (see Berry *et al.*, 1990).

Northeast Australia

Feary *et al.* (1991) estimate that sea surface temperatures in northeast Australia decrease to 10–13 °C, but remained slightly above those recorded after *ca* 35 Ma. This cooling is seen as moderate compared to high latitudes: temperatures remained approximately constant at the equator.

No reliable latest Eocene–earliest Oligocene palynofloras are recorded from Queensland (see Foster, 1982). *Nothofagidites* spp. are abundant in Early Oligocene (or older) oil shales in the Plevna area of north Queensland (Grimes, 1980), consistent with relatively cool conditions.

Central regions

Lignitic sediments continued to accumulate within the Murray Basin (Moorlands Lignite), St Vincent Basin (Clinton Formation) and Pirie Basin (Kanaka Beds) during the latest Eocene and earliest Oligocene. A contrasting depositional environment is provided by the Buccleuch Forma-

tion, a sequence of shallow marine clays and limestones deposited near the seaway linking the Murray Basin to the Southern Ocean. Foraminifera and dinoflagellates confirm that these sediments are correlatives of the Upper *N. asperus* Zone.

Palynofloras recovered from the Buccleuch Formation (M. K. Macphail, unpublished data) resemble contemporaneous assemblages from the central-west Murray Basin, except that Araucariaceae (*Araucariacites*; 24%) are more abundant and other gymnosperms less common. *Nothofagus* (*Brassospora*) (36%) greatly exceed Casuarinaceae (8%). Rare taxa include *Nothofagus* (*Fuscospora*, *Nothofagus*), *Lagarostrobos* (4%), *Beauprea* (including *Beaupreaidites verrucosus*), *Canthium*, Convolvulaceae, Cupanieae, *Dysoxylum* type (*Tetracolporites palynius*), Ericales, Euphorbiaceae (including *Micrantheum*), *Gunnera*, Gyrostemonaceae, Loranthaceae (*Gothanipollis bassensis*), Myrtaceae, *Phyllocladus*, Proteaceae (including *Embothrium* and *Agastachys* type), *Quintinia*, Sapotaceae, Trimeniaceae and a variety of cryptogams (e.g. *Hemitelia*, *Lygodium*, *Microsorium* and *Lycopodium laterale* type).

It is improbable that these would have been derived from the one vegetation type, but overall they are consistent with the presence of mesothermal rainforest in the southwest of the basin. There are few if any indications of sclerophyll associations, whilst the virtual absence of reeds and sedges suggests that woody species formed the riparian vegetation. On the basis of pollen recovered from the Moorland Lignite, the peat-swamp vegetation in the same area included *Nothofagus* (*Fuscospora*) (8%), *Microcachrys* (13%), *Proteacidites rectomarginis* and variable populations of Casuarinaceae (6–30%). There is no compelling evidence that *Nothofagus* (*Brassospora*) grew locally or that *Lagarostrobos* was part of any local swamp forest association.

Pollen data from the St Vincent Basin and the northwest Murray Basin confirm the (variable) presence of *Nothofagus* (*Brassospora*) and general scarcity of Araucariaceae and Myrtaceae in the regional vegetation. Isolated instances exist where peat-swamp communities were wholly dominated by Casuarinaceae (67–80%) or *Laga-*

rostrobos (29%) and (one sample from the Lochiel Coal Pit) by *Eugenia*. *Eucalyptus* pollen comprises 3% of the same palynoflora. Cunoniaceae pollen, if present, are extremely rare. In both areas percentages of Sparganiaceae and/or Typhaceae are sufficiently high (4–7%) to show that reed swamps were present, although probably not extensive.

Harris & Twidale (1991) cited an 'Upper *N. asperus* Zone equivalent' palynoflora from the Ayers Rock area as containing Malvaceae (*Malvacearumpollis*), *Quintinia*, Gyrostemonaceae, an abundance of gymnosperm pollen and a low diversity of Proteaceae pollen.

Western Australia

No palynofloras of latest Eocene–earliest Oligocene age have been published for this region.

Early Oligocene–late Early Miocene *ca* 34–18.5 Ma (35–18.5 Ma)

During this period Australia drifted northwards through five degrees of latitude, resulting in the northern margin reaching *ca* 20° S by the Middle Miocene. Although global climates continued to cool, this trend was punctuated by episodes of rapid cooling and limited warming. Shackleton (1986) related cooling events at *ca* 31 and 24 Ma to intense glacial events in Antarctica; warming episodes include the local warming of shelf waters along the southern margin in the Early Oligocene, and a more general reversal of the cooling trend during the early Late Oligocene and Early Miocene (Feary *et al.*, 1991). To date, few, if any, palaeobotanical events can be linked to a particular climatic excursion, although broad directions of change correlate well with long-term climatic trends.

A major drop in relative sea level at 30 Ma, followed by marine transgression during the Late Oligocene and Early Miocene, has resulted in a regional unconformity separating Early Oligocene and Late Oligocene–Early Miocene shelf deposits. At about this time the Bassian Rise was submerged, isolating Tasmania (then at *ca* 57° S) from the southern mainland except during periods of glacio-eustatic low sea levels.

Southeast Australia

A. *Proteacidites tuberculatus* Zone (Early Oligocene)

Due to widespread erosion of shelf sediments and nondeposition elsewhere, there are few confirmed deposits of Early Oligocene *P. tuberculatus* Zone age. These are (bathyal) calcareous to noncalcareous shales, principally deposited up to 50–100 km away from the palaeocoastline in the Bass Strait region. Apart from confirming that *Lophosoria* (*Cyatheacidites annulatus*), a Filicopsidan fern now restricted to upland tropical America, was part of the Bassian flora by *ca* 35 Ma, the sparse palynofloras are little more than a record of well-dispersed types in the regional flora. The few known onshore correlatives tend to reflect local or specialised floras.

An example of the latter is the Childers Formation underlying the Thorpdale Volcanics in the Moolamoona Coal Mine, onshore Gippsland Basin (Partridge, 1971). Palynofloras recovered from lignites and associated clays correspond to the Lower *P. tuberculatus* Zone assemblage recognised by Stover & Partridge (1973). The earliest records of Chenopodiaceae-Amaranthaceae (*Chenopodipollis chenopodiaceoides*) and *Eucalyptus* pollen in the Latrobe Valley Depression occur in this zonule: *Beaupreaidites verrucosus*, *Embothrium*, *Epacris virgata* type, *Gillbea* type and *Halocarpus bidwillii* type are last recorded in the zonule. The assemblages are dominated by *Nothofagus* (*Brassospora*) spp. and include frequent numbers of *Proteacidites rectomarginis* and a *Lycopodium* sp. (*Foveotriletes crater*).

Three apparently contemporaneous Early Oligocene assemblages occur in northern Tasmania – at Lemonthyme Creek and Little Rapid River in the northwest and at Cethana in the central north (M. K. Macphail, unpublished data). *Cyatheacidites annulatus* provides a confident *P. tuberculatus* Zone maximum age, but the upper age limits are less certain due to possible diachronism in the times of extinction of accessory species such as *Beaupreaidites verrucosus*, *Granodiporites nebulosus* and *Triporopollenites ambiguus*. In each case the vegetation is interpreted as a variant of microthermal–mesothermal rainforest.

All sites are located within mountainous terrain and to that extent are unrepresentative of ecological conditions occurring in the relatively subdued topography of the mainland. Nevertheless three developments are of general significance:

1. Two of the sites (Cethana, Little Rapid River) preserve macrofossil evidence of *Nothofagus* (*Brassospora*) spp. (Hill, 1987; Carpenter, 1991). Thus it is certain that *Brassospora* was in fact part of mesothermal rainforest associations growing around the foreshore at these sites. This calls into question the widely held assumption, first expressed by Duigan (1966), that *Nothofagus* (*Brassospora* spp.) were restricted to upland sites beyond the macrofossil source areas.

2. The presence of *Acmopyle* (a podocarp now restricted to New Caledonia and Fiji), *Papuacedrus*, *Libocedrus* (Cupressaceae restricted to New Guinea and New Zealand/New Caledonia, respectively) and a shrub or liane species of the Asteraceae tribe Mutisieae (largely confined to South America) confirm that this rainforest and probably most other early Tertiary communities in southern Australia simply have no close living analogues.

3. Fewer subcanopy angiosperms are represented than is the case with either Eocene or Late Oligocene–Early Miocene assemblages in northern Tasmania. This sequence is ecologically consistent with a temporary reversal of the global cooling trend during the Early Miocene but the link remains speculative at present.

On present indications, cryptogams and possibly some herbs had extended their dominance at the expense of small trees and shrubs during the Early Oligocene: there is no pollen evidence for a corresponding shift in canopy dominance away from *Nothofagus* spp. (chiefly *Brassospora*: 30–53%) and conifers (chiefly *Dacrydium/Podocarpus*, with lesser numbers of *Lagarostrobos*, *Dacrycarpus*, *Phyllocladus* and Araucariaceae). This development is most clearly recorded at Lemonthyme Creek, located within a valley deeply incised into mountains adjoining the northwest margin of the Central Plateau. Samples taken

at 1–1.5 m intervals through a 34 m thick sequence of deep-water clays and silts yielded essentially identical palynofloras. Palynological dominance is shared broadly between conifers (*Dacrydium/Podocarpus*, *Lagarostrobos*), *Nothofagus* (*Brassospora*) and cryptogams (Cyatheaceae, Dicksoniaceae, lycopsids), with the maximum value reached by any one species being less than 30%. Araucariaceae, *Microcachrys* (*Podosporites parvus*), *Microstrobos*, *Dacrycarpus*, *Dacrydium* Group B, *Nothofagus* (*Fuscospora*, *Lophozonia*, *Nothofagus*) and Casuarinaceae are represented by low pollen values (< 1–6%). The herb flora includes Cyperaceae, Restionaceae, Droseraceae, Gentianaceae (*Striasyncolpites laxus*), *Gunnera* (*Tricolpites reticulatus*) and the earliest record to date of Stylidiaceae (*Forstera* type).

On present indications the affinities of this rainforest community are more likely to be microtherm than mesotherm. This impression is reinforced by the persistent presence of taxa that are characteristic of *Nothofagus cunninghamii* cool temperate rainforest in Tasmania – *Microstrobos*, *Anopterus* (Escalloniaceae), epacrids and two Proteaceae (*Agastachys*, *Telopea* type) – and the extreme rarity of thermophilous taxa (*Beaupreaidites verrucosus*, Cupanieae, Sapotaceae).

The diversity of subcanopy angiosperms represented by pollen at Cethana is even lower, with Casuarinaceae (1%), Proteaceae (including *Embothrium*) and Euphorbiaceae being the only taxa able to be identified with certainty. This may be an artefact of preservation, since the sites preserve a rich macrofossil flora, including cycads, *Gymnostoma*, three genera of Cunoniaceae (*Callicoma*, *Vesselowskya*, *Weinmannia/Cunonia*), Elaeocarpaceae, Lauraceae, Myrtaceae, Proteaceae (including *Banksia/Dryandra*, *Lomatia*) and Sterculiaceae (*Brachychiton*). However the range of conifer and *Nothofagus* spp. represented by macrofossils accords well with the pollen data, e.g. *Lophozonia* (9%) and *Nothofagus* (7%): the herb (*Gunnera*, Droseraceae, Restionaceae) and the cryptogam flora is similar to that of Lemonthyme Creek except that macrofossils confirm the presence of *Lygodium* and Gleicheniaceae. Casuarinaceae is the only subcanopy angiosperm represented by frequent–common pollen (16%)

at Little Rapid River, although trace values demonstrate that *Ascarina* and *Quintinia* were also present in the general area. The site differs from Cethana and Lemonthyme Creek in that two possible sclerophyllous taxa *Dodonaea triquetra* and Gyrostemonaceae occurred within the pollen source area.

B. *Proteacidites tuberculatus* Zone (Late Oligocene–late Early Miocene)

For various reasons, including tectonic ponding, damming of valleys by lava in the Eastern Highlands and the existence of ephemeral lakes and sluggish rivers in the interior, Late Oligo-Miocene sediments are well preserved but scattered throughout much of the lower eastern half of the continent.

Irrespective of location, these palynostratigraphies give the impression of containing fewer first appearances and extinctions than Eocene–earliest Oligocene palynofloras. To some extent this reflects the lack of taxonomic attention paid to the less distinctive tricolpate and tricolporate pollen taxa, but there can be little doubt that many apparently different vegetation types shared large numbers of fossil (pollen) species and that shifts in community composition were diachronous within southeast Australia (see Macphail & Truswell, 1989). For possibly the first time during the Cenozoic, the pollen data allow the ecological spectrum of rainforest associations to be described in some detail.

1. Lowland peat-swamp/lake complex, Gippsland Basin

Proteacidites tuberculatus Zone sediments in the Latrobe Valley Depression consist of thick brown coal measures that accumulated inland of a sand barrier complex marking the maximum point of Cenozoic marine transgression in the basin. Whilst not wholly isolated from a marine influence during the Late Oligocene–Miocene (Holdgate & Sluiter, 1991), these coal measures were deposited in freshwater environments ranging from open-water lakes (intermittently affected by widespread fluvial incursions) to extensive peat bogs and swamps. Changes in depositional environments are linked to colour of the sediments, with light lithotypes indicating open-water conditions and dark lithotypes peat-swamp environments (Luly *et al.*, 1980). *Nothofagus* pollen becomes relatively less common in the darker lithotypes, although overall *Brassospora* spp. dominate (56–62%) all assemblages (Sluiter & Kershaw, 1982).

Macrofossils (Blackburn & Sluiter, Chapter 14, this volume) and pollen data indicate that Aquifoliaceae, Casuarinaceae, possibly *Nothofagus* (*Lophozonia* but probably not *Brassospora* or *Fuscospora*), were part of the peat-swamp vegetation. Whether or not all *Brassospora* spp. were wholly confined to the adjacent uplands or extended on to freely draining sites within the Latrobe Valley Depression is uncertain but wide fluctuations in the relative representation of *Nothofagidites deminutus-vansteenisii* (see Holdgate & Sluiter, 1991) suggests that one ecotype was part of vegetation fringing the peat swamps.

Quantitative pollen analysis (Sluiter & Kershaw, 1982) has identified a number of plant associations (Figure 10.6) and allowed several angiosperms to be associated with a particular habitat. For example, the relatively higher values of *Quintinia*, Cunoniaceae (tricolpate) and Sapotaceae pollen in open-water lithotypes are consistent with a regional distribution but do not preclude a local presence. Three main associations are recognised within the wetland vegetation (maximum pollen values in parentheses). These are:

1. *Lagarostrobos* (40%), *Phyllocladus* (12%), dicolpate Cunoniaceae (8%), *Quintinia* (5%), Sapotaceae (2%) and Aquifoliaceae (2%);
2. Myrtaceae (30%) and Elaeocarpaceae (15%), possibly associated with Cunoniaceae (*Ceratopetalum*-type; 20%), *Xylomelum occidentale*-type (8%) and *Dacrydium* Group B species (20%);
3. *Dacrydium/Podocarpus* (8%), Epacridaceae (6%) and Gleicheniaceae (15%) probably associated with *Banksia/Dryandra* (2%) and Restionaceae (4–10%).

The last association occurs in peat lithotypes

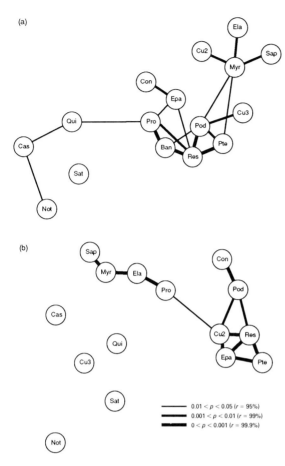

(a)

(b)

━━━ 0.01 < p < 0.05 (r = 95%)
━━━ 0.001 < p < 0.01 (r = 99%)
━━━ 0 < p < 0.001 (r = 99.9%)

Figure 10.6 A comparison of probable plant associations occupying the Latrobe Valley depression during (*a*) the late Early–Middle Miocene, *T. bellus* Zone and (*b*) Late Oligocene–Early Miocene *P. tuberculatus* Zone. Associations are inferred from correlation coefficients between relative pollen frequencies of major taxa from the Yallorn and Morwell coal formations. Ban, *Banksia*; Cu2, dicolpate Cunoniaceae; Cu3, tricolpate Cunoniaceae; Cas, Casuarinaceae; Con, other Coniferae; Ela, Elaeocarpaceae; Epa, Epacridaceae; Myr, Myrtaceae; Not, *Nothofagus*; Pod, *Podocarpus*; Pte, Pteridophytes; Qui, *Quintinia*; Sap, Sapindaceae; Sat, Sapotaceae; Res, Restionaceae. (Adapted from Sluiter & Kershaw, 1982.)

which contain abundant charcoal, indicating repeated wildfires. The community is interpreted as relatively open heath/scrub on highly acidic, infertile soils. Conversely, the Myrtaceae–Elaeocarpaceae association is correlated with kaolin-rich interseam sediments deposited,

according to Holdgate & Sluiter (1991), during periods of relative sea level rise and therefore with warmer temperatures.

2. Alluvial fan, Pioneer, Northeast Tasmania

Proteacidites tuberculatus Zone palynofloras are preserved in siltstones in a small cut-off channel at 90 m elevation in an alluvial fan extending from granite uplands in north-east Tasmania (Hill & Macphail, 1983). These deposits probably represent short-term flooding events in what was a high energy environment. K–Ar dating fixes the upper age limit at about the Early/Middle Miocene boundary, although the palynoflora may be as old as Late Oligocene.

In spite of the abundance of *Nothofagus* (*Brassospora*) relative to *Lophozonia* pollen (49% versus 4%), the macrofossil data indicate that the local rainforest was in fact formed by an extinct member of the latter group, *N. johnstonii*, possibly associated with *Fuscospora* sp. (4%). Conifers appear to have formed a sparse overstorey, although the only taxa that can be shown to be local are *Athrotaxis*, *Phyllocladus* and ?*Dacrydium* Group B (4%). Unlike the earlier Lemonthyme Creek sequence, a rich cryptogam flora (chiefly Cyatheaceae and Blechnaceae types; up to 15%) is associated with a diverse subcanopy angiosperm flora including *Agastachys* type, tricolpate Cunoniaceae (1%), Cupanieae, *Eucalyptus* type, Euphorbiaceae, *Ilex*, Loranthaceae, *Quintinia* (2%), *Rhodomyrtus* type, Sapotaceae, Trimeniaceae, Winteraceae and *Xylomelum occidentale* type. Casuarinaceae values (1%) are too low to confirm a local presence. On present indications the affinity of this rainforest lies with the mesotherm group.

The absence of *Nothofagus* (*Brassospora*) macrofossils led Hill & Macphail (1983) to suggest that stands of this taxon were located upslope on the granite batholith. If correct, then the local vegetation was altitudinally or, less likely, edaphically zoned. This conclusion must be considered suspect in the light of the Cethana and Little Rapid River evidence (see above).

3. Highlands, Southern New South Wales and Central Tasmania

Proteacidites tuberculatus Zone palynofloras and associated macrofossil assemblages occur in fresh-water pond sediments from within the present-day upper subalpine/alpine tension zone – at Kiandra (elevation 1600 m) on the Southern Highlands of New South Wales (Owen, 1988) and Monpeelyata (elevation 920 m) on the Central Plateau of Tasmania (Macphail *et al.*, 1991). Both sites are associated with Tertiary volcanics, although volcanism has had no detectable impact on either palaeoflora. K–Ar dates demonstrate the upper age limit is Early Miocene at each site. The Early Miocene lower age limit for the Kiandra assemblages is based on an anomalously early record of the Middle–Late Miocene indicator *Canthiumidites bellus* (*Randia*: Rubiaceae) and the absence of significant erosion between the sediments and the overlying basalts: a Late Oligocene age remains possible for the Monpeelyata assemblage.

Although the precise chronological relationship between the two floras cannot be determined, the data are sufficient to demonstrate that temperature-forced impoverishment of the late early Tertiary rainforest flora was time transgressive along latitudinal and altitudinal gradients:

1. The Kiandra flora (which closely resembles the Pioneer assemblage) is significantly more diverse than that at Monpeelyata. Except for a single grain of *Archidendron* type (*Acaciapollenites miocenicus*, misidentified as *A. myriosporites* by Macphail *et al.* (1991), the Monpeelyata flora lacks taxa with megatherm–mesotherm affinities. This is not the case at Kiandra where the flora includes, in addition to *Eugenia*, Lauraceae and *Syzygium* (see below), *Beauprea* (*Beaupreaidites verrucosus*), Cupanieae and Sapotaceae. In contrast, gymnosperms, especially *Lagarostrobos* (45%), and herbs, including Apiaceae, Droseraceae, Gentianaceae, Poaceae and Restionaceae, are distinctly more common at Monpeelyata.
2. The dominance (88.5%) of the macrofloras at Monpeelyata by very small angiosperm leaves

almost certainly reflects an adaptive response to global cooling. No such evidence of cool climates has been demonstrated at Kiandra, where the macrofossil flora consists mainly of podocarps (*Dacrycarpus*) and notophyllous taxa including Lauraceae and Myrtaceae (chiefly lowland–montane genera such as *Acmena*, *Austromyrtus*, *Eugenia*, *Syzygium*, *Xanthomyrtus* and Leptospermoidae). Silicified stumps and wood of *Nothofagus*, *Acacia* and Myrtaceae (aff. *Eucalyptus*) occur in Early Miocene volcanoclastic deposits within the upper Lachlan Valley to the north of Kiandra (Bishop & Bamber, 1985). None of the fossils is definitely in growth position, but the diameters of the remains are consistent with the relative warm conditions implied by the low (maximum 350 m) elevation; the *Nothofagus* specimen has distinct annual rings, suggesting seasonality, but no information is available on the width of the annual growth increments.

Macrofossils confirm that *Microstrobos*, and at least three species each of *Araucaria* and *Dacrycarpus* were part of the Monpeelyata flora. Suprisingly, in view of its high pollen values (43%), *Lagarostrobos* remains were not found. A similar mismatch exists for *Nothofagus*. *Fuscospora* and *Lophozonia* spp. are represented by macrofossils but only by trace amounts of pollen: *Brassospora* pollen is abundant (43%), but macrofossils are absent. Like Pioneer, pollen and macrofossils confirm that Casuarinaceae were not part of the local flora. This can be viewed as an early indication of the divergence between the vegetation of Tasmania and the mainland (where the majority of Oligo-Miocene macrofossil floras include either *Casuarina* or *Gymnostoma*).

4. Riverine plain, Murray Basin

Oligo-Miocene transgression of the Murray Basin has resulted in a laterally extensive and diverse range of marginal marine environments (Figure 10.7) that preserve not only spores and pollen produced by plants restricted to specialised shoreline communities but also palynomorphs from a wide

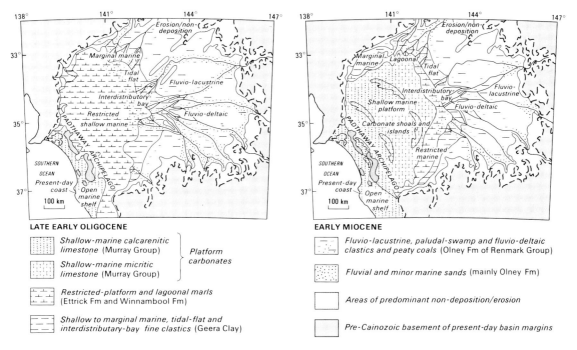

Figure 10.7 Changes in depositional environment during the Late Oligocene–Early Miocene marine transgression of the Murray Basin. (From Brown, 1989.)

range of communities upstream. Pollen and spores recovered from fluvio-lacustrine sediments represent a much narrower range of plant communities, chiefly plants growing along the riverbanks or in peat swamps infilling cut-off channels within the floodplain (Martin, 1987; Macphail & Truswell, 1989).

As elsewhere on the southeastern mainland, all palynofloras are dominated by *Nothofagus* (*Brassospora*), Casuarinaceae and gymnosperms (chiefly *Dacrydium/Podocarpus* and Araucariaceae). Taxa that are better represented in shoreline, strand and tidal-flat than in fluvio-lacustrine sediments include ?*Acacia*, reeds (*Aglaoreidia qualumis*, *Sparganiaceaepollenites* spp.), and sedges (Cyperaceae, Restionaceae), halophytic shrubs such as Chenopodiaceae–Amaranthaceae (*Chenopodipollis chenopodiaceoides*), bryophytes (*Rudolphisporis rudolphi*, *Ricciaesporites kawaraensis*) and algae (*Botryococcus*, Desmidaceae, *Pediastrum*). Occasionally, fluvio-lacustrine sediments may be dominated by one or more of Cupanieae, Euphor-

biaceae, Myrtaceae (noneucalypt types), *Nothofagus (Fuscospora)*, Proteaceae and tricolporate types.

The diversity of riverine environments within the basin is matched by a regional pollen and spore flora which greatly exceeds contemporaneous floras in the Latrobe Valley and Tasmania in terms of species richness. Angiosperm affinities range from nonseasonal megathermal to mildly seasonal mesothermal rainforest and sclerophyllous vegetation types: *Alangium*, *Alyxia*, Arecaceae species, *Ascarina*, Caesalpinoidae, Convolulaceae, Droseraceae, Euphorbiaceae (including *Austrobuxus* spp. and *Micrantheum*), Gyrostemonaceae, Loranthaceae (*Cranwellia costata*, *Gothanipollis*), Malpighiacieae, Malvaceae, Mimosaceae (including *Acacia* and *Archidendron* type), Myrtaceae (including *Eugenia*, *Eucalyptus* and *Rhodomyrtus* types), Onagraceae (*Epilobium* and *Fuchsia*), Polygonaceae (*Glencopollis ornatus*), Portulacaceae (*Lymingtonia*), Proteaceae (including *Agastachys* type, *Banksia/Dryandra*, *Gevuina/Hicksbeachia*,

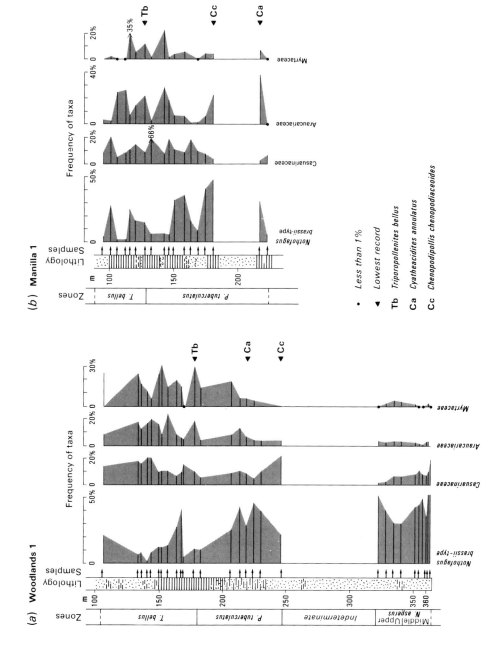

(a) Woodlands 1

(b) Manilla 1

Frequency of taxa

Zones Lithology Samples

• Less than 1%
• Lowest record
▼ Tb Triporopollenites bellus
Ca Cyatheacidites annulatus
Cc Chenopodipollis chenopodiaceoides

(d) **Hatfield 1**

(c) **Piangil West 2**

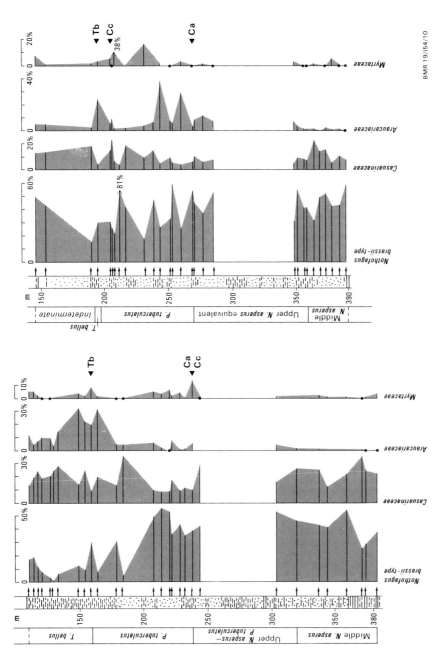

Figure 10.8 Relative pollen data illustrating the (diachronous) replacement of *Nothofagus* (*Brassospora*)-dominated evergreen rainforest by drier vegetation types in the central-west Murray Basin during the Late Oligocene–Early Miocene. (From Macphail & Truswell, 1989.)

Hakea/Grevillea, *Isopogon*, *Telopea*-type, *Xylomelum occidentale*-type and numerous extinct species such as *Proteacidites rectomarginis*, *P. pachypolus*), Rubiaceae (including *Canthium*, *Coprosma*, *Gardenia* and *Guettarda* types), Trimeniaceae and Winteraceae. Ferns include Blechnaceae (*Stenochlaena* sp. cf. *S. papuanum*) and Pteridaceae. One conifer (*Microstrobos*) has microtherm affinities.

Unlike the Bassian region, there is a distinct but time-transgressive shift away from *Nothofagus* (*Brassospora*) to palynological dominance by Myrtaceae and Casuarinaceae via a phase (usually in the Early Miocene) in which Araucariaceae are prominent (Figure 10.8). This development almost certainly represents the replacement of evergreen rainforest communities by drier, more open forest types as regional precipitation became more seasonally distributed. Because the shifts in relative dominance do not conform to any predictable pattern, at least within the central west sector, it has not been possible to determine what, if any, gradients in precipitation existed across the basin (see Macphail & Truswell, 1989). *Lagarostrobos* appears to have become progressively restricted to the southern (coastal) sector of the basin.

5. Coastal lowlands, New South Wales

Lignitic sediments outcrop at a number of locations along the New South Wales coastline. Many are Holocene but two deposits have been shown to preserve generalised Oligo-Miocene palynofloras – the Long Beach Formation near Merimbula on the south coast and the 'leaf beds' at Little Bay between Sydney Heads and Botany Bay. Both sites are within tens of metres of the ocean but only one sample (at Little Bay) contains evidence for marginal marine conditions at the time of deposition. Zone index species are absent, although neither site appears to be older than Oligocene or younger than Miocene on the basis of accessory species. Determinations based on projections of palaeomagnetic poles on the Cenozoic Polar Wander Path for Australia (Nott *et al.*, 1991) reveal a minimum Early Miocene age for deposition of the Long Beach Formation: *Rugulatisporites cowrensis* near the base of the Little

Bay sequence is consistent with an Early Miocene maximum age limit (M. K. Macphail, unpublished data). In both instances the source of the palynofloras is interpreted as floristically depauperate *Nothofagus* mesothermal rainforest, possibly growing on nutrient-poor soils.

(a) Long Beach Formation

Regional stream aggradation, related to uplift of the Southeastern Highlands as well as eustacy, covered much of the southern New South Wales coastal lowlands with a thin (< 150 m) blanket of terrestrial sediments during the Tertiary. The Long Beach Formation, now forming a 10–30 m high sea-scarp, is one of a number of strongly weathered remnants of these deposits preserved at and below present-day sea level.

The stratigraphic relationship of two samples of lignites preserving spore–pollen is unknown (see Morgan, 1974) but pollen aggregates indicate a very local origin for much of the pollen and spores, including possibly *Nothofagus* (*Brassospora* spp.) (13–16%) and *Lophozonia* (3–5%). In one instance the depositional environment almost certainly was a peat-swamp supporting Gleicheniaceae (*Dictyophyllidites arcuatus*: 42%), an unidentified Rutaceae-like species (10%) and, less certainly, Sparganiaceae/Typhaceae (3%) and *Dacrycarpus* (4%). The other habitat is less clearly defined but Casuarinaceae (18%), an unidentified tricolpate species (11%) and *Dictyophyllidites arcuatus* (48%) appear to have been part of the peat-forming vegetation. Otherwise *Dacrydium/Podocarpus* (5–7%) is the only taxon to be frequent in both palynofloras.

Rare taxa provide hints of a richer flora elsewhere: Cupanieae, *Eucalyptus* type, *Halocarpus bidwillii* type, *Lagarostrobos*, *Nothofagus* (*Fuscospora*, *Nothofagus*), *Xylomelum occidentale* type and, less certainly, Meliaceae and *Nymphoides* (Menyanthaceae).

(b) Little Bay

Tertiary sediments at Little Bay consist of a *ca* 14 m thick sequence of sandy clays and lignitic sands. Overall the sequence coarsens upwards, terminating in an indurated black sand similar in appearance to 'coffee rock' (Bird, 1968, p. 135) and yielding low numbers of a marine dinoflagellate and (rare) Chenopodiaceae pollen.

Clays at the base of the sequence are overwhelmingly dominated by *Nothofagus* (*Brassospora*) spp. (81%, including 9% *Nothofagidites falcatus*). At the top of the sequence, values have decreased to 59%, whereas those of *Lophozonia* and Casuarinaceae have increased (from 6% to 9% and 2% to 7%, respectively). *Eucalyptus* (2%) is the predominant Myrtaceae and a type resembling *Mallotus* (Euphorbiaceae) is prominent (2%) amongst the tricolporate pollen. The only gymnosperms recorded are Araucariaceae (1%), *Dacrydium* Group B, *Dacrydium/Podocarpus* (2%) and *Phyllocladus*.

As in the Long Beach Formation, rare pollen types suggest a much more diverse (mesothermal) regional flora, including a palm (*Dicolpopollis* cf. *metroxylonoides*), *Acacia*, *Ascarina*, Caesalpinoidae, tricolpate Cunoniaceae, Cupanieae, Elaeocarpaceae, Euphorbiaceae (*Malvacipollis*), Malvaceae, Meliaceae, Myrtaceae (including *Eugenia* and *Rhodomyrtus* types), *Quintinia*, Sapotaceae (possibly *Planchonella*) and *Ungeria*.

Northeast Australia

Carbonate build-ups did not occur in northeastern Australia during the Oligocene in spite of the continent being at palaeolatitudes as low as 23° S. Sea surface paleotemperatures reached minimum values during the Late Oligocene (see Table 10.1). Subtropical build-ups, succeeded by tropical coral reefs first developed in the northernmost part of the area during the latest Oligocene or Early Miocene and later developed further south, with continued warming in the Late Miocene and Pliocene (Figure 10.4).

One Late Oligocene–Early Miocene palynosequence is available, from fluvio-deltaic sediments intersected in a borehole at Sandy Cape on Fraser Island near Brisbane (Wood, 1986). This sequence confirms that southern Queensland shared a large number of fossil species with southeast Australia but shows that the coastal rainforest was relatively lacking in *Nothofagus* (*Brassospora*)

(< 20%) and all gymnosperms except Araucariaceae (up to 15%). The impression given is that the vegetation represents the warmer end of an ecological continuum of mesothermal (rather than megathermal) rainforest types in which *Nothofagus* spp., except possibly *Nothofagidites falcatus* (8%), were competitively unsuccessful: because much of the Araucariaceae pollen is of the *Dilwynites* type, it is premature to suggest its irregularly high values reflect mildly seasonal climates.

High spore values (up to 60%) imply that many palynomorphs have been water-transported to the site. If this is correct, the dryland component should be biassed toward species lining the river banks. These appear to have included Myrtaceae (up to 10%), *Mallotus* type (up to 4%), Proteaceae (3–6%) and a plethora of unidentified tricolpate and tricolporate types (up to 12%). At the fossil species level, the Sandy Cape palynofloras are very similar to contemporaneous assemblages in the Murray Basin. The more distinctive of these are *Alangium villosum* type, *Archidendron* type, Convolvulaceae, Caesalpinoidae, *Epilobium*, *Guettarda* type, Malvaceae, and Proteaceae spp., including *Proteacidites pachypolus*. Both this flora and that of Kiandra in the southeastern highlands include early records of *Randia* (*Canthiumidites bellus*). Most of these taxa are absent or very rare in the Gippsland Basin and Tasmania.

Central regions

Palynological records of the Oligo-Miocene flora and vegetation are confined to onshore basins in the centre and south of South Australia. In many instances these can be independently dated using forams, e.g. restricted marine to paralic facies in the southwest Murray Basin (Upper Buccleuch, Ettrick and Geera Clay Formations) and in the Barossa Basin (Rowland Flat Sands). Contemporaneous fluvio-lacustrine sediments inland include the Etadunna and Namba Formations in the Lake Eyre Basin.

Early–early Late Oligocene palynofloras from the southwest of the Murray Basin are usually dominated by *Nothofagus* (*Brassospora*) spp. (14–67%) but differ from the Upper *N. asperus*

Zone assemblages in several other ways (N. F. Alley, unpublished data):

1. The representation of *Fuscospora* and *Lophozonia* types is sufficiently high (5–13%) to indicate that both were local.
2. Myrtaceae and, by the Late Oligocene, Araucariaceae (up to 35%) had replaced Casuarinaceae (4–8%) as the most abundant pollen types apart from *Nothofagus*.
3. Proteaceae are both rare (< 1%) and of restricted diversity.

As in the centre-west of the basin, all developments are likely to be part of a precipitation-forced trend towards drier, more open forest types. Very rare occurrences of *Eucalyptus spathulata*-type pollen in some Late Oligocene assemblages are consistent with drier climates, since eucalypts producing this pollen type are now confined to warm, subhumid to semiarid regions in the far southwest of Western Australia (see Martin & Gadek, 1988). The records are of considerable phytogeographical interest because the pollen type has not been recorded in southeast Australia before the Late Miocene.

Widespread but minor taxa include Arecaceae, *Beauprea*, Cupanieae, Cupressaceae/Taxodiaceae, Cyperaceae (1–7%), *Dacrycarpus*, *Epilobium*, Ericales, *Fuchsia*, *Halocarpus bidwillii* type, *Ilex*, *Lagarostrobos*, Malvaceae, *Micrantheum*, *Microcachrys*, Poaceae (up to 2%), Polygalaceae, Restionaceae (< 1%), Sapotaceae and Sparganiaceae/Typhaceae (1–2%). *Eucalyptus* pollen may form up to 2% of some palynofloras. Asteraceae are first recorded in the Middle Oligocene. Casuarinaceae (5–21%) and Cunoniaceae (1–5%) are better represented in the northwest than in the southwest of the basin within South Australia.

Latest Oligocene–Early Miocene palynofloras in the Barossa Basin (N. F. Alley, unpublished data) are dominated by *Nothofagus* (*Brassospora*) spp. (18–60%) and include significant amounts of *Fuscospora* spp. (up to 12%), although *Lophozonia* are rare (< 1%). Cupanieae and *Quintinia* pollen are infrequent (*ca* 2%). The assemblages resemble contemporaneous palynofloras from the southwest Murray Basin in that Myrtaceae

(4–28%) are usually more abundant than Casuarinaceae (3–20%). Araucariaceae, however, are relatively rare (< 6%) with *Dilwynites* occurring only in trace amounts. Rare taxa include palms, *Anacolosa* (*Anacolosidites sectus*), *Ascarina*, Euphorbiaceae, *Ilex*, Loranthaceae, Malvaceae, Poaceae, Restionaceae, *Santalum* type, Sapotaceae and Winteraceae.

On present indications, the source vegetation more closely resembled rainforest present in the region during the Late Eocene than the contemporaneous rainforest communities in the adjoining Murray Basin. The reasons for this almost certainly include warm sea surface waters, since pollen of the mangrove palm *Nypa* (*Spinizonocolpites baculatus*) are present in the sequence. As with *Anacolosidites sectus*, this represents the youngest record to date of the taxon in southern Australia.

Little information is available for the interior of the continent during the Oligo-Miocene. Martin (1990) has proposed that a palynoflora from the Callabonna Basin near Lake Frome is Late Oligocene–Early Miocene. If this is correct, then populations of *Nothofagus* (*Brassospora*) (< 5%) had undergone a massive contraction during the Oligocene. Significantly, *Nothofagidites falcatus* and *N. deminutus-vansteenisii* constitute 50% of the total *Nothofagus* pollen count: all *Fuscospora* and *Lophozonia* types are absent. Gymnosperms are relatively prominent, in particular *Dacrydium/ Podocarpus* (9%), *Dacrydium* Group B (5%), *Lagarostrobos* (3%) and *Phyllocladus* (3%). Because the assemblage is dominated by locally sourced pollen (Restionaceae, 39%, Sparganiaceae/Typhaceae, 5%), it is probable that all dryland taxa, including rare (< 1%) types such as Araucariaceae, Cupressaceae/Taxodiaceae and Myrtaceae, were in fact more frequent than the relative pollen values suggest.

Whether *Nothofagus* had become restricted to the adjacent Flinders Ranges to the west, or were merely very rare elements in the local dryland vegetation, is unknown. In either case, the logical explanation is a reduced/more seasonal rainfall, although levels were adequate to support (?gallery) wet forest communities of Cyatheaceae (*Cyathidites paleospora*; 10%), Casuarinaceae (10%) and rainforest gymnosperms at, or upstream from, the site (a sedgeland with some indications of Sparganiaceae/Typhaceae reed swamp). Restionaceae values are far in excess of values recorded at any other Oligocene–Miocene site. Whilst it is premature to conclude from this that sedgeland had become widespread along rivers in the interior of the continent, the development would be consistent with increasingly seasonal/sluggish stream flow regimes. The pollen spectra do not confirm the presence of the grasslands suggested by Callen & Tedford (1976).

Western Australia

The only palynofloras possibly of Late Oligocene–Early Miocene age come from the Weelumurra Beds, Fortesque River area, in northwest Western Australia (Truswell, 1987b). The depositional environment appears to have been a freshwater lake, impounded behind a bar of Proterozoic dolomite where the ancestral Weelumurra Creek discharged on to the Fortesque Plain from its source in the Hamersley Range. The age determination is based on *Acaciapollenites myriosporites* and an absence of Asteraceae and Poaceae pollen: the maximum age is unlikely to be older than Late Eocene on the basis of the presence of *Nothofagidites falcatus* and *Proteacidites truncatus*.

The palynosequence is dominated by *Eucalyptus* (22–31%), Casuarinaceae (11–24%), Gleicheniaceae (14–37%) and *Nothofagus* (*Fuscospora*: 5–13% and *Brassospora*: 3–9%), with significant numbers of Restionaceae (1–3%). Gymnosperms are rare, with *Dacrydium* Group B species (3% in one sample) and *Dacrydium/Podocarpus* (up to 2%) being the only taxa represented. The diversity of angiosperms represented by low to trace amounts of pollen is not much greater, the only species able to be associated with extant taxa being *Acacia*, *Isopogon* (*Proteacidites truncatus*), *Micrantheum* and *Xylomelum occidentale* type.

From the data available it is likely that the palynofloras represent shoreline communities around the lake, one of which appears to have included *Nothofagus* (*Fuscospora* spp.). Casuarinaceae and *Eucalyptus* almost certainly were part of the local vegetation, but whether this was analogous to any modern sclerophyllous formation is unknown. Pollen values of *Nothofagus* (*Brassospora*) spp. are insufficient to indicate whether the source(s) were rare elements in a local (?gallery) wet forest or more extensive upstream on the Hamersley Ranges.

While data from onshore areas of Western Australia remain sparse for this period, a sequence from DSDP Site 254 on the southern end of Ninetyeast Ridge (Kemp, 1974; Kemp & Harris, 1975, 1977) may provide some indication of floristic developments in the general region during the Oligocene. The occurrence of cuticle, as well as pollen still retained in tetrads and larger aggregates in a lignitic sediment, demonstrates that at least some of the source plants were growing on islands on the now wholly submerged ridge.

Angiosperms dominate all palynofloras, the most abundant taxon being a species of *Rhoipites* (*Rhoipites grandis*) that broadly resembles pollen of the Vitaceae (up to 19%). Other prominent taxa are a possible palm (*Arecipites* cf. *waitakiensis*; up to 9%), Myrtaceae including *Myrtaceidites* cf. *mesonesus* (up to 6%) and *M. oceanicus* (up to 5%), *Ascarina*, Casuarinaceae and Cupanieae (all up to 3%), a porate type resembling Moraceae or Urticaceae pollen (up to 8%) and an unidentified tricolporate type (*Nyassapollenites* cf. *endobalteus*; up to 6%). The herb flora includes *Gunnera* (up to 3%), Asteraceae, Poaceae, Sparganiaceae and a tricolpate species similar to modern Scrophulariaceae pollen (*Tricolpites asperamarginis*). Only three gymnosperms are recorded: Araucariaceae (*Araucariacites australis*; up to 8%), *Dacrydium/Podocarpus* (up to 11%) and *Microcachrys* (< 1%). Of these, Araucariaceae, and *Dacrydium/Podocarpus* had established themselves on Ninetyeast Ridge during the Oligo-Eocene, whereas *Dacrydium* Group B species, *Phyllocladus* and *Podocarpus* (*Podocarpidites ellipticus*) appear to have become extinct during the same period. Of phytogeographical interest are (1) the presence of the mangrove palm *Nypa* and (2) the complete absence of *Nothofagus*. The latter suggests that the prevailing winds were westerly at the palaeolatitude of Ninetyeast Ridge (ca 30° S) during the Late Oligocene, since *Nothofagus* pollen is able to be transported very long distances across the Southern Ocean.

Whether the source vegetation was megathermal rainforest is unknown but there is no doubt that part of the flora (including *Microsorium* and *Pteris*) has megatherm–mesotherm affinities. If, as noted by Kemp & Harris (1977), high pollen densities imply that the island at DSDP Site 254 was high and large, then vegetation similar in physiognomy if not in composition to rainforest on islands in the Bismark Sea is a distinct possibility. Kemp (1974) has noted that present-day oceanic islands at the palaeolatitude of Ninetyeast Ridge have floras that are considerably less diverse and thermophilous than the DSDP Site 254 flora, consistent with Neogene cooling of the southern Indian Ocean.

PHYTOGEOGRAPHICAL CONSIDERATIONS

In the 130 years since the publication of J. D. Hooker's influential essay in his '*Flora Tasmaniae*' (1860), the Australian flora has been seen as comprising three principal elements. These are (1) an Australian (autochthonous) element characteristically forming open, sclerophyllous communities; (2) an Indo-Malaysian element prominent in the tropical (megathermal seasonal) monsoon forests and (megathermal, nonseasonal) rainforests; and (3) an Antarctic element restricted to subtropical–temperate (mesothermal–microthermal) rainforest and (microthermal) subalpine–alpine communities. All three elements were seen as initially intrusive, originating from the invasion of the Australian continent by different floras at different times and from different directions, although no agreement was reached on the importance (if any) of long-distance dispersal (see Barlow, 1981).

This 'invasion' hypothesis has been strongly contested by plant geographers on the basis of, for example, the typically low vagility of rainforest species and relatively high levels of regional endemism in Australasian rainforest floras (see Dawson, 1986; Hartley, 1986; Morat *et al.*, 1986; Thorne, 1986; Webb *et al.*, 1986). Blurring the distinction between the Indo-Malasian and Ant-

arctic floristic elements is evidence that a number of the terranes within the Southeast Asian region (see Burrett *et al.*, 1991; Daly *et al.*, 1991) have a Gondwanan/Australian origin. Disjunct distributions are viewed as evidence for shared ancestry rather than as evidence of long-distance dispersal (see Nix, 1982): the Australian flora is interpreted as being overwhelmingly autochthonous, evolved from Gondwanic stock under conditions of increasing isolation prior to the Middle Miocene collision with Southeast Asia.

The fossil pollen and spore data presented in this chapter strongly support a 'Gondwanic' origin for many taxa but challenge the limited role attributed to long-distance dispersal during the early Tertiary at the generic and family levels (Table 10.2).

Many Gondwanic taxa may have been dispersed via northern Australia during the Late Cretaceous but there are few confirmed examples. At present the strongest candidates are: (1) *Lophosoria*, whose highly distinctive spores (*Cyatheacidites tectifera*) first appear in the mid-Aptian in northern Australia, some 10 Ma earlier than along the southern margin (Turonian) although Berriasian (Late Jurassic–earliest Cretaceous) records exist on the Antarctic Peninsula (Dettmann, 1989); and (2) *Anacolosa*, whose pollen occurs in probable Campanian sediments in the Bonaparte Basin in northern Australia (A.D. Partridge, personal communication). Other families that may have reached Australia during the Late Cretaceous have not been recorded prior to the Oligo-Miocene (e.g. Goodeniaceae) or remain unidentified in the pre-Quaternary pollen record (e.g. Pittosporaceae).

Contrary to the phytogeographical evidence for autochthony, there is a growing body of spore and pollen evidence for floristic interchange between Australia and regions to the north and west during its early Tertiary isolation phase (Table 10.2). For example, on the Ninetyeast Ridge in the Indian Ocean, during the Oligocene approximately two thirds of the pollen species were shared with Australia. Taxa established there during the Oligo-Eocene included Araucariaceae, Asteraceae, Cupanieae, Loranthaceae, Poaceae, *Nypa*,

Table 10.2. *Late Cretaceous–Early Tertiary migration. A comparison of the time of first appearance of selected taxa in southeast Australia and earliest known fossil records elsewhere*

Family	Genus [Tribe]	Earliest record[a]	First appearance in SE Australia
Late Cretaceous			
Ulmaceae	*Celtis* type	N America	Miocene
Chenopodiaceae–Amaranthaceae		N America	Oligocene
Onagraceae	*Epilobium*	N America	Late Eocene
Arecaceae	*Nypa*	S America	Early Eocene
Restionaceae		W Africa	Early Eocene
Sapotaceae		SE Asia	Early Eocene
Sapindaceae	[Cupanieae]	Africa/India	Late Paleocene
Symplocaceae		N America	Early Miocene
Paleocene			
Apocynaceae	*Alyxia*	SE Asia	Late Eocene
Didymelaceae		Indian region	Early Paleocene
Poaceae		S America/Africa	Early Eocene
Polygalaceae		S America	Late Paleocene
Polygonaceae	*Polygonum*	Europe	Early Miocene
Sparganiaceae		Europe	Middle Eocene
Tiliaceae		Europe	Late Paleocene
Verbenaceae	*Avicennia* type	W Australia	?Eocene
Early Eocene			
Droseraceae		Europe	Early Eocence
Loranthaceae	(*Gothanipollis*)	Europe	Middle Eocene
Poaceae		Africa/S America	Middle Eocene
Middle–Late Eocene			
Convolvulaceae		Africa/S America	Late Eocene
Cyperaceae		Europe	Late Eocene
Fabaceae	[Caesalpinoidae]	S America	Late Eocene
Malvaceae		S America	Late Eocene
Meliaceae	*Dysoxylum*	New Zealand	Late Eocene
Mimosaceae	*Acacia*	Africa	Middle Oligocene
Mimosaceae	*Archidendron* type	Africa	Early Oligocene
Rubiaceae	*Gardenia*	Europe	Early Miocene
Malpighiaceae		S America	Late Eocene
Scrophulariaceae		Europe	Late Eocene
Vitaceae	*Cissus* type	?Australia	Late Eocene
Oligocene			
Asteraceae		Europe/USA	Oligocene
Goodeniaceae	*Scaevola*	Europe/New Zealand	?Early Miocene
Stylidiaceae	*Forstera* type	Australia	Early Oligocene

[a]After Muller (1981).

Restionaceae, *Rhodomyrtus* type, Sapotaceae, Scrophulariaceae type (*Tricolpites asperamarginis*) and Sparganiaceae. Several of these appear to have undergone local diversification during the same period, e.g. *Myrtaceidites oceanicus* (*Rhodomyrtus* lineage) and Sparganiaceae. Over the same period, *Dacrydium* Group B species, *Phyllocladus* and other podocarps were replaced by Araucariaceae and different species of *Podocarpus* and/or *Dacrydium*.

Obviously it is premature to nominate this now submerged ridge as a major plant dispersal route into and from Australia but the palynosequences are good evidence for ongoing, very long-distance dispersal of a broad range of phanerogams and cryptogams across a major ocean. Comparisons of Eocene, Oligocene and Early Miocene palynofloras from Southeast Asia (e.g. Muller, 1968; Morley, 1978; Watanasak, 1990) reveal much the same phenomenon, although the absence of moist montane habitats along the northwest margin of Australia apears to have restricted the entry of taxa such as *Abies*, *Alnus*, *Picea* and *Pinus* (cf. Morley, 1990). Other examples are discussed by Truswell *et al.* (1987).

Studies by Kemp & Barrett (1975), Dettmann & Thomson (1987), Dettmann (1989), Askin (1990; Askin *et al.*, 1991), Dettmann *et al.*, (1990), Dettman & Jarzen (1990, 1991) and Truswell (1990) confirmed that the Antarctic region shared major floristic elements with Australia and may have been a centre of origin and a dispersal corridor for plants entering Australia during the Late Cretaceous and Tertiary. However, the number of taxa with identifiable modern descendants is low (time of first appearance in southeast Australia in square brackets): Araucariaceae, Podocarpaceae, *Ascarina*, *Ilex*, Gunneraceae, Myrtaceae, *Nothofagus*, Epacridaceae, several lineages of Proteaceae including *Grevillea robusta* type and *Telopea*, Trimeniaceae and Winteraceae [Late Cretaceous]; Casuarinaceae [Danian]; *Beauprea* (*Beaupreaidites verrucosus*) and Cupanieae [latest Thanetian]; Bombacaceae [Early Eocene]; and *Ascarina* (*Clavatipollenites glarius*), *Anacolosa* (*Anacolosidites sectus*), Loranthaceae (*Cranwellia striata*) [Late Eocene] and Chenopodiaceae [Late Oligocene].

To what extent Antarctica exchanged taxa with Australia during the early Tertiary is uncertain due to the lack of *in situ* spore and pollen assemblages (see Truswell, 1983) in the East Antarctic sector adjacent to Australia, and to the fact that recycling in the glacial environment confuses the chronostratigraphic record (see Truswell, 1990). Macrofossils indicate that a moderately diverse flora was present during the Eocene, although data are only available from the Antarctic Peninsula. Pollen evidence suggests podocarps and *Nothofagus* forests persisted into the Oligocene on the Ross Sea margins (Mildenhall, 1989). This conclusion is borne out by the discovery of a leaf of a Late Oligocene *Nothofagus* sp., provisionally referred to the shrub *N. gunnii*, now endemic to the upper subalpine–alpine zone in Tasmania (Hill, 1989*b*). Another *Nothofagus* sp. survived (or was re-established) in Antarctica during the Pliocene (Hill, Chapter 16, this volume).

One probable example of long-distance dispersal from Antarctica is *Lophosoria*, which (as *Cyatheacidites tectifera*) had become extinct in Australia during the Campanian (see Dettmann, 1986: Helby *et al.*, 1987). The fern was present on the Antarctic Peninsula during the Eocene (Shen, 1990) but does not appear in Australia until the Early Oligocene and then only in the relatively cool, moist southeast. From here it appears to have rapidly extended west into eastern South Australia and north on to the Northern Tablelands of New South Wales (McMinn, 1989). This almost certainly reflects its good vagility and a preference for cool climates, since *Lophosoria* spores occur in ?Middle Miocene lignites on the Kerguelen Islands (Cookson, 1947), although they are also present in Early Miocene sediments on Mt Tambourine in Queensland (Cookson, 1957; Dettmann, 1989). The taxon became extinct again in Australia during the Late Pliocene.

Because of the limited (published) pollen data for Southeast Asia, the role of Australia as a corridor for plant migration into the Pacific region can be clearly demonstrated with respect to only New Zealand (Table 10.3). In this case it is certain that the pathways have involved dispersal across the Tasman Ocean, since New Zealand has been isolated from the rest of Gondwana for the past 80 million years. For establishment to be successful, suitable environments must have existed and in this context the broad similarity in vegetation types dominating southeast Australia and New Zealand during the early Tertiary is noted, i.e

Table 10.3. *A comparison of first appearances of early Tertiary palynomorphs common to southeast Australia and New Zealand*

Modern affinity		Fossil species	First appearance in	
Family	Genus		SE Australia	New Zealand
Bryophytes				
Ricciaceae	—	*Ricciaesporites kawaraensis*	Early Eocene	Early Miocene
?	—	*Bryosporis anisopolaris*	Late Oligocene	Quaternary
Pteridophytes				
Azollaceae	*Azolla* type	Undescribed sp.	Campanian	Quaternary
?Blechnaceae	—	*Monolites alveolatus*	?Early Eocene	Late Oligocene
?Blechnaceae	—	*Peromonolites densus*	Campanian	Maastrichtian
Cyatheaceae	*Hemitelia*	*Kuylisporites waterbolkii*	Early Eocene	Paleocene
Dennstaedtiaceae	*Hypolepis*	*Hypolepis spinyspora*	Early Eocene	Early Miocene
Dicksoniaceae	*Dicksonia*	*Matonisporites ornamentalis*	Early Cretaceous	Maastrichtian
Isoetaceae	*ªIsoetes*	—	Oligo-Miocene	Quaternary
Lycopodiaceae	*Lycopodium*			
Lycopodiaceae	*L. australianum* type	*Foveotriletes palaequetrus*	Middle Eocene	Pliocene
Lycopodiaceae	*L. laterale*	*Latrobosporites marginis*	Late Eocene	Early Miocene
Lycopodiaceae	—	*Verrucosisporites kopukuensis*	Paleocene	Middle Eocene
Polypodiaceae	*Microsorium*	*Polypodiisporites* spp.	Late Paleocene	Paleocene
Psilotaceae	*ªTmesipteris*	—	Late Eocene	Early Miocene
Pteridaceae	*Histeopteris*	*Polypodiisporites histeopteroides*	Late Eocene	Oligocene
Pteridaceae	*Pteris*	*Polypodiaceoisporites* spp.	Early Eocene	?Middle Eocene
Schizaeaceae	*Lygodium*	*Crassiretitriletes vanraadshoovenii*	Early Eocene	Middle Miocene
Thyrsopteridaceae	*Culcita*	*Rugulatisporites mallatus*	Maastrichtian	Paleocene
Thyrsopteridaceae	*Culcita*	*R. cowrensis*	Early Miocene	Middle Miocene
Thyrsopteridaceae	*Culcita*	*R. trophus*	Middle Eocene	Early Miocene
Conifers				
Araucariaceae	—	*Dilwynites*	Turonian	Maastrichtian
Podocarpaceae	—	*Dacrycarpidites australiensis*	Campanian	Maastrichtian
Podocarpaceae	—	*Lygistepollenites florinii*	Turonian	Maastrichtian
Podocarpaceae	*Halocarpus bidwillii* type	*Parvisaccites catastus*	Paleocene	Middle Eocene
Podocarpaceae	*Microstrobos*	*Podosporites erugatus*	Early Eocene	Late Miocene
Podocarpaceae	*Phyllocladus*	*Microalatidites palaeogenicus*	Senonian	Senonian
?Podocarpaceae	*Microcachrys* type	*Podosporites parvus*	?Oligocene	Oligocene
Angiosperms (monocots)				
Agavaceae	*Phormium*	*Luminidites phormoides*	Late Eocene	Late Eocene
Arecaceae	*Calamus* type	*Dicolpopollis* spp.	Late Oligocene	Miocene
Cyperaceae	—	*Cyperaceaepollis neogenicus*	Early Eocene	Oligocene
Liliaceae	*Astelia*	—	Quaternary	Middle Eocene
Pandanaceae	*Freycinetia*	*Lateropora glabra*	Oligo-Miocene	Late Eocene
	?*Freycinetia*	*Dryptopollenites semilunatus*	Late Paleocene	Early Eocene
Poaceae	—	*Graminidites* spp.	Middle Eocene	Oligocene
Restionaceae	—	*Milfordia* spp.	Early Eocene	Oligocene
Sparganiaceae/ Typhaceae	—	*Aglaoreidia qualumis*	Late Eocene	Early Miocene

Table 10.3. (*cont.*)

Modern affinity		Fossil species	First appearance in	
Family	Genus		SE Australia	New Zealand
Angiosperms (Dicots)				
Alangiaceae	*Alangium villosum* type	—	Early Eocene	Not recorded
Apocynaceae	*Alyxia*	*Psilodiporites* sp.	Late Eocene	Not recorded
Asteraceae	Tubuliflorae	*Tubulifloridites* spp.	Oligocene	Late Oligocene
Bombacaceae	*?Bombax*	*Bombacadidites bombaxoides*	Early Eocene	Paleocene
Callitrichaceae	—	*Australopollis obscurus*	Cenomanian	?Not recorded
Callitrichaceae	—	*Retistephanocolpites* spp.	Early Eocene	Late Miocene
Casuarinaceae	—	*Haloragacidites harrisii*	Paleocene	Paleocene
Chenopodiaceae	—	*Chenopodipollis chenopodiaceoides*	Oligocene	Early Miocene
Convolvulaceae	—	*Lymingtonia* sp.	Late Eocene	Late Oligocene
Cunoniaceae	—	Di/tricolpate spp.	?Paleocene	Oligocene
Cunoniaceae	*Gillbeea* type	*Concolpites leptos*	Maastrichtian	Maastrichtian
Didymelaceae	—	*Schizocolpus marlinensis*	Paleocene	Paleocene
Droseraceae	—	Undescribed spp.	Early Eocene	Early Miocene
Elaeocarpaceae	—	Undescribed spp.	?Late Eocene	Early Miocene
Epacridaceae	*Epacris virgata* type	*Paripollis orchesis*	Middle Eocene	Middle Eocene
Ericales	—	*Ericipites* spp.	Late Cretaceous	Late Cretaceous
Euphorbiaceae	*Austrobuxus* type	*Malvacipollis* spp.	Paleocene	Paleocene
Euphorbiaceae	*Micrantheum*	*Malvacipollis spinyspora*	Middle Eocene	Late Miocene
Fabaceae	Caesalpinoidae	*Margocolporites vanwijhei*	Late Eocene	Late Oligocene
Fagaceae	(*Nothofagus* subgenera)			
Fagaceae	*Lophozonia*	*Nothofagidites asperus*	Paleocene	Late Eocene
Fagaceae	*Fuscospora*	*N. brachyspinulosus*	Maastrichtian	Paleocene
Fagaceae	*Brassospora*	*N. deminutus-vansteenisii*	Early Eocene	Middle Eocene
Fagaceae	*Nothofagus*	*N. flemingii*	Campanian	Middle Eocene
Fagaceae	*Brassospora*	*N. falcatus*	Middle Eocene	Oligocene
Fagaceae	*Brassospora*	*N. longispinosa*	Middle Eocene	Pliocene
Gentianaceae	*Liparophyllum* type	*Striasyncolpites laxus*	?Early Oligocene	Middle Miocene
Geraniaceae	*Pelargonium*	*Tricolporopollenites pelargonioides*	Oligocene	Quaternary
Grossulariaceae	*Quintinia*	*Quintiniapollis psilatispora*	Middle Eocene	Middle Eocene
Gyrostemonaceae	—	*Gyropollis psilatus*	Late Eocene	Early Miocene
Haloragaceae	*Gonocarpus*	*Haloragacidites halaragoides*	Early Miocene	Early Miocene
Haloragaceae	*Haloragodendron*	*Stephanocolpites oblatus*	Middle Miocene	Middle Miocene
Loranthaceae	—	*Cranwellia costata*	Late Eocene	Middle Miocene
Loranthaceae	—	*C. striata*	Late Eocene	Middle Eocene
Loranthaceae	—	*Gothanipollis bassensis*	Middle Eocene	Middle Eocene
Malvaceae	—	*Malvacearumpollis* spp.	Latest Eocene	Early Eocene
Mimosaceae	*Acacia*	*Acaciapollenites myriosporites*	Early Oligocene	Oligocene
Mimosaceae	*Archidendron* type	*A. miocenicus*	Early Oligocene	Middle Miocene
Myrtaceae	*?Metrosideros*	*Myrtaceidites parvus-mesonesus*	Maastrichtian	Paleocene
Myrtaceae	*Rhodomyrtus* type	*M. verrucosus*	Early Eocene	Early Oligocene
Olacaceae	*Anacolosa*	*Anacolosidites luteoides*	Campanian	Early Eocene
Onagraceae	*Epilobium*	*Corsinipollenites* spp.	Late Eocene	Late Oligocene
Onagraceae	*Fuchsia*	*Diporites aspis*	Early Oligocene	Middle Oligocene
Polygalaceae	—	*Polycolporopollenites esobalteus*	Late Paleocene	Middle Eocene

Table 10.3. (*cont.*)

Modern affinity			First appearance in	
Family	Genus	Fossil species	SE Australia	New Zealand
Polygonaceae	*Polygonum*	*Glencopollis ornatus*	Early Miocene	Early Miocene
Proteaceae	*Agastachys* type	*Proteacidites obscurus*	Late Paleocene	Paleocene
Proteaceae	*Banksia/Dryandra*	*Banksieaeidites elongatus*	Paleocene	Not recorded
Proteaceae	*Beauprea*	*Beaupreaidites elegansiformis*	Maastrichtian	Maastrichtian
Proteaceae	*Beauprea*	*B. verrucosus*	Paleocene	Maastrichtian
Proteaceae	*Embothrium*	*Granodiporites nebulosus*	Latest Eocene	Not recorded
Proteaceae	*Hakea/Grevillea*	*Hakeidites* spp.	Late Eocene	Not recorded
Proteaceae	*Gevuina/ Hicksbeachia*	*Proteacidites reticuloscabratus*	Late Paleocene	Not recorded
Proteaceae	*Grevillea robusta* type	Undescribed sp.	Campanian	Not recorded
Proteaceae	*Knightia* type	Undescribed sp.	Campanian	?Eocene
Proteaceae	*Isopogon*	*Proteacidites isopogoniformis*	Late Eocene	Late Oligocene
Proteaceae	Musgraveinae	*Banksieaeidites arcuatus*	Late Paleocene	Not recorded
Proteaceae	*Persoonia/Toronia*	Undescribed sp.	Maastrichtian	Early Quaternary
Proteaceae	*Telopea* type	*Triporopollenites ambiguus*	Paleocene	Late Oligocene
Proteaceae	*Xylomelum occidentale* type	*Proteacidites annularis*	Late Paleocene	Paleocene
Proteaceae	—	*Proteacidites crassus*	Campanian	?Early Eocene
Proteaceae	—	*P. grandis*	Campanian	Early Eocene
Proteaceae	—	*P. incurvatus*	Late Paleocene	?Early Eocene
Proteaceae	—	*P. rectomarginis*	Late Eocene	Late Eocene
Proteaceae	—	*P. rectus*	Campanian	Middle Eocene
Proteaceae	—	*P. spiniferus*	Not recorded	Middle Eocene
Rubiaceae	*Canthium*	*Rubipollis oblatus*	Late Eocene	Late Miocene
Rubiaceae	*Coprosma* type	*Palaeocoprosmadites zelandiae*	Late Eocene	Late Oligocene
Rubiaceae	*Randia*	*Canthiumidites bellus*	Early Miocene	Late Miocene
Santalaceae	*Santalum* type	*Santalumidites cainozoicus*	Early Eocene	Middle Eocene
Sapotaceae	—	*Sapotaceoidaepollenites rotundus*	Early Eocene	Middle Eocene
Sapindaceae	Cupanieae	*Cupanieidites orthoteichus*	Late Paleocene	Early Eocene
Sapindaceae	*Dodonaea triquetra* type	*Nuxpollenites* sp.	Late Eocene	Early Miocene
Scrophulariaceae	—	*Tricolpites trioblatus*	Late Eocene	Late Miocene
Sterculiaceae	*Ungeria*	*Reevesiapollis reticulatus*	Late Eocene	Late Eocene
Strasburgeriaceae	*Strasburgeria*	*Bluffopollis scabratus*	Late Paleocene – Early Eocene	Late Eocene
Stylidiaceae	*Forstera* type	Undescribed sp.	?Early Oligocene	Quaternary
Symplocaceae	*Symplocos*	*Symplocoipollenites austellus*	Early Miocene	Early Miocene
Tiliaceae	—	*Intratriporopollenites notabilis*	Early Eocene	Paleocene
Umbelliferae	—	Undescribed spp.	?Oligocene	?Early Miocene
Vitaceae	*Cissus*	Undescribed sp.	Late Eocene	Not recorded
Winteraceae	—	*Pseudowinterapollis* spp.	Campanian	Late Cretaceous
Unknown	—	*Dryadopollis retequetrus*	Late Eocene	?Late Eocene
Unknown	—	*Tricolpites phillipsii*	Paleocene	Paleocene

*a*Macrofossil record only.
Data from Stover & Partridge, 1973; Mildenhall, 1980; Pocknall & Mildenhall, 1984; Raine, 1984; Mildenhall & Pocknall, 1989; Macphail & Truswell, 1989; Dettmann & Jarzen, 1991; and unpublished data from M. K. Macphail, D. C. Mildenhall & A. D. Partridge.

gymnosperm-dominated rainforest during the Paleocene, a coastal/lowland vegetation dominated by Casuarinaceae, Euphorbiaceae and Proteaceae during the Early to Middle Eocene, and *Nothofagus*-dominated rainforest during and since the Late Eocene (Raine, 1984; Pocknall, 1989, 1990).

Trans-Tasman dispersal

Bryophyta

Precise identification of, and dispersal patterns within, the cryptogams are blurred by the considerable morphological variation within most species. Nevertheless, at least two bryophytes have appeared in New Zealand substantially later than in Australia: *Bryosporis anisopolaris/problematicus* (Bryophyta) and *Ricciaesporites kawaraensis* (?Ricciaceae) are found in southeast Australia from the Oligocene and late Early Eocene, respectively (Macphail & Truswell, 1989; M. K. Macphail, unpublished data), but are not recorded in New Zealand until the mid-Pleistocene and Early Miocene, respectively. The same pattern is likely to be true of *Rudolphisporis rudolphi* (Anthocerotae).

Ferns and fern allies

From the data available, the only Tertiary fern recorded in New Zealand before Australia is *Hemitelia* (*Kuylisporites waterbolkii*). Two genera that are part of the Early Eocene cryptogam flora and survive in Australian warm temperate–tropical rainforest (*Lygodium*, *Pteris*) appear to have reached New Zealand during the Early–Middle Eocene: *Lycopodium* spp. that tend to colonise exposed or disturbed areas (*L. australianum*, *L. laterale*) are recorded in Australia during the Paleogene but did not reach New Zealand until the Neogene. It is unlikely that the wholly Quaternary records of *Azolla* (Azollaceae) in New Zealand reflect its actual residency because the fern occurs in Campanian sediments in the Otway Basin (M. K. Macphail, unpublished data).

Gymnosperms

Any early Tertiary diversification and trans-Tasman dispersal of gymnosperms tends to be obscured by their conservative pollen morphologies. For example, none of the early Tertiary genera of Araucariaceae, Cupressaceae and Taxodiaceae species described by Carpenter *et al.* (Chapter 12, this volume) can be distinguished through their pollen and it is unknown whether affinities with South American, New Zealand and Melanesian taxa represent diversification of a common Gondwanic stock or subsequent long-distance dispersal. In contrast, fossil lineages within the Podocarpaceae can be assigned to modern genera or groups of species with a high degree of confidence, viz. *Halocarpus bidwillii* type, *Microcachrys*, *Microstrobos* and *Phyllocladus*.

Halocarpus bidwillii type, which remains present in New Zealand (*H. bidwillii*, *H. biforme*, *H. kirkii*) first appears in southeast Australia during the Late Paleocene but appears in New Zealand only in the Middle Eocene. *Microcachrydites antarcticus* (*sensu* Dettmann, 1973) is present in Gondwana from the latest Jurassic onwards, but pollen virtually identical with that of living *Microcachrys tetragona* becomes the predominant morphotype only during the Tertiary. The related species *Podosporites parvus* (see Mildenhall & Pocknall, 1989) becomes prominent only during the Late Paleocene in Australia and New Zealand. Pollen identical with those of living *Microstrobos* occur in Middle Eocene sediments in the Gippsland Basin, some 20–30 Ma earlier than the first (Late Miocene) record in New Zealand (Mildenhall & Pocknall, 1989). Macrofossils confirm that this shrub was present in eastern Tasmania during the Early Eocene and part of a highland cool-climate flora in central Tasmania during the Late Oligocene–Early Miocene. Apart from an isolated population of *Microstrobos* on the Blue Mountains west of Sydney, this conifer and *Microcachrys* survive only in the upper subalpine–alpine flora in Tasmania. *Phyllocladus* reached New Zealand during the Late Eocene, a period in which the Australian flora included species distinct from, as well as similar to, living lineages (Hill, 1989*a*).

Angiosperms

Within the limitations of international correlation, there is a considerable degree of parallelism between Australia and New Zealand in the pattern of first appearances of taxa, although, as with the cryptogams and gymnosperms, in almost all cases first appearances in southeast Australia are older than in New Zealand.

Consistent with Antarctica being the centre of dispersal, there are a relatively high number of taxa appearing in both regions during the Paleocene, e.g. *Agastachys* type, Casuarinaceae, Epacridaceae (*Paripollis orchesis*) and Euphorbiaceae (*Malvacipollis subtilis*). Significantly all are shrubs and/or trees. Only Bombacaceae and, less certainly, Tiliaceae are recorded in New Zealand earlier than in southeast Australia.

The same bias exists in the Eocene–Early Miocene, where out of *ca* 55 taxa referable to modern plants, only five appear first in New Zealand: Malvaceae and *Ungeria* in the Early Eocene, *Astelia* and Loranthaceae (*Cranwellia striata*) in the Middle Eocene and *Freycinetia* (*Lateropora glabra*) in the Late Eocene. One Proteaceae (*Proteacidites rectomarginis*) appears to have become established at approximately the same time (Late Eocene) in both southeast Australia and New Zealand. From the data presented in Table 10.3, approximately two thirds of the taxa appearing in southeast Australia during the Late Eocene do not appear in New Zealand until the Late Oligocene–Early Miocene and one *Nothofagus* sp. (*Nothofagidites longispinosa*) not until the Pliocene. This preferential west to east dispersal across the Tasman Sea continued into the later Neogene, e.g. Thymeleaceae (*Pimelea*) during the Pliocene.

SPECIALISED PLANT COMMUNITIES

The pollen data allow the evolution of several specialised vegetation types to be traced back into the early Tertiary. These are (1) halophytic communities, (2) sclerophyllous formations and (3) cool–cold climate communities.

Halophytic communities

The extensive mangrove swamps and salt marshes found in other countries at similar latitudes are absent from much of the Australian coastline because of the relatively arid and sandy nature of many estuaries, high wave energies and a low to moderate tidal range (Jennings & Bird, 1967; Love, 1981). Only in eastern Queensland, protected by the Great Barrier Reef, and along the northern and northwestern margins are low energy coastlines associated with high tidal ranges and/or seasonally high rainfall. Here 29 mangrove species are recorded. South of 21° S the numbers are reduced to nine species: only *Exoecaria* (Euphorbiaceae), *Aegiceras* (Myrsinaceae) and *Avicennia* (Verbenaceae) range south of 33° S. The southernmost location of mangroves (*Avicennia*) is at Corner Inlet (38° 45′ S) on Wilson's Promontory, Victoria.

On present indications, early Tertiary climates would have been adequate to support a diverse range of mesothermal–megathermal mangroves (see Woodroffe & Grinrod, 1991). However, pollen evidence appears to show that few mangrove species were present and that most of these are only sporadically represented by high relative pollen values. The pollen data are clear that this is not an artefact of preservation or limited sampling, at least along the southern margin. Possible ecological explanations are that mean sea surface temperatures were below the minimum level required to support the development of extensive mangrove stands (suggested to be 24° C by Barth (1982)) or that tidal regimes were unsuitable. Alternatively, pollen of a range of mangroves are present in the fossil record but have not been identified as such.

The latter is a distinct possibility on two accounts. (1) The pollen of many mangroves are difficult to distinguish from those of freshwater/dryland species within the same taxon. For example, *Osbornia*, which still occurs in northern Australia, produces pollen that closely resembles that of other Myrtaceae genera. Tiliaceae pollen is equally conservative and it is possible that the mangrove genus *Brownlowia* was present during the Early Eocene. Other mangrove taxa that would be difficult to identify in Australian early Tertiary palynofloras are *Phoenix* (Arecaceae), *Laguncularia* (Combretaceae), *Cynometra* (Caesalpinoidae), *Aegiceras* (Myrsinaceae) and *Heritiera* (Sterculiaceae) (see Thanikaimoni, 1987). (2) Because of altered ecological preferences, some now wholly dryland taxa may have included mangrove species during the early Tertiary. A. D. Partridge (personal communication) has proposed that the Proteaceae included salt-tolerant species during the Early Eocene. This has limited support in that one species (*Proteacidites pachypolus*) is restricted to marine sediments during the Early Eocene *Apectodinium* (*Wetzeliella*) *hyperacanthum* transgression in southeastern Australia and an ?analogous salt-influenced environment in the Lake Eyre Basin. The species disappears from the fossil record during the Middle *M. diversus* Zone but, on the basis of subsequent records, had diversified away from salt-influenced habitats by the late Early Eocene Upper *M. diversus* Zone.

The earliest record to date of a probable halophyte in southern Australia is pollen of an undescribed *Spinizonocolpites* sp. occurring in Turonian–earliest Santonian sediments in the offshore Otway Basin (M. K. Macphail, unpublished data). This record pre-dates occurrences of *Nypa* in South America and Africa (Muller, 1981) but is consistent with (1) a Gondwanic centre of origin for the taxon and (2) the growing body of evidence for an early Late Cretaceous marine influence along the southern margin of Australia (see Marshall, 1988). It is probable that any mangrove communities formed during the Maastrichtian and Paleocene in southern Australia included mangrove species of *Acrostichum* (*Cyathidites splendens* vars. ?*C. gigantis*). Otherwise the earliest reliably dated record of a mangrove in southern Australia is earliest Early Eocene. At this time *Nypa* communities were present and occasionally extensive on the west coast of Tasmania and in marginal marine environments within the Bass, Gippsland, Otway, Eucla and ?Perth Basins.

Churchill (1973) noted that pollen closely resembling *Avicennia* and Rhizophoraceae, and

wood closely resembling that of *Barringtonia* (Barringtoniaceae), occur in marine sediments (Plantagenet Beds) in southwest Western Australia. These records, which have yet to be confirmed (H. Martin, personal communication), remain the only early Tertiary records of the mangroves in southern Australia. Pollen resembling *Sonneratia* (*Florschuetzia levipoli*) occurs very rarely in late Early Eocene sediments in the Gippsland Basin (M. K. Macphail & A. D. Partridge, unpublished data) but it is certain that none of the other mangroves producing highly distinctive pollen is present, e.g. *Acanthus* (Acanthaceae), *Aegialitis* (Plumbaginaceae) and Rhizophoraceae. Hekel (1972) recorded Rhizophoraceae and *Nypa* pollen in probable Eocene sediments in southern Queensland, an association also found in the northern hemisphere during the early Tertiary, e.g. in the Eocene London Clays (Chandler, 1951). The youngest record of *Nypa* in southern Australia occurs in latest Oligocene–Early Miocene sediments in the Barossa Basin (N. F. Alley, unpublished data).

Whether the early Tertiary palaeogeography would have allowed the development of regionally extensive mangrove communities in southern Australia is uncertain. In the very flat-lying Murray Basin at least, the development of tidal flat and other estuarine environments during the Oligocene and Early Miocene is associated with the appearance of salt-marsh plants, notably Chenopodiaceae–Amaranthaceae, not with any identifiable mangrove species or community (Macphail & Truswell, 1989). Since chenopods are first recorded in the Maastrichtian in North America, it is more likely that the appearance of this taxon in southern Australia during the Early Oligocene reflects slow migration rather than environmental trends. Salt-marsh associations are recorded in the Gippsland Basin at about the same time, although quantitative data (A. D. Partridge & M. K. Macphail, unpublished data) suggest that these did not become widespread until the Late Miocene. It is possible that the Asteraceae, Poaceae and Scrophulariaceae also included halophytic species because pollen of these families also appears in coastal sediments at about the same time. As with mangroves, the majority of palynologically distinctive genera found in modern salt-marsh and brackish water communities have not been recorded in southern Australia during the early Tertiary. Examples are *Disphyma* (Aizoaceae), *Apium* (Apiaceae), *Wilsonia* (Convolvulaceae), *Baumea* (Cyperaceae), *Selliera* (Goodeniaceae), *Triglochin* (Juncaginaceae), *Limonium* (Plumbaginaceae), *Samolus* (Primulaceae) and *Ruppia* (Ruppiaceae).

Sclerophyllous communities

Sclerophyllous plant formations dominate the Australian landscape: in most instances the canopy stratum is formed by shrub or tree species of Casuarinaceae, Chenopodiaceae, Epacridaceae, Mimosaceae, Myrtaceae, Proteaceae and occasionally by Asteraceae on disturbed sites. Not surprisingly, a considerable effort has been made to trace the evolution of these communities as well as the origins of the more important canopy genera such as *Acacia*, *Casuarina* and *Eucalyptus* or some distinctive minor elements, e.g. *Dodonaea triquetra* (Lange, 1978; Truswell & Harris, 1982; Martin & Gadek, 1988; Singh, 1988; Martin, 1989*b*).

Fossil pollen have contributed to this search but are not good indicators of the presence of sclerophyllous communities because of the conservative nature of the pollen produced, and the wide spectrum of habitats occupied, by these families. On present indications there is little reason to doubt that sclerophylly per se originated during the early Tertiary as an adaptation to infertile soils (Barlow, 1981) or more locally, cool–cold climates (Hill & Gibson, 1986; Macphail *et al.*, 1991). What can be demonstrated from the fossil pollen record is that (1) the Australian sclerophyllous flora has evolved largely from Gondwanic rainforest stock (Table 10.1) and (2) that the modern identification of sclerophyllous communities with a significant (seasonal) water deficit or repeated wildfires is a post-Early Miocene development. Prior to this, taxa that are now largely sclerophyllous/xerophytic were part of mesophytic plant communities. Examples are:

1. **Casuarinaceae**: Because it is uncertain whether fossil pollen represent rainforest (*Gymnostoma*) or sclerophyllous (*Allocasuarina*/*Casuarina*) species, the nature of the vegetation including Casuarinaceae can only be deduced from the associated species or macrofossils. During the early Tertiary, this was almost always rainforest, consistent with the source species being *Gymnostoma* (see Hill, Chapter 16, this volume). Since Casuarinaceae pollen are equally well represented at times when *Nothofagus* (*Brassospora*) spp. are rare–absent (Early Eocene, Miocene) or abundant (Middle Eocene–Oligocene), any association with *Nothofagus* was a loose one. There is some evidence from central and western Australia that the abundance of Casuarinaceae was inversely related to the Cunoniaceae and Myrtaceae and, in coastal situations, directly related to an increasing marine influence.

On present indications, the Casuarinaceae included species that were (i) adapted to low photoperiods during the Paleocene, (ii) part of the riparian vegetation in tide-influenced deltas during the Early Eocene, and (iii) widely distributed along rivers and around freshwater lakes, but not part of peat-swamp vegetation during the Middle Eocene–Early Miocene. On the basis of abundant *Eucalyptus*-type pollen, the earliest evidence to date of a possible sclerophyllous community including Casuarinaceae is Early–Middle Miocene, at Weelumurra Creek, northwest Western Australia.

2. *Acacia*: Like the Casuarinaceae, the Mimosaceae include rainforest as well as sclerophyllous species but at present are dominant in open forests, woodlands and shrublands only in drier areas of the continent (see Johnson & Burrows, 1981).

The earliest pollen record of the family (*Polyadopollenites* sp. cf. *P. granulosus*, *Acaciapollenites miocenicus*) occurs in the latest Eocene–Early Oligocene in the Murray and Eucla Basins. Both fossil species represent mesophytic taxa such as *Archidendron* and, possibly significant, both are first recorded in marginal marine sediments. *Acacia* (*Acaciapollenites myriosporites*) shows a similar preference for riparian habitats when it first appears in the Gippsland and Murray Basins during the Middle Oligocene. If correct, then *Acacia* spp. have evolved from coastal/shoreline species. This thesis accords with a global appearance of *Acacia* polyads in the Oligocene. Pollen data from the Lake Frome area of central Australia confirm that *Acacia* was part of a sclerophyllous community (including *Eucalyptus* and *Grevillea/Hakea*) lining watercourses by the Late Miocene–Pliocene (Martin, 1990). Asteraceae may have had a similar history in southeast Australia during the early Tertiary.

3. *Eucalyptus*: Pollen resembling *Eucalyptus* (*Myrtaceidites eucalyptoides*) are recorded very sporadically in central and southeast Australia from the Early Eocene onwards but in all cases are associated with low to frequent amounts of *Nothofagus* (chiefly *Brassospora*) spp. This association may not be fortuitous, based on the widespread presence of eucalypts in infrequently burnt *Nothofagus* (*Lophozonia*) forest in southeast Australia and the association of *Nothofagus* (*Brassospora*) spp. with the primitive eucalyptoid *Arilastrum gumiferum* in New Caledonia (G. S. Hope, personal communication). From the pollen data, eucalypts did not become prominent relative to other Myrtaceae until the Early Miocene. Examples include Weelumurra Creek (above) and Little Bay on the central New South Wales coast.

The earliest definite record of a eucalypt able to be referred to an extant species is Late Oligocene: *Eucalyptus spathulata* type from the southwest Murray Basin (N. F. Alley, unpublished data). *Myrtaceidites tenuis*, which occurs in southeast Australia from the late Early into the Middle Eocene, may have been a eucalyptoid species inhabiting a range of coastal/lowland and upland environments. Both species are strongly associated with mesothermal rainforest types.

4. *Dodonaea triquetra*: Pollen closely resembling that of the widespread dry sclerophyllous shrub *Dodonaea triquetra* (Sapindaceae) is widespread in Middle–Late Eocene sediments in central Australia and in Oligo-Miocene sediments in the

Murray Basin and are sporadic in basins in the southeast.

Like the Late Oligocene source of *Eucalyptus spathulata*-type pollen, there is no reason to suppose that the early Tertiary sources of *Dodonaea triquetra*-type pollen had the same (sub-humid to semiarid) climatic preferences as the extant species because, in the Murray Basin at least, the taxon first appears some 9 Ma before the first (Late Oligocene) indication of mildly seasonal climates (Macphail & Truswell, 1989). It is a persistent element in the flora during the Late Oligocene–Early Miocene replacement of *Nothofagus* (*Brassospora*) rainforest by drier (Araucariaceae, Myrtaceae) types, but appears to be absent during the Pliocene when open woodlands and grasslands were established across the basin (data from Martin, 1987).

5. Poaceae: Martin (1990) has shown that pollen evidence for extensive grasslands in central Australia as suggested by Harris (in Callen & Tedford, 1976) is incorrect. On present indications, grasses were a very rare element in southern Australia during the Middle Eocene–late Early Miocene. The type is recorded in a wide range of depositional environments. An apparent increase in the number of trace records during the Oligo-Miocene in the Murray Basin is probably due to the large size of pollen source area, although, if correctly identified, Poaceae values of 2–11% may imply that grasses were locally prominent in or around a sedge-reed swamp community in the northwest of the basin during the Early Miocene (see Truswell *et al.*, 1985).

Cool–cold climate communities

Cool–cold climate vegetation types are now restricted to the upper slopes and summits of mountains on the southeastern Highlands and in Tasmania (see Costin, 1981). Apart from the shrub conifers *Microstrobos* and *Microcachrys* and liliaceous taxa such as *Astelia* and *Isophysis*, few taxa that can be specifically identified using pollen are confined to these vegetation types in Australia (see Macphail, 1986). There is no reason to suppose that circumstances were different during the early Tertiary.

Although this makes it difficult to use pollen to detect the impact of cooling global climates, except via the elimination of megathermal taxa, the fossil pollen record demonstrates the following.

1. The majority of the present-day *woody* cold climate species have evolved from lowland/rainforest taxa. Predictably all are Gondwanic elements, although many are now endemic to Tasmania (see Macphail, 1986) or New Zealand. Conifers include: (i) *Athrotaxis*, present in the eastern lowlands of Tasmania during the Early Eocene (Townrow, 1965*b*); (ii) *Microstrobos*, widely distributed throughout southeast Australia during the Late Paleogene and Neogene but to date not recorded in any abundance before the Pliocene (western Tasmania); and (iii) *Microcachrys*, present in Gondwanan lowland floras since the Jurassic. Similarly the angiosperm families prominent in highland shrubberies, e.g. Epacridaceae, Myrtaceae, Proteaceae and Winteraceae, are widespread in Late Cretaceous and early Tertiary coastal/lowland vegetation types. One exception is *Coprosma* (Rubiaceae), a shrub now prominent at high elevations in Tasmania (and New Zealand) but not recorded in southeast Australia before the Late Eocene.

2. Apart from *Gunnera*, Cyperaceae, Poaceae and Restionaceae, the majority of herb families now prominent in the subalpine–alpine flora are unlikely to have reached Australia before the Oligocene, e.g. Apiaceae, Asteraceae, Campanulaceae, Gentianaceae, Geraniaceae, Lamiaceae, Lobeliaceae, Loganiaceae, Plantaginaceae, Ranunculaceae and Scrophulariaceae. Relatively few have reliable earlier fossil records elsewhere (see Muller, 1981), but a northern hemisphere centre of origin seems probable for many. Exceptions are (i) Stylidiaceae, to date only recorded in ?Early Oligocene sediments in northwest Tasmania, and (ii) *Astelia* (Liliaceae), a robust tufted herb whose highly distinctive echinate monocolpate pollen

occurs in New Zealand from the Late Eocene onwards but which is not recorded in Australia before the Quaternary. In New Zealand, the genus includes epiphytic species in lowland wet forests and turf-forming colonies in coastal areas and it seems reasonable to assume that Neogene analogues were the source of highland *Astelia* ecotypes in Australia and New Zealand.

Not surprisingly, macrofossils are better evidence for the evolution of cool–cold-adapted floras during the Tertiary and, by good fortune, the three most intensively studied palynosequences from highland areas in New South Wales and Tasmania are associated with well-preserved macrofossil remains. The combined evidence shows that by the late Early Miocene cool climate floras were present at high elevations in Tasmania (palaeolatitude 57° S), but not at similar or higher elevations on the Southeastern Highlands (palaeolatitude 48° S).

Although seasonally cool conditions existed in the Southeastern Highlands during the Paleocene (Taylor *et al.*, 1990) this is not indicated, for example, by small leaf sizes (see Christophel, Chapter 11, this volume). Reinforcing the nexus between cool climate floras and Tertiary global cooling rather than high topographical relief and high latitudes *per se* is pollen evidence showing that there are few differences between the gymnosperm-dominated associations in the Southeastern Highlands of New South Wales (60° S) and contemporaneous associations in the coastal/lowlands at *ca* 65° S during the Paleocene.

Macrofossils show that the Early Miocene forest vegetation in the Southeastern Highlands (Kiandra) was dominated by notophyllous taxa, including the Lauraceae, a family with no members in the modern highland flora. By this time, the Tasmanian highlands (Monpeelyata) supported at least two taxa with strong modern cold climate affinities, viz. *Microstrobos* and *Nothofagus gunnii*. The associated leaf remains are wholly dominated by microphyllous taxa (Hill & Gibson, 1986). This remains the earliest evidence to date for a Tertiary cool-adapted flora.

CONCLUSIONS

Palynological research in Australia has reached the stage where the fossil data are useful as a primary source of evidence *at the family and generic levels* for the evolution of the Australian Cenozoic flora.

One common thread linking the early Tertiary palynofloras is that the first appearances of many angiosperms and some gymnosperms and cryptogams are being pushed back into the Paleocene and Late Cretaceous. We anticipate that, as knowledge of the palynofloras improves, many species will be found to have appeared earlier in inland and northwest Australia than in the southeast. Much information must become available from the Arafura and Timor Sea regions and Southeast Asia before it will be possible to determine which taxa have invaded Australia from the north; however, there is no reason to disagree with Barlow's (1981) conclusion that the history of the Tertiary flora is largely one of differentiation from the original Gondwanic stock.

What is debatable is the current lack of emphasis given to long-distance dispersal. The data suggest that the geographical isolation of Australia during the early Tertiary did not pre-empt exchange of rainforest and other taxa across sea barriers to the north and Antarctica to the south. This was certainly the case with respect to New Zealand. Possible routes for biotic exchanges are expected to become better defined as knowledge of the palaeotectonic and palaeogeographical interrelationships of Australia and adjacent land masses improves. In this respect, the limited 'invasion' hypothesis as expounded by Barlow (1981) appears to be the more appropriate model for the early Tertiary flora than the vicariance model supported by, for example, Webb *et al.* (1986).

As with the Neogene, the usefulness of pollen and spores is limited by taxonomic constraints and difficulties in differentiating between local and distant plant sources. A comparison of the macro- and microfossil data confirm that these are a partial record of past floras and individually are an inadequate basis for identifying past plant associ-

ations or for reconstructing the past vegetation except in broad terms. It is unfortunate that where sites are abundant, e.g. in southeast Australia, the sheer volume of the data and magnitude of local variations tend to mask developments of regional significance: elsewhere the paucity of sites increases the danger that local developments will be given an unwarranted regional emphasis.

What can be concluded with certainty is that the relatively well-studied, often *Nothofagus*-dominated, palynosequences in southeast Australia are a poor guide to the composition of the early Tertiary vegetation elsewhere in Australia. There is no compelling evidence for the existence of a 'pan-Australian' flora (whether dominated by *Nothofagus* or any other taxon) at any time during this period. This almost certainly is an artefact of the limited fossil species base and the even more limited number of species sharing the pollen dominance in the early Tertiary palynofloras. As in the Quaternary, there is no reason to assume that these were the only canopy taxa involved in the massive reassortments characterising the early Tertiary vegetation.

Conversely, there is much that supports the concept of a flora undergoing progressive differentiation into a spectrum of ecological associations. With few exceptions, shifts in species distributions and trends in community dominance conform well with independently established patterns of warming and cooling during the early Tertiary operating at the regional scale.

Paleocene

Plant communities occupying lowland/coastal and highland areas in southeast Australia are interpreted as conifer-dominated wet forests able to withstand periods of winter darkness.

The majority of species were elements within the Cretaceous floras that had extended their dominance at the expense of angiosperms, in particular species occupying the freshwater-swamp niche. The angiosperm element increases both in diversity and abundance throughout the Danian: 'ancestral' *Nothofagus* spp. become locally dominant in the Bassian region during the latest

Thanetian. The number of taxa with well-defined mesotherm–megatherm affinities is low. Not surprisingly, all are of Gondwanic origin, with some migrating into, and others evolving within, the region.

Data from the Lake Eyre Basin suggest that a more varied and variable mesothermal rainforest vegetation existed in inland Australia, with Cunoniaceae and/or Proteaceae, not *Nothofagus*, being the most prominent angiosperms present. As in southeast Australia, gymnosperm dominance is linked to freshwater lake or swamp environments, although the mix of taxa is different. For example, *Lygistepollenites balmei* is consistently rare, whereas Cupressaceae/Taxodiaceae are sporadically abundant.

Whether conifers or angiosperms dominated Paleocene forest communities in northern and western Australia is unknown. Pollen evidence from Ninetyeast Ridge suggests that the floras in both regions will have shared many genera with eastern Australia.

Early Eocene

In contrast to the Paleocene, rainforests present on the coastal/lowland regions in southeast Australia are distinctly megathermal in character. This is reflected in the diversity and palynological dominance of angiosperms (including mangroves now restricted to tropical regions) and the rarity of *Nothofagus* spp. Whilst there is little doubt that these communities evolved in response to global warming and regional year-round high humidity, at least part of the development can be associated with repeated marine inundation of low-lying coastal plains. It seems likely that similar types of rainforest extended along the northwest and northeast Australian coastlines in spite of the relatively low estimated sea surface temperatures (maximum < 18° C) in the latter region.

Away from the coastal/lowland areas, rainforest communities appear to have been mesotherm in character although many species are shared with megathermal rainforest. Marked differences in pollen dominance at and between inland sites are seen as early manifestations of the strong ecologi-

cal differentiation that characterises the later Cenozoic vegetation. On present indications, *Nothofagus*-dominated communities were largely confined to Tasmania and possibly the Southeastern Highlands, but it is premature to nominate these areas as refugia for *Brassospora* spp. during the Early Eocene thermal maximum. Vegetation in inland central Australia appears to have been a mosaic of rainforest communities in which dominance varied between gymnosperms (including Cupressaceae and/or Taxodiaceae), Casuarinaceae, Cunoniaceae and Myrtaceae.

A number of distinctive Proteaceae species may have evolved in inland regions during the Early Eocene and subsequently migrated into coastal southeast Australia. It is tempting to associate this with more variable and extreme climatic conditions away from the coast, although eustacy also is seen as a factor behind the progressive diversification of the flora in the Bassian region. In neither instance was the rate of change sufficiently high to allow the development of extensive low/open vegetation types such as sedgeland even though taxa such as Poaceae and Restionaceae were present.

Middle–Late Eocene

If the Early Eocene can be categorised as the period of maximum development of nonseasonal megathermal rainforests in Australia, it is equally valid to view the Middle–Late Eocene as the period of maximum development of nonseasonal mesothermal rainforest.

To a large degree this trend is expressed by a locally variable but very widespread expansion of *Nothofagus* (*Brassospora* spp.). The apparent rapidity of this expansion may be partly an artefact of the geological record, but there is no reason not to associate the phenomenon with global cooling. For example the mangrove palm *Nypa* was eliminated from the Bassian region but maintained a presence along with other tropical mangrove taxa in southern Queensland and, because of locally warm shelf waters, possibly also on the South Australian coast. Consistent with this climatic trend is the first appearance of modern, now xerophytic genera, e.g. *Dodonaea triquetra* type and *Banksia serrata* type as well as a time-transgressive expansion of 'open habitat' taxa such as sedges and rushes: local conditions seem to have permitted the immigration into southern Australian of megathermal lineages within, for example, the Fabaceae (Caesalpinoidae) and Malpighiaceae. These appear to have dispersed around the coast since there is no record to date of these taxa in the continental interior.

It is highly probable that the diversity of Middle–Late Eocene floras is strongly linked to the competitive ability and gregarious behaviour of *Nothofagus*. Thus, in the cooler southeastern corner of Australia, *Nothofagus* appears to have successfully captured all but swamp habitats within the coastal/lowland regions, resulting in species-poor rainforest communities. In ?warmer coastal areas to the west and north, this was not the case and diverse mesothermal–megathermal rainforest associations are recorded. Within the limits of pollen resolution, the flora of the Murray Basin closely resembled that on the south coast of Queensland, although there is no reason to suppose that the source (rainforest) vegetation was identical in terms of physiognomy or species dominance. Few consistent patterns have been established for the continental interior except that *Nothofagus* (*Brassospora*) spp. were part of a mosaic of mesothermal rainforest associations in which the (shifting) pollen dominance was shared with Casuarinaceae, Cunoniaceae, gymnosperms and Myrtaceae.

Gondwanic groups, including *Nothofagus* and the Proteaceae continue to diversify both in terms of the number of species and in ecological amplitude (ecotypes). Examples include *Proteacidites pachypolus* and *Nothofagidites falcatus*. The latter species, which first appears in southeast Australia during the Middle Eocene, shows rapid extension through eastern and some areas of central Australia, reaching northwest Australia and New Zealand during the Oligocene. *Proteacidites pachypolus*, which first appears in saltwater-influenced environments during the Early Eocene, had become widely established around coastal/lowland swamps, in uplands and the continental interior by Late Eocene times.

All factors contributed to the trend – already apparent in the Early Eocene – towards ecological diversification of, and geographical provincialism in, the Australian flora. High diversity of Proteaceae within the southwest sector is one example of such provincialism.

Latest Eocene–earliest Oligocene

The Terminal Eocene is widely seen as a critical period in the history of Cenozoic climates. For this reason, latest Eocene–earliest Oligocene palynofloras have been discussed separately in order to determine whether this time also was a biological turning point in the history of the Australian vegetation. There are some indications that this was the case in southern Australia:

1. A significant number of trees and shrub taxa which had been typical of, even if uncommon in, Eocene mesothermal–megathermal rainforest are last recorded during the Late Eocene–earliest Oligocene. One, *Sphagnum* spp., that had been a feature of the flora since the Maastrichtian became extinct at this time.
2. In the overwhelming majority of areas for which data are available, palynological dominance became increasingly restricted to *Nothofagus* (*Brassospora*) and only one or two other taxa, usually Casuarinaceae or a podocarp, less frequently Araucariaceae.

Whilst macrofossil data show that this floristic impoverishment is partly an artefact of specialised peat-swamp depositional environments, there is little doubt that the expansion of *Nothofagus* outside of the southeastern corner of Australia was a real phenomenon and one that can be related to climatic trends that began during the Early Eocene.

As with the late Cenozoic flora, it is to be anticipated that the responses and resiliences of the various early Tertiary rainforest species to any climatic excursion will have varied, and therefore changes at the plant community level will be time-transgressive or masked by local factors. An example is the relatively high surface temperatures along the southern margin that appear to have allowed megathermal species to survive in the Murray Basin flora after their demise in the Bassian region to the east. Another is the better representation of sedges and reeds in inland situations.

Whether the transitional nature of the rainforest flora was mirrored by developments elsewhere in northwest Australia during the latest Eocene–earliest Oligocene is unknown. What can be predicted is that any shifts in community dominance in inland regions are likely to reflect precipitation as well as temperature trends. A corollary is that coastal regions in the northeast and along the northern margins may have begun to act as refugia for megathermal elements. The data are emphatic that conditions most favourable to the dominance of *Nothofagus* (*Brassospora*) were restricted to southeast Australia, although one species (*Nothofagidites falcatus*) had the necessary ecological attributes to thrive at lower latitudes in both coastal and inland environments.

Early Oligocene–late Early Miocene

The Oligocene–Miocene probably has become the best-described period in the history of the Australian pre-Quaternary vegetation, at least in southeast Australia, but one that is difficult to summarise because of poor biostratigraphic control, particularly within the Late Oligocene–Early Miocene, and also because of the local origin for most spores and pollen. Both phenomena point to an increasingly provincial and heterogeneous vegetation in which local factors have dictated plant community composition and, presumably, structure. What can be stated with a moderate degree of confidence is:

1. The relatively low number of first appearances and extinctions of spore and pollen species within the Oligo-Miocene flora is unlikely to reflect a period of stability within the Australian flora; rather it is more probable that ongoing diversification is 'concealed' by conservative pollen and spore morphologies, in particular tricolpate and tricolporate types.
2. Global cooling, associated with increasingly seasonal rainfall patterns, had crossed the bio-

logical thresholds leading to the evolution of cool–cold climate floras, araucarian 'dry' rainforest types and low/open communities such as sedgeland and reed swamp. Not unexpectedly, these are first recorded at high elevation/high latitude and continental interior sites, and *Nothofagus* (*Brassospora* spp.) and thermophilous angiosperms are amongst the taxa whose distributions were most affected.

On present indications, *Nothofagus* mesothermal rainforest was increasingly restricted to Tasmania and southeastern Victoria, although some *Brassospora* spp. including *Nothofagidites falcatus* maintained a much wider distribution. Macrofossil data confirm the presence of *Nothofagus* (*Brassospora*) spp. in lowland rainforest communities but challenge its importance in this vegetation, e.g. *Lophozonia* spp. appear to have been the dominant taxon bordering lakes and rivers in Tasmania. Whether this applied to upland situations is less clear, but by the Early Miocene *Nothofagus* microthermal rainforest appears to have evolved at mid to high elevations. On mainland Australia, there are some indications that the *Brassospora* spp. had become increasingly restricted to upland situations or at best were minor elements in ?gallery wet-forest dominated by one or more of Casuarinaceae, Cunoniaceae, Lauraceae and Myrtaceae along the inland rivers and in the northwest of Australia.

The associated decline in thermophilous angiosperms in southeastern Australia (and Casuarinaceae in Tasmania) was offset by increasingly diverse herb and cryptogam floras. Gymnosperms such as Araucariaceae become increasingly prominent in the Murray Basin and along the coast in South Australia, but these floras also appear to have retained a greater proportion of minor taxa with well-defined mesotherm–megatherm affinities. To what extent similarities between palynofloras in southern Queensland and inland areas such as the Murray Basin are an artefact of limited pollen resolution is unknown but, whereas rainforest in Queensland is likely to be nonseasonal, megathermal, those in the Murray Basin are more clearly seasonal, mesothermal.

The strong ecological differentiation of the

M. K. MACPHAIL *et al*

flora during the Oligocene–Early Miocene and the appearance of families such as the Asteraceae, Chenopodiaceae, Mimosaceae and Poaceae give some plant communities a modern appearance.

That this is deceptive is confirmed by macrofossil data that demonstrate (1) not only is the actual diversity of the flora rather greater than indicated by pollen and spores (a conclusion that applies to the Cenozoic in general), but also (2) the phytogeographical affinities of the flora are much wider than has been suspected. Examples are records of *Acmopyle* (Podocarpaceae) and *Libocedrus* (Cupressaceae) in Tasmania. A surprising number of taxa first recorded in the Oligocene have South American affinities. All taxonomic groups are

represented, e.g. *Lophosoria* (Filicopsida), *Austrocedrus* (Cupressaceae), Mutiseae (Asteraceae) and *Embothrium* (Proteaceae). Whilst the gymnosperms could have been derived from the inherited stock of Gondwanic taxa in Australia, *Lophosoria* and the two angiosperms appear to be immigrants. Logically these entered Australia via Antarctica, although to date only *Lophosoria* has been confirmed as part of the Tertiary flora in this region.

By the end of the Early Miocene a marked contrast between coastal and inland environments was well established, although not to the degree required to permit the development of sclerophyllous plant formations or grasslands on the regional scale.

Figure 10.9 Photomicrographs of commonly occurring early Tertiary spore and pollen types from southeast Australia (modern analogue in parentheses). Except where stated otherwise, the magnification is ×950: (*a*) *Cyathidites splendens* (*Acrostichum aureum* type); (*b*) *Gleicheniidites* (Gleicheniaceae ×599); (*c*) *Rugulatisporites mallatus* (Culcita ×599); (*d*) *Araucariacites australis* (Araucariaceae); (*e*) *Dilwynites granulatus* (Araucariaceae); (*f*) *D. tuberculatus* (Araucariaceae); (*g*) *Dacrycarpites australiensis* (*Dacrycarpus*); (*h*) *Ephedripites notensis* (*Ephreda*); (*i*) *Lygistepollenites florinii* (*Dacrydium* Group B spp.); (*j*) *Microcachrydites antarcticus* (*Microcachrys*); (*k, l*) *Phyllocladidites mawsonii* (*Lagarostrobos*); (*m*) *Podosporites microsaccatus* (extinct Podocarpaceae); (*n*) *Podocarpidites* (*Dacrydium/Podocarpus*); (*o*) *Trisaccites* sp. cf. *Microcachrydites antarcticus* (extinct? *Microcachrys*); (*p*) *Australopollis obscurus* (Callitrichaceae); (*q, r*) *Ericipites scabratus* (Ericales); (*s*) *Gambierina edwardsii* (extinct angiosperm); (*t*) *G. rudata* (extinct angiosperm); (*u, v*) *Myrtaceidites parvus-mesonesus* (Myrtaceae); (*w*) *Concolpites leptos* (*Gillbeea* type); (*x*) *Nothofagidites brachyspinulosus* (*Nothofagus* subgenus *Fuscospora*); (*y*) *N. flemingii* (*Nothofagus* subgenus *Nothofagus*); (*z*) *Periporopollenites demarcatus* (Trimeniaceae); (*aa*) *P. polyoratus* (?Caryophyllaceae).

Figure 10.10 Photomicrographs of commonly occurring early Tertiary spore and pollen types from southeast Australia (modern analogue in parentheses) Except where stated otherwise, the magnification is ×1088: (*a*) *Matonisporites ornamentalis* (Dicksoniaceae ×686); (*b*) *Parvisaccites catastus* (*Halocarpus*); (*c*) *Anacolosidites acutullus* (*Anacolosa*); (*d*) *Banksieaeidites arcuatus* (Musgraveinae); (*e*) *Haloragacidites harrisii* (Casuarinaceae); (*f*) *Cupanieidites orthoteichus* (Cupanieae); (*g*) *C. reticularis* (Cupanieae); (*h*) *Malvacipollis subtilis* (*Austrobuxus*); (*i–k*) *Malvacipollis* (Euphorbiaceae); (*l*) *Nothofagidites asperus* (*Nothofagus* subgenus *Lophozonia*); (*m–o*) *N. emarcidus-heterus* (*Nothofagus* subgenus *Brassospora*); (*p*) *Polycolpites langstonii* (extinct angiosperm); (*q*) *Proteacidites annularis* (*Xylomelum occidentale*-type); (*r*) *Proteacidites grandis* (extinct Proteaceae

×686); (*s*) *Proteacidites reticuloscabratus* (*Gevuina/Hicksbeachia*); (*t*) *Schizocolpus marlinensis* (Didymelaceae); (*u*) *Triporopollenites ambiguus* (*Telopea*).

Figure 10.11 Photomicrographs of commonly occurring early Tertiary spore and pollen types from southeast Australia (modern analogue in parentheses). Except where stated otherwise, the magnification is ×1100: (*a*) *Kuylisporites waterbolkii* (*Hemitelia*); (*b*) *Beaupreaidites verrucosus* (*Beauprea* ×693); (*c*) *Clavatipollenites glarius* (*Ascarina*); (*d, e*) *Malvacipollis* (Euphorbiaceae); (*f*) *Milfordia hypolaenoides* (Restionaceae); (*g*) *Myrtaceoipollenites australis* (? *Normapolles*); (*h*) *Myrtaceidites tenuis* (? eucalyptoid Myrtaceae); (*i, j*) *Nothofagidites deminutus-vansteenisii* (*Nothofagus* subgenus *Brassospora*); (*k*) *Polycolporopollenites esobalteus* (Polygalaceae); (*l*) *Spinizonocolpites prominatus* (*Nypa*); (*m*) *Proteacidites leightonii* (extinct Proteaceae ×693); (*n*) *Proteacidites pachypolus* (extinct Proteaceae); (*o*) *P. pachypolus* (geographical variant); (*p*) *Proteacidites ornatus* (extinct Proteaceae ×693); (*q,r*) *Santalumidites cainozoicus* (*Santalum* type); (*s,t*) *Sapotaceoidaepollenites rotundus* (Sapotaceae); (*u*) *Intratriporopollenites notabilis* (Tiliaceae ×693).

Figure 10.12 Photomicrographs of commonly occurring early Tertiary spore and pollen types from southeast Australia (modern analogue in parentheses). Except where stated otherwise, the magnification is ×1250: (*a*) *Anacolosidites sectus* (*Anacolosa*); (*b*) *Aglaoreidia qualumis* (Sparganiaceae/Typhaceae); (*c*) *Gothanipollis bassensis* (Loranthaceae); (*d*) *Nothofagidites falcatus* (*Nothofagus* subgenus *Brassospora*); (*e*) *Proteacidites asperopolus* (extinct Proteaceae); (*f*) *Proteacidites rectomarginis* (extinct Proteaceae); (*g*) *Proteacidites reticulatus* (extinct Proteaceae); (*h*) *Psilastephanocolporites micus* (extinct? angiosperm); (*i*) *Reevesiapollis reticulatus* (*Ungeria*); (*j*) *Triorites magnificus* (extinct Proteaceae ×788); (*k*) *Cyathidites subtilis* (Cyatheaceae ×788); (*l*) *Cyatheacidites annulatus* (*Lophosoria* ×788). (*m*) *Granodiporites nebulosus* (*Embothrium*); (*n*) *Tubulifloridites antipoda* (Asteraceae).

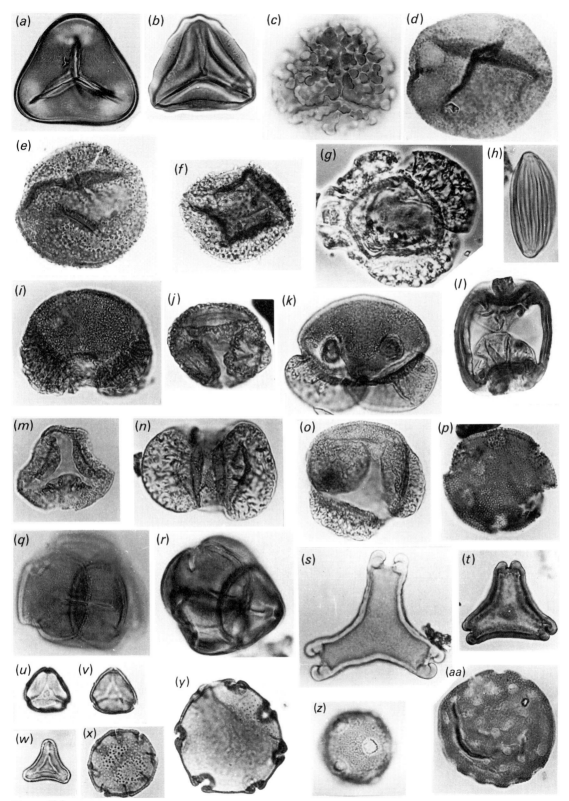

Figure 10.9

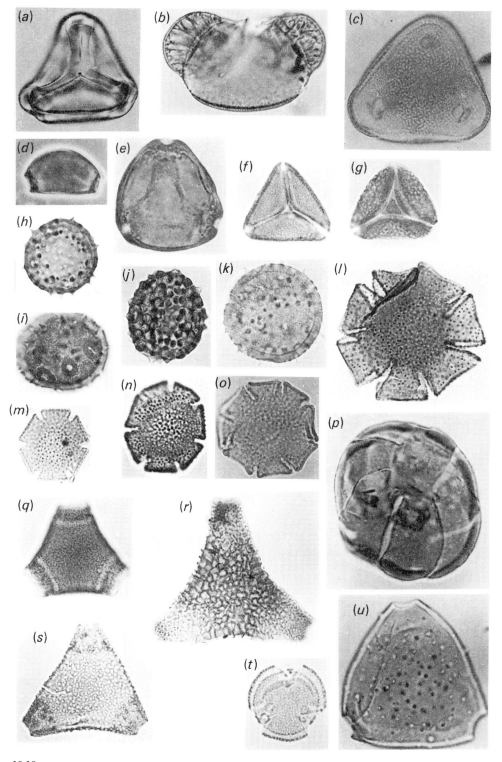

Figure 10.10

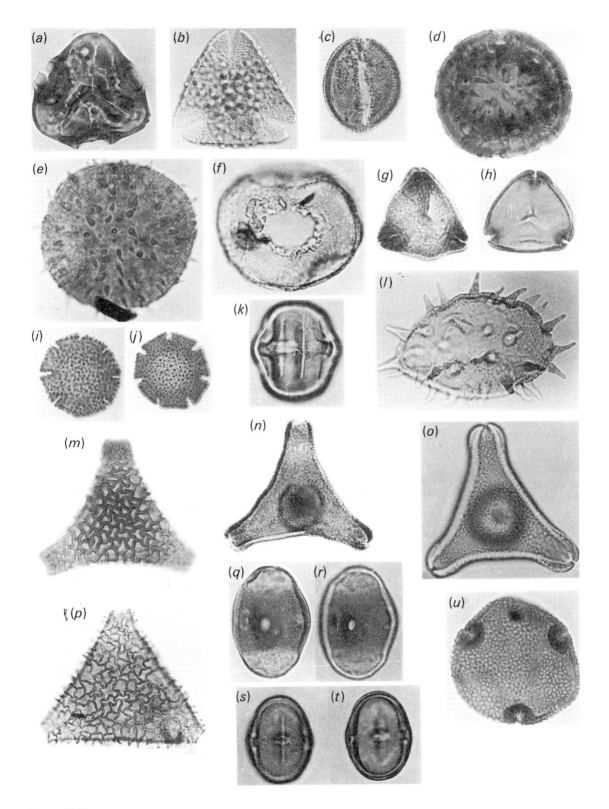

Figure 10.11

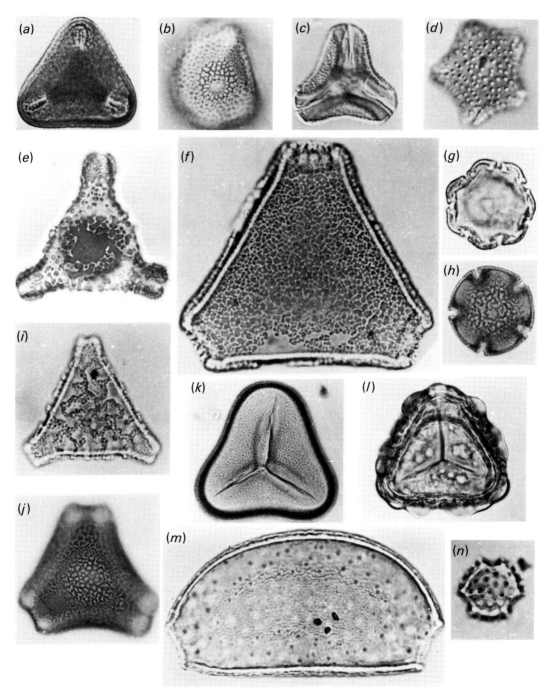

Figure 10.12

Appendix 10.1. *Suggested modern affinities of the fossil spores and pollen (selected spores and pollen grains are illustrated in Figures 10.9–10.12)*

Fossil taxon	Modern analogue	Habit(s)	Modern distribution
Bryophytes			
Bryosporis anisopolaris	?Anthocerotae	?Liverwort	Unknown, but includes Australia and New Zealand
Ricciaesporites kawaraensis	Ricciaceae	Liverworts	Includes Australia and New Zealand
Rudolphisporis rudolphi	*Anthoceros*	Liverworts	Includes Australia, New Zealand and Europe
Pteridophytes			
Asseretospora	*Pteris* (Pteridaceae)	Terrestrial ferns	Tropical and warm temperate, including Indo-malesia and Australia
Camarozonosporites bullatus	Lycopsida	Unknown	?Extinct
Crassiretitriletes vanraadshoovenii	*Lygodium* (Schizaeaceae)	Climbing ferns	Tropical and warm temperate, including Indo-malesia and Australia
Cyatheacidites annulatus	*Lophosoria* (Lophosoriaceae)	Small tree fern	Tropical America (mountains)
Cyathidites splendens	Includes *Acrostichum* (Adiantaceae)	Mangrove ferns	Pan-tropical including Indo-malesia and Australia
Cyathidites spp.	Includes *Cyathea* (Cyatheaceae)	Tree ferns to 24 m tall	Widespread, mainly tropical and warm temperate
Foveotriletes crater	*Lycopodium* sp.	Terrestrial lycopod	Unknown, ?extinct
Foveotriletes lacunosus	*Lycopodium varium type*	Terrestrial lycopod	Eastern Australia and New Zealand; mainly cool temperate
Foveotriletes palaequetrus	*L. australianum type*	Terrestrial lycopod	Eastern Australia, New Zealand and Indonesia
Gleicheniidites spp.	Gleicheniaceae	Thicket-forming ground fern	Tropical to cold climates
Herkosporites elliotii	cf. *Lycopodium deuterodensum*	Terrestrial lycopod	Eastern Australia, New Zealand and New Caledonia
Hypolepis spinyspora	*Hypolepis* (Dennstaedtiaceae)	Terrestrial fern	Mainly tropical and warm temperate
Kuylisporites waterbolkii	*Hemitelia* (Cyatheaceae)	Tree ferns	Tropical South America
Latrobosporites marginis	*Lycopodium laterale*	Terrestrial lycopod	Eastern Australia, New Zealand and New Caledonia
Matonisporites ornamentalis	*Dicksonia* (Dicksoniaceae)	Tree fern	Tropical America, Malesia (mountains), Australia, New Zealand, New Caledonia
Polypodiaceoisporites spp.	*Pteris* (Pteridaceae)	Terrestrial ferns	Tropical and warm temperate, including Indo-malesia and Australia
Polypodiisporites histeopteroides	*Histeopteris incisa* (Pteridaceae)	ground fern	Australia (except WA), New Zealand, southern Africa, South America, West Indies
Polypodiisporites spp.	Includes *Microsorium* (Polypodiaceae)	Ground ferns	Mainly tropical, includes Australia and New Zealand
Rugulatisporites mallatus	*Culcita* (Thyrsopteridaceae)	Terrestrial fern	Tropical America, Melanesia, eastern Australia
Gymnosperms			
Araucariacites australis	Araucariaceae	Tall trees	SW Pacific (especially New Caledonia), South America

Appendix 10.1. (*cont.*)

Fossil taxon	Modern analogue	Habit(s)	Modern distribution
Dacrycarpites australiensis	*Dacrycarpus, Podocarpus*	Tall trees	SE Asia to W Pacific including New Zealand
Dilwynites spp.	Araucariaceae	?Trees	?Extinct
Ephredripites notensis	*Ephredra* (Ephredraceae)	Shrubs	Mediterranean and warm temperate America, China
Lygistepollenites balmei	?Podocarpaceae	?Trees	Extinct
Lygistepollenites florinii	*Dacrydium* Group B spp. of Florin, 1931 (Podocarpaceae)	Trees	SE Asia to New Zealand
Microalatidites palaeogenicus	*Phyllocladus* (Podocarpaceae)	Trees and shrubs	Borneo to New Guinea, New Zealand, Tasmania
Microcachrydites antarcticus	*Microcachrys tetragona* (Podocarpaceae)	Shrub	Tasmania (mountains)
Parvisaccites catastus	*Halocarpus bidwillii* type (Podocarpaceae)	Shrubs, trees	New Zealand (mainly on mountains)
Phyllocladidites mawsonii	*Lagarostrobos franklinii* type (Podocarpaceae)	Shrub to tall tree	Tasmania
Phyllocladidites reticulosaccatus	*Lagarostrobos* (Podocarpaceae)	?Tree	Extinct
Podocarpidites spp.	*Dacrydium* and *Podocarpus* (Podocarpaceae)	Mainly tall trees	Southern temperate through tropical highlands to West Indies and Japan
Podosporites erugatus	*Microstrobos* (Podocarpaceae)	Shrubs	Tasmania (mountains), NSW (Blue Mts)
Podosporites microsaccatus	Extinct ?Podocarpaceae	?Shrubs	Extinct
Podosporites parvus	*?Microcachrys* (Podocarpaceae)	?Shrubs	Extinct
Angiosperms: monocots			
Aglaoreidia qualumis	*?Typha* (?Typhaceae)	Bull-rushes	Cosmopolitan
Arecipites	?Arecaceae	?Palms	Tropical and warm temperate
Cyperaceaepollis	Cyperaceae	Sedges	Cosmopolitan, especially temperate regions
Dicolpopollis metroxylonoides	*Metroxylon* (Arecaceae)	Palms	Thailand to Fiji
Graminidites	Poaceae	Grasses	Cosmopolitan
Dryptopollenites semilunatus	*?Freycinetia* (Pandancaeae)	Lianes	Sri Lanka to New Zealand and Polynesia
Florschuetzia levipoli	*Sonneratia* (Sonneratiaceae)	Mangrove trees	Coastal areas, Indian and Pacific Oceans
Lateropora glabra	*Freycinetia* (Pandanaceae)	Lianes	Sri Lanka to New Zealand and Polynesia
Liliacidites	Liliaceae and ?Iridaceae	Lilies, ?irises	Cosmopolitan, especially dryish and warm temperate
Luminidites phormoides	*Phormium* (Agavaceae)	Flax	New Zealand
Milfordia	Restionaceae	Herbs	Southern hemisphere, especially Africa and Australia
Sparganiaceaepollenites	Sparganiaceae	Bull-rushes	North temperate to Australia and New Zealand
Spinizonocolpites prominatus	*Nypa* (Arecaceae)	Mangrove palm	India to Indo-malesia
Angiosperms: dicots			
Acaciapollenites miocenicus	*Archidendron/Pithecellobium* (Mimosaceae)	Trees	Indo-malesia and tropical America

Appendix 10.1. (*cont.*)

Fossil taxon	Modern analogue	Habit(s)	Modern distribution
Acaciapollenites myriosporites	*Acacia* (Mimosaceae)	Trees, shrubs	Tropical and warm temperate, especially Australia
Alangiopollis sp.	*Alangium villosum* (Alangiaceae)	Shrubs	Tropical Africa, China to NE Australia
Anacolosidites	*Anacolosa* (Olacaceae)	Trees, shrubs, lianes	Tropical and southern Africa
Australopollis obscurus	Callitrichaceae	Aquatic herbs	Cosmopolitan
Banksieaeidites arcuatus	Musgraveinae (Proteaceae)	Trees	Queensland
Banksieaeidites spp.	*Banksia* (Proteaceae)	Small trees, shrubs	Australia and New Guinea
Basopollis	Unknown	Unknown	?Extinct
Beaupreaidites	*Beauprea* (Proteaceae)	Shrubs	New Caledonia
Bluffopollis scabratus	*Strasburgeria*	Tree	New Caledonia
Bombacacidites bombaxoides	?*Bombax* (Bombacaceae)	Trees, shrubs	Tropical
Canthiumidites bellus	*Randia* (Rubiaceae)	Shrubs	Tropical
Chenopodipollis chenopodiaceoides	Chenopodiaceae–Amaranthaceae	Halophytic shrubs, herbs	Cosmopolitan, especially arid areas, and salt marshes
Clavatipollenites glarius	*Ascarina* (Chloranthaceae)	Small tree	Madagascar, Melanesia, Polynesia and New Zealand
Concolpites leptos	*Gillbea* type (Cunoniaceae)	Trees	Queensland and New Guinea
Corsinipollis	*Epilobium* (Onagraceae)	Herbs	Cosmopolitan
Cranwellia	Loranthaceae	Hemiparasites	Tropical and temperate
Cupanieidites	Sapindaceae (Cupanieae)	Trees, shrubs, climbers	Tropical and warm temperate
Diporites aspis	*Fuchsia* (Onagraceae)	Trees, shrubs	Central and south America, Tahiti, New Zealand
Dodonaea sphaerica	*Dodonaea* (Sapindaceae)	Trees, shrubs	Tropical and warm temperate, especially Australia
Ericipites	Ericales (?Epacridaceae)	Small trees, shrubs	Predominantly Australia
Gambierina	Unknown	Unknown	Extinct
Glencopollis ornatus	*Polygonum* (Polygonaceae)	Herbs	Cosmopolitan
Gothanipollis spp.	Loranthaceae	Hemiparasites	Tropical and temperate
Granodiporites nebulosus	*Embothrium* (Proteaceae)	Shrub	Central and southern Andes, South America
Guettardidites	*Guettarda* (Rubiaceae)	Trees, shrubs	New Caledonia, tropical America
Gyropollis psilatus	Gyrostemonaceae	Trees, shrubs	Australia
Hakeidites	*Grevillea/Hakea* (Proteaceae)	Trees, shrubs	Australia
Haloragacidites harrisii	Casuarinaceae	Trees, shrubs	Indo-malesia and Australia
Ilexpollenites	*Ilex* (Aquifoliaceae)	Trees, shrubs, climbers	Cosmopolitan, especially tropical and temperate Asia and America
Intratriporopollenites notabilis	Tiliaceae	Trees, shrubs, herbs	Subcosmopolitan
Lymingtonia	Portulacaceae	Shrubs, herbs	Cosmopolitan, especially western America
Malvacearumpollis	Malvaceae	Trees, shrubs, herbs	Cosmopolitan, especially tropical
Malvacipollis diversus	Euphorbiaceae	?Shrubs	Malesia to Fiji, Australia
Malvacipollis spinyspora	*Micrantheum* (Euphorbiaceae)	Shrubs	Australia

Appendix 10.1. (*cont.*)

Fossil taxon	Modern analogue	Habit(s)	Modern distribution
Malvacipollis subtilis	Euphorbiaceae	?Shrubs	Malesia to Fiji, Australia
Margocolporites vanwijhei	Caesalpinoidae (Fabaceae)	Trees, shrubs	Mainly tropical
Myrtaceidites eucalyptoides	*Eucalyptus* (Myrtaceae)	Trees, shrubs	Mainly Australia
Myrtaceidites eugenioides	*Eugenia* type (Myrtaceae)	Trees	Tropical, especially America, 1 sp. in Australia
Myrtaceidites oceanicus	*Rhodomyrtus* type (Myrtaceae)	?Shrubs	?Extinct
Myrtaceidites parvus-mesonesus	?*Metrosideros* (Myrtaceae)	Trees, shrubs, ?climbers	Malesia, South Africa, Pacific region including New Zealand
Myrtaceidites tenuis	?*Eucalyptus* (Myrtaceae)	Trees, ?shrubs	Extinct
Myrtaceidites verrucosus	*Rhodomyrtus* type	Trees, shrubs	Indo-malesia, Australia
Nothofagidites asperus	*Nothofagus* (*Lophozonia*) spp. (Fagaceae)	Trees, shrubs	Eastern Australia, New Zealand, South America
Nothofagidites brachyspinulosus	*Nothofagus* (*Fuscospora*) spp.	Trees, shrubs	Tasmania, New Zealand, South America
Nothofagidites deminutus-vansteenisii	*Nothofagus* (*Brassospora*) spp.	Trees, ?shrubs	Extinct
Nothofagidites emarcidus-heterus	*Nothofagus* (*Brassospora*) spp.	Trees, shrubs	New Guinea, New Caledonia
Nothofagidites endurus	'Ancestral' *Nothofagus*	Trees, ?shrubs	Extinct
Nothofagidites falcatus	*Nothofagus* (*Brassospora*) spp.	Trees, ?shrubs	?New Caledonia, ?New Guinea
Nothofagidites flemingii	*Nothofagus* (*Nothofagus*) spp.	Trees, ?shrubs	South America
Nothofagidites goniatus	*Nothofagus* (*Brassospora*) spp.	Trees, ?shrubs	Extinct
Nothofagidites longispinosa	*Nothofagus* (*Brassospora*) spp.	Trees, ?shrubs	Extinct
Nuxpollenites sp.	*Dodonaea triquetra* (Sapindaceae)	Shrub	Eastern Australia
Palaeocoprosmadites	*Coprosma/Opercularia* (Rubiaceae)	Shrubs	Malesia, Australia and Pacific including New Zealand
Paripollis orchesis	*Epacris virgata* type (Epacridaceae)	Shrubs	Australia
Perfotricolpites sp. cf. *P. digitatus*	*Merremia* type (Convolvulaceae)	Herbs, climbers	Mainly tropical
Periporopollenites demarcatus	Trimeniaceae	Trees, shrubs	Malesia to SE Australia
Periporopollenites polyoratus	?Caryophyllaceae	?Herbs	Cosmopolitan
Perisyncolporites pokornyi	Malpighiaceae	Trees, shrubs, lianes	Tropical and warm temperate, especially South America
Polyadopollenites sp. cf. *P. granulosus*	*Archidendron* type (Mimosaceae)	Trees, ?shrubs	Tropical and warm temperate
Polycolpites langstonii	Unknown	Unknown	Extinct
Polycolporopollenites esobalteus	Polygalaceae	Trees, shrubs, lianes	Subcosmopolitan (not in west Pacific)
Polyorificites oblatus	Euphorbiaceae	?Shrubs	Malesia to Fiji, Australia
Proteacidites annularis	*Xylomelum occidentale* type (Proteaceae)	Shrub	SW Australia?
Proteacidites isopogoniformis	*Isopogon* (Proteaceae)	Shrubs	Eastern and SW Australia
Proteacidites obscurus	*Agastachys* type (Proteaceae)	Shrub	Tasmania
Proteacidites truncatus	*Isopogon* (Proteaceae)	Shrubs	Eastern and SW Australia
Proteacidites spp.	Proteaceae	Trees, shrubs	Tropical to temperate, mainly Australia and South Africa
Psilastephanocolporites micus	Unknown	Unknown	?Extinct
Psilodiporites sp.	*Alyxia* (Apocynaceae)	Shrubs	Indo-malesia to West Pacific, Australia

Appendix 10.1. (*cont.*)

Fossil taxon	Modern analogue	Habit(s)	Modern distribution
Quintiniapollis psilatispora	*Quintinia* (Grossulariaceae)	Trees, shrubs	Malesia to Australia and New Zealand
Reevesiapollis reticulatus	*Ungeria* (Sterculiaceae)	?Shrub	Norfolk Island
Rhoipites sp.	*Cissus* (Vitaceae)	Lianes	Tropical and warm temperate
Rubipollis oblatus	*Canthium* (Rubiaceae)	Trees, shrubs	Tropical
Santalumidites cainozoicus	*Santalum* (Santalaceae)	Tree	Indo-malesia to Australia, Pacific Is.
Sapotaceoidaepollenites rotundus, S. latizonatus	Sapotaceae	Trees, shrubs	Mainly tropical
Simpsonipollis	Unknown	Unknown	?Extinct
Stephanocolpites oblatus	*Haloragodendron*	Herbs, ?shrubs	Southern Australia
Striasyncolpites laxus	*Liparophyllum* type (Gentianaceae)	Herbs	Tasmania and New Zealand
Tetracolporites palynius	*Dysoxylum* (Meliaceae)	Trees, shrubs	Indo-malesia to New Zealand and Tonga
Tricolpites reticulatus	*Gunnera* (Gunneraceae)	Herbs	Tropical and South Africa, Malesia, Tasmania, Antarctic Is., Hawaii, South America
Tricolpites simatus	Loranthaceae	?Hemiparasites	Tropical and temperate
Tricolpites thomasii	Loranthaceae	?Hemiparasites	Tropical and temperate
Trocolporopollenites endobalteus	*Macaranga/Mallotus* (Euphorbiaceae)	Trees, shrubs	Tropical, Indo-malesia to East Asia, Fiji, eastern Australia, Africa
Triporopollenites ambiguus	*Telopea* type (Proteaceae)	Shrubs	Australia
Triporotetradites sp.	*Gardenia* (Rubiaceae)	Trees, shrubs	Tropical and warm temperate
Tubulifloridites spp.	Tubulifloridae (Asteraceae)	Trees, shrubs, herbs	Cosmopolitan

Distribution data after D. J. Mabberly (1989), *The Plant Book*, Cambridge University Press.

ACKNOWLEDGEMENTS

It is difficult to overstate the debt owed by Australian paleopalynologists in general and the authors in particular to Alan Partridge (Esso Australia Ltd, Sydney) and his colleagues Lew Stover (Exxon Corporation, Houseton) and Robin Helby (Consultant palynologist, Sydney). Without their palynological zonation schema, this chapter would have consisted of little more than isloated observations.

REFERENCES

ALLEY, N. F. (1987). Middle Eocene age of the megafossil flora at Golden Grove, South Australia: preliminary report and comparison with the Maslin Bay flora. *Transactions of the Royal Society of South Australia*, **111**, 211–12.

ASKIN, R. A. (1990). Campanian to Paleocene spore and pollen assemblages of Seymour Island, Antarctica. *Review of Palaeobotany and Palynology*, **65**, 105–13.

ASKIN, R. A., ELLIOT, D. H., STILWELL, J. D. & ZINSMEISTER, W. J. (1991). Stratigraphy and palaeontology of Campanian and Eocene sediments, Cockburn Island, Antarctic Peninsula. *Journal of South American Earth Sciences*, **4**, 99–117.

BAKER, G. (1973). Dr Isabel Clifton Cookson. *Geological Society of Australia Special Publication*, **4**, i–x.

BALME, B. E. & CHURCHILL, D. M. (1959). Tertiary sediments at Coolgardie, Western Australia. *Journal of the Royal Society of Western Australia*, **42**, 37–43.

BARLOW, B. A. (1981). The Australian flora: its origin and evolution. In *Flora of Australia*, vol. 1, pp. 25–75. Canberra: Australian Government Publishing Service.

BARRON, E. J. (1985). Explanations of the Tertiary global cooling trend. *Palaeogeography, Palaeoclimatology, Palaeoecology*, **50**, 45–61.

BARRON, E. J. & PETERSON, W. H. (1991). The Cenozoic ocean circulation based on ocean General Circulation Model results. *Palaeogeography, Palaeoclimatology, Palaeoecology*, **83**, 1–28.

BARTH, H. (1982). The biogeography of mangroves. *Tasks in Vegetation Science*, **2**, 35–60.

BERGGREN, W. A. (1969). Cainozoic chronostratigraphy, planktonic foraminiferal zonation and the radiometric time scale. *Nature*, **224**, 1072–5.

BERGGREN, W. A., KENT, D. V., FLYNN, J. J. & COUVERING, J. A. VAN (1985). Cenozoic geochronology. *Geological Society of America, Bulletin*, **96**, 1407–18.

BERRY, P. E., SKVARLA, J. J., PARTRIDGE, A. D. & MACPHAIL, M. K. (1990). *Fuchsia* pollen from the Tertiary of Australia. *Australian Systematic Botany*, **3**, 739–44.

BINT, A. N. (1981). An Early Pliocene pollen assemblage from Lake Tay, south-western Australia and its phytogeographic implications. *Australian Journal of Botany*, **29**, 277–91.

BIRD, E. C. F. (1968). *Coasts*. Canberra: Australian National University Press.

BIRKS, H. J. B. & BIRKS, H. H. (1980). *Quaternary Palaeoecology*. London: Edward Arnold.

BISHOP, P. & BAMBER, R. K. (1985). Silicified wood of Early Miocene *Nothofagus, Acacia* and Myrtaceae (aff. *Eucalyptus* B) from the upper Lachlan valley, New South Wales. *Alcheringa*, **9**, 221–8.

BLACKBURN, D. T. (1981). Tertiary megafossil flora of Maslin Bay, South Australia: numerical taxonomic study of selected leaves. *Alcheringa*, **5**, 9–28.

BLOW, W. H. (1969). Late Middle Eocene to Recent planktonic foraminiferal biostratigraphy. *Proceedings, First International Conference on Planktonic Microfossils, Geneva, 1967*, I, 199–422.

BLOW, W. H. (1979). *The Cainozoic Globigerinida. A Study of the Morphology, Taxonomy, Evolutionary Relationships and the Stratigraphical Distribution of some Globigerinida (Mainly Globigerinacea)*. Leiden: E. J. Brill.

BODDARD, J. M., WALL, V. J. & KANEN, R. A. (1986). Lithostratigraphic and depositional architecture of the Latrobe Group, offshore Gippsland Basin. In *Second South-Eastern Australia Oil Exploration Symposium*, ed. R. C. Glenie, pp. 113–36. Melbourne: Petroleum Exploration Society of Australia, Victorian and Tasmanian Branch.

BOWMAN, D. M. J. S., MACLEAN, A. R. & CROWDEN, R. K. (1986). Vegetation–soil relationships in the lowlands of south-west Tasmania. *Australian Journal of Ecology*, **11**, 141–53.

BROWN, C. M. (1989). Structural and stratigraphic framework of groundwater occurrence and surface discharge in the Murray Basin, southeastern Australia. *BMR Journal of Australian Geology and Geophysics*, **11**, 127–46.

BURBIDGE, N. T. (1960). The phytogeography of the Australian region. *Australian Journal of Botany*, **8**, 75–212.

BURRETT, C., DUHIG, N., BERRY, R. & VARNE, R. (1991). Asian and south-western Pacific continental terranes derived from Gondwana, and their biogeographic significance. *Australian Systematic Botany*, **4**, 13–24.

CALLEN, R. A. & TEDFORD, R. H. (1976). New Late Cainozoic rock units and depositional environments, Lake

Frome area, South Australia. *Transactions of the Royal Society of South Australia*, **100**, 125–67.

CARPENTER, R. (1991). Palaeovegetation and environment at Cethana, Tasmania. Ph.D. thesis, University of Tasmania.

CHANDLER, M. E. J. (1951). Notes on the occurrence of mangroves in the London Clay. *Proceedings Geological Association*, **62**, 271–2.

CHRISTOPHEL, D. C. (1980). Occurrence of *Casuarina* megafossils in the Tertiary of south-eastern Australia. *Australian Journal of Botany*, **28**, 249–59.

CHRISTOPHEL, D. C., HARRIS, W. K. & SYBER, A. K. (1987) The Eocene flora of the Anglesea locality, Victoria. *Alcheringa*, **11**, 303–23.

CHURCHILL, D. M. (1973). The ecological significance of tropical mangroves in the Early Tertiary floras of southern Australia. *Geological Society of Australia Special Publication*, **4**, 55–72.

COOKSON, I. C. (1947). Plant microfossils from lignites of Kerguelen Archipelago. *BANZ Antarctic Research Expedition 1929–1931 Reports Series A*, **2**, 127–42.

COOKSON, I. C. (1957). On some Australian Tertiary spores and pollen grains that extend the geological and geographical distribution of living genera. *Proceedings of the Royal Society of Victoria*, **69**, 41–53.

COOKSON, I. C. & EISENACK, A. (1967). Some Early Tertiary microplankton and pollen grains from a deposit near Strahan, western Tasmania. *Proceedings of the Royal Society of Victoria*, **80**, 131–40.

COSTIN, A. B. (1981). Alpine and sub-alpine vegetation. In *Australian Vegetation*, ed. R. H. Groves, pp. 361–75. Cambridge: Cambridge University Press.

CROCKER, R. L. & WOOD, J. G. (1947). Some historical influences on the development of South Australian vegetation communities and their bearing on concepts and classification in ecology. *Transactions of the Royal Society of South Australia*, **71**, 91–136.

DALY, M. C., COOPER, M. A., WILSON, I., SMITH, D. G. & HOOPER, B. G. D. (1991). Cenozoic plate tectonics and basin evolution in Indonesia. *Marine Petroleum Geology*, **8**, 2–21.

DAWSON, J. W. (1986). Floristic relationships of lowland rainforest phanerogams of New Zealand. *Telopea*, **2**, 681–96.

DETTMANN, M. E. (1963). Upper Mesozoic microfloras from south-eastern Australia. *Proceedings of the Royal Society of Victoria*, **77**, 1–148.

DETTMANN, M. E. (1986). Significance of the Cretaceous–Tertiary spore genus *Cyatheacidites* in tracing the origin and migration of *Lophosoria* (Filicopsida). *Special Publications in Palaeontology*, **35**, 63–94.

DETTMANN, M. E. (1989). Antarctica: Cretaceous cradle of austral temperate rainforests? *Geological Society of Australia Special Publication*, **47**, 89–105.

DETTMANN, M. E. & JARZEN, D. M. (1990). The Antarctic/Australian rift valley: Late Cretaceous cradle of northeastern Australasian relicts? *Review of Palaeobotany and Palynology*, **65**, 131–44.

DETTMANN, M. E. & JARZEN, D. M. (1990). Pollen evidence for Late Cretaceous differentiation of Proteaceae in southern polar forests. *Canadian Journal of Botany*, **69**, 901–6.

DETTMANN, M. E., POCKNALL, D. T., ROMERO, E. J. & ZAMALOA, M. A. DEL C. (1990). *Nothofagidites* Erdtman ex Potonié, 1960; a catalogue of species with notes on the palaeogeographic distribution of *Nothofagus* Bl. (Southern Beech). *New Zealand Geological Survey Paleontological Bulletin*, **60**.

DETTMANN, M. E. & THOMSON, M. R. A. (1987). Cretaceous palynomorphs from the James Ross Island area, Antarctica – a pilot study. *British Antarctic Survey Bulletin*, **77**, 13–59.

DUDGEON, M. J. (1983). Eocene pollen of probable proteaceous affinity from the Yaamba Basin, central Queensland. *Memoirs of the Association of Australasian Palaeontologists*, **1**, 339–62.

DUIGAN, S. L. (1966). The nature and relationships of the Tertiary brown coal flora of the Yallourn area in Victoria, Australia. *Palaeobotanist*, **14**, 191–201.

FARLEY, M. B. (1990). Vegetation distribution across the Early Eocene depositional landscape from palynological analysis. *Palaeogeography, Palaeoclimatology, Palaeoecology*, **79**, 11–27.

FEARY, D. A., DAVIES, P. J., PIGRAM, C. J. & SYMONDS, P. A. (1991). Climatic evolution and control on carbonate deposition in northeast Australia. *Palaeogeography, Palaeoclimatology, Palaeoecology (Global and Planetary Change Section)*, **89**, 341–61.

FORSYTH, S. M. (1989). Tamar Graben. *Geological Society of Australia Special Publication*, **15**, 358–9.

FOSTER, C. B. (1982). Illustrations of Early Tertiary (Eocene) plant microfossils from the Yaamba Basin, Queensland. *Queensland Department of Mines Publication*, **381**, 1–33.

GRIMES, K. G. (1980). The Tertiary geology of north Queensland. In *The Geology and Geophysics of Northeastern Australia*, ed. R. A. Henderson & P. J. Stephenson, pp. 329–47. Brisbane: Geological Society of Australia.

HAQ, B. U., HARDENBOL, J. & VAIL, P. R. (1987). Chronology of fluctuating sea levels since the Triassic. *Science*, **235**, 1156–67.

HARLAND, W. B., ARMSTRONG, R. L., COX, A. V., CRAIG, L. E., SMITH, A. G. & SMITH, D. G. (1990). *A Geological Time Scale*. Cambridge: Cambridge University Press.

HARRIS, W. K. (1965a). Basal Tertiary microfloras from the Princetown area, Victoria, Australia. *Palaeontographica B*, **120**, 76–106.

HARRIS, W. K. (1965b). Tertiary microfloras from Brisbane, Queensland. *Geological Survey of Queensland Report*, **10**.

HARRIS, W. K. (1971). Tertiary stratigraphic palynology,

Otway basin. *Special Bulletin, Geological Surveys of South Australia, Victoria*, pp. 67–87.

HARRIS, W. K. (1974). Palynology of Paleocene sediments at Site 214, Ninetyeast Ridge. *Initial Reports, Deep Sea Drilling Project*, **22**, 503–19.

HARRIS, W. K. (1985). Middle to Late Eocene depositional cycles and dinoflagellates zones in southern Australia. *Special Publication, South Australia Department of Mines*, **5**, 133–44.

HARRIS, W. K. & TWIDALE, C. T. (1991). Revised age for Ayers Rock and the Olgas. *Transactions of the Royal Society of South Australia*, **115**, 109.

HARTLEY, T. G. (1986). Floristic relationships of the rainforest flora of New Guinea. *Telopea*, **2**, 619–30.

HEKEL, H. (1972). Pollen and spore assemblages from Queensland Tertiary sediments. *Geological Survey of Queensland Palaeontological Paper*, **30**, 1–34.

HELBY, R., MORGAN, R. & PARTRIDGE, A. D. (1987). A palynological zonation of the Australian Mesozoic. *Memoir of the Association of Australasian Palaeontologists*, **4**, 1–134.

HILL, R. S. (1982). Rainforest fire in western Tasmania. *Australian Journal of Botany*, **30**, 583–9.

HILL, R. S. (1987). Discovery of *Nothofagus* fruits corresponding to an important Tertiary pollen type. *Nature*, **327**, 56–8.

HILL, R. S. (1989a). New species of *Phyllocladus* (Podocarpaceae) macrofossils from southeastern Australia. *Alcheringa*, **13**, 193–208.

HILL, R. S (1989b). Fossil leaf. *DSIR Bulletin*, **245**, 143–4.

HILL, R. S. (1991). Leaves of *Eucryphia* (Eucryphiaceae) from Tertiary sediments in south-eastern Australia. *Australian Systematic Botany*, **4**, 481–97.

HILL, R. S. & GIBSON, N. (1986). Macrofossil evidence for the evolution of the alpine and subalpine vegetation of Tasmania. In *Flora and Fauna of Alpine Australasia: Ages and Origins*, ed. B. A. Barlow, pp. 204–17. Melbourne: CSIRO/Australian Systematic Botany Society.

HILL, R. S. & MACPHAIL, M. K. (1983). Reconstruction of the Oligocene vegetation at Pioneer, northeast Tasmania. *Alcheringa*, **7**, 281–99.

HILL, R. S. & READ, J. (1991). A revised infrageneric classification of *Nothofagus* (Fagaceae). *Botanical Journal of the Linnean Society*, **105**, 37–72.

HOLDGATE, G. R. & SLUITER, I. R. K. (1991). Oligocene–Miocene marine incursions in the Latrobe Valley Depression, onshore Gippsland Basin: evidence, facies relationships and chronology. *Geological Society of Australia Special Publication*, **18**, 137–57.

HOOKER, J. D. (1860). Introductory essay. In *Botany of the Antarctic Voyage of HM. Discovery ships 'Erebus' and 'Terror', in the years 1839–1843*. III. *Flora Tasmaniae*. London: Lovell Reeve.

HOS, D. (1975). Preliminary investigations of the palynology of the upper Eocene Werrillup Foramation,

Western Australia. *Journal and Proceedings of the Royal Society of Western Australia*, **58**, 1–14.

JACKSON, W. D. (1968*a*). Fire, air, water and earth – an elemental ecology of Tasmania. *Proceedings of the Ecological Society of Australia*, **3**, 9–16.

JACKSON, W. D. (1968*b*). Fire and the Tasmanian flora. *Tasmanian Yearbook*, 2/1968, 50–5.

JACOBSON, G. L. & BRADSHAW, R. H. W. (1981). The selection of sites for palaeoenvironmental studies. *Quaternary Research*, **16**, 80–96.

JARZEN, D. M. & DETTMANN, M. E. (1989). Taxonomic revision of *Tricolpites reticulatus* Cookson ex Couper, 1953 with notes on the biogeography of *Gunnera* L. *Pollen et Spores*, **31**, 97–112.

JENNINGS, J. N. & BIRD, E. C. F. (1967). Regional geomorphological characteristics of some Australian estuaries. In *Estuaries*, ed. G. H. Lauff, pp. 121–8. Washington: American Association for the Advancement of Science.

JOHNSON, R. W. & BURROWS, W. H. (1981). *Acacia* open-forests, woodlands and shrublands. In *Australian Vegetation*, ed. R. H. Groves, pp. 198–226. Cambridge: Cambridge University Press.

KAMP, P. J. J., WAGHORN, D. B. & NELSON, C. S. (1990). Late Eocene–Early Oligocene integrated isotope stratigraphy and biostratigraphy for paleoshelf sequences in southern Australia: paleoceanographic implications. *Palaeogeography, Palaeoclimatology, Palaeoecology*, **80**, 169–208.

KEMP, E. M. (1974). Preliminary palynology of samples from site 254, Ninetyeast Ridge. *Initial Reports, Deep Sea Drilling Project*, **26**, 815–23.

KEMP, E. M. (1976). Early Tertiary pollen from Napperby, Central Australia. *BMR Journal of Australian Geology and Geophysics*, **1**, 109–14.

KEMP, E. M. (1978). Tertiary climatic evolution and vegetation history in the southeast Indian Ocean region. *Palaeogeography, Palaeoclimatology, Palaeoecology*, **24**, 169–208.

KEMP, E. M. (1981). Pre-Quaternary fire in Australia. In *Fire and the Australian Biota*, ed. A. M. Gill, R. H. Groves & I. R. Noble, pp. 1–21. Canberra: Australian Academy of Science.

KEMP, E. M. & BARRETT, P. J. (1975). Antarctic glaciation and early Tertiary vegetation. *Nature*, **258**, 507–8.

KEMP, E. M. & HARRIS, W. K. (1975). The vegetation of Tertiary islands on Ninetyeast Ridge. *Nature*, **258**, 303–7.

KEMP, E. M. & HARRIS, W. K. (1977). The palynology of Early Tertiary sediments, Ninetyeast Ridge, Indian Ocean. *Special Paper in Palaeontology*, **19**, 1–69.

KENNETT, J. P. (1982). *Marine Geology*. Englewood Cliffs, NJ: Prentice Hall.

KENNETT, J. P., BORCH C. VAN DER, BAKER, P. A. *et al*. (1985). Palaeotectonic implications of increased late Eocene–early Oligocene volcanism from South Pacific DSDP sites. *Nature*, **316**, 507–11.

KERSHAW, A. P. (1988). Australasia. In *Vegetation History*,

ed. B. Huntley & T. Webb III, pp. 237–306. New York: Kluwer Academic Publishers.

KIRKPATRICK, J. B. (1977). Native vegetation of the West Coast region of Tasmania. In *Landscape and Man*, ed. M. R. Banks & J. B. Kirkpatrick. Proceedings of a Symposium of the Royal Society of Tasmania 1977, 55–80.

KNIGHTON, A. D., MILLS, K. & WOODROFFE, C. D. (1991). Tidal creek extension and saltwater intrusion in northern Australia. *Geology*, **19**, 831–4.

LAING, S., DEE, C. N. & BEST, P. W. (1989). The Otway Basin. *APEA Journal*, **29**, 417–29.

LANGE, R. T. (1978). Carpological evidence for fossil *Eucalyptus* and other Leptospermae (subfamily Leptospermoideae of Myrtaceae) from a Tertiary deposit in the South Australia arid zone. *Australian Journal of Botany*, **26**, 221–33.

LANGE, R. T. (1982). Australian Tertiary vegetation. In *A History of Australasian Vegetation*, ed. J. M. B. Smith, pp. 44–89. Sydney: McGraw-Hill.

LOVE, L. D. (1981). Mangrove swamps and salt marshes. In *Australian Vegetation*, ed. R. H. Groves, pp. 319–24. Cambridge: Cambridge University Press.

LULY, J., SLUITER, I. R. K. & KERSHAW, A. P. (1980). Pollen studies of Tertiary brown coals: preliminary analyses of lithotypes within the Latrobe Valley, Victoria. *Monash Publications in Geography*, **23**, 1–78.

MACPHAIL, M. K. (1980). Fossil and modern pollen of *Beilschmiedia* (Lauraceae) in New Zealand. *New Zealand Journal of Botany*, **18**, 149–52.

MACPHAIL, M. K. (1986). Over the top: pollen-based reconstructions of past alpine floras and vegetation in Tasmania. In *Flora and Fauna of Alpine Australasia: Ages and Origins*, ed. B. A. Barlow, pp. 173–204. Melbourne: CSIRO/Australian Systematic Botany Society.

MACPHAIL, M. K. (1991). Cooking up our data? *Palynological and Palaeobotanical Association of Australasia Newsletter*, **22**, 5–8.

MACPHAIL, M. K., COLHOUN, E. A., KIERNAN, K. & HANNAN, D. (1993). Glacial climates in the Antarctic region during the late Paleogene: evidence from northwest Tasmania, Australia. *Geology*, **21**, 145–8.

MACPHAIL, M. K. & HILL, R. S. (1983). Cool temperate rainforest in Tasmania: a reply. *Search*, **14**, 186–7.

MACPHAIL, M. K., HILL, R. S., WELLS, P. M. & FORSYTH, S. M. (1991). A Late Oligocene–Early Miocene cool climate flora in Tasmania. *Alcheringa*, **15**, 87–106.

MACPHAIL, M. K. & PARTRIDGE, A. D. (1991). *Australopollis obscurus. Palynological and Palaeobotanical Association of Australasia Newsletter*, **21**, 6–7.

MACPHAIL, M. K. & TRUSWELL, E. M. (1989). Palynostratigraphy of the central west Murray Basin. *BMR Journal Australian Geology Geophysics*, **11**, 301–31.

MARSHALL, N. G. (1988). A Santonian dinoflagellate assemblage from the Gippsland Basin, southeastern Australia. *Memoir of the Association of Australasian Palaeontologists*, **5**, 195–215.

MARSHALL, N. G. & PARTRIDGE, A. D. (1988). The Eocene acritarch *Tritonites* gen. nov. and the age of the Marlin Channel, Gippsland Basin, southeastern Australia. *Memoir of the Association of Australasian Palaeontologists*, **5**, 239–57.

MARTIN, A. R. H. (1982). Proteaceae and the early differentiation of the central Australian flora. In *Evolution of the Flora and Fauna of Arid Australia*, ed. W. R. Barker & P. J. M. Greenslade, pp. 77–83. Frewville: Peacock Publications.

MARTIN, H. A. (1973). Upper Tertiary palynology in southern New South Wales. In *Mesozoic and Cainozoic Palynology: Essays in Honour of Isabel Cookson*, ed. J. E. Glover & G. Playford. *Geological Society of Australia Special Publication*, **4**, 35–54.

MARTIN, H. A. (1974). The identification of some Tertiary-pollen belonging to the Family Euphorbiaceae. *Australian Journal of Botany*, **22**, 271–91.

MARTIN, H. A. (1977). The history of *Ilex* (Aquifoliaceae) with special reference to Australia: evidence from pollen. *Australian Journal of Botany*, **25**, 655–73.

MARTIN, H. A. (1978). The Tertiary flora. In *Ecological Biogeography of Australia*, ed. A. Keast, pp. 393–406. The Hague: W. Junk.

MARTIN, H. A. (1982). Changing Cenozoic barriers and the Australian paleobotanical record. *Annals of the Missouri Botanical Garden*, **69**, 625–67.

MARTIN, H. A. (1987). Cainozoic history of the vegetation and climate of the Lachlan River region, New South Wales. *Proceedings of the Linnean Society of New South Wales*, **109**, 213–57.

MARTIN, H. A. (1989*a*). Vegetation and climate of the late Cenozoic in the Murray Basin and their bearing on the salinity problem. *BMR Journal of Australian Geology and Geophysics*, **11**, 291–9.

MARTIN, H. A. (1989*b*). Evolution of mallee and its environment. In *Mediterranean Landscapes in Australia*, ed. J. C. Noble & R. A. Bradstock, pp. 83–92. Canberra: CSIRO.

MARTIN, H. A. (1990). The palynology of the Namba Formation in the Wooltana-1 bore, Callabonna Basin (Lake Frome), South Australia, and its relevance to Miocene grasslands in central Australia. *Alcheringa*, **14**, 247–55.

MARTIN, H. A. & GADEK, P. A. (1988). Identification of *Eucalyptus spathulata* pollen and its presence in the fossil record. *Memoir of the Association of Australasian Palaeontologists*, **5**, 311–27.

MARTIN, H. A., WORRALL, L. & CHALSON, J. (1987). The first occurrence of the Paleocene *Lygistepollenites balmei* Zone in the Eastern Highlands Region, New South Wales. *Australian Journal of Earth Sciences*, **34**, 359–65.

McGOWRAN, B. (1979). The Tertiary of Australia: foraminiferal overview. *Marine Micropalaeontology*, **4**, 235–64.

McGOWRAN, B. (1989). The later Eocene transgressions in southern Australia. *Alcheringa*, **13**, 45–68.

McGOWRAN, B., HARRIS, W. K. & LINDSAY, J. M.

(1970). The Maslin Bay flora, South Australia. 1. Evidence for an early Middle Eocene age. *Neue Jahrbuch für Geologie und Paläontologie, Monatschrifte*, **8**, 481–5.

McMINN, A. (1989). Tertiary palynology in the Inverell area. *NSW Geological Survey Quarterly Notes*, **76**, 2–10.

MELVILLE, R. (1975). The distribution of Australian relict plants and its bearing on angiosperm evolution. *Botanical Journal of the Linnean Society*, **71**, 67–88.

MILDENHALL, D. C. (1980). New Zealand Late Cretaceous and Cenozoic plant biogeography: a contribution. *Palaeogeography, Palaeoclimatology, Palaeoecology*, **31**, 197–233.

MILDENHALL, D. C. (1989). Tertiary palynology. *DSIR Bulletin*, **245**, 119–27.

MILDENHALL, D. C. & POCKNALL, D. T. (1989). Miocene–Pleistocene spores and pollen from Central Otago, South Island, New Zealand. *New Zealand Geological Survey Palaeontological Bulletin*, **59**, 1–128.

MILNE, L. A. (1988). Palynology of a late Eocene lignitic sequence from the western margin of the Eucla Basin, Western Australia. *Memoir of the Association of Australasian Palaeontologists*, **5**, 285–310.

MORAT, Ph., VEILLON, J.-M. & MACKEE, H. S. (1986). Floristic relationships of New Caledonian rainforest phanerogams. *Telopea*, **2**, 631–80.

MORGAN, R. (1974). Probable Early Miocene microfloras from the south coast, New South Wales. *Geological Survey of New South Wales Palaeontological Report*, 1974/16.

MORLEY, R. J. (1978). Palynology of Tertiary and Quaternary sediments in Southeast Asia. *Proceedings of the Indonesian Petroleum Association, Sixth Annual Conference*, pp. 255–76.

MORLEY, R. J. (1990). Tertiary stratigraphic palynology in Southeast Asia; current status and new directions. *Geological Society of Malaysia Petroleum Geology Seminar, Kuala Lumpur*, pp. 1–41.

MOUNT, A. B. (1979). Natural regeneration processes in Tasmanian forests. *Search*, **10**, 180–6.

MULLER, J. (1968). Palynology of the Pedawan and Plateau Sandstone Formations (Cretaceous–Eocene) in Sarawak, Malaysia. *Micropalaeontology*, **14**, 1–37.

MULLER, J. (1981). Fossil pollen records of extant angiosperms. *Botanical Review*, **47**, 1–146.

NIX, H. (1982). Environmental determinants of biogeography and evolution in Terra Australis. In *Evolution of the Flora and Fauna of Arid Australia*, ed. W. R. Barker & P. J. M. Greenslade, pp. 47–66. Frewville: Peacock Publications.

NOTT, J. F., IDNURM, M. & YOUNG, R. W. (1991). Sedimentology, weathering, age and geomorphological significance of Tertiary sediments on the far south coast of New South Wales. *Australian Journal of Earth Sciences*, **38**, 357–72.

OWEN, J. A. K. (1988). Miocene palynomorph assemblages from Kiandra, New South Wales. *Alcheringa*, **12**, 269–97.

PARTRIDGE, A. D. (1971). Stratigraphic palynology of

the onshore Tertiary sediments of the Gippsland Basin, Victoria. M.Sc. Thesis, University of New South Wales.

PARTRIDGE, A. D. (1976). The geological expression of eustacy in the Early Tertiary of the Gippsland Basin. *APEA Journal*, 73–9.

PICKETT, J. W., SMITH, N., BISHOP, P. M., HILL, R. S., MACPHAIL, M. K. & HOLMES, W. B. K. (1990). A stratigraphic evaluation of Ettinghausen's New England Tertiary plant localities. *Australian Journal of Earth Sciences*, 37, 293–303.

POCKNALL, D. T. (1989). Late Eocene to Early Miocene vegetation and climates of New Zealand. *Journal of the Royal Society of New Zealand*, 19, 1–18.

POCKNALL, D. T. (1990). Palynological evidence for the early to middle Eocene vegetation and climate history of New Zealand. *Review of Palaeobotany and Palynology*, 65, 57–69.

POCKNALL, D. T. & MILDENHALL, D. S. (1984). Late Oligocene–early Miocene spores and pollen from Southland, New Zealand. *New Zealand Geological Survey Palaeontological Bulletin*, 51, 1–66.

QUILTY, P. G. (1974). Tertiary stratigraphy of Western Australia. *Journal of the Geological Society of Australia*, 21, 301–18.

QUILTY, P. G. (1978). The Late Cretaceous–Tertiary section in Challenger No. 1 (Perth Basin) – details and implications. *Bulletin Bureau Mineral Resources Geology Geophysics Australia*, 192, 109–35.

RAHMANIAN, V., MOORE, P. S., MUDGE, W. J. & SPRING, D. E. (1990). Sequence stratigraphy and the habitat of hydrocarbons, Gippsland Basin, Australia. *Geological Society of London Special Publication*, 50, 525–41.

RAINE, J. I. (1984). Outline of a palynological zonation of Cretaceous to Paleogene terrestrial sediments in West Coast region, South Island, New Zealand. *New Zealand Geological Survey Report*, 109, 1–82.

RAJ, B. & VAN DER WERFF, H. (1988). A contribution to the pollen morphology of neotropical Lauraceae. *Annals of the Missouri Botanical Gardens*, 75, 130–67.

READ, J. (1990). Some effects of acclimation temperature on net photosynthesis in some tropical and extra-tropical Australasian *Nothofagus* species. *Journal of Ecology*, 78, 100–12.

READ, J., HOPE, G. S. & HILL, R. S. (1990). Integrating historical and ecophysiological studies in *Nothofagus* to examine the factors shaping the development of cool rainforest in southeastern Australia. *Proceedings, Third IOP Conference Melbourne*, ed. J. G. Douglas & D. C. Christophel, pp. 97–106. Melbourne: A–Z Printers.

RHEA, D. K., ZACHOS, J. C., OWEN, R. M. & GINGERICH, P. D. (1990). Global change at the Paleocene–Eocene boundary: climatic and evolutionary consequences of tectonic events. *Palaeogeography, Palaeoclimatology, Palaeoecology*, 79, 117–28.

ROBERT, C. & CHAMLEY, H. (1991). Development of early Eocene warm climates, as inferred from clay mineral variations in oceanic sediments. *Palaeogeography, Palaeoclimatology, Palaeoecology*, 89, 315–31.

SHACKLETON, N. J. (1986). Paleogene stable isotope events. *Palaeogeography, Palaeoclimatology, Palaeoecology*, 57, 91–102.

SHEN, Yan-bin (1990). Progress in stratigraphy and palaeontology of Fildes Peninsula, King George Island, Antarctica. *Acta Palaeontologica Sinica*, 29, 136–9.

SINGH, G. (1988). History of aridland vegetation and climate: a global perspective. *Biological Reviews*, 63, 159–95.

SLUITER, I. R. K. (1991). Early Tertiary vegetation and climates, Lake Eyre region, northeastern South Australia. *Geological Society of Australia, Special Publication*, 18, 99–118.

SLUITER, I. R. & KERSHAW, A. P. (1982). The nature of Late Tertiary vegetation in Australia. *Alcheringa*, 6, 211–22.

SLUITER, I. R. & MACPHAIL, M. K. (1985). Quantitative analysis of Paleocene *L. balmei* palynofloras in selected Gippsland wells. Esso Australia Ltd Palaeontological Report 1984/10 (unpublished).

SMITH, J. M. B. (1975). Living fragments of the flora of Gondwanaland. *Australian Geographical Studies*, 13, 3–12.

SOLOMON, A. M. & SILKWOOD, A. B. (1986). Spatial patterns of atmospheric pollen transport in a montane region. *Quaternary Research*, 25, 150–62.

SPECHT, R. A. (1970). Vegetation. In *The Australian Environment*, ed. G. W. Leeper, pp. 44–67. Melbourne: CSIRO/Melbourne University Press.

STOVER, L. E. & EVANS, P. R. (1973). Upper Cretaceous–Eocene spore–pollen zonation, offshore Gippsland Basin, Australia. *Geological Society of Australia Special Publication*, 4, 55–72.

STOVER, L. E. & PARTRIDGE, A. D. (1973). Tertiary and Late Cretaceous spores and pollen from the Gippsland Basin, southeastern Australia. *Proceedings of the Royal Society of Victoria*, 85, 237–86.

STOVER, L. E. & PARTRIDGE, A. D. (1982). Eocene spore–pollen from the Werillup Formation, Western Australia. *Palynology*, 6, 69–95.

SUC, J.-P. & DRIVALIARI (1991). Transport of bisaccate coniferous pollen grains to coastal sediments: an example from the earliest Pliocene Orb ria (Languedoc, southern France). *Review of Palaeobotany and Palynology*, 70, 247–53.

TAYLOR, D. J. (1971). Foraminifera and the Cretaceous and Tertiary depositional history in the Otway Basin in Victoria. *Special Bulletin of the Geological Surveys of South Australia and Victoria*, pp. 217–33.

TAYLOR, G., TRUSWELL, E. M., McQUEEN, K. G. & BROWN, M. C. (1990). Early Tertiary palaeogeography, landform evolution and palaeoclimates of the Southern Monaro, NSW, Australia. *Palaeogeography, Palaeoclimatology, Palaeoecology*, 78, 109–34.

THANIKAIMONI, G. (1987). Mangrove palynology. *Institut Français de Pondichery*, 24, 1–100.

THOMPSON, B. R. (1986). The Gippsland Basin – development and stratigraphy. In *Second South-Eastern Australia Oil Exploration Symposium*, ed. R. C. Glenie, pp. 57–64. Melbourne: Petroleum Exploration Society of Australia, Victorian and Tasmanian Branch.

THORNE, R. F. (1986). Antarctic elements in Australasian rainforests. *Telopea*, 2, 611–18.

TOWNROW, J. A. (1965a). Notes on some Tasmanian pines. I. Some Lower Tertiary podocarps. *Papers and Proceedings of the Royal Society of Tasmania*, 99, 87–107.

TOWNROW, J. A. (1965b). Notes on some Tasmanian pines. II. *Athrotaxis* from the Lower Tertiary. *Papers and Proceedings of the Royal Society of Tasmania*, 99, 109–13.

TRUSWELL, E. M. (1981). Pre-Cenozoic palynology and continental movements. *Palaeoreconstruction of the Continents Geodynamic Series*, 2, 13–25.

TRUSWELL, E. M. (1983). Recycled Cretaceous and Tertiary pollen and spores in Antarctic marine sediments: a catalogue. *Palaeontographica B*, 186, 16–174.

TRUSWELL, E. M. (1987a). Early Tertiary sediments from the Edmund 1:250,000 Sheet Area, Western Australia. *BMR Professional Opinion*, 87/001.

TRUSWELL, E. M. (1987b). Palynological age determination of the Weelumurra Beds, western Fortesque Plain, Western Australia. *BMR Professional Opinion*, 87/003, 1–3.

TRUSWELL, E. M. (1990). Cretaceous and Tertiary vegetation of Antarctica: a palynological perspective. In *Antarctic Paleobiology*, ed. T. N. Taylor & E. L. Taylor, pp. 71–88. New York: Springer-Verlag.

TRUSWELL, E. M. & HARRIS, W. K. (1982). The Cainozoic palaeobotanical record in arid Australia: fossil evidence for the origins of an arid-adapted flora. In *Evolution of the Flora and Fauna of Arid Australia*, ed. W. R. Barker & P. J. M. Greenslade, pp. 67–76. Frewville: Peacock Publications.

TRUSWELL, E. M., KERSHAW, A. P. & SLUITER, I. R. (1987). The Australian–South-East Asian connection: evidence from the palaeobotanical record. In *Biogeographic Evolution of the Malay Archipelago*, ed. T. C. Whitmore, pp. 32–49. Oxford: Oxford University Press.

TRUSWELL, E. M. & MARCHANT, N. G. (1986). Early Tertiary pollen of probable Droseracean affinity from central Australia. *Special Papers in Palaeontology*, 35, 163–78.

TRUSWELL, E. M. & OWEN, J. A. (1988). Eocene pollen from Bungonia, New South Wales. *Memoir of the Association of Australasian Palaeontologists*, 5, 259–84.

TRUSWELL, E. M. & PLANE, M. (1983). Dating Cenozoic deposits in inland Australia. *BMR 82 – Yearbook of the Bureau of Mineral Resources, Geology and Geophysics, Canberra*, pp. 25–32.

TRUSWELL, E. M., SLUITER, I. R. & HARRIS, W. K. (1985). Palynology of the Oligocene–Miocene

sequence in the Oakvale-1 corehole, western Murray Basin, South Australia. *BMR Journal of Australian Geology and Geophysics*, 9, 267–95.

TWIDALE, C. R. & HARRIS, W. K. (1977). The age of Ayers Rock and the Olgas, central Australia. *Transactions of the Royal Society of South Australia*, 101, 45–50.

VEEVERS, J. J., POWELL, C. McA. & ROOTS, S. R. (1991). Review of seafloor spreading around Australia. 1. Synthesis of patterns of spreading. *Australian Journal of Earth Sciences*, 38, 373–89.

WALKER, D. & FLENLEY, J. R. (1979). Late Quaternary vegetation of the Enga Province of upland Papua New Guinea. *Philosophical Transactions of the Royal Society of London B*, 286, 265–344.

WALKER, D. & SINGH, G. (1981). Vegetation history. In *Australian Vegetation*, ed. R. H. Groves, pp. 26–43. Cambridge: Cambridge University Press.

WATANASAK, M. (1990). Mid Tertiary palynostratigraphy of Thailand. *Journal of Southeast Asian Earth Sciences*, 4, 203–18.

WEBB, L. J. (1959). A physiognomic classification of Australian rainforests. *Journal of Ecology*, 47, 551–70.

WEBB, L. J. (1968). Environmental relationships of the structural types of Australian rainforest vegetation. *Ecology*, 49, 296–311.

WEBB, L. J., TRACEY, J. G. & JESSUP, L. W. (1986). Recent evidence for the autochthony of Australian tropical and subtropical rainforest floristic elements. *Telopea*, 2, 575–90.

WEBB, L. J., TRACEY, J. G. & WILLIAMS, W. T. (1972). Regeneration and pattern in the subtropical rainforest. *Journal of Ecology*, 60, 675–95.

WEBB, L. J., TRACEY, J. G. & WILLIAMS, W. T. (1984). A floristic framework of Australian rainforests. *Australian Journal of Ecology*, 9, 169–98.

WEI, W. (1991). Evidence for an earliest Oligocene abrupt cooling in the surface waters of the Southern Ocean. *Geology*, 19, 780–3.

WEI, W. & WISE, S. W. (1990). Biogeographic gradients of middle Eocene–Oligocene calcareous nannoplankton in the South Atlantic Ocean. *Palaeogeography, Palaeoclimatology, Palaeoecology*, 79, 29–61.

WILLIAMS, E. (1989). Summary and synthesis. In *Geology and Mineral Resources of Tasmania*, ed. C. F. Burrett & E. L. Martin. *Geological Society of Australia Special Publication*, 15, 468–99.

WOOD, G. R. (1986). Late Oligocene to Early Miocene palynomorphs from GSQ Sandy Cape 1–3R. *Geological Survey of Queensland Publication*, 387, 1–27.

WOODROFFE, C. D. & GRINROD, J. (1991). Mangrove biogeography: the role of Quaternary environmental and sea-level change. *Journal of Biogeography*, 18, 479–92.

WOPFNER, H., CALLEN, R. A. & HARRIS, W. K. (1974). The lower Tertiary Eyre Formation of the southwestern Great Artesian Basin. *Journal of the Geological Society of Australia*, 2, 17–51.

11   The early Tertiary macrofloras of continental Australia

D. C. CHRISTOPHEL

Plant macrofossil deposits have been recorded from Australia for well over 100 years. Early reports were primarily from localities in Victoria and New South Wales; these are population centres and hence excavations likely to produce fossil material were commonest in those two states. Early work included reports by von Ettingshausen (1883, 1886, 1888), Deane (1895, 1900, 1902a,b,c, 1904) and von Mueller (1874, 1883), and was generally concerned with relating the fossils to extant plants on the basis of gross morphological features. Our current knowledge of continental drift throws doubt on the suggested affinities, found in the earlier papers, of the Australian fossils with northern hemisphere plants. Even those taxa with suggested Australian affinities unfortunately cannot be confirmed, as much of the material is either no longer available or is preserved in such a manner as to render it useless (Hill, 1988b). Therefore, these older reports and their deposits are not considered in this chapter. The reader interested in following them up can consult Hill (1988b) for general references or Christophel (1992) for specific references on Victoria. The locality that is an exception is Vegetable Creek, New South Wales. Hill (1988b) has reported that a number of the fossils from the original von Ettingshausen work on this deposit are still available and are excellently preserved. This flora, considered to be Late Eocene, is particularly important, as it contains the earliest

macrofossils attributable to *Nothofagus* from continental Australia (Hill, 1988b).

Theoretically, the scope of this chapter is the entire early Tertiary, which includes the Paleocene, Eocene and Oligocene epochs. The reality of the situation, however, is that the chapter is basically an essay on the Eocene floras of Australia – primarily the Middle Eocene of southeastern Australia. This is due solely to the fact that most of the published accounts of early Tertiary macrofossil floras from continental Australia deal with Eocene localities. Indeed, only two Paleocene and two Oligocene floras (published or under current study) are known to the author, and none of these is extensive. It is the opinion of one biostratigrapher (B. McGowran, personal communication) that continental Australia has no Early Oligocene plant fossil assemblages, and hence this period, potentially of great evolutionary import after the global terminal Eocene climatic deterioration (McGowran, 1992), may never be documented.

In this chapter, therefore, I first present a summary of the published floristics of each of the various major Eocene deposits, followed by a brief summary of some of the less well-documented macrofossil floras that are awaiting further work, or at least are incompletely known. The fossil deposits discussed here are shown in figure 11.1. I conclude with an overview of the macrofossil floras and Australian early Tertiary on the basis

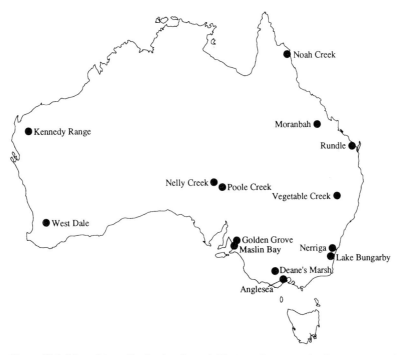

Figure 11.1 Map of Australia showing the early Tertiary plant macrofossil sites mentioned in this chapter.

of continental macrofossil floras and consider the ramifications for evolution later in the Tertiary and for the modern flora. While it may be convincingly argued that in the early Tertiary a separation of the continental and Tasmanian floras is somewhat artificial, the work on the evolution of the Tasmanian flora through the Tertiary forms a compact and elegant story, and seems best treated as a separate unit (Carpenter *et al.*, Chapter 12, this volume).

It is important to realise while considering these various floras, particularly the rather large group that are Middle Eocene, that even those which are palynologically indistinguishable could still in reality be separated by several million years. Even within one locality, individual input events (fossil leaf horizons) might well be separated by thousands of years, and the lack of precise dating guarantees that comparisons between floras must be done with caution. To expand an analogy used by Greenwood (Chapter 4, this volume), it is better to consider each of these floras as a snapshot of

a particular point in an ecological and evolutionary continuum rather than frames on a continuous roll of film.

MIDDLE EOCENE MACROFLORAS

Nerriga

The Nerriga locality occurs approximately 140 km east of Canberra in New South Wales, and is exposed by the erosional action of Titringo Creek. Raine (1967) interpreted the carbonaceous siltstones as part of the depositional sequence of a freshwater lake or slow-moving river. The age, based on the interpretation of several, not all together consistent, radiometric datings of basalt, is at least 45 million years (Ma) old (Wellman & McDougall, 1974). Palynological dating by Owen (1975) was consistent with the radiometric dating. This makes the Nerriga locality the oldest of the well-documented early Tertiary macrofossil floras.

Hill (1982) described 25 broad-leaved angiosperm taxa from the Nerriga locality on the basis of a numerical taxonomic study. At the time, none of these was identified; however Christophel (1980) recognised the microphyllous *Gymnostoma* (Casuarinaceae). In addition, three cycads – *Bowenia papillosa* (Hill, 1978), *Lepidozamia foveolata* and *Pterostoma anastomosans* (Hill, 1980) – as well as a fern with possible affinities to *Sticherus* (Gleicheniaceae) were described.

Since that first description, Hill has re-examined some components of the flora and placed them taxonomically. Specifically he has described 12 species of Lauraceae and placed them in the organ genus *Laurophyllum* (Hill, 1986). Later, Hill (1989) described *Menispermaphyllum tomentosum* (Menispermaceae).

There are several interesting aspects of the vegetation composition of the Nerriga flora. Firstly, as mentioned by Hill (1982), it is unusual (relative to other Eocene floras) that over 90% of the broad-leaved angiosperm specimens can be placed in only 5 of the 25 taxa, thus indicating clear dominants within the flora. Two of these dominants have since been placed by Hill (1986) in the Lauraceae, while another of them was placed in the Menispermaceae (Hill, 1989). This latter taxon might effectively be eliminated from the consideration of dominants within the standing vegetation, as there is a strong likelihood that it represents a vine. This fact in itself is highly interesting, in that there is a very poor documentation of vines in the Australian early Tertiary. This general lack of vines is unexpected, considering the proposed megathermal–mesothermal rainforest assignment of many of the deposits, and that many of Australia's rainforest formations are characterised by the presence of lianas (Webb, 1959). Leaves collected from litter traps placed along Noah Creek in northern Queensland rainforest include a significant quantity of vine leaves, suggesting that if they were present they could be expected to be found in fossil leaf-litter deposits.

Another interesting feature of the Nerriga flora is that the 12 Lauraceae species constitute 46% of the total macrofossil species diversity. Only one North American flora has documented that high a percentage (Hill, 1986), although both the Anglesea (Victoria) flora (Christophel *et al.*, 1987) and the Maslin Bay (South Australia) flora (L. Scriven, personal communication) probably have five or more species each. Although not formally described, some aspects of the familial composition of the Nerriga flora are interesting. There are no members of the Proteaceae identified as macrofossils, and, considering that all taxa described by Hill (1982) as having a paracytic subsidary cell arrangement have now been placed by him in the Lauraceae (Hill, 1986), there is none remaining with the typical Proteaceae subsidiary cell arrangement. The presence of Proteaceae in the Nerriga region has been confirmed by dispersed cuticle fragments identified as belonging to that family (R. S. Hill, personal communication) Similarly, no leaves are described that fit either with the typical intramarginal-veined, high angle, secondary vein pattern found in the majority of *Syzygium*-like Myrtaceae, nor have any of the taxa described been reported to have lid cells. Thus, two families that have a prominent palynological place in early Tertiary floras, and, as will be seen in the following sections, have a prominent place in the Middle Eocene macrofossil floras of southeastern Australia, are absent from Nerriga.

Anglesea

Plant remains from this locality in Victoria were first described by Douglas (1977) and since then many additional reports have appeared. These have fallen broadly into two categories: those concerned with the overall flora or the mechanics of deposition (e.g. Christophel, 1981, 1989; Christophel *et al.*, 1987; Christophel & Greenwood, 1988, 1989; Barrett & Christophel, 1990; Rowett & Christophel 1990), and those concerned with the floristic elements of the deposit (e.g. Hill 1978, 1980; Christophel, 1981, 1984; Christophel & Basinger, 1982; Basinger & Christophel, 1985; Christophel & Lys, 1986; Greenwood, 1987; Hill & Christophel, 1988; Scriven & Christophel, 1990; Hill & Pole, 1992).

All information published to date on the Anglesea flora shows it to be a highly complex system. The plant fossils occur in clay lenses in the overburden of a brown coal mine currently operated by Alcoa of Australia. Six individual lenses have been documented to date, and an additional one was uncovered in the early part of 1991. Christophel *et al.* (1987) considered all lenses to represent oxbow lakes or otherwise isolated channels in a meandering stream deposit. They placed the age as Middle Eocene on the basis of palynostratigraphy, and this age (approximately 40–43 Ma) is supported by McGowran (1992), who sees it as perhaps slightly younger than the Maslin Bay and Golden Grove floras, which are discussed in the next section.

Several of the lenses described from Anglesea have distinctive compositions (Christophel *et al.*, 1987). The first from which fossil plants were described (Christophel, 1980) was dominated by remains of *Gymnostoma* (Casuarinaceae). Christophel *et al.* (1987) suggested that this could have represented a discrete population of *Gymnostoma*, limited in lateral extent by its small size and indicative of autochthonous deposition. A second possibility is now suggested by observation of an extant population of *Gymnostoma australianum* in Noah Creek, northern Queensland. Although over 200 trees occur on an island in Noah Creek, and produce numerous infructescences, little evidence of the presence of this taxon is seen in downstream litter deposits, which contain a mix of many other streamside and island taxa. However, high water during the summer wet season has on occasion scoured the island and banks of *Gymnostoma* material and other fruits and twigs have been found deposited more than 1 km downstream from the island. A similar event in the Anglesea depositional cycle provides an alternative explanation for the *Gymnostoma*-dominated lens found there. Such beds, which are the result of high energy deposition of primarily woody material, are also known from modern depositional systems in the northern hemisphere (Spicer, 1980).

Only one other lens at Anglesea (the *Brachychiton* lens) has low diversity, with *Brachychiton* (Sterculiaceae) dominating an assemblage consisting of fewer than ten taxa. Within the remaining described lenses (Mesophyll Lens, *Austrodiospyros* lens, Pit II Lens A and Pit II Lens B; Christophel *et al.*, 1987), and in the recently discovered lens,

diverse mixtures of taxa (usually 30–40) are found. Elements identified (and numbered as taxa by Christophel *et al.* (1987)) include three Lauraceae, five Proteaceae (including two species of *Banksieaephyllum*; Hill & Christophel, 1988), two Myrtaceae (*Myrtaciphyllum*; Christophel & Lys, 1986), at least four Podocarpaceae (Greenwood, 1987), and at least two Zamiaceae (Hill, 1978, 1980). Recent work by J. Conran & D. C. Christophel (unpublished data) suggests that in addition to one palm already known, at least three other monocotyledonous taxa are present. One species of Ebenaceae, *Austrodiospyros eocenica*, has been described (Basinger & Christophel, 1985), and single species of Casuarinaceae, Elaeocarpaceae and Escalloniaceae have been reported (Christophel *et al.*, 1987).

When this collection of taxa is examined and compared to vegetation types found in the modern Australian flora, the general pattern is found to be common in many rainforest types. Indeed, forests with the approximate numbers of Proteaceae, Myrtaceae and Lauraceae species as seen in the common lenses at Anglesea can be found in at least six of the extant rainforest types listed by Webb (1959) in his classification of Australian rainforests. These six range from temperate microphyll mossy forests to the warmest of tropical mesophyll vine forests.

The presence at Anglesea of taxa extinct at the generic or familial level guarantees that no exact modern homologue can be found for the flora. Such extinct taxa include one common angiosperm leaf type, the cycad *Pterostoma eocenica* (Hill, 1980) and several of the Podocarpaceae (Hill & Carpenter, 1991; Hill & Pole, 1992). Despite the presence of these extinct taxa, extant communities do exist with reasonable numbers of related taxa to Anglesea. One such locality, a patch of simple notophyll vine forest (*sensu* Webb, 1959) growing on an island in Noah Creek, Queensland (16° 05′ S, 145° 30′ E), has been discussed in some detail (Christophel *et al.*, 1987; Christophel, 1992).

A list of the major components of the Noah Creek Island flora is shown in Table 11.1 along with a similar list for the Anglesea flora. While many of the taxa, including the Lauraceae and Myrtaceae, are found in similar proportions in

Table 11.1. *Number of species in a range of taxa for the Middle Eocene Anglesea locality and five extant vegetation types*

	Locality					
Taxa	Anglesea	Noah Creek Island	Baileys Creek Island	Wongabel (CMVF)	Mt Lewis (CNVF)	Mt Rescue (heath)
Casuarinaceae	1	1	0	0	0	3
Gymnostoma	1	1	0	0	0	0
Proteaceae	5+	5	3	0	6	10
Musgraveinae	1	1	2	0	1	0
Oriteae	1	1	0	0	0	0
Myrtaceae	2	5	9	2	7	22
Podocarpaceae	5	1	0	0	2	0
Sterculiaceae	1	3	3	1	2	2
Brachychiton	1	0	0	0	0	0
Lauraceae	5+	5	18	7	5	3
Ebenaceae	1	2	2	0	1	0
Elaeocarpaceae	2	3	2	1	2	0
Rutaceae	?	4	7	6	4	4
Euphorbiaceae	?	4	9	4	4	2
Sapindaceae	?	3	8	5	4	1
Total woody taxa	100+	96	113	61	52	125

CMVF, complex mesophyll vine forest; CNVF, complex nothophyll vine forest
Data for the three rainforest types (three columns after Noah Creek) were taken from lists compiled by Webb & Tracey (1982), while the data for the sclerophyllous community represents an unoffical species list compiled for the sclerophyllous vegetation at Mt Rescue National Park, South Australia as an undergraduate teaching exercise.

many rainforest assemblages in Australia and montane New Guinea today, certain combinations of taxa – for example the combination of *Gymnostoma*, *Orites* sp. nov., *Diospyros*-like Ebenaceae, members of the Musgraveinae, *Bowenia* and large-leafed *Podocarpus* – are unique to Noah Creek Island and Anglesea.

Several other taxa from Anglesea have been assigned to extant genera or tribes, including *Lygodium* (Rozefelds *et al.*, 1991), *Banksieaephyllum* (extant tribe Banksieae) (Hill & Christophel, 1988), and *Elaeocarpus/Sloanea* (O'Dowd *et al.*, 1991). However the majority of known taxa (60–70 for all lenses) at Anglesea are represented by architecturally indistinct, elliptical to ovate notophylls that are as yet undescribed. The full picture of the Anglesea flora, therefore, is far from complete.

Although many of the extant taxa to which the described Anglesea taxa are closely related are known from lowland mesophyll vine forests (e.g. Musgraveinae, *Bowenia*), the physiognomic signature suggests that the assemblage from the best-known of the Anglesea lenses would be classified as simple notophyll vine forest (Christophel & Greenwood, 1989; Greenwood, Chapter 4, this volume). This is consistent with the Noah Creek Island flora, which contains a mixture of lowland mesophyllous and microphyllous taxa (e.g. *Gymnostoma*) from cooler forest types. The Noah Creek Island flora is also classified as simple notophyll vine forest.

Although it is rare for plant and animal remains to be preserved together because the acidic depositional environment required to preserve plants is often destructive to bone, the chiton of arthropods is not often affected. Indeed, the Anglesea locality has yielded several species of mite (O'Dowd *et al.*, 1991) and has provided evidence for the earliest record of mites in leaf domatia.

Maslin Bay

The Maslin Bay locality in South Australia was discovered in the mid-1960s when sand-quarrying operations approximately 30 km south of Adelaide unearthed a fine-grained, carbonaceous clay lens that was rich in fossil plant remains. Stratigraphically the deposit is found within the North Maslin Sands, and on the basis of palynological analysis, has been dated as Middle Eocene (McGowran et al., 1970; Alley, 1987). More recently, the age has been suggested as upper Middle Eocene (McGowran, 1992), in the range of 42–44 Ma; however, the sequence is actively being reassessed (B. McGowran, personal communication).

The Maslin Bay fossils have been the subject of several papers (Lange, 1969, 1970; Lange & Smith, 1971; Southcott & Lange, 1971; Christophel & Blackburn, 1978; Blackburn, 1981), which have either been preliminary in nature or have addressed minor specific components of the flora. A broad systematic treatment of the flora has never been published.

Early accounts of the flora have described it as highly diverse, with approximately 200 macrofossil taxa (Christophel & Blackburn, 1978; Blackburn, 1981). Current work being undertaken as a Ph.D. study (L. Scriven, personal communication) supports the high diversity of the flora, but suggests a diversity of nearer 180 taxa. Using either estimate, the Maslin Bay flora is the most diverse of any of the known Australian Eocene localities, exceeding even the combined six lenses at Anglesea.

Four species of plant were formally named and described from the Maslin Bay locality by Blackburn (1981), and include one species of Araliaceae (Parafatsia), one species of Podocarpaceae (Decussocarpus, now Willungia (Hill & Pole, 1992)) and two species of Proteaceae (Maslinia and Banksieaephyllum). An additional species of Proteaceae (Banksieaeformis decurrens) was described by Hill & Christophel (1988). Other recognisable taxa have been reported from the locality, and include Gymnostoma (Christophel, 1980; Scriven & Christophel, 1990), Lygodium (Rozefelds et al.,

1992) and Agathis (Christophel & Blackburn, 1978). Christophel & Blackburn (1978, their Fig. 2C) illustrated a specimen which they called a five-petalled angiosperm flower. That specimen has since been interpreted as the fruit of Ceratopetalum (Cunoniaceae), which retains a persistent calyx as a flotation aid to water dispersal (A. Floyd, personal communication) and that assignment is supported here. Ceratopetalum is a rainforest/open forest tree or shrub genus with members found from the New South Wales simple notophyll vine forest (C. apetalum) to the simple notophyll, complex notophyll and complex mesophyll vine forests of northern Queensland (e.g. C. macropetalum).

The general composition of the Maslin Bay flora is similar to many other Eocene deposits in that Proteaceae and Lauraceae have quite diverse representation and other Gondwanic families, such as the Casuarinaceae (Gymnostoma), Araucariaceae (Agathis), Podocarpaceae (Willungia), Sterculiaceae (Brachychiton) and Ebenaceae (Diospyros/Austrodiospyros) are present. The Proteaceae are particularly diverse, with possibly as many as 20 taxa (L. Scriven, personal communication). Interestingly, no Myrtaceae leaves were reported from an early sampling (Christophel & Blackburn, 1978), a situation now considered unusual for Eocene localities. However, L. Scriven (personal communication) reports a number of specimens from that family, with possibly as many as five taxa. If confirmed, this would give Maslin the highest diversity of all the Eocene macrofossil floras for that family.

Several other interesting features of the Maslin Bay flora are worth noting. Firstly, Christophel & Blackburn (1978) figured one of the compound leaves found at the deposit. L. Scriven (personal communication) now believes that compound leaf taxa were an important part of the flora. Since the Maslin Bay fossils are usually preserved as compressions, there is a better opportunity to prove the presence of compound leaves than at most of the localities, where the mummified leaves are usually preserved in isolation with little clue to their phyllotaxy.

A second interesting feature of the Maslin Bay

flora is the source of nearest living relatives for some of the taxa. Whereas most of the taxa identified from other localities appear to have close extant relatives within Australia, this may not be the case for a number of the Maslin Bay fossils. Blackburn (1981) described *Parafatsia* (Araliaceae) and noted its total dissimilarity to any Australian taxa. Indeed, the only modern genus bearing close resemblance (*Fatsia*) comes from Japan. Similarly, a taxon from the Proteaceae shows no close similarity to any modern Australian species, and at least superficially appears much more similar to *Beauprea* from New Caledonia or to some of the South American members of the family. Considering the diversity of the flora, the amount of time available for evolution to occur, the drastic changes in climate that have occurred in Australia over the last 40 Ma and the Gondwanic nature of many of the families represented in the Maslin Bay flora, such biogeographically diverse relationships should not be surprising. However, one significance of that diversity is that any attempts to determine the affinities of the numerous taxa collected will have to include comparisons with a very wide range of modern floras.

Finally, the interpretation of the climate at the time of Maslin Bay fossil deposition is also interesting. Greenwood (Chapter 4, this volume) has confirmed the comparison with modern complex mesophyll vine forest (*sensu* Webb, 1959) made originally by Christophel (1981) and later by Christophel & Greenwood (1988). This would be consistent with a mean annual temperature (MAT) of 20–27 °C and makes the Maslin Bay locality the warmest documented Middle Eocene deposit. This is even more interesting if the flora is considered to be possibly coeval with the Golden Grove flora described below (Alley, 1987; Greenwood, Chapter 4, this volume) as it suggests the Adelaide region to have been a tension zone with at least two forest types represented, and with MAT values therefore at the lower end of the range for complex notophyll vine forest.

Golden Grove

The Golden Grove fossil plant deposit was discovered in 1985 during sand quarrying operations at Monier's Golden Grove Quarry, approximately 25 km north of Adelaide, South Australia. The plant fossils occur in a carbonaceous clay lens that is interpreted as an oxbow lake within a meandering stream system (Christophel & Greenwood, 1987; Barrett & Christophel, 1990). The lens is considered to be a part of the North Maslin Sands and is palynologically placed in the same zone as the Maslin Bay locality (Alley, 1987). Thus the two deposits may be coeval.

Fossil leaves, flowers and fruits are preserved either as mummified structures, or as oxidised impressions. Although the fossil plant remains appear to be homogeneously mixed vertically within a given lens at both the Anglesea and Maslin Bay localities, at Golden Grove the fossils occur as three distinct carbonaceous bands, each of approximately 1 cm thickness (Barrett & Christophel, 1990). This deposition suggested the possibility of separate input events for each of the layers. Mapping and careful collecting demonstrated that two of the three leaf horizons probably represent seasonal accumulations of the same vegetation, differing only in frequency of given taxa. However, the third leaf horizon is sufficiently different to suggest a separate input – representing either a separate plant association within a mosaic forest or an evolutionary continuum of the same flora (Barrett & Christophel, 1990). Floristically, Barrett & Christophel (1990) isolated 24 leaf taxa. In addition, *Musgraveinanthus* is known from the deposit as is the fern *Lygodium aureonemorosum* (Rozefelds *et al.*, 1992). Initial studies were based primarily on the mummified flora, but in 1989 it was discovered that many of the oxidised 'impression' fossils would yield cuticle, giving far more scope to their study. A realistic estimate for the diversity of the Golden Grove flora may therefore be around 40–45 taxa. This diversity equates to one of the more species-rich Anglesea lenses, but is far short of the estimate for Maslin Bay.

In addition to *Lygodium*, only one other species has been formally described from Golden Grove.

Banksieaephyllum cuneatum (Hill & Christophel, 1988) is relatively common in those samples from the locality which have been analysed quantitatively (Barrett & Christophel, 1990). The presence of a large-leaved species of a member of the Banksineae subtribe is not expected within the normal floristics of an Australian subtropical or tropical rainforest flora, and yet they are present at three of the Middle Eocene sites considered here. Additionally, even if such structures were to be expected, the delicate preservation of flowers occurring in all deposits considered thus far would lead to the expectation of Banksineae reproductive structures in the same deposits. These, however, have not been reported. One possible explanation for this is that the *Banksieaephyllum* leaves described from the Anglesea, Golden Grove, Maslin Bay and Deane's Marsh, Victoria, deposits are actually associated with the *Musgraveinanthus* inflorescences reported thus far from Anglesea and Golden Grove. The Musgraveinae are the sister subtribe of the Banksineae within the tribe Banksieae, and the juvenile leaves of extant *Musgravea* bear a resemblance to the fossil *Banksieaephyllum* leaves described from the various Middle Eocene deposits. This explanation is further supported by the lack of any other leaves in either the Anglesea or the Golden Grove deposit that might reasonably be associated with the *Musgraveinanthus* inflorescences (Christophel, 1992).

Although as yet not formally described, the affinities of several of the Golden Grove taxa are known. One of the most striking is the compound-leaved Proteaceae that appears to be closely related to extant *Neorites* (Barrett & Christophel, 1990). It is one of the ten most frequently encountered taxa within the deposit, and has not been reported from any other locality. Several other Proteaceae taxa are known, as are several belonging to the Lauraceae, and at least two belonging to the Myrtaceae (Christophel & Greenwood, 1987), giving a floristic profile to this flora similar to that of the two Middle Eocene Floras discussed earlier. One of the two commonest taxa has affinities with extant *Sloanea* or *Elaeocarpus* in the Elaeocarpaceae, a family also occurring commonly at both Anglesea and Maslin Bay. Two or

three species of *Brachychiton* (Sterculiaceae) are also known from Golden Grove, and they appear to be different from those species from Anglesea and Maslin Bay. *Gymnostoma* (Casuarinaceae), which is common at Anglesea and present at Maslin Bay (as well as the two other major Middle Eocene floras discussed below) has not been found in any of the numerous collections made from the Golden Grove deposit.

Earlier in this chapter (see also Greenwood, Chapter 4, this volume) it was mentioned that stratigraphically and palynologically the Maslin Bay and Golden Grove deposits are generally indistinguishable (Alley, 1987). In Chapter 4, Greenwood concluded that the two were physiognomically different, but recognised that two equally physiognomically different forest types occurring within a 60 km radius is common today in the mosaic forests of northern Queensland, and hence in itself does not negate the possibility of the two deposits being coeval. Floristically, at first glance the two deposits share common occurrences of several families and genera. However, it is important to note that the similarity is as great between Anglesea and Golden Grove as it is between Golden Grove and Maslin Bay. In addition, several distinctive taxa, such as *Gymnostoma*, *Neorites* and *Agathis* occur in one deposit but have not been recovered from the other. However, the same three taxa (or their nearest living relatives) could be found in only one of two extant forests sampled within a 50 km radius in the Daintree region of far north Queensland. Therefore, it must be concluded that, whereas two floras from different localities that shared numerous specific taxa might be considered to be proved coeval, the differences and similarities exhibited between the macrofossil floras at Golden Grove and Maslin Bay cannot really be used either to support or to reject an isochronous relationship.

Nelly Creek and other floras of the Lake Eyre Basin

The Middle Eocene floras dealt with thus far all occurred near the coast. However, an extensive set of inland floras have been known for more than 50 years (Chapman, 1937), and are collectively named the silcrete floras. They consist of silicified blocks on whose surface appear finely detailed imprints of leaves and moulds of fruits, seeds and twigs. Until recently their stratigraphic placement has been questionable, and their lack of organic remains has precluded them from palynological analysis as well. Thus, they have figured little in interpretations of the evolution of the Australian flora (Christophel, 1989).

In 1987 a carbonaceous clay lens bearing mummified leaves was discovered at Nelly Creek in northern South Australia. It has been dated as Middle Eocene (Alley, 1987), and a preliminary report of the macrofossil flora has now been published (Christophel *et al.*, 1992). Its far inland position, and the fact that Greenwood *et al.* (1990) have assigned a Middle Eocene age to one of the nearby silcrete floras (Poole Creek) make this macrofossil flora particularly interesting. The first examination of the Nelly Creek macrofossil flora, based on 269 specimens, yielded 16 taxa (Christophel *et al.*, 1992). These include one species of Myrtaceae leaf (*Myrtaciphyllum eremeaensis*), two species of *Gymnostoma* (Casuarinaceae), four taxa assigned to the Proteaceae, one to *Brachychiton* (Sterculiaceae), one to *Agathis* (Araucariaceae), one to the Podocarpaceae and one to the Lauraceae. While 11 of the 16 taxa are identified to the family level, more than 50% of the flora consist of leaves of one microphyllous angiosperm leaf whose affinities are as yet unknown. However, this apparent dominance could well be an artefact of the small sample analysed.

Comparisons of the Nelly Creek flora with the other southern Australian Middle Eocene deposits described above provide little information. The common microphyll from Nelly Creek is not known from any of the other deposits. *Myrtaciphyllum erermeaensis* is different from the members of that family described from Anglesea and those known from Golden Grove. Neither the imbricate Podocarpaceae taxon nor the commonest, lobed Proteaceae taxon are known from the other Middle Eocene deposits. At least one of the

Gymnostoma spp. is different from those occurring at the other Middle Eocene deposits, and the ten other taxa from Nelly Creek are sufficiently rare as to not make meaningful interlocality comparisons at present.

Ignoring the specific matches (or lack of matches) with other floras, the overall nature of the vegetation at Nelly Creek seems very different from the other Middle Eocene floras described above. Physiognomically the flora is distinctive (Greenwood, Chapter 4 this volume), with many small, more or less sclerophyllous leaves. Although only 269 leaves have been considered, the appearance of only one specimen of Lauraceae at Nelly Creek, given the family's importance at all other Middle Eocene localities, is noteworthy. Dispersed cuticle examined from Nelly Creek (A. Rowett, personal communication) confirms the rarity of the Lauraceae and highlights the importance of *Myrtaciphyllum* at the locality.

In combination, the silcrete floras show quite a high diversity and a strong range of leaf forms. However, the Poole Creek flora, highlighted because of the superior stratigraphic control on it, has a reported diversity of 18 leaf taxa (Greenwood *et al.*, 1990). While Greenwood *et al.* stressed the preliminary and stratigraphic nature of the taxa created by them, it is nonetheless instructive to note several features. Firstly, several representatives of both Myrtaceae and Proteaceae are described. The Proteaceae taxa include two and possibly three assignable to *Banksieaeformis*. One large leaved, multilobed taxon with possible affinities to *Brachychiton* is also described, as are two types of Casuarinaceae cone with likely affinities to *Gymnostoma*. One last particular taxon described by Greenwood *et al.* (1990) is worth further comment. This is an apparently woody (?possibly reproductive) structure that is occasionally encountered in the silcretes. A solitary, poorly preserved specimen has been recovered from the Nelly Creek clay lens. The radially arranged structures around the central axis appear to be pairs of woody bracts. This allows the possibility of it being some form of Proteaceae reproductive structure or infructescence. In particular, however, the presence of a structure unknown from any other Middle Eocene locality in both the Poole Creek and Nelly Creek floras supports a link between them.

In general, the macrofossil flora reported from the Poole Creek locality includes a component of leaves much larger than the Nelly Creek locality. Greenwood *et al.* (1990) commented that this could well represent a mosaic of gallery forest and more sclerophyllous thickets, and current work on Nelly Creek does not contradict that. The silcrete floras are demonstrably mixtures of large, mesophytic leaves and smaller sclerophyllous ones, while anecdotal evidence suggests that a large, mesophytic-leaved component of the Nelly Creek clay lens was also once collected (Christophel *et al.*, 1992). Whatever the ultimate floristic interpretation of these inland deposits, it is already clear that they represent a vegetation at least partially different from the other known Middle Eocene macrofossil floras, and that they also provide support for the idea that the Middle Eocene flora of Australia was neither taxonomically nor ecologically uniform.

THE LATE EOCENE VEGETABLE CREEK MACROFLORA

The original Vegetable Creek locality described by von Ettingshausen was considered by him to be one site (von Ettingshausen, 1888). However, an addendum by Etheridge stated that von Ettingshausen's flora was actually the combination of two or more localities (Hill, 1988*b*). The putative localities are found in northeastern New South Wales, and the fossils occur in fine-grained clays associated with deep leads unearthed during tin-mining excavations. Recent palynological investigation of carbonaceous samples of the clays by Macphail (in Hill, 1988*b*) places the clays as Late Eocene.

The original flora described by Ettingshausen was diverse, with many elements identified as relating to modern northern hemisphere plants and others having affinities with southern hemisphere extant taxa. The obvious incorrectness of the former identifications requires that all of von Ettingshausen's determinations be re-examined.

Three taxa have been re-examined by Hill (1988*b*) and Hill & Carpenter (1991). A trinerved leaf placed by von Ettingshausen in *Cinnamomum* is incorrectly placed generically but is correctly placed in the Lauraceae. Podocarpaceae specimens assigned to a single taxon by von Ettingshausen are now assigned to species of *Dacrycarpus* and *Podocarpus*. Most importantly, leaves identified by von Ettingshausen as *Fagus* (Fagaceae) (*Nothofagus* was not uniformly recognised as distinct at that time) are indeed *Nothofagus*, and represent the earliest foliar record of that genus on the continent (Hill, 1988*b*). Blackburn (1981) in his description of *Parafatsia* (Araliaceae) noted that two of von Ettingshausen's Araliaceae spp. probably belonged in that family, but in a genus different from that used by von Ettingshausen. Finally, I have seen one of Ettingshausen's type specimens, placed by him in *Darlingia* (Proteaceae) which I consider to be correct, at least at the family level. Unfortunately, no cuticle appears to remain on the specimen, and hence formal confirmation is unlikely.

The relative accuracy of the identification of those taxa checked, and the quality of the material makes further investigation of those specimens remaining very important. Hill (1988*b*) suggested that the Vegetable Creek *Nothofagus* represented a taxon that would be expected in a flora more temperate than those Eocene floras just described. This increases the significance of the flora. In 1978 I collected material from a spoil heap at a tin mine near the original Vegetable Creek locality, and, although subsequent trips have failed to produce additional material, further collection of the locality must remain a possibility (D. C. Christophel, unpublished data).

MISCELLANEOUS FLORAS

There are several early Tertiary floras that are either reported with few taxa or are still in the stage of being studied. The small amount of published information on them does not warrant their being discussed as separate floras, but a brief mention will at least give recognition to their existence. The first of these is the Merlinleigh flora in the Kennedy Range of Western Australia. It is important for two reasons: firstly, our knowledge of the early Tertiary of the western half of the continent is very poor and hence any knowledge is valuable; and, secondly, from it is described the earliest record of infructescences assignable to *Banksia*. The only published record from this flora to date is this report of *Banksia* cones, and the deposit is dated as Early Eocene on the basis of the stratigraphic position of the sediments (MacNamara & Scott, 1983).

A second Western Australian flora from West Dale contains a diverse, extremely well preserved impression flora (Hill & Merrifield 1993). Hill & Merrifield described 35 taxa from the Middle Eocene–Oligocene sediments, including ferns, conifers (*Agathis, Dacrycarpus, Retrophyllum*) and a diverse angiosperm flora including *Gymnostoma, Nothofagus*, Cunoniaceae, Lauraceae and a predominance of Myrtaceae and Proteaceae. This flora is particularly important in tracing the development of the sclerophyllous vegetation.

An additional Early Eocene flora is known from Deane's Marsh, Victoria. The fossiliferous clays are found in the Demon's Bluff Formation and the flora is known only from compressions and a few rare mummified leaves from 50 year old spoil heaps of a now-flooded brown coal mine (Christophel, 1992). The few specimens recovered include lobed Proteaceae and Lauraceae leaves. The leaves are generally smaller than those at Anglesea and apparently are from quite different taxa.

Another poorly documented flora with much potential is from Moranbah in Queensland. Large numbers of *Gymnostoma* infructescences have been recovered from the locality (Scriven & Christophel, 1990), and tentative identifications of leaves include *Agathis* (Araucariaceae), Proteaceae and Lauraceae. Pollen recovered from the clay lens, and a radiometric date of a basalt capping the clay, suggest an Eocene age.

Finally, a leaf flora from Lake Bungarby near the New South Wales/Victoria border has been reported (Hill, 1991; Hill & Carpenter, 1991). This Late Paleocene macroflora contains at least 40 taxa including *Eucryphia*, Proteaceae,

Lauraceae, Cunoniaceae, *Acmopyle* (and various other Podocarpaceae), possibly *Nothofagus* and Cupressaceae (R. S. Hill, personal communication).

DISCUSSION

In considering the above series of 'snapshots' of specific floras at specific points in time during the early Tertiary, it is possible to make some generalisations without incurring the dangers of direct comparisons between floras that cannot be proved to be coeval. The Middle Eocene is a period during which several macrofossil floras are available for consideration. From both physiognomic and systematic analysis of these floras, it can be generalised that mesothermal and megathermal forests (*sensu* Nix, 1982) were common in the southeastern quarter of the continent at that time. However, they were neither uniform, nor even highly similar, either at one time across the region, or in one region through time. Evidence from localities such as Anglesea (Christophel *et al.*, 1987; Barrett & Christophel, 1990) demonstrates that even through relatively short intervals (some lenses are possibly even coeval) a mosaic pattern of vegetation was evident that is reminiscent of the extant rainforest of northern Australia.

It is also worth while examining the comparative floristic composition of the Middle Eocene localities. Certain taxa are common or prevalent at most of the sites studied. The Lauraceae form a significant part of the flora at four of the five localities considered in detail. They are particularly prevalent at Nerriga and Anglesea, and reasonably common at Golden Grove and Maslin Bay. Only at Nelly Creek are they conspicuous in their rarity. The Proteaceae are particularly diverse and abundant at Maslin Bay and Anglesea, reasonably so at Golden Grove and Nelly Creek, while apparently missing from the macrofossil flora at Nerriga. Specifically within the Proteaceae, *Musgraveinanthus* (Musgraveinae) is present as a macrofossil at Golden Grove and Anglesea and the presence of *Banksieaeidites arcuatus*, the pollen grain found in *Musgraveinanthus alcoensis* (Christophel, 1984) at all sites suggests its widespread occurrence. *Banksieaephyllum* and/or *Banksieaeformis* are also reported from Anglesea, Maslin Bay and Golden Grove, and while not reported from Nelly Creek, are known as common elements of the neighbouring silcrete deposits (Greenwood *et al.*, 1990). *Gymnostoma* is reported from all sites except Golden Grove. Specimens attributed to *Elaeocarpus/Sloanea* (Elaeocarpaceae) are reported as common from Golden Grove and Anglesea, and are clearly present at Maslin Bay and possibly at Nerriga. *Brachychiton* is known from all localities. Finally, Myrtaceae and *Lygodium* are reported from a majority of sites and the Zamiaceae are known from Anglesea, Nerriga and Moranbah.

Perhaps as important as comparisons between individual Eocene floras are comparisons of them with various extant vegetation types. Table 11.1 shows the distribution of the macrofossil taxa discussed above as well as the constituents of five major extant plant community types. It can be seen that the Proteaceae and Myrtaceae form an important part of all of the extant communities listed, as they do for the fossil communities. The Lauraceae are present in all of the extant rainforest floras listed, and are also present in the sclerophyllous community if the leafless parasite *Cassytha* is included.

It is apparent from the data that there are some commonly occurring taxa that have not been mentioned in the context of the fossil floras. However, the total diversity of most of the fossil floras shows that a majority of taxa have not been identified as yet, allowing for the possibility of some of the common extant families to be found there still. Indeed, the leaves of many of those families, such as the Euphorbiaceae, Rutaceae and Sapindaceae, have relatively indistinct venation and cuticle patterns (or at least studies have so far failed to determine distinctive features), and hence their presence or absence cannot safely be commented on at this time. Only one or two taxa are distinctive enough that their absence is both apparent and conspicuous. The first of these is the Mimosaceae, and in particular *Acacia*. All extant *Acacia* spp. have either conspicuous phyllodes or bipinnate compound leaves. No fossils that could be

interpreted as having either of those foliar forms are known from our Eocene deposits. This is particularly strange in that members of that genus are found in extant rainforest situations, and also the strong Gondwanic distribution of the group should preclude its pre-Miocene absence from the gene pool, although it must be noted that *Acacia* pollen has a pre-Miocene record (see Macphail *et al.*, Chapter 10, this volume).

The second taxon conspicuous by its absence is the Moraceae, particularly *Ficus*. The much lower angle of the basal pair of secondary veins of *Ficus* makes its leaf architecture conspicuous, and yet no definite occurrences have been reported from the Eocene. The overall flavour of the Middle Eocene macrofossil floras, then, is one of diversity, with the major families, and in some cases genera, prevalent in today's forest systems already present and diversifying. These Middle Eocene plant communities, despite their similarities to some modern systems, were unique. They were composed of mixtures of plants, some of which ultimately became extinct, or retreated to refugial locations in pockets of rainforest, or evolved into some of the extant taxa unreported from the Eocene.

REFERENCES

ALLEY, N. F. (1987). Middle Eocene age of the megafossil flora at Golden Grove, South Australia: preliminary report, and comparison with the Maslin Bay Flora. *Transactions of the Royal Society of South Australia*, 111, 211–12.

BARRETT, D. J. & CHRISTOPHEL, D. C. (1990). The spatial and temporal components of two Australian Tertiary plant megafossil deposits. In *Proceedings of the Third IOP Conference*, ed. J. G. Douglas & D. C. Christophel, pp. 43–9. Melbourne: A–Z Press.

BASINGER, J. F. & CHRISTOPHEL, D. C. (1985). Fossil flowers and leaves of the Ebenaceae from the Eocene of southern Australia. *Canadian Journal of Botany*, 63, 1825–43.

BLACKBURN, D. T. (1981). Tertiary megafossil flora of Maslin Bay, South Australia: numerical taxonomic study of selected leaves. *Alcheringa*, 5, 9–28.

CHAPMAN, F. (1937). Descriptions of Tertiary plant remains from central Australia and from other Australian localities. *Transactions of the Royal Society of South Australia*, 61, 1–16.

CHRISTOPHEL, D. C. (1980). Occurrence of *Casuarina*

megafossils in the Tertiary of southeastern Australia. *Australian Journal of Botany*, 28, 249–59.

CHRISTOPHEL, D. C. (1981). Tertiary megafossil floras as indicators of floristic associations and palaeoclimate. In *Ecological Biogeography of Australia*, ed. A. Keast, pp. 379–90. The Hague: W. Junk.

CHRISTOPHEL, D. C. (1984). Early Tertiary Proteaceae: the first floral evidence for the Musgraveinae. *Australian Journal of Botany*, 32, 177–86.

CHRISTOPHEL, D. C. (1989). Evolution of the Australian Flora through the Tertiary. *Plant Systematics and Evolution*, 162, 63–78.

CHRISTOPHEL, D. C. (1992). The prehistory of the flora of Victoria. In *The Flora of Victoria*, vol. I, ed. D. Foreman. Melbourne: Inkata Press, in press.

CHRISTOPHEL, D. C. & BASINGER, J. F. (1982). Earliest floral evidence for the Ebenaceae in Australia. *Nature*, 296, 439–41.

CHRISTOPHEL, D. C. & BLACKBURN, D. T. (1978). The Tertiary megafossil flora of Maslin Bay, South Australia: a preliminary report. *Alcheringa*, 2, 311–19.

CHRISTOPHEL, D. C. & GREENWOOD, D. R. (1987). A megafossil flora from the Eocene of Golden Grove, South Australia. *Transactions of the Royal Society of South Australia*, 111, 155–62.

CHRISTOPHEL, D. C. & GREENWOOD, D. R. (1988). A comparison of Australian tropical rainforest and Tertiary fossil leaf-beds. In *The Ecology of Australia's Wet Tropics*, ed. R. Kitching. *Proceedings of the Ecological Society of Australia*, 15, 139–48. Chipping Norton, NSW: Surrey Beatty & Sons.

CHRISTOPHEL, D. C. & GREENWOOD, D. R. (1989). Changes in climate and vegetation in Australia during the Tertiary. *Review of Palaeobotany and Palynology*, 58, 95–109.

CHRISTOPHEL, D. C., HARRIS, W. K. & SYBER, A. K. (1987). The Eocene flora of the Anglesea locality, Victoria. *Alcheringa*, 11, 303–23.

CHRISTOPHEL, D. C. & LYS, S. D. (1986). Mummified leaves of two new species of Myrtaceae from the Eocene of Victoria, Australia. *Australian Journal of Botany*, 34, 649–62.

CHRISTOPHEL, D. C., SCRIVEN, L. J. & GREENWOOD, D. R. (1992). An Eocene megaflora from Nelly Creek, South Australia. *Transactions of the Royal Society of South Australia*, 116, 65–76.

DEANE, H. (1895). President's address. *Proceedings of the Linnean Society of New South Wales*, 10, 619–67.

DEANE, H. (1900). Observations on the Tertiary flora of Australia, with special reference to Ettingshausen's theory of the Tertiary cosmopolitan flora. *Proceedings of the Linnean Society of New South Wales*, 25, 463–75.

DEANE, H. (1902a). Notes on fossil leaves from the Tertiary deposits of Wingello and Bungonia. *Records of the Geological Survey of New South Wales*, 7, 59–65.

DEANE, H. (1902b). Notes on the fossil flora of Pitfield and Mornington. *Records of the Geological Survey of Victoria*, 1, 15–20.

DEANE, H. (1902c). Notes on the fossil flora of Berwick. *Records of the Geological Survey of Victoria*, 1, 21–32.

DEANE, H. (1904). Further notes on the Cainozoic flora of Sentinel Rock, Otway Coast. *Records of the Geological Survey of Victoria*, 1, 212–16.

DOUGLAS, J. G. (1977). A new fossil plant assemblage from the Eastern View Coal measures. Geological Survey of Victoria, unpublished report 1977/31.

ETTINGSHAUSEN, C. VON (1883). Beiträge zur Kenntniss der Tertiar-flora Australiens. Part 1. *Denkschriften der Österreichischen Akademie der Wissenschaften, Mathematisch-Naturwissenschaftliche Klasse*, 47, 101–48.

ETTINGSHAUSEN, C. VON (1886). Beiträge zur Kenntniss der Tertiar-flora Australiens. Part 2. *Denkschriften der Österreichischen Akademie der Wissenschaften, Mathematisch-Naturwissenschaftliche Klasse*, 53, 81–142.

ETTINGSHAUSEN, C. VON (1888). Contributions to the Tertiary flora of Australia. *Memoirs of the Geological Survey of New South Wales, Palaeontology*, 2, 1–189.

GREENWOOD, D. R. (1987). Early Tertiary Podocarpaceae: megafossils from the Eocene Anglesea location, Victoria. *Australian Journal of Botany*, 35, 111–33.

GREENWOOD, D. R., CALLEN, R. A. & ALLEY, N. F. (1990). *Tertiary macrofloras and Tertiary Stratigraphy of Poole Creek Palaeochannel, Lake Eyre Basin*. Australian Geological Congress, Hobart, February 1990, Abstract.

HILL, R. S. (1978). Two new species of *Bowenia* Hook. ex Hook. f. from the Eocene of Eastern Australia. *Australian Journal of Botany*, 26, 837–46.

HILL, R. S. (1980). Three new Eocene cycads from Eastern Australia. *Australian Journal of Botany*, 28, 105–22.

HILL, R. S. (1982). The Eocene megafossil flora of Nerriga, New South Wales, Australia. *Palaeontographica Abt. B*, 181, 44–77.

HILL, R. S. (1986). Lauraceous leaves from the Eocene of Nerriga, New South Wales. *Alcheringa*, 10, 327–51.

HILL, R. S. (1988a). Australian Tertiary angiosperm and gymnosperm remains – an updated catalogue. *Alcheringa*, 12, 207–19.

HILL, R. S. (1988b). A re-investigation of *Nothofagus muelleri* (Ett.) Patterson and *Cinnamomum nuytsii* Ett. from the Eocene of Vegetable Creek. *Alcheringa*, 12, 221–31.

HILL, R. S. (1989). Early Tertiary Leaves of the Menispermaceae from Nerriga, New South Wales. *Alcheringa*, 13, 37–44.

HILL, R. S. (1991). *Eucryphia* (Eucryphiaceae) leaves from Tertiary sediments in southeastern Australia. *Australian Systematic Botany*, 4, 481–97.

HILL, R. S. & CARPENTER, R. (1991). Evolution of *Acmopyle* and *Dacrycarpus* (Podocarpaceae) foliage as inferred from macrofossils in south-eastern Australia. *Australian Systematic Botany*, 4, 449–79.

HILL, R. S. & CHRISTOPHEL, D. C. (1988). Tertiary leaves of the tribe Banksieae (Proteaceae) from south-eastern Australia. *Botanical Journal of the Linnean Society*, **97**, 205–27.

HILL, R. S. & MERRIFIELD, H. E. (1993). An Early Tertiary macroflora from West Dale, south-western Australia. *Alcheringa*. **17**, 285–326.

HILL, R. S. & POLE, M. S. (1992). Leaf and shoot morphology of extant *Afrocarpus*, *Nageia* and *Retrophyllum* (Podocarpaceae) species, and species with similar leaf arrangement from Tertiary sediments in Australasia. *Australian Systematic Botany*, **5**, 337–58.

LANGE, R. T. (1969). Recent and fossil epiphyllous fungi of the *Manginula-Shortensis* group. *Australian Journal of Botany*, **17**, 565–74.

LANGE, R. T. (1970). The Maslin Bay flora, South Australia, 2. The assemblage of fossils. *Neues Jahrbuch für Geologie und Paläontologie*, **8**, 486–90.

LANGE, R. T. & SMITH, P. H. (1971). The Maslin Bay flora, South Australia. 3. Dispersed fungal spores. *Neues Jahrbuch für Geologie und Paläontologie*, **11**, 663–81.

MACNAMARA, K. J. & SCOTT, J. K. (1983). A new species of *Banksia* (Proteaceae) from the Eocene Merlinleigh sandstone of the Kennedy Range, Western Australia. *Alcheringa*, **7**, 185–93.

McGOWRAN, B. (1992). Maastrichtian and Early Cainozoic, Southern Australia: planktonic foraminiferal Biostratigraphy. In *Pre Cenozoic of Australia*, ed. M. Willams & P. De Deckker. *Geological Society of Australia Special Publication*, in press.

McGOWRAN, B., HARRIS, W. K. & LINDSAY, J. M. (1970). The Maslin Bay flora, South Australia, 1. Evidence for an Early Middle Eocene age. *Neues Jahrbuch für Geologie und Paläontologie*, **8**, 481–5.

MUELLER, F. VON (1874). *Observations on New Vegetable Fossils of the Auriferous Drifts*. 1 December. Geological Survey of Victoria.

MUELLER, F. VON (1883). *Observations on New Vegetable Fossils of the Auriferous Drifts*. 2 December. Geological Survey of Victoria.

NIX, H. (1982). Environmental determinants of biogeography and evolution in Terra Australis. In *Evolution of the Flora and Fauna of Arid Australia*, ed. W. R. Barker & P. J. M. Greenslade, pp. 47–66. Frewville: Peacock Publications.

O'DOWD, D. J., BREW, C. R., CHRISTOPHEL, D. C. & NORTON, R. A. (1991). Mite–Plant associations from the Eocene of Southern Australia. *Science*, **252**, 99–101.

OWEN, J. A. (1975). Palynology of some Tertiary deposits from New South Wales. Ph. D. thesis, Australian National University.

RAINE, J. I. (1967). Geology of the Nerriga area. Hons. thesis, Australian National University.

ROWETT, A. I. & CHRISTOPHEL, D. C. (1990). The dispersed cuticle profile of the Eocene Anglesea clay lenses. In *Proceedings of the Third IOP Conference*, ed. J. G. Douglas & D. C. Christophel, pp. 115–21. Melbourne: A – Z Press.

ROZEFELDS, A. C., CHRISTOPHEL, D. C. & ALLEY, N. F. (1992). Tertiary occurrence of the fern *Lygodium* (Schizaeaceae) in Australia & New Zealand. *Memoirs of the Queensland Museum*, **32**, 203–22.

SCRIVEN, L. J. & CHRISTOPHEL, D. C. (1990). A numerical study of extant and fossil *Gymnostoma*. In *Proceedings of the Third IOP Conference*, ed. J. G. Douglas & D. C. Christophel, pp. 137–47. Melbourne: A – Z Press.

SOUTHCOTT, R. V. & LANGE, R. T. (1971). Acarine and other micro-fossils from the Maslin Eocene, South Australia. *Records of the South Australian Museum*, **16**, 1–21.

SPICER, R. A. (1980). The importance of depositional sorting to the biostratigraphy of plant megafossils. In *Biostratigraphy of Fossil Plants. Successional and Paleoecological Analyses*, ed. D. L. Dilcher & T. N. Taylor, pp. 171–83. Pennsylvania: Dowden, Hutchinson and Ross.

WEBB, L. J. (1959). A physiognomic classification of Australian rainforests. *Journal of Ecology*, **47**, 551–70.

WEBB, L. J. & TRACEY, J. G. (1982). *Rainforests: Data on Floristics and Site Characteristics*. Canberra: Biogeographic Information System, Bureau of Flora and Fauna.

WELLMAN, P. & McDOUGALL, I. (1974). Potassium–argon ages of the Cainozoic volcanic rocks of New South Wales. *Journal of the Geological Society of Australia*, **21**, 247–72.

12 Cenozoic vegetation in Tasmania: macrofossil evidence

R. J. CARPENTER, R. S. HILL & G. J. JORDAN

The Tasmanian Cenozoic macrofossil record is relatively rich, and changes that have occurred in the vegetation of the region are becoming increasingly well understood. The record is essentially one of rainforest elements, especially in the Paleogene, but taxa that are now common in sclerophyllous heathlands and woodlands are increasingly prevalent in Quaternary sediments.

Extant Tasmanian rainforest is renowned for its beauty, and botanists have long recognised its marked taxonomic and structural similarity to other southern hemisphere 'cool temperate' forests of New Zealand and Chile. These are generally dominated by *Nothofagus* trees, their boughs laden with lichens and verdant shrouds of bryophytes. Other links are often made by phytogeographers to similar forests in high altitude regions of northern New South Wales and the much more species-rich vegetation of the generally montane regions of New Guinea and New Caledonia where *Nothofagus* also grows. A striking aspect of these forests is the presence of a variety of conifers, principally Podocarpaceae, but also Cupressaceae and Araucariaceae. In Tasmania the Araucariaceae are extinct, but the region is unique in the southern hemisphere in having a genus of Taxodiaceae, *Athrotaxis*. *Athrotaxis* spp. are often associated with Australia's only winter deciduous plant, *Nothofagus gunnii*, in montane regions of the island. The macrofossil record shows conclusively that the current diversity of Tasmania's woody rainforest flora is very much lower than at any other time during the Cenozoic. It confirms that there are strong floristic links to regions as widespread as eastern and southwestern mainland Australia, southern South America, New Zealand and New Guinea. In fact, Tasmanian Paleogene floras contain a wealth of taxa that are closely related to plants now confined to these regions.

Apart from the relatively large tracts of rainforest in Tasmania, closed forest lacking eucalypts is now confined to small patches along the east coast of Australia. In contrast to mainland Australia, Tasmania is relatively mountainous and has a well-developed woody alpine vegetation, dominated by shrubs of the Asteraceae, Epacridaceae, Myrtaceae and Proteaceae. A physiognomically (and often taxonomically) similar vegetation is typical of low altitude sedgeland heaths of Tasmania and coastal areas of other parts of Australia. Further, in contrast to much of mainland Australia, Tasmania lacks truly arid regions. The Cenozoic macrofossil record of taxa now present in alpine, eucalypt woodland, coastal heathland and arid regions is not as complete as it is for rainforest. This can be attributed partly to the fact that these vegetation types are often not found near depositional sites, but it must also be a reflection of unsuitable environmental factors during much of the Cenozoic.

The palaeobotanist's aim is to extract the maximum information from each fossil deposit.

This can be achieved by identifying at least some of the fossil plant organs, evaluating the size and shape (physiognomy) of the leaves, and interpreting the nature of the fossil site and how the material came to be preserved within it (taphonomy). Using a uniformitarian approach, the interrelated aspects of ecology, physiology, physiognomy and community relations of living plants are then considered, and this is particularly relevant when taxa with close affinity to the fossils are identified. In addition, the task of reconstructing past environments offers some scope for the imagination.

Collections of Tasmanian Cenozoic plant macrofossils were initiated by Darwin (1860), who obtained specimens from the Hobart region in 1836. In the late nineteenth century, Robert Johnston, the Government geologist, produced a text on the geology of the island (Johnston, 1888) in which he referred to numerous plant-bearing Tertiary sediments, including those from the ancient river and lake systems of the Derwent and Tamar estuaries. He described and illustrated fossils as new species or referred them to existing taxa. This material is of little use in current research because much was poorly preserved and/or has deteriorated in the century since its collection. Nevertheless, there is no doubt that at least some of Johnston's determinations were sound, and among the more interesting records are leaves of *Nothofagus*, Lauraceae and *Araucaria* (Johnston, 1885, 1888). The Swedish palaeobotanist Olof Selling described *Araucaria derwentensis* from Tertiary beds near Hobart and *Podocarpus brownei* (now considered to be a podocarpaceous genus of uncertain generic affinity by Hill & Pole (1992)) from Burnie (Selling, 1950). Townrow contributed two papers on the Buckland flora (which has since been dated as Early Eocene by Macphail in Hill & Carpenter (1991*a*)), in which he described numerous podocarps (Townrow, 1965*a*) and what he believed to be a species of *Athrotaxis* (Townrow, 1965*b*). These studies were significant in that Townrow realised there was an association of taxa that have modern relatives now widely distributed latitudinally. Macrofossil studies of Tasmanian Cenozoic floras have recently escalated, and this

research forms the basis of this review, in which the vegetation of the Tasmanian region is discussed chronologically. Elements of the extant Tasmanian flora that appear in the macrofossil record are highlighted. The Tasmanian record should be considered in conjunction with that of mainland Australia (Christophel, Chapter 11, this volume) and with the Cenozoic palynological record (Macphail *et al.*, Chapter 10; Kershaw *et al.*, Chapter 13, this volume). Also, investigation of many fossil sites has only recently begun, and consequently some of our interpretations are preliminary.

The Tasmanian Tertiary fossil sites and environmental reconstructions discussed in this chapter are shown in Figure 12.1. They range in age from Early Eocene to earliest Miocene, but most are Oligocene, and this epoch is therefore the best known. The current lack of macrofossil deposits of Cretaceous–Paleocene or Late Miocene–Pliocene age in the region is unfortunate since, in the former period, the global rise to dominance of the angiosperms occurred, and in the latter there was an apparent decrease in Tasmanian rainforest diversity in response to climatic changes. Several Tasmanian Pleistocene

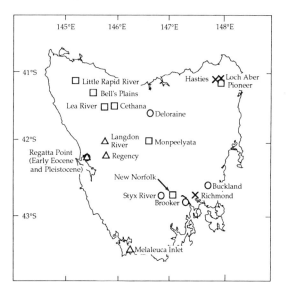

Figure 12.1 Location of Tasmanian fossil sites discussed in this chapter: (○) Early Eocene, (X) Middle–Late Eocene, (□) Oligocene–Early Miocene, (△) Pleistocene.

macrofossil floras have also been studied (Figure 12.1).

The known Cenozoic deposits were formed in a variety of sedimentological environments. Most have been dated by palynological correlation, using the stratigraphy of Stover & Partridge (1973), but more precise K–Ar dates have been ascertained for those sites at which the sediments have an associated basalt layer. Radiocarbon ages have been used for dating Late Pleistocene assemblages, although this method is unreliable beyond about 30 thousand years ago, because of potential contamination (Colhoun, 1986).

CENOZOIC CLIMATE AND LANDFORMS

An overview of the climate and physical geography of the Tasmanian region during the Cenozoic is presented, but detailed discussion of these topics appears elsewhere (Wilford & Brown, Chapter 2; Quilty, Chapter 3; Taylor, Chapter 5, this volume). Palaeoclimatological evidence, based in part on oxygen isotope ratios obtained from foraminifera recovered from deep-sea cores, indicates that the climate in the southern hemisphere near Tasmania during the Early Eocene was highly equable, with warm, wet and very humid conditions prevailing. These conditions persisted until the Middle–Late Eocene, when a general cooling began. This temperature decline was most marked around the Eocene/Oligocene boundary, and the remainder of the Oligocene was also cool, with an apparent increase in the seasonality of precipitation and seasonal and diurnal temperature range. Associated with the accumulation of polar ice, the Miocene is seen as a period of further development of this trend. Superimposed on a general trend toward decreasing temperatures and increased aridity, the Pliocene and Pleistocene were periods of major and frequent climatic fluctuations. At least four periods of glaciation of large parts of Tasmania (glacials) interspersed with relatively short, warmer, wetter, ice-free periods (interglacials) occurred in the Pleistocene, although the record of other glacials

may have been eroded by subsequent glaciers (Fitzsimons & Colhoun, 1991).

An important feature of the Tasmanian environment, past and present, is its physical geography. Mountainous country has existed in the region since at least the start of the Cenozoic (Taylor, Chapter 5, this volume). The latitude of southern Tasmania in the Early Eocene was about 60° S (Wilford & Brown, Chapter 2, this volume), about the same as much of coastal Antarctica today. This implies that Tasmania experienced almost continuously dark conditions during winter. As Australia moved toward the Equator these periods of darkness lessened, but by the Oligocene Tasmania was still at about 55° S. Hence, throughout the Paleogene, the combination of a low incident sunlight angle and rugged topography would have created a much broader range of physical niches than exist currently in Tasmania. The Pleistocene glaciations caused major landform disruptions, altering not only the high country but also the lowland riverine systems. Sea levels were much lower at these times, and Bass Strait was closed.

Edaphic and pyric factors also influence the nature of vegetation. Soils in much of western Tasmania are derived from Cambrian and Precambrian rocks and are extremely oligotrophic. Most of the rest of Tasmania has moderately nutrient poor soils derived from igneous rocks and Permian and Triassic sediments. Presumably the nutrient status of soils in these regions has been similar throughout the Cenozoic. However, in the Oligocene and Miocene Tasmania experienced much volcanic activity, which resulted in the development of much richer basaltic soils in much of the north and parts of the southeast. Acid peats are now widespread in western Tasmania, but there is little or no evidence for their existence prior to the Late Pleistocene. Fire has a major influence on modern Tasmanian ecosystems. Lightning strikes have probably caused rare fires throughout the Cenozoic, but frequent burning of vegetation became common only with the arrival of Aborigines in the Late Pleistocene.

VEGETATION RECONSTRUCTION AND FLORISTICS

Early Eocene

The Regatta Point, Buckland, Styx River, Deloraine and Brooker floras are characterised by an abundance of gymnosperms and (where determinable) angiosperm leaves that are relatively large (Figure 12.2(*a*)). *Nothofagus* has been recorded only from Deloraine. Some of the taxa are now extinct at the generic level, e.g. *Araucarioides* (Araucariaceae; Bigwood & Hill, 1985), from Regatta Point, which had leaves at least 16 cm long. Cycads have been recovered from two deposits – a serrate-leaved species of *Bowenia* similar to the extant *B. serrulata* (Regatta Point), and the extinct genus *Pterostoma* (Hill, 1980) (Buckland). The two living species of *Bowenia* are confined to eastern Queensland, in lowland megathermal rainforests or more open forest types. *Bowenia* has also been described from three Eocene deposits in southeastern mainland Australia (Hill, 1978; Christophel, Chapter 11, this volume). *Pterostoma* occurs in two of these floras and Hill (1980) noted that its closest relationships are not with extant cycads, but rather with extinct Cretaceous taxa. In living cycads the fronds are not shed when they die, but slowly decay while attached to the parent plant. However, in *Pterostoma* the rachis bases are expanded and have the appearance of an abscission zone (Hill, 1980). It is possible that *Pterostoma* periodically discarded its fronds, perhaps in response to long dark periods in winter or due to less regular periods of stress. The *Nothofagus* leaves from Deloraine are morphologically very similar to those of modern deciduous species from South America and probably adopted the same habit. The ability to shed foliage during the dark season would have been ecophysiologically advantageous in high latitude, everwarm (and wet) Eocene environments, because it is a means of circumventing the risk of excessively high dark respiration rates by the leaves during winter. It is possible that *Pterostoma*

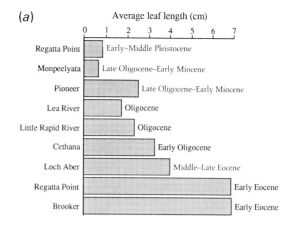

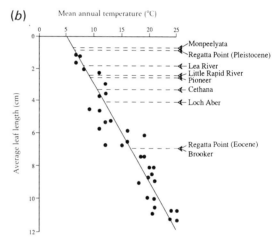

Figure 12.2 (*a*) Leaf length data from Tasmanian fossil sites. Average leaf lengths are based on measurements of all available angiosperm leaves (except Casuarinaceae). (*b*) Plot of average leaf length versus mean annual temperature (MAT) for rainforest sites in Tasmania and eastern Australia. Leaves were obtained from forest floor leaf-litter. For each site MAT was obtained by using climate surfaces derived by M. F. Hutchinson (e.g. Hutchinson & Bischoff, 1983) as implemented in BIOCLIM Version 2.0, written by J. R. Busby (e.g. Busby, 1988).

and other taxa were able to persist into the Paleogene in the most southerly part of Australia, when elsewhere they were outcompeted by plants that could better exploit the increased growing seasons created by the northward-drifting continent.

The Styx River sediments contain many conifers, including an extinct large-leaved *Araucaria* (Araucariaceae) and extinct taxa that are probably podocarpaceous. Townrow (1965*a*) also recorded an extinct putative podocarp species, *Coronelia molinae*, from Buckland. Podocarps are now a familiar component of the wet western and highland regions of Tasmania, where five (four endemic) species from five genera (*Lagarostrobos, Microcachrys, Microstrobos, Phyllocladus* and *Podocarpus*) are found. Pollen from the Early Eocene is referable to each of these genera, and macrofossils of the last three are known (Townrow, 1965*a*; R. S. Hill, unpublished data). Several other podocarpaceous genera have been recorded from the Early Eocene in Tasmania, and these now have representatives that occur in northeastern Queensland and/or other regions of Australasia or Malesia. These are *Acmopyle, Prumnopitys* and *Dacrycarpus* from Buckland (Townrow, 1965*a*; Hill & Carpenter, 1991*a*), *Acmopyle* and *Dacrycarpus* from Regatta Point (Hill & Carpenter, 1991*a*) and *Dacrydium* from Styx River (R. S. Hill, unpublished data). Foliage of *Libocedrus* (Cupressaceae) and several species of *Araucaria* have also been recovered from Regatta Point (Hill & Bigwood, 1987; R. S. Hill, unpublished data). The former genus is now confined to New Zealand and New Caledonia.

The diversity of angiosperms in the Early Eocene deposits is generally quite low. For instance, at Styx River rare dispersed cuticle of only one species has been recovered. However, many more taxa occur at Regatta Point, including the evergreen *Eucryphia microstoma* (Hill, 1991*a*). *Eucryphia* is now restricted to microthermal rainforests of Tasmania, Chile, southern New South Wales and Mt Bartle Frere in northeastern Queensland. Leaves or dispersed cuticles of the Lauraceae and Proteaceae are common components of almost all sites. Unlike *Eucryphia* and *Nothofagus*, no extant species of these families are deciduous. Many of the fossil lauraceous leaves are more than 10 cm long, and if these plants were evergreen, then such broad-leaved forms must have been able to survive the prevailing dark seasons. The southerly extent of the

Lauraceae (excluding the leafless parasite *Cassytha*) in Australia is now southern New South Wales, and the vast majority of species occur in northeastern Queensland (Hyland, 1989). The microphyllous casuarinaceous genus *Gymnostoma* is common at Regatta Point, and this genus is currently restricted to tropical regions, with only one small population occurring in northeastern Australia. An interesting Regatta Point record is a palm frond (M. S. Pole, personal communication), which may be related to *Nypa*, which is common in the pollen flora at that site (Macphail *et al.*, Chapter 10, this volume). *Nypa* is a monotypic mangrove genus that is now confined to the tropics. Similarly aged *Nypa* macrofossils are often quite abundant in the northern hemisphere (e.g. Collinson & Hooker, 1987), where they are regarded as good indicators of very warm climates. It is likely that the Regatta Point sediments were deposited in an estuary or lagoon.

The floristic data accumulated to date emphasise that the Early Eocene vegetation of Tasmania was growing under climatic conditions quite different from those of any modern environment; there are many taxa that do not have close affinities to any extant plants, others that have nearest living relatives in microthermal environments, and still others that have nearest living relatives in mesothermal–megathermal environments. These forests clearly have no modern structural or floristic analogue.

Middle–Late Eocene

Only two Middle–Late Eocene Tasmanian macrofloras are known, at Hasties and Loch Aber in the northeast. Palynologically, this period is significant in having a marked increase in the abundance of *Nothofagus* pollen (see Macphail *et al.*, Chapter 10, this volume), and these floras are the oldest in which evergreen *Nothofagus* macrofossils (leaves similar to extant *N. moorei*) have been detected (Hill, 1990*a*; M. S. Pole, personal communication). Another deposit at Richmond, in southeastern Tasmania, is not easily datable, but may be of Middle–Late Eocene age. This flora contains the deciduous gymnosperm *Ginkgo* and a species of *Nothofagus* that was probably also deciduous. This provides further evidence of the unique suitability of the Tasmanian region for a diversity of deciduous plants in the Paleogene, since the nearest temporal (and spatial) record of *Ginkgo* in Australia is from Cretaceous sediments in Victoria (Douglas, Chapter 9, this volume).

As discussed earlier, a general climatic cooling commenced around the Late Eocene and this probably produced conditions favourable for the spread of microthermal elements. These floras also exhibit a relatively high diversity of gymnosperms, including species of at least six genera of Podocarpaceae from Hasties (Pole, 1992) and *Araucaria hastiensis*, which belongs to the now endemic South American section *Columbea* (Bigwood & Hill, 1985; Hill, 1990*b*). The Loch Aber flora also includes specimens of Araucariaceae, *Phyllocladus*, *Acmopyle*, *Falcatifolium* and *Willungia* (Podocarpaceae) (Hill, 1989*a*, and unpublished data; Hill & Carpenter, 1991*a*; Hill & Pole, 1992). Hill (1989*a*) has shown that phylloclades of *P. aberensis* are morphologically very similar to those of *P. aspleniifolius*, which is a common tree in extant Tasmanian rainforest. *Phyllocladus* is extinct from mainland Australia but was present in the Late Eocene at least as far north as Vegetable Creek (von Ettingshausen, 1888). The angiosperm component of the Hasties flora includes leaves of *Nothofagus* subgenus *Lophozonia*, Lauraceae, microphylls with pronounced drip-tips comparable to *Xanthomyrtus* (Myrtaceae), *Gymnostoma* (Casuarinaceae) and possibly Cunoniaceae and *Cenarrhenes* (Proteaceae) (Pole, 1992). The Proteaceae is represented by *Banksieaephyllum attenuatum* at Loch Aber (Hill & Christophel, 1988). Hill (1991*a*) demonstrated that *Eucryphia aberensis* from Loch Aber is most closely related to the extant Tasmanian species *E. lucida* and *E. milliganii*.

The average leaf length of angiosperm specimens recovered from these two floras is similar, and shows a decrease from the Early Eocene (Figure 12.2*a*). This is probably a reflection of a general temperature decline.

Oligocene–Early Miocene

Five Oligocene–Early Miocene macrofloras from Tasmania have been studied in some detail; Pioneer, Cethana, Little Rapid River, Lea River and Monpeelyata. They offer an excellent opportunity to reconstruct the Oligocene vegetation and environment of what is now northern Tasmania. Although their relative ages are uncertain, Cethana is likely to be Early Oligocene and anyway older than the other four deposits; Little Rapid River and Lea River are Oligocene, and are older than Pioneer and Monpeelyata, which are Late Oligocene–Early Miocene (Macphail *et al.*, Chapter 10, this volume). These deposits were formed in quite varied physical environments. The Pioneer sediments were deposited in a braided river at low altitude (Hill & Macphail, 1983). In contrast, the probable fluvial sediments of the Lea River assemblage are currently 800+ m above sea level (a.s.l.) and were probably deposited at similar altitude in the Oligocene. Cethana was probably a small lake at moderate altitude (about 300 m a.s.l.), and sedimentological features suggest that it was located adjacent to a steep slope in a topographically diverse landscape (Carpenter, 1991*a*). Its macrofossils exist as compressions and impressions, in contrast to common mummifications at the other sites. Basal sediments in the moderate altitude (about 350 m a.s.l.) Little Rapid River deposit reflect relatively quiet water deposition, but an upper layer of coarse organic material (including logs) suggests a much more rapid input, possibly associated with a local catastrophic event (Wells & Hill, 1989). The Monpeelyata site currently occurs at 920 m a.s.l., and its sediments were accumulated in a lake at this or at a lower altitude (Macphail *et al.*, 1991). Other Tasmanian sediments, of probable Oligocene age, from which macrofossils have been recovered are Bell's Plains (Wells & Hill, 1989) and New Norfolk (Hill, 1983*a*). The Oligocene macrofossil floras are characterised by abundant and diverse gymnosperms and *Nothofagus*, and by their overall microphyllous or nanophyllous nature. At both Cethana and Little Rapid River at least 20 conifer species have been identified, and it is likely that up to three times as many angiosperms occurred at each site. The following composite scenario is presented for the Tasmanian region during the Oligocene to earliest Miocene.

Pteridophytes

Hill & Macphail (1983) concluded that vegetation at Pioneer included a diverse assemblage of ferns, and although this was based principally on microfossil evidence there are numerous unidentified sporangia preserved in these sediments. Sporangia have also been found at Little Rapid River and Cethana, but it is only in the latter flora that well-preserved foliar remains occur. These include macrofossils of *Lygodium* (Schizaeaceae), Hymenophyllaceae, *Gleichenia* and *Sticherus* (Gleicheniaceae) and probably *Blechnum* (Blechnaceae) and Cyatheaceae (Figure 12.3(*a*)) (Carpenter, 1991*a*). A single specimen was assigned to *Tmesipteris* (Psilotaceae) by Carpenter (1988). With the exception of *Lygodium*, these taxa are characteristic of the high latitude regions of Australasia, and all occur in and/or marginal to *Nothofagus* forest. Most notably, the *Sticherus* specimens are indistinguishable from *S. tener*, which is now confined to southeastern Australia, and is abundant in Tasmanian forests (Duncan & Isaac, 1986). All of the taxa recognised from Cethana also now occur in lower latitude regions of northeastern Queensland, the north island of New Zealand, New Guinea and New Caledonia, most frequently in montane regions. The climbing fern *Lygodium* is more typical of lowland rainforest margins in these regions. However, *Lygodium* is not confined to this habitat, as exemplified by the fact that it is abundant near the summit of New Caledonia's highest peak, Mt Panié (about 1640 m a.s.l.). Hill (1988) has described leaves and megaspores of the freshwater hydrophyte *Isoetes*, a fern ally, from Monpeelyata. Species of this genus are generally found in temperate regions of the world, including alpine and sub-alpine lakes of Tasmania (Duncan & Isaac, 1986).

Gymnosperms

A diverse conifer component has been found in Oligocene–Early Miocene sediments, and if taphonomic factors are considered (where usually only waterside plants contribute significantly to the sediment), it can reasonably be expected that the identified taxa represent only a fraction of the total regional flora. As for the Eocene, there are numerous podocarps and araucarians, but for the first time in the macrofossil record of the Tasmanian region definite Taxodiaceae are present, including both reproductive and vegetative structures. One (or possibly more) species of *Athrotaxis*, distinct from the three extant species, occurs at Pioneer, Little Rapid River and Lea River. Hill & Carpenter (1989) described cupressaceous macrofossils from three closely related genera, *Libocedrus*, *Papuacedrus* and *Austrocedrus*, which are now confined to New Zealand/New Caledonia, New Guinea/Moluccas and southern South America, respectively. Species from the former two genera have been found at Pioneer and Cethana, and well-preserved ovulate cones and foliage of *Libocedrus* from Little Rapid River and Lea River. Ovulate cones of *Austrocedrus* have been recovered from Monpeelyata.

Foliage types referred to numerous species of extant podocarpaceous genera are listed in Table 12.1. The fossil *Lagarostrobos* spp. are clearly distinct from the extant Tasmanian species *Lagarostrobos franklinii*. The genus is now confined to Tasmania and New Zealand. *Falcatifolium* is now restricted to New Caledonia and Malesia. Some Oligocene Podocarpaceae have been referred to extinct genera. Wells & Hill (1989) referred imbricate foliage from Pioneer and Little Rapid River to the extinct genus *Mesibovia*, but it has since been transferred to the Taxodiaceae and is now considered to be an extinct species of the extant genus *Athrotaxis* (Hill *et al.*, 1993). Hill & Pole (1992) described the extinct genera *Willungia* and *Smithtonia* spp. from Little Rapid River. At least one now extinct, possibly podocarpaceous taxon occurs at Cethana (Carpenter 1991*a*).

Several species of *Araucaria* and at least one of

Figure 12.3 Fossils from Tasmanian Paleogene deposits. Scale bars in (*a*), (*f*) and (*g*) represent 1 cm, (*b*), (*c*) and (*e*) 5 mm, and (*d*) 20 μm. (*a*) Section of a fern frond from Cethana, possibly belonging to the Cyatheaceae. (*b*) Section of *Araucaria* section *Eutacta* stem from Lea River, morphologically very similar to species now confined to New Caledonia. (*c*) *Agathis* cone-scale from Cethana. (*d*) Scanning electron micrograph of the abaxial cuticle surface of a xeromorphic ? *Lomatia* species from Cethana. The stomates are surrounded by raised rims of cuticle and a trichome base is associated with each epidermal cell. (*e*) Microphyllous Myrtaceae leaf from Cethana. Note the impression of an insect leaf mine. (*f*) *Banksieaephyllum* leaf from Cethana, architecturally very similar to leaves of *Banksia grandis*, a species now confined to southwestern Western Australia. (*g*) Leaf of Lauraceae from Little Rapid River.

Table 12.1. *Accepted records of Tasmanian Oligocene–Early Miocene Podocarpaceae from extant genera. The designation 'sp.' may or may not imply distinct status from a described species*

Species	Deposit	Reference
Acmopyle glabra	Cethana	Hill & Carpenter (1991*a*)
Dacrycarpus mucronatus	Cethana	Hill & Carpenter (1991*a*)
D. mucronatus	Little Rapid River	Wells & Hill (1989)
D. linifolius	Little Rapid River	Wells & Hill (1989)
D. arcuatus	Little Rapid River	Wells & Hill (1989)
D. cupressiformis	Little Rapid River	Wells & Hill (1989)
D. linearis	Little Rapid River	Wells & Hill (1989)
D. involutus	Monpeelyata	Wells & Hill (1989)
D. acutifolius	Monpeelyata	Wells & Hill (1989)
D. lanceolatus	Monpeelyata	Wells & Hill (1989)
D. crenulatus	Pioneer	Wells & Hill (1989)
Dacrycarpus spp. (1 ? or more)	Lea River	R. J. Carpenter & R. S. Hill (unpublished data)
Dacrydium tasmanicum	Little Rapid River	Wells & Hill (1989)
D. aciculare	Little Rapid River	Wells & Hill (1989)
D. sinuosum	Pioneer	Wells & Hill (1989)
Dacrydium spp. (2)	Cethana	Carpenter (1991*a*)
Falcatifolium sp.	Little Rapid River	R. S. Hill (unpublished data)
Lagarostrobos marginatus	Little Rapid River	Wells & Hill (1989)
Lagarostrobos sp.	Cethana	Carpenter (1991*a*)
Microstrobos microfolius	Little Rapid River	Wells & Hill (1989)
M. microfolius	Monpeelyata	Wells & Hill (1989)
Phyllocladus annulatus	Pioneer	Hill (1989*a*)
P. aberensis	Little Rapid River	Hill (1989*a*)
P. lobatus	Little Rapid River	Hill (1989*a*)
Phyllocladus spp. (2)	Cethana	Carpenter (1991*a*)
Podocarpus spp. (2)	Cethana	Carpenter (1991*a*)
Podocarpus sp.	Little Rapid River	R. S. Hill (unpublished data)
Podocarpus sp.	Lea River	R. J. Carpenter & R. S. Hill (unpublished data)

Agathis, the other extant genus of the Araucariaceae, are also known from the Oligocene–Early Miocene (Hill & Bigwood, 1987; Hill, 1990*b*; Carpenter, 1991*a*, and unpublished data). However, *Araucarioides* has not been recovered and may have become extinct by this time, possibly due to climatic change. The *Araucaria* fossils include species with close affinity to modern species from New Caledonia (Figure 12.3(*b*)) and Chile. *Agathis* spp. are now found in a variety of ecological situations in Queensland, Malesia, New Zealand, Fiji, Vanuatu, New Caledonia and the Solomon Islands. *Agathis* macrofossils, including cone scales (Figure 12.3(*c*)) and leaves, have been found at Little Rapid River (Hill & Bigwood, 1987) and Cethana (Carpenter, 1991*a*).

The cycads *Pterostoma* and *Macrozamia* (Zami-aceae) have been recovered from Cethana (Carpenter, 1991*b*). The likelihood that Cethana is the oldest of the sites and the apparent lack of cycads elsewhere suggests that climatic factors led to their extinction from the region by the Late Oligocene. Alternatively, their occurrence at Cethana could be attributable to the presence of suitable habitats related to the unique physical attributes of the site (Carpenter, 1991*a,b*).

Angiosperms

Although many angiosperm macrofossils are difficult to identify, there is no doubt that Tasmanian Oligocene rainforest was complex and contained a relatively high angiosperm diversity compared to that of modern vegetation, even in the tropics.

Some of these fossil taxa have close affinities to modern Tasmanian plants. However, the regions with the greatest taxonomic and physiognomic (and by inference climatic) similarity to the various Oligocene forests of Tasmania include the lower montane rainforests of New Guinea and, at least in terms of their angiosperm components, the *Nothofagus moorei* and associated microthermal–mesothermal vegetation of montane northernmost New South Wales. So far, taxa from the families Proteaceae, Casuarinaceae, Cunoniaceae, Fagaceae, Elaeocarpaceae, Lauraceae, Myrtaceae and Epacridaceae have been identified. The highest diversities occur at Cethana and Little Rapid River.

Fagaceae is represented by leaves and sometimes cupules of *Nothofagus* at all sites. A review of this genus (including taxonomic, evolutionary and ecological aspects of its fossil history) is presented by Hill (Chapter 16, this volume).

Macrofossils from three of the four subgenera of *Nothofagus* as recognised by Hill & Read (1991) occur at Little Rapid River and Cethana (Carpenter, 1991*a*; Hill, 1991*b*). There are species at both sites that have close affinity to living species now separated by over 40° of latitude and occur across the entire geographical range of the genus. The fossils from Little Rapid River are relatively well preserved and have been studied in detail (Hill, 1987, 1991*b*). There are cupules and leaves of more than one species from subgenus *Brassospora*, and at least one each from subgenera *Nothofagus* and *Lophozonia*. It is significant that these include the first macrofossils of subgenus *Brassospora* recorded, as the pollen type produced by the species of this group dominates Late Eocene to Miocene deposits of southeastern Australia, and the subgenus is now restricted to New Guinea and New Caledonia (see Macphail *et al.*, Chapter 10; Kershaw *et al.*, Chapter 13, this volume). Hill (1990*a*, 1991*b*) has discussed this phenomenon and attributes the unusually common presence of subgenus *Brassospora* macrofossils at Little Rapid River to a local disturbance event. At Cethana, there are macrofossils closely resembling extant species from Tasmania, New Guinea, northern New South Wales–southernmost Queensland

and Chile. These species belong to subgenera *Brassospora*, *Lophozonia* and *Fuscospora* of *Nothofagus*. The latter subgenus has not been recovered from Little Rapid River, but is diverse at Cethana, where *N. gunnii* and leaves with close morphological similarity to the Chilean species *N. alessandri* and the New Zealand species *N. fusca* and *N. truncata* have been recorded (Hill, 1984; Carpenter, 1991*a*). The first two species are winter deciduous. *Nothofagus gunnii* leaves also occur at Lea River and Monpeelyata, and Hill (1989*b*) has recorded the species from Antarctic sediments of similar age.

Nothofagus macrofossils of subgenus *Lophozonia* are abundant in all Oligocene deposits studied (Hill, 1991*b*), especially in the Pioneer, Lea River and Monpeelyata assemblages, where they dominate the entire angiosperm floras. The species described are considered to be intermediate between the two extant Australian species of this subgenus, with closer relationship to either *N. cunninghamii* or *N. moorei* (Hill, 1983*a*,*b*, 1984, 1990*a*, 1991*b*, Chapter 16, this volume). *Nothofagus cunninghamii* and *N. moorei* are restricted to, and dominate, microthermal rainforests of Tasmania–southern Victoria and the high altitude regions of northern New South Wales–southernmost Queensland, respectively.

The results of recent ecophysiological research help to explain the fact that now widely distributed *Nothofagus* subgenera were able to coexist in the past. Read & Hope (1989) found that the tropical species have a lower frost resistance than *N. moorei*, which in turn has a lower frost resistance than *N. cunninghamii*. In addition, Read (1990) found that the southern species have a photosynthetic tolerance of a wider range of temperatures around their optima than do the New Guinean species. This evidence supports the hypothesis proposed by Hill (1990*a*, 1991*b*) and Read *et al.* (1990) that species of the subgenus *Lophozonia* dominated the more exposed forest margin habitats, whereas those of subgenus *Brassospora* occurred in the forest proper, where temperatures were more constant. Deciduous species were most probably found on sites that received ample light in the growing season, but were in prolonged shade or darkness during winter. Read & Hill (1985) found that *N. gunnii* has a high light requirement for photosynthesis and a high light compensation point and dark respiration rate. A favourable site for deciduous *Nothofagus* spp. may therefore have been on the steep slopes near the lake at Cethana (Carpenter, 1991*a*).

The Cunoniaceae is well represented in the rainforests of Australasia, although at present (excluding *Bauera*) only the monotypic genus *Anodopetalum* is found in Tasmania. In the Oligocene the diversity of this family was much higher. For example, Carpenter & Buchanan (1993) recognised species from at least five extant genera from Cethana, and these have close affinity to species now represented in microthermal–mesothermal forests from New South Wales to New Guinea, and possibly South America. One of these fossil taxa has foliage indistinguishable from extant species of the closely related genera, *Weinmannia* and *Cunonia*. These genera are now absent from Australia, although the former is by far the largest and most widespread genus (over 130 species) in the family, being found in South and Central America, New Zealand, Pacific Islands, Malesia and Madagascar. *Cunonia* has a remarkable extant distribution, being found in South Africa and New Caledonia. Leaves and fruits of a *Weinmannia*-like plant also occur in the Lea River sediments. A cuticle-bearing leaf and an infructescence with seeds from Cethana are identical with those from the extant species *Callicoma serratifolia*. In addition, leaves of *Vesselowskya* and ?*Ceratopetalum* occur in this deposit, and flowers of *Acsmithia* and *Schizomeria*. Species of four of these genera (*Callicoma*, *Vesselowskya*, *Schizomeria* and *Ceratopetalum*) may now be found together in the rainforests of northeastern New South Wales, particularly in regions where the warm and cool temperate rainforest associations described by Baur (1957) intergrade on the poorer soil types. The presence of these taxa at Cethana may therefore be a reflection of local edaphic factors. *Callicoma serratifolia* is restricted to New South Wales and southern Queensland and *Vesselowskya* is confined to upland scarp regions of the New South Wales

Great Divide, usually in association with *Nothofagus moorei* in microthermal rainforest. *Nothofagus tasmanica*, one of the common taxa in the Cethana flora, has been shown by Hill (1983*b*) to be virtually identical in leaf morphology with *N. moorei*. Therefore, a distinct element of the Cethana flora can be regarded as having strong relationships with the *N. moorei* forests of regions such as the Dorrigo escarpment and Mt Banda Banda. These data therefore substantiate the prediction of Hill *et al.* (1988) that *Ceratopetalum apetalum* and *Vesselowskya rubifolia*, or their immediate ancestors, once occurred in Tasmania but were unable to adapt or evolve to the changing conditions in Tasmania from the mid-Tertiary onwards. These authors could have included *Callicoma serratifolia* with this group because it is frequently associated with *N. moorei*, although it does occur in other vegetation types. *Ceratopetalum*, *Schizomeria* and *Weinmannia* are often common elements of lower–upper montane rainforest in New Guinea (Johns, 1982) and the family is also well represented in northeastern Queensland, New Caledonia and Fiji.

At present Tasmania has three monotypic endemic genera of the Proteaceae, *Cenarrhenes*, *Agastachys* and *Bellendena*, as well as species from nine other, more widespread, genera. The family has a diverse representation at Cethana, where at least 20 species occur (Carpenter, 1991*a*). Apart from the unidentified and probably extinct taxa, these leaf fossils belong to at least three of the seven tribes of the subfamily Grevilleoideae as recognised by Johnson & Briggs (1975). The fossils exhibit a generally high level of scleromorphy and in some cases obvious xeromorphy, and some are also of great biogeographical interest. Modern species with affinity to these plants occur in a range of habitats from rainforests to sclerophyllous heathlands.

The tribe Embothrieae is represented at Cethana by up to four species of *Lomatia*. *Lomatia xeromorpha* has foliage almost identical with that of *L. tinctoria*, a Tasmanian endemic shrub typical of fire-prone dry sclerophyllous vegetation (Carpenter & Hill, 1988). Another undescribed species conforms closely to the north Queensland rainforest species *L. fraxinifolia*. The two other taxa exhibit an affinity to the genus, but not to any extant species. Both exhibit a reduced leaf form and one in particular has a markedly xeromorphic cuticle (Figure 12.3*d*). *Lomatia* is now found in a variety of habitats along the east coast of Australia and in southern South America. The subtribe Embothriinae is also represented by a leaf exhibiting strong macromorphological and cuticular similarity to leaves of *Telopea truncata*, which is now endemic to the wetter parts of Tasmania, particularly subalpine regions.

Another fossil leaf probably belongs to the subtribe Gevuininae of the tribe Macadamieae. This group is of great biogeographical interest, having closely related taxa in South America, New Caledonia, Vanuatu, Fiji, New Guinea and Queensland.

Six species have been assigned to the tribe Banksieae, and placed in *Banksieaephyllum* or *Banksieaeformis*, depending on their preservation. This number of species is as great as that which may now be found at locations in the centres of diversity of *Banksia*, in southwestern Australia and the Hawkesbury sandstone region of New South Wales. Only two species of *Banksia* are extant in Tasmania. The high diversity of *Banksia/ Dryandra*-like species at Cethana, including fossils with architectural similarity to that of species now confined to southwestern Australia (Figure 12.3*f*), strongly suggests that the high species numbers now found in the modern centres of diversity is a relictual situation, and not the result of recent speciation. Hopper (1979) discussed several reasons for the high level of endemism in the southwest of Western Australia and there is no doubt that the survival of relictual forms could be associated with the presence of ancient landscapes with deeply weathered soils, continuous relatively high rainfall and the lack of glaciations in the region.

Among extant species of *Banksia* and *Dryandra*, the common presence of stomata in deep depressions in the leaf is interpreted as an adaptation for limiting water loss (e.g. Hill, 1990*c*). None of the *Banksieaephyllum* species so far described from sediments older than Late Oligocene (Latrobe Valley) exhibits this feature (Blackburn & Sluiter,

Chapter 14, this volume), although many, including specimens from Cethana, have very sclerophyllous small leaves with hairy surfaces. These are also characters that can be termed xeromorphic. This may be interpreted as evidence for developing seasonal aridity at Cethana, or perhaps more likely (given the abundance of associated obligate rainforest taxa), the presence of variably drained sites imposing a variety of water stress levels.

Hill & Macphail (1983) illustrated a fruit from the Pioneer flora that they believed to be probably referrable to the tribe Banksieae. An associated species of *Banksieaephyllum*, *B. regularis* (Hill & Christophel, 1988) may have been produced by the same plants. These fruits are much smaller than, but morphologically very similar to, those of extant *Banksia*.

Microphyllous proteaceous leaves occur at Monpeelyata and were probably derived from sclerophyllous alpine shrubs (Macphail *et al.*, 1991). Their affinities have not been further investigated. However, their presence shows that plants with the sclerophyllous, xeromorphic leaf form, typical of modern subalpine Tasmanian vegetation, were growing at high altitude over 20 million years (Ma) ago.

No other angiosperm macrofossils have been formally described from Tasmanian Oligocene sediments, but they include taxa that are common in mainland Eocene floras (discussed in detail by Christophel, Chapter 11, this volume), and which are now found in a variety of mainland eastern Australian rainforest types. They include leaves with affinity to *Elaeocarpus* (Elaeocarpaceae), *Brachychiton* (Sterculiaceae), *Gymnostoma* (Casuarinaceae), Myrtaceae (Figure 12.3(*e*)) and Lauraceae (Figure 12.3(*g*)). In addition, microphyllous epacridaceous leaves of uncertain affinity have been recovered from the Little Rapid River and Monpeelyata floras (Hill, 1990*a*; Macphail *et al.*, 1991).

Physiognomy

Foliar physiognomic analysis has been discussed in detail by Greenwood (Chapter 4, this volume).

Tasmanian Oligocene–Early Miocene floras are dominated by microphylls and nanophylls of less than half the length of leaves in Early Eocene deposits (Figures 12.2(*a*) and (*b*)). Also, the high altitude sites of Lea River and Monpeelyata have much smaller leaves than near contemporaneous sites at Pioneer, Cethana and Little Rapid River. A high correlation ($r = 0.88$) (and presumably a causal relationship) exists between the mean annual temperature (MAT) at diverse extant Tasmanian and eastern Australian rainforest sites and the mean length of leaves obtained from litter collected from these sites (Figure 12.2(*b*)). The smallest leaves occur at high altitude in Tasmania and the largest at low altitude in northeastern Queensland. Leaves of larger intermediate lengths occur at high altitude in northeastern Queensland and shorter intermediate lengths at high altitude in New South Wales–southernmost Queensland and low altitude in Victoria and Tasmania. Greenwood (Chapter 4, this volume) discusses the taphonomic factors that generally bias a sample set of fossil leaves to smaller leaf sizes than those typical of the source vegetation. We have further reservations about the applicability of using fossil leaf lengths for determining precise MAT estimates of past depositional regions, and in particular the merit of discussing modern vegetational analogues of past environments (*sensu* Christophel & Greenwood, 1989). In particular, on the basis of floristic affinities, we are doubtful that floras such as that of Cethana are entirely composed of rainforest taxa, and there are several other unknowns, such as the influence of seasonal light regimes not now present in rainforests. Nevertheless, bearing in mind the bias toward smaller leaves in fossil floras, extant sites with leaves of lengths similar to those of the Lea River and Monpeelyata floras occur at high altitude in Tasmania, where the MAT is of the order of 7.5 °C. Sites with leaves of lengths similar to those of the lower altitude Oligocene sites now occur at low altitude in Tasmania and southern Victoria and high altitude in New South Wales, such as at Mt Banda Banda (MAT = 12 °C), a region that exhibits some floristic similarities to Cethana (see previous discussion).

It is well known that the leaves of angiosperms from modern lowland (mesothermal and megathermal) tropical rainforests characteristically have entire margins and often have drip-tips, whereas those of montane tropical and high latitude regions are usually serrate-margined and lack drip-tips. Tasmanian Oligocene–Early Miocene floras are composed mostly of leaves of the latter form. Therefore, several lines of evidence indicate that microthermal rainforests were typical in the region.

Greenwood (Chapter 4, this volume) has also reviewed the evidence for climatic interpretation provided by epiphyllous fungal germlings. All of the Tasmanian Tertiary floras so far studied have abundant Grade V forms (Lange, 1978) and although the proportions of the grades and the density of germlings vary somewhat from site to site, there seems little doubt that climates during this time were very wet and humid.

Miocene–Pliocene

There are no known Middle–Late Miocene or Pliocene macrofossil deposits in Tasmania. This is unfortunate, since palaeoclimatological records indicate that during the Miocene climatic seasonality similar to that experienced at present developed. The most geographically proximal Miocene floras occur in southern Victoria (e.g. Blackburn & Sluiter, Chapter 14, this volume). Taxa that probably became extinct in Tasmania during the Miocene–Pliocene, at least in part because of climatic factors, include many conifers, including Araucariaceae, some Cupressaceae and Taxodiaceae, and many Podocarpaceae spp. Apart from changes in climate, the effects of altered biotic and edaphic/physiographical factors should not be discounted when one attempts to determine the reasons for these plant extinctions. For instance, research into rainforest population dynamics indicates that extant species of taxa that formerly occurred in Tasmania have a disturbance-based ecology. Studies of *Weinmannia* spp. in New Zealand (Stewart, 1986) and Chile (Veblen *et al.*, 1981) indicate that regeneration of these species is largely dependent on the presence of major openings in the forest canopy caused by events such as earthquake-triggered landslides, volcanic eruptions and floods. Our observations in montane regions of New Guinea and Vanuatu suggest that other species of *Weinmannia* have similar ecological requirements. Also it is well known that the Andes, where the majority of *Weinmannia* spp. occur, is a region of great tectonic activity. In contrast, Tasmania (and the rest of Australia) is currently geologically quiescent, and lacks the types of large mountain that are prone to massive disturbance. Probably this has also been so for long periods during the Cenozoic. Some species of *Nothofagus* and certain conifers such as *Libocedrus* also develop even-aged stands in large canopy gaps caused by major disturbance (e.g. Veblen *et al.*, 1981; Veblen & Stewart, 1982). It is possible that such plants were outcompeted in the Tertiary because suitable habitat for their regeneration was formed too infrequently (Hill, 1987; Hill & Carpenter, 1989, 1991*b*; Hill, 1990*a*).

Pleistocene

Several Tasmanian Pleistocene deposits containing macrofossils are known (Figure 12.1), two of which are Early–Middle Pleistocene, the rest are all Late Pleistocene. They are all from the west and southwest of the state, the regions that now have the largest areas of rainforest. The macrofossil treatments of all these sites have been made in association with microfossil analyses.

Sediments at Regatta Point are Early–Middle Pleistocene (Hill & Macphail, 1993), and contain a rich and well-preserved flora. This flora has strong floristic similarity to extant lowland wet forest of western Tasmania (Hill & Macphail, 1985), but there are many taxa present that are now totally or regionally extinct (Jordan, 1992). These include an extinct podocarpaceous taxon probably allied to *Dacrydium* or *Dacrycarpus*. Almost all woody species of Tasmania's modern lowland rainforest appear in these sediments, with many taxa making their first macrofossil appearances, e.g. *Nothofagus cunninghamii, Lagarostrobos franklinii* (Figure 12.4(*b*), *Atherosperma moschatum, Cenarrhenes nitida, Trochocarpa* spp., *Agastachys odorata* and *Tasmannia lanceolata*. Sclerophyllous elements of the modern vegetation are also common in the sediments, mainly for the first time as macrofossils, e.g. *Cyathodes, Acacia*

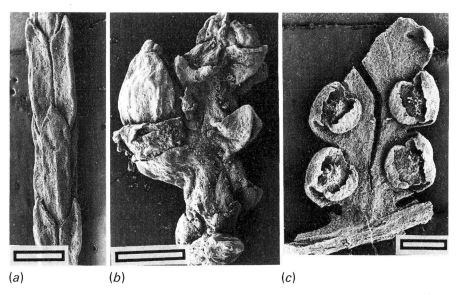

(*a*)   (*b*)   (*c*)

Figure 12.4 Scanning electron micrographs of fossils from the Regatta Point Pleistocene deposit. Scale bars represent 1 mm. (*a*) Section of stem of an extinct species of *Callitris*. (*b*) Female cone of *Lagarostrobos franklinii*. (*c*) Pinnule from a fertile frond of *Dicksonia antarctica*, with attached sori.

melanoxylon and *Richea. Eucalyptus* and *Dicksonia antarctica* (Figure 12.4(*c*)) also occur in these sediments.

The Regatta Point sediments, however, also contain elements now associated with both significantly warmer and significantly cooler climates. The warmer climate plants include taxa allied to species now occurring in the rainforests of New South Wales and further north, e.g. *Quintinia, Rubus* (aff. *moorei*) (Figure 12.5(*c*)), broad-leaved Lauraceae and Myrtoideae (not *Acmena*). Sclerophyllous species closely related to, and morphologically similar to, extant species from mainland Australia are also present, including *Banksia strahanensis* (related to *B. spinulosa*) and another species of *Banksia*. In contrast, *Microcachrys tetragona* is common in the sediments but is now restricted to subalpine and alpine areas of Tasmania, and other species present as fossils are now usually associated with subalpine habitats. A number of other extinct sclerophyllous species are

present in the sediments, but are more difficult to associate with modern vegetation types, notably species of *Callitris* (Figure 12.4(*a*)), *Acacia* and *Allocasuarina* (Figure 12.5(*a*)) and several species of the Proteaceae (e.g. Figure 12.5(*b*)). While the different elements of this curious flora do not all co-occur in any single stratum, anomalous combinations of types of taxon do occur, and it is clear that the vegetation at this time was distinct from any modern vegetation. Also individual species had ecological roles different from those of the present.

All subsequent Tasmanian macrofloras are composed almost entirely of species that now occur in western Tasmania, although there are indications of community structures that no longer occur in Tasmania. The Regency sediments are clearly more recent than Regatta Point (although they are both dated as Early – Middle Pleistocene), and contain essentially modern rainforest assemblages (Fitzsimons *et al.*, 1990),

(*a*)   (*b*)   (*c*)

Figure 12.5 Fossils of extinct species from the Regatta Point Pleistocene deposit. Scale bars represent 1 cm. (*a*) Cone of *Allocasuarina* sp. with affinity to extant *A. verticillata* from dry sclerophyllous woodlands in southeastern Australia. (*b*) Leaf of *Orites* sp. with affinity to extant *O. diversifolia* from Tasmania. (*c*) Leaf of *Rubus* sp. with affinity to extant *R. moorei* from Queensland. Note the many thorns present on the veins.

although some sclerophyllous elements, including *Eucalyptus*, occur as macrofossils (Jordan, 1992). Among the Late Pleistocene sites are Langdon River, which shows a relatively long sequence of mixed rainforest/sclerophyllous vegetation (Colhoun *et al.*, 1989), and Melaleuca Inlet, which shows the presence of sclerophyllous scrub and sedgeland/heath now typical of southwest Tasmania (Jordan *et al.*, 1991). Melaleuca Inlet is the only site postdating Regatta Point that contains macrofossils of a now extinct species, in this case *Banksia kingii* (Jordan & Hill, 1991).

DIVERSITY AND CHANGE IN THE TASMANIAN CENOZOIC

Apart from the fact that the overall diversity of past rainforests was higher than today, remarkably high numbers of genera per family and species per genus have been recovered from several Tasmanian fossil assemblages. This is of interest, because it relates to recent speculation and debate about the establishment and maintenance of high species diversity in modern ecosystems. Generalities regarding diversity in tropical forests have been widely discussed (e.g. Richards, 1952; Ashton, 1969; Connell, 1978; Gentry, 1988; Currie, 1991). There is evidence that a wet climate, which is stable over time and without diurnal or seasonal extremes, can be expected to produce a highly species rich ecosystem. Indeed, Gentry (1988) has compiled much data which indicate that (at least in the lowland neotropics) a direct relationship between annual precipitation and plant diversity exists, and that species richness in 0.1 ha plots reaches an asymptote of about 250 species of $\geqslant$ 2.5 cm diameter at breast height at about 4000 mm of annual precipitation. Currie & Paquin (1987) found that a relationship between diversity and actual evapotranspiration (AET), which is itself a function of solar radiation and precipitation, is more precise. Thus, angiosperm richness is strongly related to capturable energy, since AET is highly correlated with global patterns of terrestrial primary productivity.

Ashton (1969) argued that, on a large scale, regional seasonal and geological stability has led to selection for mutual avoidance, where great complexity and high productivity ensue as organisms become increasingly specialised to occupy increasingly narrow niches. Thus, although closely allied rainforest trees of the same genus may be sympatric, or even appear to occupy the same niche, hybrid survival is strongly selected against. Connell (1978) has stressed that competitive exclusion over time would be retarded where local moderate disturbance events in the forest lead to the creation of heterogeneous physical habitats. Hyndman & Menzies (1990) studied an area in the Ok Tedi headwaters of New Guinea that has a rich and diverse flora and fauna and appears to exemplify the type of region outlined above. Here the rainfall is continuously high, temperatures are mild, and there is much cloudiness and a high relative humidity. Further, the region is predisposed to periodic local disturbances in the form of landslips. These authors concluded that the heavy rainfall and persistent cloudiness acted to compress ecological zonation to produce extremely diversified biota in a comparatively small area. It may be no coincidence that Currie (1991) found that North American tree species diversity is richest in high energy, moist, mountainous areas of the southeast of that region. The study of Hyndman & Menzies (1990) is also significant in that the vegetation of their study site in New Guinea, and montane regions of the island in general, have some floristic similarity to that of the Paleogene in Tasmania.

The region of northern New South Wales and southernmost Queensland known as the Macpherson-Macleay Overlap has long been recognised for its important rainforested massifs (Burbidge, 1960). It includes environments suitable for each of microthermal, mesothermal and megathermal vegetation in close proximity, and also has quite a high and even distribution of annual rainfall (Nix, 1982). Hill & Macphail (1983), Hill *et al.* (1988) and Carpenter (1991*a*) have noted a floristic and physiognomic similarity of the *Nothofagus moorei* forests of this region to that of the Pioneer and Cethana floras. This suggests that a similar compressed ecological zonation occurred at these sites in the Tasmanian Oligocene–Early Miocene.

High species diversity has also been documented for nonforested regions. Of particular note are the sclerophyllous heathlands (kwongan vegetation) of southwestern Australia and the South African southwestern Cape flora. Linder (1991) found that total precipitation is the best predictor of species richness for the Cape flora, but soil and landform type appear to be more significant influences on kwongan (Lamont *et al.*, 1984). Both of these regions have tracts of undulating sandy soils and floras that feature diverse Proteaceae. The high diversity of these floras can also be attributed to the ancient and stable nature of the landform and the mosaic effects of fire history and microhabitat differences (e.g. Hopper 1979; Cowling, 1987). Also, Lamont *et al.* (1984) suggested that the extreme lack of nutrients and consequent reduced plant growth rates and sizes make it unlikely that any one species can dominate at the expense of others.

The botanical history of a region should not be considered in isolation of its faunal component. Indeed, the high diversity of plant forms and taxa in tropical ecosystems exists in interaction with a proportionate multitude of animal species, particularly insects. Studies have also highlighted remarkable examples of coevolution, where, for example, a single plant species may rely upon a single animal species for pollination of its flowers. Although Australian Tertiary faunas are little known, recent palaeontological work on the Late Oligocene–Early Miocene Riversleigh fauna of north Queensland demonstrates the presence of a much higher diversity of marsupials (often at the subgeneric level) than is found anywhere today, in what was a lowland rainforest (Archer *et al.*, Chapter 6, this volume).

To what degree these hypotheses and observations are relevant to the Tasmanian Paleogene is uncertain, particularly since it has been demonstrated that plant communities were different from those of today. However, as discussed above, there is independent evidence that the climate at the time was everwet and there would have been a broad range of niches in topographically diverse, high latitude regions. Importantly, the incidence of solar energy and evapotranspiration would have

been relatively very high in the summer months, and in turn so would the primary productivity (see Chaloner & Creber, 1989). The degree of the limiting effect of the winter period on species diversity and the potential effects of the low sun angle on vegetation structure are unknown.

Although there is no doubt that mainland Paleogene floras were also highly diverse (Christophel, Chapter 11, this volume), there is no evidence of high diversity in any one genus, given that the 12 species of *Laurophyllum* described by Hill (1986) from Nerriga probably come from several natural genera. In fact, the genera that show great diversity (three or more species per deposit) in Tasmanian Paleogene deposits – *Banksieaephyllum/ Banksieaeformis*, *Lomatia* and *Nothofagus* at Cethana, *Nothofagus* and *Dacrycarpus* at Little Rapid River, *Dacrycarpus* and *Araucaria* at Monpeelyata and *Araucaria* at Regatta Point – are at most represented by one species in any single mainland deposit, with the exception of *Banksieaephyllum* in the spatially and temporally extensive Latrobe Valley coal floras in Victoria (Blackburn & Sluiter, Chapter 14, this volume). How and why did the high sympatric diversity of these particular taxa become established in Tasmania? No doubt, aspects of the environmental factors that appear to enable the development of high species diversity in modern ecosystems were significant, as discussed previously. More particularly, there is evidence that there was a radiation of certain taxa (especially *Nothofagus*) by the Oligocene. In accordance with Hill (1990*a*) we hypothesise that at high latitudes many elements of the mesothermal rainforests of the Eocene were led to extinction or were forced to migrate northwards or to lower altitudes when the relatively sudden cooling event occurred near the Eocene/Oligocene boundary. Subsequent vacant niches were filled by taxa derived from a pool with a preference for microthermal environments. The source of these pool taxa can only have been from cooler higher latitudes and/or altitudes. If Australia (Tasmania) had already separated from Antarctica (South Tasman Rise) (Wilford & Brown, Chapter 2, this volume), then this implies that these taxa were growing in the Tasmanian uplands, and/or were those pre-adapted to cooler conditions by virtue of their sclerophyllous leaf form, having evolved in low nutrient conditions. This hypothesis explains the apparent proliferation of sclerophyllous taxa at Cethana in particular, where there is evidence of a Precambrian substrate that would have been highly weathered. It would be interesting to test this hypothesis by examining a high altitude flora contemporaneous with the Regatta Point and other Early Eocene floras.

Holistic palaeoecological interpretation of past ecosystems (incorporating plant and animal records) is an exciting field for research in the future, and clearly depends upon increased knowledge of the taxonomic make-up of more fossil assemblages. Palaeobotanists generally seek explanations for plant evolution in terms of climate and landform change alone, but perhaps biotic factors should be given more emphasis (Wing & Tiffney, 1987). Already, there are questions to be answered regarding flowering and fruiting phenology, the modes of pollination and dispersal of propagules, and the nature and degree of herbivory. Knoll (1984) and Traverse (1988) argued convincingly that throughout time plant taxa in general are much less subject to the type of short-term catastrophic events that result in the extinction of animals. Certainly, our studies show that many plant species have persisted in the Australasian region since the Paleogene with little or no apparent change. Is the same true of the fauna that interacts with them? Unfortunately the Paleogene record of the most relevant animal groups (birds, insects and mammals) in Australia is very poor (see e.g. Archer et al., Chapter 6, this volume). However, there has almost certainly been a marsupial fauna since the Cretaceous, although its Cenozoic record is one of numerous extinctions. The coevolutionary relationship between small nectar-feeding pollinator species such as the honey possum *Tarsipes spenserae* of Western Australia and certain flowering plants is hypothesised to have dated from the Cretaceous (Sussman & Raven, 1978). Keast (1972) has suggested that it may be significant that *T. spenserae* occurs in the kwongan region, where there is a deficiency of nectar-feeding birds

compared with similar habitats in eastern Australia, yet it is exceptionally rich in flowering shrubs.

Flower-visiting among avian families is believed to have become widespread only in the Miocene (Sussmann & Raven, 1978; Armstrong, 1979). In Australia, although there is considerable doubt, the radiation of the honeyeaters, which are now characteristic pollinators of flowers of such sclerophyllous genera as *Eucalyptus*, *Banksia*, *Grevillea* and *Melaleuca*, may not have occurred until this time (see Paton, 1986). Frugivorous bats are also important pollinators of a variety of native plants from Gondwanic families in northern and eastern Australia (Armstrong, 1979). However, the impressive fossil record of bats from Riversleigh is entirely one of insectivorous taxa, and the larger fruit-eating species almost certainly arrived from Southeast Asia (Archer et al., Chapter 6, this volume). Of course, the other principal pollinators of flowers throughout the Cenozoic must have been insects. To what extent did changing insect, mammal and bird assemblages affect the make-up of vegetation communities?

Perhaps the high plant diversity was in fact a reflection of reduced niche breadth, caused by increased competition for pollinators during the growing, flowering and fruiting season. For instance, are the completely fertilised *Banksia*-like cones at Pioneer (Hill & Macphail, 1983) a reflection of a very active, diverse pollinator suite, or perhaps the result of a highly efficient, specific coevolutionary relationship? It is possible that the difficulty in explaining the low fruit : flower ratios commonly observed in the Proteaceae (e.g. Collins & Rebelo, 1987; Walker & Whelan, 1991) can be attributed to this being a relatively recent phenomenon associated with the extinction of certain pollinators, and an evolutionary lag exists.

How were ecosystems maintained during the dark winters, when productivity must have nearly ceased? If there were many pollinating birds present during the Paleogene in the Tasmanian region, then most presumably migrated away from the high latitude winters. In fact, it may not have been efficient for them to act as pollinators in these regions during this time. On the other hand, their radiation may have been associated with the cooling event around the Eocene/Oligocene boundary, as Stiles (1978) noted that cold, rainy weather inhibits the activity of most insects, whereas birds, being homeothermic, can continue foraging and Paton (1986) presented data which demonstrates that normally insectivorous birds resort to nectar-feeding during cold weather. In turn, the success of honeyeaters may have been related to the extinction of nectivorous marsupials in eastern Australia, and many coevolved plant species. Other species were outcompeted as the birds tended to facilitate the regeneration of the plants with the most suitable inflorescences.

There is evidence that atmospheric CO_2 concentrations have changed during time. If levels in the Paleogene were significantly higher than those of today, then it is possible that plants could sacrifice the relative amount of area open to the environment (through reduced stomatal density and/or pore size), with the concomitant benefit of reducing water loss. There are numerous taxa from Cethana with markedly xeromorphic cuticles. It is possible that a diverse range of species could exploit the marginal (?intermittently arid) environments at this site because they were able to protect their stomata without encountering a problem in obtaining sufficient CO_2.

The Regatta Point sediments show the persistence into the Early Pleistocene of some elements of the highly diverse Tertiary rainforests. The Regency and Regatta Point floras make a good comparison because they are both very well preserved, represent moderately long time periods, are spatially similar, and contain both sclerophyllous and rainforest components. The presence of numerous extinct species in the Regatta Point sediments and no macrofossils of extinct species in the Regency sediments therefore suggest that many extinctions occurred during the Early and Middle Pleistocene, and that relatively few have occurred since. The major fluctuations in climate typical of the Pleistocene started to develop in the Pliocene, and it is likely that the final decline of the diverse rainforest floras of the Tertiary happened during the Late Pliocene and finally in the Early Pleistocene due to these climatic changes. It is possible

that it was seasonal dryness, rather than the colder climate during the coolest times, that directly or indirectly caused the extinction of the 'Tertiary' species, since some of the taxa involved have relatives occurring relatively far south in the cold, more everwet parts of New Zealand. Also, the Regatta Point sediments were laid down after several of these climatic perturbations, and after times that were considerably cooler than at present. The vegetation that contributed to the fossil flora must somehow have survived these severe conditions. The decline of the Tertiary vegetation was probably stepwise, in response to each major climatic fluctuation until the mid-Pleistocene, with the species remaining in the Tasmanian region being those that could survive cold and dry periods. The current distributions of plant species thus may not represent the limits of their potential distribution, or indeed their preferred habitats.

It is likely that some groups typical of sclerophyllous vegetation also declined. Two species of *Banksia* and several other species of the Proteaceae that occur in the Regatta Point sediments are now extinct. There is now only one species of *Banksia* in southern and western Tasmania, *B. marginata*. Its habitat range is much smaller on mainland Australia than in Tasmania, even though it is widely distributed. Other species of *Banksia* occur in habitats similar to those occupied by *B. marginata* in Tasmania, e.g. *B. robur* and *B. ericifolia* in dystrophic swamps, *B. aemula* and *B. serrata* in coastal woodlands, *B. integrifolia* var. *integrifolia* on the coast and *B. spinulosa* in cool, wet woodlands. It is likely that the range of *B. marginata* has expanded to fill niches left unoccupied by the extinction of other species of *Banksia* (or possibly other sclerophyllous species). An alternative hypothesis, that there is some unknown factor associated with warmer climates that has created more niches for *Banksia* species on mainland Australia can be discounted because two extinct species of *Banksia* grew at Regatta Point well after periods of cooler climate than at present and *Banksia* spp. are relatively poorly dispersed. Other species have similarly wide habitat ranges in Tasmania, and much more restricted ecological but not geographical ranges on mainland Australia, e.g. *Grevillea australis*, *Leptospermum scoparium* and *Acacia melanoxylon*. Each of these has many related species on the mainland, which occupy a wide range of niches. Two *Acacia* spp. and several *Leptospermum*-like spp. occur at Regatta Point. Overall, this evidence indicates that there has been considerable extinction of sclerophyllous taxa from Tasmania, although there may have been recent speciation of some groups, e.g. *Eucalyptus*, *Acacia* and some Asteraceae, to take up some of the 'ecological slack' left by extinctions.

The extinct species that occur in the Regatta Point sediments are likely to show a decline in diversity, rather than natural turnover in species, since almost all modern Tasmanian rainforest taxa are found in the sediments, and the closest modern Tasmanian relatives of many of the extinct sclerophyllous taxa have distributions that suggest that they are likely to be relatively old (e.g. *Banksia marginata* is poorly dispersed but is widely distributed in south-eastern Australia).

Even though the species present in the Pleistocene sediments after the time of Regatta Point are extant, the communities present were probably quite different from modern ones. *Allocasuarina* spp. are dominant, or at least very prominent, often in otherwise rainforest floras, in virtually all the Pleistocene macrofossil floras. This strongly suggests that *Allocasuarina* spp. and rainforest species formed mixed forest communities, much in the way that *Eucalyptus* spp. now form mixed forest communities with rainforest. Such communities are unknown at present, but a more dominant role of the Casuarinaceae has been long assumed for much of Australia before the increase in fire frequency that occurred about 40 thousand years ago (see Hope, Chapter 15, this volume).

CONCLUSION

Obtaining a complete picture of the changes in the vegetation of a region is limited by the number, age and quality of fossil deposits that are available. However, several features of the floristic and community make-up of the vegetation of the Tasmanian region through time are clear. In particular, microthermal rainforests with *Nothofagus*

and conifers as dominant elements have been growing in the region for at least 35 Ma (i.e. since the Early Oligocene). However, these forests have relatively recently lost the bulk of their floristic diversity, probably as a consequence of climatic changes that resulted in greater seasonal and diurnal variation in rainfall and temperature. An interesting phenomenon evident from the macrofossil record is the presence of now extinct or unfamiliar community types in the past, and this supports the notion that plant species often exhibit quite individualistic responses to climate change (Jablonski, 1991).

Prior to the Late Eocene, the nature of the vegetation is more obscure, since it contained many elements that have no modern descendants, and grew under conditions totally unlike those found at very high latitudes today.

There is a meagre Tertiary macrofossil record for the typical coastal sclerophyllous heathland and eucalypt-dominated woodlands containing abundant and diverse Myrtaceae, Epacridaceae and legumes which are particularly common in the relatively dry eastern and nutrient-deficient wet southwestern regions of Tasmania today. However, this can be attributed partly to taphonomic factors, and the apparent spread of these vegetation types during the Quaternary, when fire frequencies increased (see Hope, Chapter 15, this volume). Nevertheless, there is some evidence for these floral elements, and possibly a similar structural vegetation type. An oligotrophic edaphic complex similar to that which is found in the modern Cethana region probably occurred there in the Paleogene. Today it supports a stunted sclerophyllous flora including *Eucalyptus*, various epacrids and legumes, *Lomatia tinctoria*, *Banksia*, *Gleichenia* and *Sticherus*. Macrofossils with affinity to the latter four taxa have been recovered from the deposit.

The modern alpine and subalpine vegetation of Tasmania typically contains small-leaved sclerophyllous shrubs or small trees of such taxa as Taxodiaceae, Myrtaceae, Proteaceae, Epacridaceae, *Microstrobos*, *Microcachrys* and *Nothofagus gunnii*. The fern ally *Isoetes* is common in the numerous small lakes of the high country. Macrofossils of most of these taxa occur in the high altitude Monpeelyata site, indicating that a similar vegetation has extended across the Tasmanian mountains for at least 22 Ma.

An important aim of macrofossil research in Tasmania will concentrate on identifying the numerous angiosperm leaves that occur at sites such as Regatta Point, Little Rapid River and Cethana. Leaf physiognomy and the ecophysiological information that can be elicited from plant cuticles and epiphyllous fungi will yield further clues as to the nature of climate and climate change through time.

REFERENCES

ARMSTRONG, J. A. (1979). Biotic pollination mechanisms in the Australian flora: a review. *New Zealand Journal of Botany*, 17, 467–508.

ASHTON, P. S. (1969). Speciation among tropical forest trees: some deductions in the light of recent evidence. *Biological Journal of the Linnean Society*, 1, 155–96.

BAUR, G. N. (1957). Nature and distribution of rainforests in New South Wales. *Australian Journal of Botany*, 5, 190–233.

BIGWOOD, A. J. & HILL, R. S. (1985). Tertiary araucarian macrofossils from Tasmania. *Australian Journal of Botany*, 33, 645–56.

BURBIDGE, N. T. (1960). The phytogeography of the Australian region. *Australian Journal of Botany*, 8, 75–211.

BUSBY, J. R. (1988). Potential impacts of climate change on Australia's flora and fauna. In *Greenhouse: Planning for Climate Change*, ed. G. I. Pearman, pp. 387–98. Melbourne: CSIRO.

CARPENTER, R. J. (1988). Early Tertiary *Tmesipteris* (Psilotaceae) macrofossil from Tasmania. *Australian Systematic Botany*, 1, 171–6.

CARPENTER, R. J. (1991a). Palaeovegetation and environment at Cethana, Tasmania. Ph.D. thesis, University of Tasmania.

CARPENTER, R. J. (1991b). *Macrozamia* from the Early Tertiary of Tasmania and a study of the cuticles of extant species. *Australian Systematic Botany*, 4, 433–44.

CARPENTER, R. J. & BUCHANAN, A. M. (1993). Oligocene leaves, fruit and flowers of the Cunoniaceae from Cethana, Tasmania. *Australian Systematic Botany*, 6, 91–109.

CARPENTER, R. J. & HILL, R. S. (1988). Early Tertiary *Lomatia* (Proteaceae) macrofossils from Tasmania, Australia. *Review of Palaeobotany and Palynology*, 56, 141–50.

CHALONER, W. G. & CREBER, G. T. (1989). The phenomenon of forest growth in Antarctica: a review. In *Origins and Evolution of the Antarctic Biota*, ed. J. A. Crame. *Geological Society of London Special Publication*, 47, 85–8.

CHRISTOPHEL, D. C. & GREENWOOD, D. R. (1989). Changes in climate and vegetation in Australia during the Tertiary. *Review of Palaeobotany and Palynology*, 58, 95–109.

COLHOUN, E. A. (1986). Field problems of radiocarbon dating in Tasmania. *Papers and Proceedings of the Royal Society of Tasmania*, 120, 1–6.

COLHOUN, E. A., VAN DER GEER, G., HILL, R. S. & BIRD, T. (1989). Interglacial pollen and plant macrofossils from Langdon River, western Tasmania. *New Phytologist*, 111, 531–48.

COLLINS, B. G. & REBELO, T. (1987). Pollination biology of the Proteaceae in Australia and southern Africa. *Australian Journal of Ecology*, 12, 387–421.

COLLINSON, M. E. & HOOKER, J. J. (1987). Vegetational and mammalian faunal changes in the Early Tertiary of southern England. In *The Origins of the Angiosperms and Their Biological Consequences*, ed. E. M. Friis, W. G. Chaloner & P. R. Crane, pp. 259–303. Cambridge: Cambridge University Press.

CONNELL, J. H. (1978). Diversity in tropical rainforests and tropical reefs. *Science*, 199, 1302–10.

COWLING, R. M. (1987). Fire and its role in the coexistence and speciation of Gondwanan shrublands. *South African Journal of Science*, 83, 106–12.

CURRIE, D. J. (1991). Energy and large-scale patterns of animal- and plant-species richness. *American Naturalist*, 137, 27–49.

CURRIE, D. J. & PAQUIN, V. (1987). Large-scale biogeographical patterns of species richness in trees. *Nature*, 329, 326–7.

DARWIN, C. (1860). *Journal of Researches into the Natural history and Geology of the Countries visited during the Voyage of HMS Beagle*. London: J. Murray.

DUNCAN, B. D. & ISAAC, G. (1986). *Ferns and Allied Plants of Victoria, Tasmania and South Australia*. Melbourne: Melbourne University Press.

ETTINGSHAUSEN, C. VON (1888). Contributions to the Tertiary flora of Australia. *Memoirs of the Geological Survey of New South Wales, Palaeontology*, 2, 1–189.

FITZSIMONS, S. J. & COLHOUN, E. A. (1991). Pleistocene glaciation of the King Valley, western Tasmania, Australia. *Quaternary Research*, 36, 135–56.

FITZSIMONS, S. J., COLHOUN, E. A., VAN DER GEER, G. & HILL, R. S. (1990). Definition and character of the Regency Interglacial and early–middle Pleistocene stratigraphy in the King Valley, western Tasmania, Australia. *Boreas*, 19, 1–15.

GENTRY, A. H. (1988). Changes in plant community diversity and floristic composition on environmental and geographical gradients. *Annals of the Missouri Botanical Garden*, 75, 1–34.

HILL, R. S. (1978). Two new species of *Bowenia* Hook. ex Hook. f. from the Eocene of eastern Australia. *Australian Journal of Botany*, 26, 837–46.

HILL, R. S. (1980). Three new Eocene cycads from eastern Australia. *Australian Journal of Botany*, 28, 105–22.

HILL, R. S. (1983a). *Nothofagus* macrofossils from the Tertiary of Tasmania. *Alcheringa*, 7, 169–83.

HILL, R. S. (1983b). Evolution of *Nothofagus cunninghamii* and its relationship to *N. moorei* as inferred from Tasmanian macrofossils. *Australian Journal of Botany*, 31, 453–65.

HILL, R. S. (1984). Tertiary *Nothofagus* macrofossils from Cethana, Tasmania. *Alcheringa*, 8, 81–6.

HILL, R. S. (1986). Lauraceous leaves from the Eocene of Nerriga, New South Wales. *Alcheringa*, 10, 327–51.

HILL, R. S. (1987). Discovery of *Nothofagus* fruits corresponding to an important Tertiary pollen type. *Nature*, 327, 56–8.

HILL, R. S. (1988). Tertiary *Isoetes* from Tasmania. *Alcheringa*, 12, 157–62.

HILL, R. S. (1989a). New species of *Phyllocladus* (Podocarpaceae) macrofossils from south eastern Australia. *Alcheringa*, 13, 193–208.

HILL, R. S. (1989b). Fossil leaf. In *Antarctic Cenozoic History from the CIROS-1 Drillhole, McMurdo Sound*, ed. P. J. Barrett. *DSIR Bulletin*, 245, 143–4.

HILL, R. S. (1990a). Evolution of the modern high latitude southern hemisphere flora: evidence from the Australian macrofossil record. In *Proceedings of the Third IOP Conference*, ed. J. G. Douglas & D. C. Christophel, pp. 31–42. Melbourne: A – Z Printers.

HILL, R. S. (1990b). *Araucaria* (Araucariaceae) species from Australian Tertiary sediments – a micromorphological study. *Australian Systematic Botany*, 3, 203–20.

HILL, R. S. (1990c). Tertiary Proteaceae in Australia: a re-investigation of *Banksia adunca* and *Dryandra urniformis*. *Proceedings of the Royal Society of Victoria*, 102, 23–8.

HILL, R. S. (1991a). *Eucryphia* (Eucryphiaceae) leaves from Tertiary sediments in southeastern Australia. *Australian Systematic Botany*, 4, 481–97.

HILL, R. S. (1991b). Tertiary *Nothofagus* (Fagaceae) macrofossils from Tasmania and Antarctica and their bearing on the evolution of the genus. *Botanical Journal of the Linnean Society*, 105, 73–112.

HILL, R. S. & BIGWOOD, A. J. (1987). Tertiary gymnosperms from Tasmania: Araucariaceae. *Alcheringa*, 11, 325–35.

HILL, R. S. & CARPENTER, R. J. (1989). Tertiary gymnosperms from Tasmania: Cupressaceae. *Alcheringa*, 13, 89–102.

HILL, R. S. & CARPENTER, R. J. (1991a). Evolution of *Acmopyle* and *Dacrycarpus* (Podocarpaceae) foliage as inferred from macrofossils in south-eastern Australia. *Australian Systematic Botany*, 4, 449–79.

HILL, R. S. & CARPENTER, R. J. (1991b). Extensive past distributions for major Gondwanic floral elements: macrofossil evidence. *Papers and Proceedings of the Royal Society of Tasmania*, 125, 239–47.

HILL, R. S. & CHRISTOPHEL, D. C. (1988). Tertiary leaves of the tribe Banksieae (Proteaceae) from southeastern Australia. *Botanical Journal of the Linnean Society*, 97, 205–27.

HILL, R. S., JORDAN, G. J. & CARPENTER, R. J. (1993). Taxodiaceous macrofossils from Tertiary and Quaternary sediments in Tasmania. *Australian Systematic Botany*, 6, 237–45.

HILL, R. S. & MACPHAIL, M. K. (1983). Reconstruction of the Oligocene vegetation at Pioneer, north-east Tasmania. *Alcheringa*, 7, 281–99.

HILL, R. S. & MACPHAIL, M. K. (1985). A fossil flora from rafted Plio-Pleistocene mudstones at Regatta Point, western Tasmania. *Australian Journal of Botany*, 33, 497–517.

HILL, R. S. & MACPHAIL, M. K. (1994). Tertiary history and origins of the flora and vegetation. In *Vegetation of Tasmania*, ed. J. B. Reid, R. S. Hill & M. J. Brown. Hobart: Government Printer, in press.

HILL, R. S. & POLE, M. S. (1992). Leaf and shoot morphology of extant *Afrocarpus*, *Nageia* and *Retrophyllum* (Podocarpaceae) species, and species with similar leaf arrangement from Tertiary sediments in Australasia. *Australian Systematic Botany*, 5, 337–58.

HILL, R. S. & READ, J. (1991). A revised infrageneric classification of *Nothofagus* (Fagaceae). *Botanical Journal of the Linnean Society*, 105, 37–72.

HILL, R. S., READ, J. & BUSBY, J. R. (1988). The temperature-dependence of photosynthesis of some Australian temperate rainforest trees and its biogeographical significance. *Journal of Biogeography*, 15, 431–49.

HOPPER, S. D. (1979). Biogeographical aspects of speciation in the southwest Australian flora. *Annual Review of Ecology and Systematics*, 10, 399–422.

HUTCHINSON, M. F. & BISCHOFF, R. J. (1983). A new method for estimating the spatial distribution of mean seasonal and annual rainfall applied to the Hunter Valley, New South Wales. *Australian Meteorological Magazine*, 31, 179–84.

HYLAND, B. P. M. (1989). A revision of Lauraceae in Australia (excluding *Cassytha*). *Australian Systematic Botany*, 2, 135–367.

HYNDMAN, D. C. & MENZIES, J. I. (1990). Rain forests of the Ok Tedi headwaters, New Guinea: an ecological analysis. *Journal of Biogeography*, 17, 241–73.

JABLONSKI, D. (1991). Extinctions: a paleontological perspective. *Science*, 253, 754–7.

JOHNS, R. J. (1982). Plant zonation. In *Biogeography and Ecology of New Guinea*, vol. I, ed. J. L. Gressitt, pp. 309–30. The Hague: W. Junk.

JOHNSON, L. A. S. & BRIGGS, B. G. (1975). On the Proteaceae – the evolution and classification of a southern family. *Botanical Journal of the Linnean Society*, 77, 83–182.

JOHNSTON, R. M. (1885). Descriptions of new species of fossil leaves from the Tertiary deposits of Mount Bischoff belonging to the genera *Eucalyptus, Laurus, Quercus, Cycadites* etc. *Papers and Proceedings of the Royal Society of Tasmania*, pp. 322–5.

JOHNSTON, R. M. (1888). *Systematic Account of the Geology of Tasmania*. Hobart: Government Printer.

JORDAN, G. J. (1992). Macrofossil evidence for Quaternary plant extinction and vegetation change in Western Tasmania. Ph.D. thesis, University of Tasmania.

JORDAN, G. J., CARPENTER, R. J. & HILL, R. S. (1991). Late Pleistocene vegetation and climate near Melaleuca Inlet, south-western Tasmania. *Australian Journal of Botany*, 39, 315–33.

JORDAN, G. J. & HILL, R. S. (1991). Two new *Banksia* species from Pleistocene sediments in western Tasmania. *Australian Systematic Botany*, 4, 499–511.

KEAST, A. (1972). Australian mammals: zoogeography and evolution. In *Evolution, Mammals, and Southern Continents*, ed. A. Keast, F. C. Erk & B. Glass, pp. 195–246. Albany: State University of New York Press.

KNOLL, A. H. (1984). Patterns of extinction in the fossil record of vascular plants. In *Extinctions*, ed. M. H. Nitecki, pp. 21–68. Chicago: University of Chicago Press.

LAMONT, B. B., HOPKINS, A. J. M. & HNATIUK, R. J. (1984). The flora – composition, diversity and origins. In *Kwongan, Plant Life of the Sandplain*, ed J. S. Pate & J. S. Beard, pp. 27–50. Nedlands: University of Western Australia Press.

LANGE, R. T. (1978). Southern Australian Tertiary epiphyllous fungi, modern equivalents in the Australasian region, and habitat indicator value. *Canadian Journal of Botany*, 56, 532–41.

LINDER, H. P. (1991). Environmental correlates of patterns of species richness in the south-western Cape Province of South Africa. *Journal of Biogeography*, 18, 509–18.

MACPHAIL, M. K., HILL, R. S., FORSYTH, S. M. & WELLS, P. M. (1991). A Late Oligocene–Early Miocene cool climate flora in Tasmania. *Alcheringa*, 15, 87–106.

NIX, H. A. (1982). Environmental determinants of biogeography and evolution in Terra Australis. In *Evolution of the Flora and Fauna of Arid Australia*, ed. W. R. Barker & P. J. M. Greenslade, pp. 47–66. Frewville: Peacock Publications.

PATON, D. C. (1986). Evolution of bird pollination in Australia. In *The Dynamic Partnership. Birds and Plants in Southern Australia*, ed. H. A. Ford & D. C. Paton, pp. 32–41. Adelaide: Government Printer.

POLE, M. S. (1992). Eocene vegetation from Hasties, north-eastern Tasmania. *Australian Systematic Botany*, 5, 431–75.

READ, J. (1990). Some effects of acclimation temperature on net photosynthesis in some tropical and extra-tropical Australasian *Nothofagus* species. *Journal of Ecology*, 78, 100–12.

READ, J. & HILL, R. S. (1985). Photosynthetic responses to light of Australian and Chilean *Nothofagus* species and some co-occurring canopy species, and their importance in understanding rainforest dynamics. *New Phytologist*, 101, 731–42.

READ, J. & HOPE, G. S. (1989). Foliar frost resistance of some evergreen tropical and extra-tropical Australasian *Nothofagus* species. *Australian Journal of Botany*, 37, 361–73.

READ, J., HOPE, G. S. & HILL, R. S. (1990). Integrating historical and ecophysiological studies in *Nothofagus* to examine the factors shaping the development of cool rainforest in southeastern Australia. In *Proceedings of the Third IOP Conference*, ed. J. G. Douglas & D. C. Christophel, pp. 97–106. Melbourne: A – Z Printers.

RICHARDS, P. W. (1952). *The Tropical Rainforest: An Ecological Study*. Cambridge: Cambridge University Press.

SELLING, O. H. (1950). Some Tertiary plants from Australia. *Svensk Botanisk Tidjskrift*, 44, 551–61.

STEWART, G. H. (1986). Forest dynamics and disturbance in a beech/hardwood forest, Fjordland, New Zealand. *Vegetatio*, 68, 115–26.

STILES, F. G. (1978). Ecological and evolutionary implications of bird pollination. *American Zoologist*, 18, 715–27.

STOVER, L. E. & PARTRIDGE, A. D. (1973). Tertiary and late Cretaceous spores and pollen from the Gippsland Basin, southeastern Australia. *Proceedings of the Royal Society of Victoria*, 85, 237–86.

SUSSMAN, R. W. & RAVEN, P. H. (1978). Pollination by lemurs and marsupials: an archaic co-evolutionary system. *Science*, 200, 731–4.

TOWNROW, J. A. (1965a). Notes on Tasmanian pines. I. Some Lower Tertiary podocarps. *Papers and Proceedings of the Royal Society of Tasmania*, 99, 87–107.

TOWNROW, J. A. (1965b). Notes on Tasmanian pines. II. *Athrotaxis* from the Lower Tertiary. *Papers and Proceedings of the Royal Society of Tasmania*, 99, 109–13.

TRAVERSE, A. (1988). Plant evolution dances to a different beat. *Historical Biology*, 1, 277–301.

VEBLEN, T. T., DONOSO, Z. C., SCHLEGEL, F. M. & ESCOBAR, R. B. (1981). Forest dynamics in south-central Chile. *Journal of Biogeography*, 8, 211–47.

VEBLEN, T. T. & STEWART, G. H. (1982). On the conifer regeneration gap in New Zealand: the dynamics of *Libocedrus bidwillii* stands on the South Island. *Journal of Ecology*, 70, 413–36.

WALKER, B. A. & WHELAN, R. J. (1991). Can andromonoecy explain low fruit : flower ratios in the Proteaceae? *Biological Journal of the Linnean Society*, 44, 41–6.

WELLS, P. M. & HILL, R. S. (1989). Fossil imbricate-leaved Podocarpaceae from Tertiary sediments in Tasmania. *Australian Systematic Botany*, 2, 387–423.

WING, S. L. & TIFFNEY, B. H. (1987). Interactions of angiosperms and herbivorous tetrapods through time. In *The Origins of Angiosperms and Their Biological Consequences*, ed. E. M. Friis, W. G. Chaloner & P. R. Crane, pp. 203–24. Cambridge: Cambridge University Press.

WRIGHT, D. H. (1983). Species-energy theory: an extension of species-area theory. *Oikos*, 41, 496–506.

13 The Neogene: a period of transition

A. P. KERSHAW, H. A. MARTIN & J. R. C. McEWEN MASON

The Neogene was a time of transition both in the development of the present vegetation and the palynological study of it. The vegetation cover changed from one dominated by rainforest, which is traditionally regarded as 'Tertiary', to one in which rainforest became very reduced in extent. The nature of this change has been difficult to document due to an increasingly arid landscape with a concomitant reduction in suitable pollen preservation sites. The difficulty has been compounded by a relative lack of palynological study on the period. Stratigraphic palynologists have focussed on the earlier part of the Tertiary and there is no formal or well dated biostratigraphy, for much of the period under consideration, that is applicable to Australian terrestrial environments. Palynologists concerned with vegetation reconstruction have largely restricted their attention to the later part of the Quaternary period and have had variable success when venturing back into the Tertiary, as the vegetation then was frequently very different from that of today. Consequently, the database from which we piece together this critical period in Australia's vegetation history is very fragmentary and of varying quality.

In keeping with the problematic documentation of vegetation, there are difficulties in defining the period itself. There is general agreement on its beginning – the Miocene began about 25 million years (Ma) ago, although this does not necessarily hold any palynostratigraphic or biogeographical significance – but there are different views on the best location of the end of the period, i.e. the Pliocene/Pleistocene boundary. Conventionally this boundary is placed at the top of the Olduvai palaeomagnetic event dated to 1.6 Ma (Berggren et al., 1985) but there is increasing pressure to reposition this close to the Gauss/Matuyama palaeomagnetic reversal boundary, around 2.4 Ma, as this reflects more closely the beginning of the substantial cooling and climatic fluctuations that characterise the Pleistocene period (Zagwin, 1985; Kukla, 1989).

PALAEOENVIRONMENTAL SETTING

Major factors influencing the whole of Australia during the Neogene include global climatic changes and the northward movement of the continent. The build up of ice on Antarctica, partly a result of the northward movement of Australia, which allowed the development of a circum-Antarctic ocean current, caused a steepening of the temperature gradient from equator to pole and development of the present atmospheric circulation pattern (Kemp, 1978). Although this was initiated earlier in the Tertiary, the major changes may have occurred during the Miocene. Both the ocean core oxygen isotope (^{18}O) record, reflecting global temperatures and ice build-up, and the eustatic sea level curve, which largely reflects ice

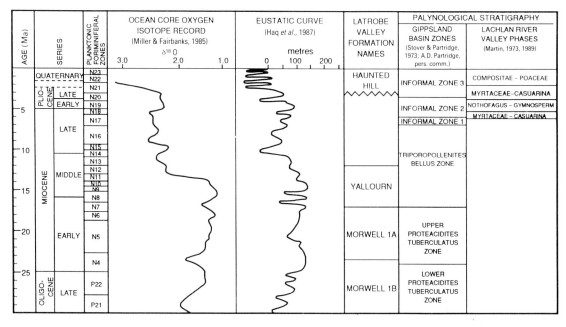

Figure 13.1 Late Cenozoic time scale, global palaeoenvironmental records and selected Australian biostratigraphies.

accumulation, show marked changes in the Middle Miocene (see Figure 13.1). The lowering of temperatures would have been offset to some degree in Australia by its northward drift, estimated to have been 50–70 mm/year during the last 30 Ma, although a combination of lower sea surface temperatures, lower sea levels and the movement of the continent into the mid-latitude high pressure belt would have resulted in reduced precipitation. Critical to moisture availability in the interior was the retreat of the seas from continental basins, particularly the Murray Basin.

Further global cooling and desiccation accompanied the expansion of continental ice sheets in the northern hemisphere within the Pliocene (Shackleton & Opdyke, 1977). In south-eastern Australia, Bowler (1982) identified 6–2.5 Ma as the period of major sustained aridity. From weathering profiles within lake sediments, he proposed that aridity was associated with the initiation and intensification of the high pressure system. This resulted in a change from a climate where rainfall was evenly distributed throughout the year to one of summer rainfall similar to that experienced in northern Australia today. Further atmospheric

intensification and movement north of the high pressure system at a greater rate than Australia's drift eventually brought southern Australia under the influence of the westerly wind belt. This change, dated around 2.5 Ma, corresponds broadly with the increasing climatic fluctuations of the Quaternary.

Unlike most other continents, Australia experienced little tectonic activity during the late Tertiary. Only the Eastern Highlands were affected, particularly by periodic volcanic activity, although they were uplands well before this period and may not have undergone further significant uplift (Williams, 1993). By contrast, earth movements associated with the collision of the Australian and Asian plates in the Middle–Late Miocene elevated the New Guinea highlands. This also brought Australia effectively into contact with the Southeast Asian region.

FOSSIL PALYNOLOGICAL SITES

All known sites or groups of sites for the Pliocene and Miocene are shown on Figure 13.2. Almost all are from the eastern part of the continent with

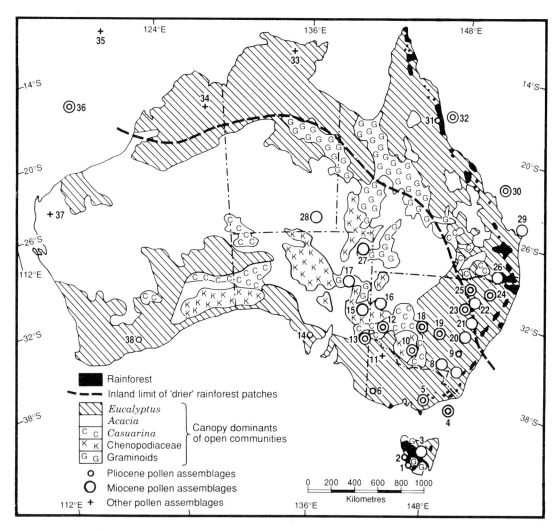

Figure 13.2 Generalised vegetation of Australia and location of Cenozoic pollen assemblages. (1) Regatta Point (Hill & Macphail, 1985), (2) Linda Valley (M. Macphail *et al.*, unpublished data), Tullabardine Dam (Colhoun & van de Geer, 1986), (3) Monpeelyata (Macphail *et al.*, 1991), (4) Gippsland Basin (Stover & Partridge, 1973), (5) Latrobe Valley (Luly *et al.*, 1980; Sluiter, 1984; Dawson, 1983; McEwen Mason & Wagstaff, unpublished data; Kershaw *et al.*, 1991*b*), (6) Grange Burn (Harris, 1971), (7) Lake Bunyan (Tulip *et al.*, 1982), (8) Kiandra (Owen, 1988), (9) Lake George (Singh & Geissler, 1985; McEwen Mason, 1989), (10) Murrumbidgee area (Martin, 1984*a*), (11) Lake Tyrrell (Luly, 1991), (12) Western Murray Basin (H. A. Martin, 1986; Macphail & Truswell, 1989), (13) SADME MC63 bore (Martin, 1991*a*), (14) Eyre Peninsula (Truswell & Harris, 1982), (15) Oakvale (Truswell *et al.*, 1985), (16) Lake Menindee region (Martin, 1988), (17) Lake Frome (W. K. Harris in Callen & Tedford, 1976; Martin, 1990*b*; Singh & Luly, 1991), (18) Hillston region (Martin, 1984*b*, 1987), (19) Lachlan Valley (Martin, 1984*b*, 1987), (20) Cadia (Owen, 1975), (21) Home Rule Kaolin Deposit (McMinn, 1981), (22) Warrumbungle Mountains (Holmes *et al.*, 1983), (23) Castlereagh River Valley (Martin, 1981), (24) Mooki Valley (Martin, 1979), (25) Namoi and Gwydir River Valleys (Martin, 1980), (26) East and West Haldon (Dudgeon, 1982), (27) Etudunna and Namba Formations (Truswell & Harris, 1982), (28) TiTree Basin (Truswell & Harris, 1982), (29) Sandy Cape (Wood, 1986), (30) Aquarius Bore (Hekel, 1972), (31) Butcher's Creek (Kershaw & Sluiter, 1982) and Lynch's Crater (Kershaw, 1986), (32) ODP Leg 133 cores (Kershaw *et al.*, 1993; Martin & McMinn, 1993), (33) South Alligator drill hole AMG 441791 (Truswell, 1982), (34) Goat Paddock (Truswell & Harris, 1982, (35) Lombok Ridge G6 4 (van der Kaars, 1991), (36) ODP Site 76S (McMinn & Martin, 1993). (37) Glen Florrie (Truswell & Harris, 1982), (38) Lake Tay (Bint, 1981).

very few from drier areas, outside the humid–subhumid fringe, identified by a present-day cover of rainforest and open eucalypt forest and woodland. The sites in drier areas are virtually restricted to the Miocene and are lake and marginal marine sediments associated with Tertiary inland basins. The Murray Basin and its catchment have been the focus of most research, generally in connection with groundwater exploration. The distribution of actual bores studied within the basin are indicated below, in Figure 13.13. Some continuous coring has been undertaken but the majority of analyses have been from bore cuttings. There is a sequence from marginal marine and lacustrine environments close to the Miocene Murray Basin embayment in the west through to fluvial and related swamp deposits in tributary valleys on the western slopes of the Eastern Highlands, in the eastern part of the catchment.

Another well-studied continental area is the Latrobe Valley of Victoria, where the extensive coal deposits have allowed detailed vegetation and environmental reconstructions (Blackburn & Sluiter, Chapter 14, this volume). Most other terrestrial sites provide very limited temporal records from lake, swamp or fluvial sediments often preserved by a capping of basalt. The only record extending back from the present to the late Tertiary is that from Lake George but, even here, core sediments are frequently barren of pollen.

Several offshore records provide, or are providing, very generalised regional vegetation histories for selected areas, and practically the only information for northern Australia. These are also extremely useful in that they provide time scales in which the pollen can be tied into the well-dated marine foraminiferal record. Neogene pollen zonations have been established for eastern Queensland (Hekel, 1972) and the Otway and Gippsland Basins of Victoria (Harris, 1971; Stover & Partridge, 1973; A. D. Partridge, personal communication). The Gippsland stratigraphy has been applied extensively within the southeast Australian region and is outlined in Figure 13.1.

In the Murray River catchment, the earlier part of the Neogene, which is represented by marginal marine sediments in the western part of the Murray Basin, has been related to the global microfaunal stratigraphy (Truswell *et al.*, 1985), while for the later Neogene a relative pollen stratigraphy is less firmly dated by comparisons with the Gippsland Basin record and inferred climatic changes from the marine oxygen isotope record.

Some time control on more isolated pollen assemblages has been provided by K–Ar dating of associated basalts, while a palaeomagnetic stratigraphy ages the Lake George sequence (McEwen Mason, 1989).

REGIONAL VEGETATION HISTORIES

Generalised pollen diagrams have been reconstructed from those areas possessing sufficient information to provide a quantitative basis for regional comparison and discussion. These diagrams have been pieced together from appropriate sequences for the Latrobe Valley (Figure 13.3), the Lake Frome Basin (Figure 13.4), the western and eastern parts of the Murray Basin catchment area (Figures 13.5, 13.6), the Southeastern Highlands (Figure 13.7), northern New South Wales (Figure 13.8), Tasmania (Figure 13.9), northeastern Queensland (Figure 13.10) and northwestern Australia (Figure 13.11). In addition, some information is presented from southern and southwestern Australia, although there are insufficient data to allow the construction of a diagram that can be usefully compared with those from the other areas.

For each sequence, data representation has been largely limited to major dryland taxa and those of indicator value, to allow a broad assessment of regional vegetation. More detailed information on community composition and evolutionary changes in taxa are considered by Blackburn & Sluiter (Chapter 14, this volume) and Hill (Chapter 16, this volume). All major taxa included have wind-dispersed pollen and the values for each reflect, to some degree, its relative importance in the vegetation. However, some variation will result from the size and nature of the depositional basin. For example, a small basin will reflect more local vegetation than a large one, while the pollen within a fluvial deposit will be biassed towards vegetation growing within the valley upstream from the site. Marine environments may reflect the vegetation growing within a very large area but most pollen is likely to have been waterborne and reflect the influence of rivers as well as the buoyancy of pollen types.

Of the major taxa recorded, *Nothofagus*, Podocarpaceae, *Elaeocarpus*-type and Cunoniaceae are almost entirely restricted to rainforest. Some shrubby species of *Podocarpus* do presently occur within open-canopied communities, particularly heath, but, like a number of taxa whose affinities are with rainforest, opinion is divided as to whether these are regarded as particular tolerant and adaptable remnants of rainforest or are true components of open communities. A similar and perhaps more difficult problem is posed by the present major canopy dominants of open forests and woodlands, *Eucalyptus* and Casuarinaceae. Within the Casuarinaceae all components in Australia today, apart from one species with very restricted distribution in northeast Queensland, are confined to open vegetation. However, many known macrofossil specimens from the Neogene can be referred to the genus *Gymnostoma*, which now inhabits rainforest in the Melanesian region. As it has not been possible to separate the different genera on their pollen morphology, it is difficult to use this pollen type on its own as an indicator of a particular vegetation type. In the case of *Eucalyptus*, it is generally assumed that, when the pollen occurs in some abundance, it is indicative of open vegetation, but there is some difficulty in distinguishing it from some other Myrtaceae. In the pollen diagrams, the categories Myrtaceae and *Eucalyptus*, where the data allow this separation, are used as an approximation of pollen from basically rainforest and open communities respectively, although many marginal rainforest taxa may be included in the former group. In one diagram, a third group, 'myrtaceous shrubs', is identified which refers to identifiable open sclerophyllous Myrtaceae.

Other essentially present-day open sclerophyllous trees and shrubs represented in some

diagrams are *Dodonaea*, Gyrostemonaceae and *Acacia*, although again it cannot be assumed that their ecology has remained consistent over time. In the case of Gyrostemonaceae, there is one species, *Codonocarpus attenuatus*, that does occur in rainforest.

It is more certain that the low-growing Poaceae, Asteraceae and Chenopodiaceae have not been associated with rainforest. All are intolerant of heavy shade and, from available fossil evidence, have their origins outside Australia. Restionaceae is an austral graminoid best represented in wetter heaths and swamps, although it also requires an open environment for extensive development.

In all diagrams, taxon values are expressed as percentages of the total pollen from what are presently considered to be dryland plants. Myrtaceous shrubs and Restionaceae are excluded. Some samples older than Miocene and averaged sample values for very recent periods are included where possible, to allow changes within the Neogene to be placed in broader temporal perspective.

The Latrobe Valley

Figure 13.3 is made up from data from several coal and overburden pollen sequences. The oldest samples from Loy Yang were selected from non-overlapping core sequences from the top of the Morwell 1B seam through the Morwell 1A seam into the Yallourn seam. The top part of the Yallourn seam is from samples taken from an exposed coal face. Only those samples from lighter coal lithotypes have been included, as these are likely to have been derived from open-water rather than vegetated-swamp environments and reflect regional rather than local vegetation (Kershaw et al., 1991a).

The Hazelwood sequence is represented by all polleniferous samples analysed and extends from the top of the Yallourn seam. The Holocene spectrum shows the averaged pollen taxon values through a swamp that formed in a fire-hole within an exposed part of the Morwell coal seam.

The Miocene samples are generally dominated by pollen of *Nothofagus* subgenus *Brassospora*, with other *Nothofagus* types, particularly the subgenus

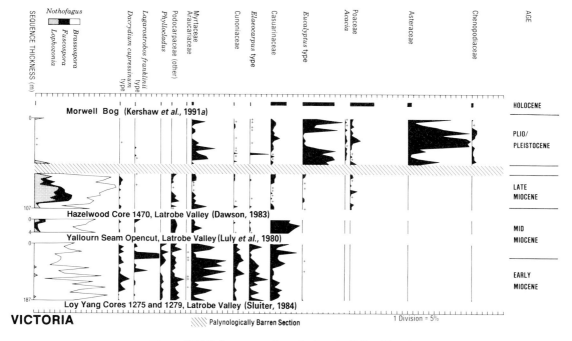

Figure 13.3 Pollen spectra from the Latrobe Valley, Victoria.

Lophozonia, becoming increasingly important from the middle part of this period. Other rainforest taxa have variable representation but generally decrease through the period. Detailed examination of the records shows systematic changes in abundance of certain rainforest taxa that reflects changing environmental conditions associated with sea level fluctuations (Holdgate & Sluiter, 1991). There is also an overall pattern of decreasing *Nothofagus* values from the Early to Middle Miocene followed by a recovery in the Late Miocene. This could reflect changing temperatures as recorded in the marine ^{18}O cores as *Nothofagus* generally dominates the cooler rainforests and rainforest assemblages become more diverse as temperatures increase.

Some opening up of the rainforest is indicated by low values of Poaceae and *Eucalyptus* pollen in the Late Miocene before an extensive period of intensive weathering that inhibited pollen presentation. After this phase the landscape changed dramatically to one dominated by open vegetation with very restricted rainforest. The high level of

Asteraceae pollen indicates that the upper part of the Hazelwood core falls into informal Zone 3 that extends from the Late Pliocene into the Pleistocene. The period of time represented by this phase cannot be determined though the pattern of change within it could represent a Pleistocene glacial cycle. Eucalypt forest was dominant in the earlier and later sections, suggesting a relatively wet and warm climate indicative of interglacial conditions. This interpretation is supported, to some extent, by maximum representation of pollen of the rainforest taxa, Cunoniaceae, *Elaeocarpus* and *Nothofagus* subgenus *Lophozonia*. The Myrtaceae also show a similar pattern, though it cannot be determined whether these were derived from rainforest or sclerophyllous forest plants. The central part of the sequence is dominated by Asteraceous herbs or shrubs, with some Poaceae and low levels of tree pollen. This suggests either dry or cool conditions: the latter is preferred as conditions were not sufficiently dry to inhibit pollen presentation.

The Holocene spectrum, representing the

present interglacial, has high eucalypt levels similar to those inferred previously, but with higher percentages of Casuarinaceae and Poaceae.

The Lake Frome and Murray Basin regions

The sequences selected from this region (Figures 13.4–13.7) are treated together as they fall along present-day precipitation and altitudinal gradients from the arid heart of the continent (Lake Frome lies at sea level and experiences about 100 mm rainfall per annum) east to the Southeastern Highlands, where Lake George, at an altitude of 700 m and receiving a mean annual rainfall of about 650 mm, is the lowest and driest of the sites from this highland area. The highland sites all occur close to the eastern divide that separates the rivers flowing into the Murray Basin from those that flow to the east coast.

The beginning of the Neogene is marked in the sequences from the Murray Basin region by a decline in pollen of the previously dominant *Nothofagus* subgenus *Brassospora*, though it remains a substantial component of the essentially rainforest vegetation through the Early–Middle Miocene. This decline could indicate an increase in temperature, although the rise in Araucariaceae pollen, along with better representation of the open vegetation taxa *Eucalyptus* and Poaceae in the western part of the Basin, could also indicate drier conditions. The existence of a precipitation gradient similar to that of today at this time is suggested by lower values for *Nothofagus* pollen in the Lake Frome Early Miocene sample (although this is compensated for to some extent by high Podocarpaceae levels) and the persistence of high levels of *Nothofagus* subgenus *Brassospora* and almost complete absence of open vegetation taxa through to the Middle Miocene in the Southeastern highlands. In contrast to the Murray Basin, Podocarpaceae, Araucariaceae, Myrtaceae and Casuarinaceae pollen percentages are low in the Southeastern Highlands, although Podocarpaceae increase in the Middle Miocene.

The existence of much drier conditions in the Late Miocene is indicated by the general absence of pollen-bearing sediments in the western

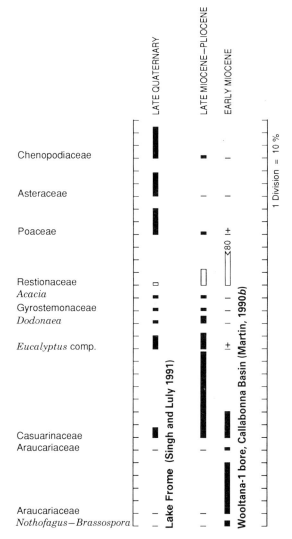

Figure 13.4 Pollen spectra from the Lake Frome Basin, South Australia.

Murray Basin and by the virtual disappearance of *Nothofagus* in the river valleys to the east of the Basin. The combination of Podocarpaceae and Araucariaceae pollen suggests some replacement of *Nothofagus* rainforest by a drier facies but the very high Myrtaceae values, which include a eucalypt component, were suggested by Martin (1988, 1991*b*) to indicate the presence of extensive wet sclerophyllous forests. This proposal will be examined later. Similar conditions appear to have continued through the Pliocene from the

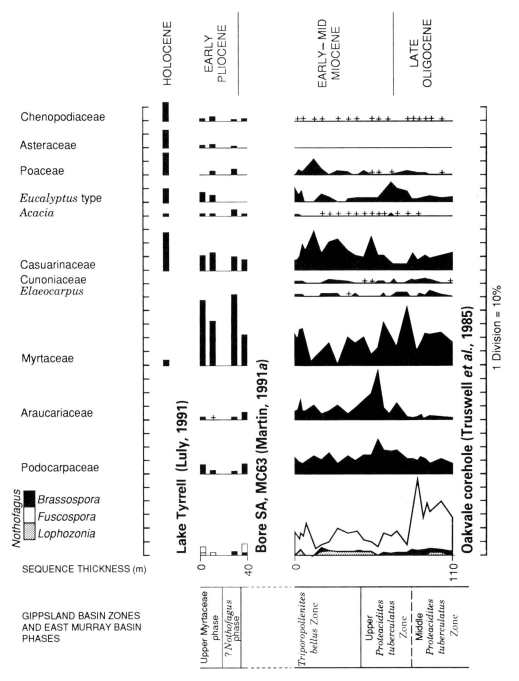

Figure 13.5 Pollen spectra from the western part of the Murray Basin.

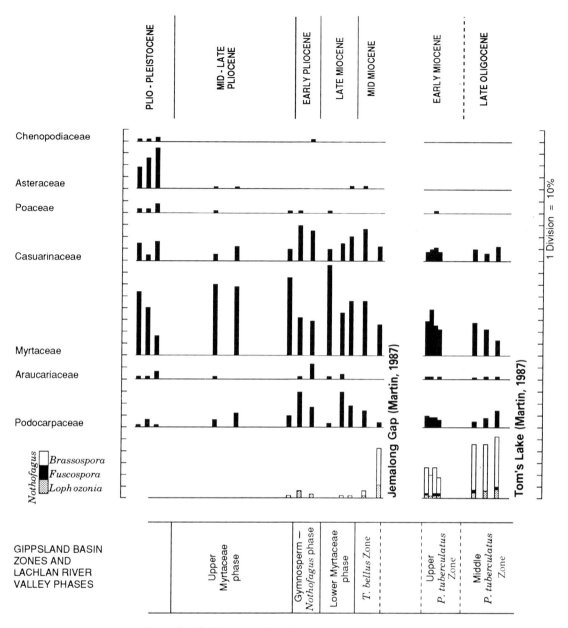

Figure 13.6 Pollen spectra from the eastern Murray catchment.

numerous sections in the river valleys and few pollen occurrences in the Murray Basin proper, except for a period of minor rainforest expansion, 'the gymnosperm–*Nothofagus* phase', but lacking subgenus *Brassospora*, tentatively correlated with the oxygen isotope temperature peak in the Early Pliocene. Casuarinaceae pollen-dominated samples from Lake Frome may also coincide with this peak.

Towards the end of the Pliocene in the river valleys, there is pollen evidence for a major expansion of herbaceous or small shrub taxa, including

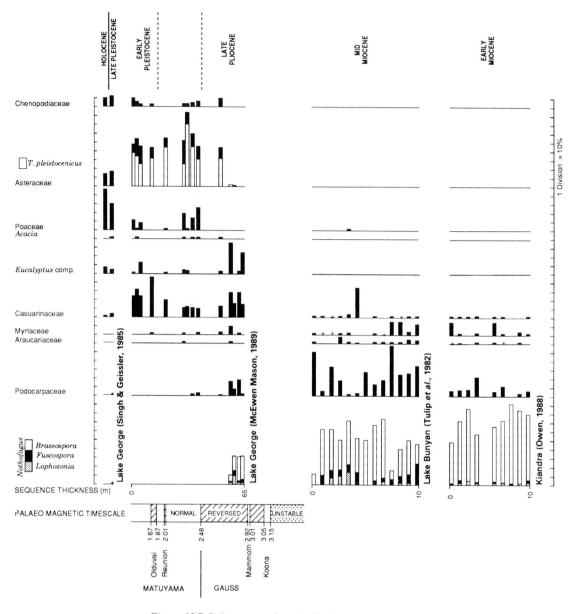

Figure 13.7 Pollen spectra from the Southeastern Highlands.

Poaceae and Chenopodiaceae but predominantly Asteraceae. The nature and timing of this change is clearly illustrated in the well-dated Lake George core in the Southeastern Highlands. Here rainforest, including *Nothofagus* subgenus *Brassospora* and Podocarpaceae, clearly survived into the Late Pliocene, although significant quantities of eucalypt and perhaps also Casuarinaceae pollen indicate the existence also of sclerophyllous vegetation, possible wet sclerophyllous forest. Wetter conditions in the highlands would have facilitated the longer survival of rainforest here than in the drier inland regions.

The change to open vegetation dominated by

Asteraceae appears to have occurred within a very short period of time around 2.7 Ma. This is effectively the end of significant rainforest representation in the area as a whole as no samples in the thoroughly analysed Lake George sequence from this time to the present day have more than traces of obligate rainforest taxa.

The nature of the change though may not have been as dramatic as it appears. A few samples just below the illustrated rainforest spectra contain small amounts of pollen composed of equal proportions of rainforest and Asteraceae pollen. Unfortunately other samples analysed, perhaps extending back into the Late Miocene (Bowler, 1982), were barren of pollen. The rainforest-dominated samples may represent then a minor phase of amelioration prior to a sustained change to open communities.

The appearance of the succeeding open vegetation is not easy to visualise as there are no obvious present-day analogues. The domination by Asteraceae with some Poaceae implies extensive steppe, perhaps existing under relatively dry and cool conditions. Those with significant levels of Casuarinaceae and *Eucalyptus* pollen (particularly in the river valleys) indicate the existence either of a woodland cover or, more likely, forest patches in more sheltered areas. Fluctuations in climate are suggested by the nonpolleniferous phases, indicating oxidation of pollen probably through periodical drying up of the lake.

The averaged pollen values for the topmost parts of the sequence, the Late Pleistocene and Holocene, representing the climatic extremes of the last glacial/interglacial cycle, are very different from those of the Late Pliocene–Early Pleistocene. Here percentages for trees, particularly Casuarinaeae, are very much reduced, whereas Poaceae is substantially more important than Asteraceae.

Northern New South Wales

The diagram from this region (Figure 13.8) contains all samples for which quantitative data are provided by Martin (1979, 1980, 1981). The samples are from a number of ground-water

exploration boreholes in the river valleys, and the correlation between bores is not sufficiently precise to allow more than an approximate ordering of samples within defined periods. The sequence is also not firmly dated.

The beginning of the sequence, in what is considered to be the Middle–Late Miocene *Triporopollenites bellus* Zone, shows high values for *Nothofagus*, Podocarpaceae and Myrtaceae pollen, which together suggest domination by rainforest under wet conditions, although significant representation by Araucariaceae indicates also the presence of a drier rainforest facies. The change to the Pliocene is marked by a substantial increase in Araucariaceae pollen, indicating the extensive development of drier rainforest, while major increases also in Poaceae and Asteraceae pollen mark the expansion of open-canopied communities, with Casuarinaceae rather than *Eucalyptus* being the canopy dominant.

The vegetation became even more open around the Pliocene/Pleistocene boundary, with the severe restriction of drier rainforest and a reduction in tree canopy cover of open communities. Casuarinaceae also declined relative to *Eucalyptus*.

Tasmania

Only a few, broadly dated pollen spectra are available for the Neogene (Figure 13.9) and these have been constructed in association with detailed macrofossil analyses considered by Carpenter *et al.* (Chapter 12, this volume). At the beginning of this period, a pollen spectrum from the Central Plateau at 920 m above sea level is dominated by the cool temperate rainforest taxa *Nothofagus* and Podocarpaceae, although the open forest taxa Poaceae, *Acacia* and possibly *Eucalyptus* are already present in minute quantities. From the Linda Valley spectrum, it might be inferred that this pattern was maintained throughout the Neogene, with only a small increase in open vegetation components, at least in the western part of the state.

The spectrum from nearby Regatta Point, originally considered to have been Late Pliocene–Early Pleistocene, is very different and has

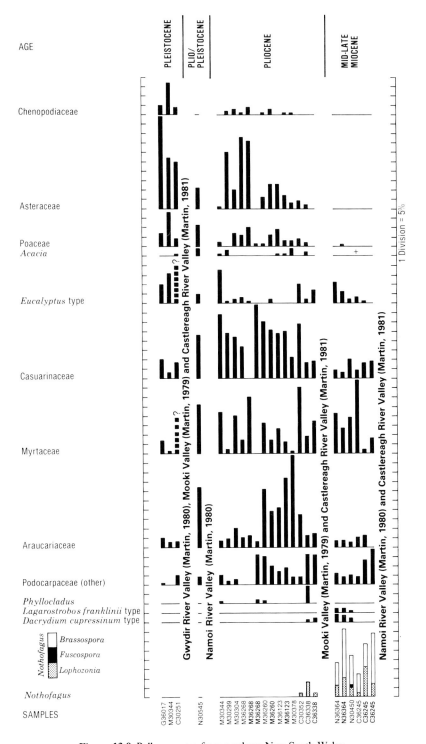

Figure 13.8 Pollen spectra from northern New South Wales.

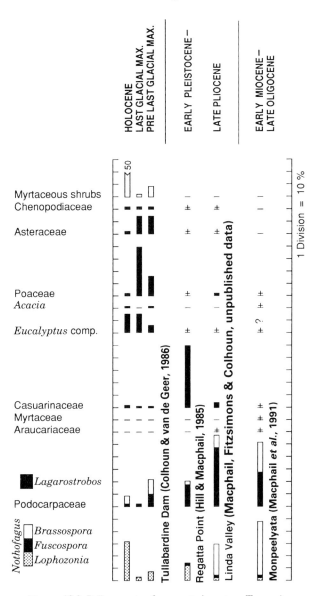

Figure 13.9 Pollen spectra from central-western Tasmania.

recently been tentatively allocated an Early or even Mid-Pleistocene age (Hill & Macphail, 1994) presumably because of these floristic differences. By this time it would appear that *Nothofagus* subgenus *Brassospora* had disappeared, although rainforest was well represented and herbs still uncommon. One very interesting feature of the site is the identification of the dominant Casuarinaceae as a

sclerophyllous, not rainforest, taxon from preserved macrofossils. This has major implications for the interpretation of vegetation containing Casuarinaceae elsewhere. The most consistent feature of all of these spectra is the high levels of pollen from the present-day Tasmanian endemic *Lagarostrobos*.

Recent pollen spectra from Tullabardine dam,

at an altitude of 230 m and presently within temperate rainforest, show substantial differences from the older spectra. *Eucalyptus* is equally or better represented than the major rainforest taxa, while Casuarinaceae has low representation, herbaceous taxa achieve high levels particularly during the last glacial period, and sclerophyllous myrtaceous shrub pollen, from plants growing on site, dominate the Holocene spectra.

Northeastern Australia

Sequences used to construct Figure 13.10 are from very different environments and are geographically separated. The Sandy Cape core is from marginal marine sediments, Aquarius 1 Well is from offshore sediments, while Butcher's Creek is from lake and swamp sediments preserved beneath basalt on the Atherton Tableland at an altitude of 700 m. The recent spectra are averaged values for the Late Pleistocene and Holocene from volcanic crater lake sediments close to Butcher's Creek. The only other available data are from ongoing studies on Ocean Drilling Programme (ODP) cores from northeast Queensland (Kershaw *et al.*, 1993).

High levels of *Nothofagus* subgenus *Brassospora* combined with significant quantities of Araucariaceae pollen appear to indicate the importance of both wet and drier rainforest vegetation in coastal areas until the Middle Miocene. However, there is some indication of a reduction in *Nothofagus* subgenus *Brassospora* before this time at Sandy Cape and it is possible that the unit considered to be Early Miocene from the central Queensland Aquarius 1 Well is in fact Eocene (Foster, 1982). If the first appearance of *Acacia* pollen is taken as the base of the Miocene (but see Chapter 10, this volume), then all but rare occurrences of *Nothofagus* pollen pre-date the Miocene in this area.

After the *Nothofagus* decline, pollen of herbaceous taxa is consistently recorded and Myrtaceae increase their representation, but the maintenance of high representation of Podocarpaceae, Araucariaceae and Casuarinaceae pollen suggests that other components of the vegetation

– probably a mix of wet and dry rainforest with some sclerophyllous forests – maintained their representation, at least until the Late Pliocene. At this time there was a substantial increase in Asteraceae and Chenopodaceae pollen, indicating a further opening up of the vegetation, most likely due to a reduction in precipitation.

Available results from the ODP cores off northeast Queensland show an actual increase in Araucariaceae pollen from the Late Miocene to the Pliocene and this taxon, together with *Podocarpus* and Casuarinaceae, dominates the spectra until the Late Pleistocene, when Araucariaceae virtually disappears. Poaceae and Asteraceae pollen increase to some extent in the Late Pliocene.

Nothofagus subgenus *Brassospora* pollen is a major component of the Butcher's Creek samples and this suggests the survival of this taxon through the whole of the Neogene on the cool wet highlands and Tablelands of Queensland. Its present absence from the state may be the result of subsequent Pleistocene climatic fluctuations or increased temperatures. It is apparently not predicted for north Queensland today from a bioclimatic analysis of its extant range in New Guinea (J. Read, personal communication) and palynological evidence suggests that temperatures may have been up to 3 deg.C cooler than present at that time (Kershaw & Sluiter, 1982).

Other changes in the vegetation from the late Neogene to the late Quaternary on the Atherton Tableland include substantial increases in *Eucalyptus* and herbaceous taxa, and also Araucariaceae until recent times. The decline of Araucariaceae pollen matches that recorded in the ODP cores.

It is interesting to note that the extensive area of complex rainforest centred on the Atherton Tableland is almost invisible in the ODP cores. Although pollen of a variety of angiosperm rainforest taxa are recorded, there is little evidence of common taxa, such as *Nothofagus*, Myrtaceae, Cunoniaceae and *Elaeocarpus*, which are known to have been present in some abundance. Part of the explanation is that pollen in the deep-sea cores is likely to have been transported by rivers and reflects mainly streamside and deltaic vegeta-

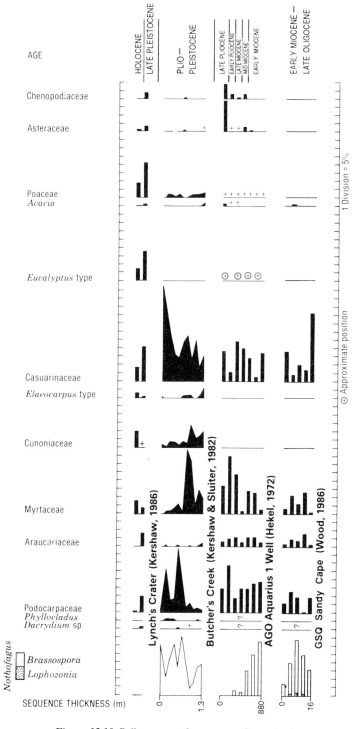

Figure 13.10 Pollen spectra from eastern Queensland.

tion, as observed elsewhere (e.g. Heusser, 1978; Chumra & Liu, 1990). Other factors may include greater buoyancy of conifer pollen and the filtering out of very small grains such as Cunoniaceae and *Elaeocarpus* during pollen preparation in the laboratory.

Northwestern Australia

Recent analyses of a core from ODP Site 765 provide the first dated evidence for vegetation changes in the northwestern part of the continent. As the core is a fair distance from land and river discharge is likely to have been low in this low-rainfall, topographically subdued region, pollen concentrations are low and several adjacent samples have been combined to produce acceptable counts (see Figure 13.11).

It is notable that rainforest pollen grains are virtually absent, even in the oldest samples from the middle Miocene. The only evidence to suggest that rainforest may have been present is the occasional grain of Podocarpaceae. Recognisable rainforest taxa are even extremely sparse in the 'Eocene' spectrum (South Alligator) from the region, a time when rainforest is considered to

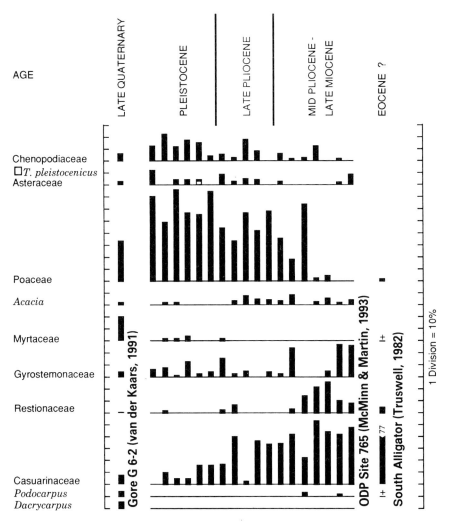

Figure 13.11 Pollen spectra from northwestern Australia.

have had its maximum extent on the continent. The fact that rainforest was once present in the region is demonstrated by Eocene spectra from Glen Florrie station and Goat Paddock (Figure 13.2) that were reviewed briefly by Truswell & Harris (1982). At Glen Florrie, buried stream deposits contain *Nothofagus* subgenus *Brassospora* and podocarpaceous pollen, similar to assemblages in central Australia (Truswell & Harris, 1982). The crater lake deposit at Goat Paddock differs in that pollen of *Nothofagus* subgenus *Brassospora* is absent, possibly because it antedates the rise in abundance of *Nothofagus* in the Middle Eocene. It does, however, contain several podocarps, *Anacolosa* and *Beauprea*, that are frequent associates of *Nothofagus* subgenus *Brassospora*.

Despite the lack of clear evidence for survival of substantial rainforest into the Miocene, it is possible that some Casuarinaceae and possibly the arid land shrub family Gyrostemonaceae were components of rainforest vegetation. The significant representation of Restionaceae, generally associated with permanent swamp environments and a common associate of rainforest assemblages in the Australian Tertiary, could suggest that climatic conditions were suitable for rainforest survival. Even today closed monsoon and dry rainforest patches occur to the east of the site in the Kimberley region (Figure 13.2). These patches, unless very extensive, can be palynologically invisible (Kershaw, 1976), lacking not only *Nothofagus* and gymnosperms, but also most other taxa that produce relatively high amounts of pollen.

At the end of the Miocene or in the Early Pliocene there are substantial increases in the pollen of grasses and chenopods, a sharp reduction in Restionaceae and some decline in Casuarinaceae. These trends continued into the Pleistocene. The Asteraceae never show very high values within the period, while the Myrtaceae are surprisingly not recorded until the very Late Pliocene. Their presence corresponds to a reduction in *Acacia* values.

Averaged values for the late Quaternary from Core G 6−2, further away from the Australian mainland, show very much higher values for Myrtaceae pollen (the majority of which is *Euca-*

lyptus). However, significant representation of conifers, particularly *Dacrycarpus*, suggests that some of the pollen was derived from islands to the north that probably included both Timor and New Guinea.

Southern and southwestern Australia

Early Pliocene vegetation at Lake Tay (Figure 13.2) had Casuarinaceae or Myrtaceae as dominants. Poaceae, Asteraceae, Restionaceae and the Chenopodiaceae-type pollen grains are present in low frequencies. There is a wealth of pollen of sclerophyllous shrub species, including *Acacia*, *Haloragis*, *Dodonaea triquetra* (as *Diplopeltis* in Bint, 1981) and the proteaceous taxa *Banksia*, *Dryandra*, *Isopogon*, *Xylomelum*, *Grevillea*, *Hakea* and *Stirlingia*. Gyrostemenaceae pollen is also present. This diversity suggests that the rich, sclerophyllous character of the southwest had developed by the Pliocene. Rare pollen grains of *Nothofagus* subgenera *Brassospora* and *Fuscospora*, and of Podocarpaceae are present (Bint, 1981) but are probably reworked, as the catchment of Lake Tay contains early Tertiary deposits (A. D. Partridge, personal communication). Relict stands of *Nothofagus* (Bint, 1981) are less likely. There is evidence, however, that rainforest with abundant *Nothofagus*, similar to that of the southeast, was widespread in the Eocene of southwestern Australia (Balme & Churchill, 1959; Hos, 1975; Milne, 1988). Early–middle Pliocene vegetation on the Eyre Peninsula was open, as indicated by common Poaceae pollen, Chenopodiceae and Asteraceae. Casuarinaceae and Cyperaceae pollen are locally abundant and Myrtaceae, including the eucalypt type, is present. Some rainforest persisted, as shown by the presence of podocarps and *Nothofagus* subgenus *Brassopora* (Truswell & Harris, 1982).

GENERAL DISCUSSION

The information presented, although of variable quality and completeness, provides a basis for examination of the general pattern of vegetation and environmental changes and their causes

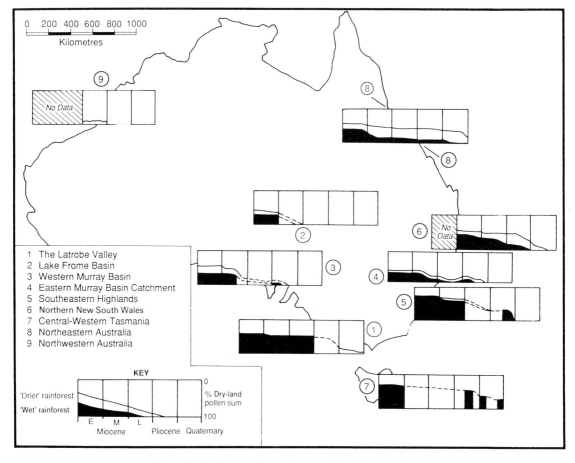

Figure 13.12 The late Cenozoic pattern of rainforest decline.

within the continent as a whole. To assist in this overview, major features of the vegetation in the different regions during the geological periods of interest are summarised in Figures 13.12 and 13.14. In Figure 13.12, an attempt has been made to separate 'wet' rainforest characterised by *Nothofagus* from 'drier' rainforest characterised by the Araucariaceae. The Podocarpaceae and rainforest angiosperms, associated with both types, have been allocated to a particular group on the known present-day ecology of their components or, where this information is lacking or where taxa are broad ranging, on the likelihood of membership of a group based on the composition of pollen assemblages from present-day rainforest types. Casuarinaceae and Myrtaceae are excluded

because they could belong to either rainforest or open-canopied vegetation. Poaceae, *Eucalyptus* and Asteraceae pollen percentages are used to characterise different open vegetation types on Figure 13.14.

The nature of Early–Middle Miocene vegetation

Rainforest containing *Nothofagus* was a major component of the vegetation, at least in the eastern part of the continent, throughout much of this period. However, evidence is restricted largely to depositional basins influenced by the Miocene marine transgressions in present-day drier areas and the record is very much biassed towards

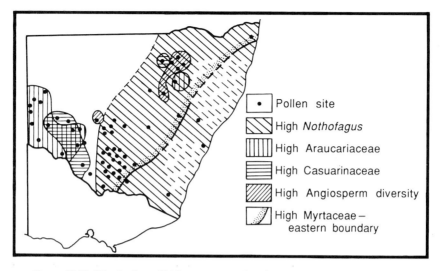

Figure 13.13 Distributions of high percentage values for selected pollen taxa in the Early–Middle Miocene of New South Wales (after Martin, 1990*a*).

wetter forest types. The *Nothofagus* component is much lower in inland areas and a precipitation gradient from coastal areas inland similar to that of today may be inferred.

There may also have been a temperature component associated with the representation of *Nothofagus*-dominated forests. Highest percentages of this taxon are recorded at higher altitudes in Tasmania and the Southeastern Highlands. In the Latrobe Valley, peaks of Myrtaceae and *Elaeocarpus*, with concomitant reductions in *Nothofagus* pollen, have been related to sea level transgressions when higher temperatures prevailed (Holdgate & Sluiter, 1991).

Very low pollen values for herbaceous plants indicate the absence of grasslands, herbfields and open sclerophyllous communities that are prevalent in drier areas today. Araucariaceae forests were important in inland southeastern Australia but it is frequently difficult to assess the community affiliations of the well-represented Myrtaceae and Casuarinaceae. It is possible that many were associated with sclerophyllous heath or scrub, the Casuarinaceae also forming relatively homogeneous stands of sclerophyllous woodland or forests. Today these communities frequently lack a significant herbaceous component.

Although some macrofossil and pollen data suggest that eucalypt-dominated communities may have been well established at this time (Ambrose *et al.*, 1979; Lange, 1978; Dudgeon, 1982; Holmes *et al.*, 1983) the bulk of pollen evidence suggests that, even if present, their distribution may have been limited.

The existence of wet heath vegetation has been demonstrated by studies on local swamp communities and is best illustrated from the combined work of Blackburn and Sluiter on the swamp precursors of the Latrobe Valley brown coals (Chapter 14, this volume). The importance of conditions during the Early–Middle Miocene to the development of heath can be gauged from a comparison of pollen assemblages from all coal seams in the Valley (see Figure 13.15, below). It is clear here that wet heath taxa such as *Banksia*, Epacridaceae/Ericaceae, Restionaceae/Centrolepidaceae and *Gleichenia* expanded at this time. These plants show their maximum expression in the dark lithotypes, which are considered to indicate swamp and raised bog environments. Prior to this time the coal basins would have been vegetated by predominantly swamp rainforest communities. This period may have witnessed the development and expansion also of dry heath communities but,

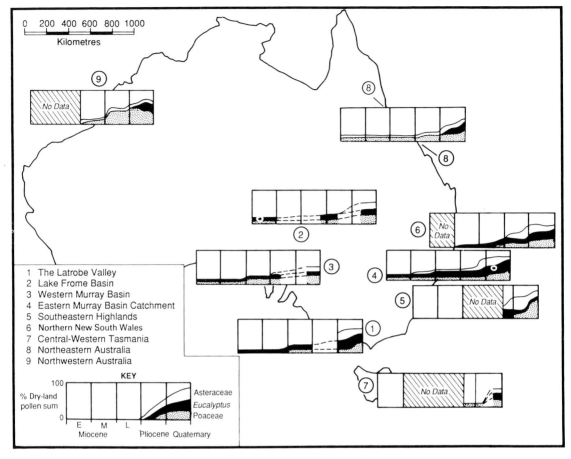

Figure 13.14 The late Cenozoic increase in selected components of open vegetation.

because of the limited pollen dispersal of characteristic heath plants, these would be poorly represented in pollen records. The development may have been related to increased climatic variability as lithotype differentiation becomes much more marked in the younger coal seams.

Although the vegetation cover of the continent has been considered to have been much more homogeneous than it is today, the large number of pollen records from New South Wales illustrate a high degree of spatial variation (Figure 13.13). One interesting feature of this pattern is that there are marked similarities in geographical variation to that of today. The *Nothofagus*-dominated area now corresponds approximately to the extent of

eucalypt-dominated sclerophyllous forest, the *Nothofagus*/Myrtaceae zone to the extent of eucalypt woodlands, and the high Casuarinaceae values in the southwest to present-day Casuarinaceae woodlands (Figure 13.2). In the latter case, especially, the possibility that there may have been some genetic continuity in Casuarinaceae populations might be worthy of investigation.

Rainforest decline

There is evidence for a rainforest decline in all sequences but the pattern varies from area to area (Figure 13.12). The sharp decline in *Nothofagus* subgenus *Brassospora* pollen towards the end of the Oligocene in the Murray Basin region has been interpreted as a reduction in rainforest area resulting from a decrease in rainfall (See Figure 13.16, below), but this may simply represent a change to a warmer facies of rainforest associated with global temperature and sea level rises. A subsequent increase in Araucariaceae pollen suggests an expansion of drier forests implying that, in this region, increasing temperatures towards the Middle Miocene maximum did result in an effective reduction in precipitation. Some reduction in rainforest relative to more open vegetation is indicated by generally higher values for *Eucalyptus*, Poaceae and probably Casuarinaceae. Available evidence from elsewhere also shows a decline in wet rainforest, particularly *Nothofagus* subgenus *Brassospora*, that could be attributed to warmer and slightly drier conditions.

The Late Miocene is characterised by a general lack of evidence, particularly for inland areas. This could indicate substantially drier conditions, but here any broad-scale reduction in precipitation may have been exacerbated by the cessation of marine conditions with the global fall in sea level. Relatively humid conditions were maintained in the Latrobe Valley where there was a resurgence of *Nothofagus*, but with a greater component of the subgenera *Fuscospora* and *Lophozonia* pollen types that are still present within the southeastern Australian region. One explanation for this change in *Nothofagus* composition is that temperatures became more variable, restricting subgenus *Brassospora* to more equable sites, while the other, more hardy, subgenera were able to expand in exposed areas and sites of cold air drainage (Read *et al.*, 1990). Some reduction in precipitation in the Latrobe Valley is suggested by the first significant presence of the open vegetation taxa *Eucalyptus* and Poaceae.

Pollen analytical evidence from terrestrial environments is totally lacking from the very Late Miocene, suggesting either very dry or extremely variable climatic conditions. However, these were not sufficiently adverse to eliminate wetter rainforest from even present-day semiarid to subhumid areas as there is evidence for a resurgence in what is presumed to have been the Early Pliocene in much of the Murray catchment area

(although *Nothofagus* subgenus *Brassospora* was absent) and the Late Pliocene at Lake George. By the Pleistocene, there is no conclusive evidence for *Nothofagus* subgenus *Brassospora* anywhere on the continent, except perhaps for the Atherton Tableland, and many of its associates had also disappeared. The surviving depauperate remnants of wetter rainforest had become restricted largely to moist east coastal areas where they still exist.

A somewhat different pattern is provided by drier rainforests, characterised by Araucariaceae. The earlier history of these communities is poorly known: they could have either expanded from drier areas in the centre of the continent where they had been contained during the mid-Tertiary, or been formulated with the onset of drier late Tertiary conditions. For the Araucariaceae, this was a return to suitable conditions, as the plants had been an important component of Australian vegetation during the Mesozoic, presumably as emergent trees above extant and extinct gymnosperms and early angiosperms rather than extant rainforest angiosperms. The drier forests appear to have progressed from the centre of the continent to the coast in response to increasingly drier conditions. They also moved northwards, presumably in response to some temperature-related variable, similar to the contraction of many taxa associated with wetter rainforests.

Although never important in the palynofloras of the southeast of the continent, Araucariaceae was present in Tasmania until the Pliocene. Araucariaceae forests maintained high representation in northeastern Australia into the Pleistocene and only suffered a substantial decline in some areas within the last 100 ka.

There has been a certain amount of debate on the underlying causes of aridity that brought about this rainforest decline. Although, in general terms, these features are related to global cooling, the patterns of change in different areas demand an examination of more local influences. Beard (1977) suggested that aridity began in the northwest of the continent as Australia moved northwards into the mid-latitude high pressure belt. This idea was challenged by Kemp (1978), who considered that the development of the present atmospheric circulation pattern, including the high pressure system, was a function of Australia's movement north away from Antarctica, which allowed the formation of a circumpolar oceanic and atmospheric circulation system. She argued that aridity began in the centre of the continent perhaps as early as the Late Eocene and expanded towards the coast. Bowler (1982), focussing on the effects of Australia's northward movement, hypothesised that this movement would have led to the development of a high pressure system to the south of the continent in the Early Miocene that would have moved north and intensified, bringing aridity first to the southern part of the continent.

The additional evidence available here allows some comparative assessment of these differing views. The Beard hypothesis was convenient in that there was virtually no fossil evidence from northwestern Australia, which itself could suggest arid conditions. The northwest shelf record does indicate that rainforest had declined early but does not really extend back far enough to determine exactly when this occurred. There is also the problem of identifying the extent of rainforest that essentially lacks plants with well-dispersed pollen during the recorded period. The fact that rainforest patches are present today in the Kimberleys could suggest that rainforest has been substantially underrepresented in the fossil record, although insufficient research has been undertaken on them to determine the extent to which the extant patches are predominantly relict or composed of recent, opportunistic arrivals (Russell-Smith, 1991).

There is both pollen and sedimentological evidence for a reduction in precipitation in the centre of the continent by the later part of the Eocene (Truswell & Harris, 1982), but the extent of drying may have been overestimated due to the incorrect identification of Poaceae pollen (Martin, 1991a). However, the general picture that rainforest declined earlier in this region than in most coastal areas is still valid.

Although it cannot be ascertained whether drying began in the northeast or centre of the continent, it was well underway by the Middle Miocene, before Bowler's hypothesis essentially comes into play. However, some subsequent patterns might be accommodated by his ideas. The lack of pollen for a substantial period at the base of the Lake George core and the barren phase in the Latrobe Valley could embrace the postulated period of intense seasonality between 6 and 2.5 Ma and the expansion northwards of the high pressure system. However, the model fails to account for the apparent increases in precipitation indicated in the Early Pliocene by the Gymnosperm/*Nothofagus* phase in the Murray Basin region and in the Late Pliocene by the rainforest phase at Lake George. Nor does it explain the evidence for dry conditions occurring much earlier in the Late Miocene within the Murray Basin. It appears most likely that intense seasonality was initiated at the end of the Middle Miocene but was punctuated by periods of amelioration, probably related to global changes in temperature and sea levels. The evidence from Lake George for survival of rainforest to practically the end of this period suggests that the switch from summer to winter rainfall as southern Australia came under the influence of the present low pressure system may have been a major cause of the disappearance of the bulk of rainforest taxa from the region. Some support for this hypothesis is provided by the records from northern New South Wales and Queensland, which are still under the influence of the high pressure system. The evidence here is that rainforest declined gradually through the period with no obvious abrupt changes, more in line with the inferred global pattern of cooling and desiccation.

The Bowler model goes some way to explaining the strong separation of present-day tropical and temperate rainforest elements that were much more mixed during the mid-Tertiary. Certainly the southern limit of drier rainforest types in northern New South Wales approximates the present summer/winter rainfall boundary (Figure 13.2).

Australian/Southeast Asian relationships

There has been a great deal of speculation about biogeographical relationships between the rainforest floras of these two regions, particularly when, and to what extent, interchange took place. Recent reviews (e.g. Truswell *et al.*, 1987; Kershaw, 1988) identified the lack of fossil evidence for the critical Miocene period in northeastern Australia, a time when the Australian and Asian continental plates came into contact, as seriously inhibiting a true assessment of the situation. The investigation of deep-sea cores from this region is being undertaken in an attempt to fill this gap in knowledge. The data from these cores tentatively support the conclusions of Truswell *et al.* (1987) that there has been no massive invasion into northern Australia since Middle Miocene contact, as the pollen assemblages are dominated by austral taxa that can be traced back to at least the early Tertiary on this continent. However, many of the possible invaders would have generally low pollen production and poor dispersal and more detailed analyses will be required before these conclusions can be verified.

The other major issue concerns the proposed expansion of austral taxa, particularly *Nothofagus* subgenus *Brassospora* and members of the Podocarpaceae, into highland New Guinea and further afield into southeast Asia. There are surprisingly few pollen grains relatable to these taxa in the ocean cores despite their wide pollen dispersal ability. It would appear that the Atherton Tableland Plio-Pleistocene assemblages dominated by *Nothofagus* subgenus *Brassospora* and Podocarpaceae indicate a local relict vegetation rather than an extensive distribution. Considering the lack of topographically suitable sites for their survival to the north of the Tableland, and the limited seed dispersal capacity of *Nothofagus* in particular, it would appear unlikely that these taxa reached New Guinea since contact and the consequent uplift of the New Guinea mountains. The most likely scenario then is that New Guinea was included within the distribution range of *Notho-fagus* and associated Podocarpaceae during the mid-Tertiary phase of maximum wet rainforest expansion. Survival of these basically temperate taxa at low tropical altitudes might have been facilitated by temperatures lower than those of today, as New Guinea would have been south of its present position. The present restriction of *Nothofagus* subgenus *Brassospora* to the New Guinea Highlands, after their formation, could have resulted from a combination of increased temperatures and competition from the southeast Asian rainforest flora that did invade the lowlands of this island. Contact and the elevation of land within the region would have allowed the more easily dispersed podocarps to expand northwards.

The development of open-canopied vegetation

Open-canopied vegetation in Australia is composed essentially of two major components: a predominantly herbaceous element, characterised by Poaceae, Asteraceae and Chenopodiaceae, which have their origins outside Australia; and an autochthonous element composed mainly of sclerophyllous trees and shrubs. Evidence for both of these elements is sparse before the Miocene, suggesting that open vegetation was very restricted. Exceptions may include some Myrtaceae- and Casuarinaceae-dominated open forests, but lacking a substantial herbaceous understorey.

The base of the Miocene may correspond with the first presence of *Acacia* (but see Macphail *et al.*, Chapter 10, this volume), which provides the most reliable evidence for the existence of open communities. Here, even very low percentages can indicate significant representation of parent plants. The Oligocene/Miocene boundary also approximately corresponds to the first real evidence for distinct sclerophyllous communities within the Latrobe Valley brown coal swamps. As previously mentioned, this has been attributed to increased climatic variability, but it also corresponds with a sharp increase in burning, which may itself be a function of the climatic conditions. The evidence for this burning is based on charcoal

322   A. P. KERSHAW *et al.*

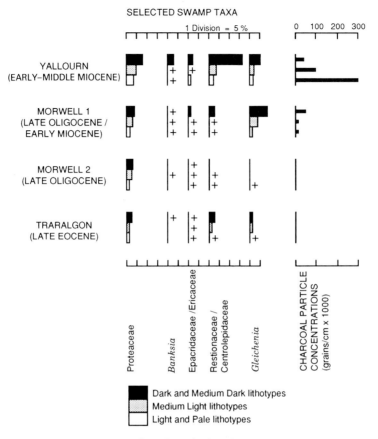

Figure 13.15 Averaged values for pollen of selected swamp taxa and charcoal concentrations for lithotype groups of major coal seams in the Latrobe Valley (after Kershaw *et al.*, 1991*a*).

particles preserved in the coal sediments (Figure 13.15).

A similar relationship between vegetation, climate and fire has been determined from the eastern part of the Murray Basin catchment (Figure 13.16). In this drier region, fire appears to have been present throughout much of the Tertiary but, from the charcoal record, its activity seems to have substantially increased with a reduction in precipitation and inferred increased seasonality of rainfall within the Late Miocene. The close correspondence between Myrtaceae phases and high charcoal levels has led to the proposal that the vegetation in these phases was wet sclerophyllous forest with the myrtaceous canopy, above an understorey containing some rainforest taxa,

requiring fire for its maintenance (Martin, 1987, 1991*b*). This vegetation is seen as analogous to some communities bordering rainforest in northeastern Australia today where the canopy can be composed of a variety of myrtaceous genera including *Eucalyptus*. The difference is that these communities today form only a narrow ecotone between rainforest and open eucalypt forest and it is difficult to visualise the maintenance of extensive wet sclerophyllous forests over the extended period of time represented by the Late Miocene and Pliocene without a drift towards the exclusion of rainforest species and the establishment of more open vegetation. On the other hand, the vegetation during this period may have been less fire promoting than the present combination of

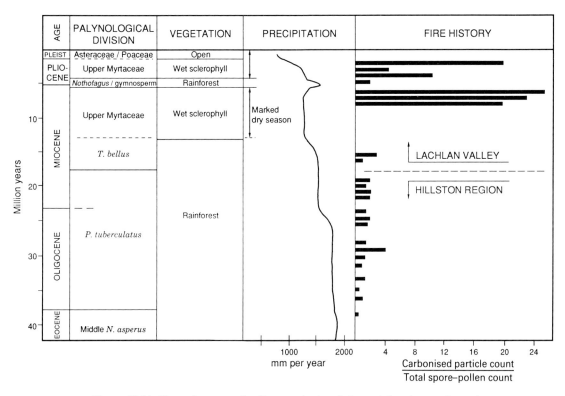

Figure 13.16 Charcoal content of pollen samples in relation to inferred vegetation and precipitation throughout the late Cenozoic of the eastern Murray catchment (after Martin, 1991*b*).

eucalypts, sclerophyllous shrubs and grasses, allowing a mixed community to survive. An alternative interpretation is that rainforest existed in the more sheltered, moister areas such as the river valleys and open vegetation, with a predominantly sclerophyllous understorey, occupying interfluves. There are no charcoal records available to allow an assessment of the role of fire in other areas, but it might be assumed from the high degree of adaptation of Australian vegetation to the survival of fire that this factor has been of broad significance for a long period of time.

The gradual trend towards the replacement of rainforest by sclerophyllous woody vegetation was dramatically altered during the Pliocene, with the expansion of herbaceous communities. This is first noted at the beginning of this period in north-western Australia, where the substantial increase in grasses provides good evidence for the develop-

ment of tropical savanna vegetation. It is likely that a range of palynologically invisible trees with a strong affinity to drier rainforests survived to form an open canopy, as they do today in some places. The Casuarinaceae probably formed dense sclerophyllous stands in more fire-protected or edaphically suitable sites.

In the southeast of the continent, it was the Asteraceae that came to dominate, with most evidence indicating a Late Pliocene–Pleistocene age. This may have corresponded with the onset of a winter rainfall regime and, from the alternations between Asteraceae and sclerophyllous tree domination in the Latrobe Valley sequence and patchy occurrence of polleniferous sediments in the Lake George core, climatic fluctuations of a magnitude characteristic of the Quaternary period might be inferred.

This expansion of herbaceous vegetation in the

southeast suggests that the climate became unsuitable for trees over much of the landscape. Very dry conditions are unlikely, as pollen preservation is generally better than throughout most of the Late Miocene and Pliocene and there is little evidence of high levels of salinity. It is therefore likely that cool conditions prevailed but temperatures were not sufficiently low to exclude trees from more sheltered areas. High winds, frosts, and incursions of cold air from the south might have been additional contributing factors. The fact that the region, for the first time, came under the influence of a winter rainfall regime might also have facilitated the expansion of opportunistic herb or small shrubs over the austral sclerophyllous flora.

There are no analogues of these assemblages in Australia today, possibly because of a lack of cool and relatively dry environments but, from a comparison with assemblages elsewhere and pollen assemblages in Australia existing during the better-documented Late Pleistocene period, the most likely vegetation type was steppe. One major component of this vegetation was a daisy producing distinctive pollen with much reduced spines in comparison to other Asteraceae. This has been given the form taxon name *Tubulifloridites pleistocenicus* (Martin, 1973). Attempts have been made to identify parent plants from the modern southeast Australian flora to define more exactly the nature of the 'Pleistocene' vegetation and its environmental controls (A. R. H. Martin, 1986; Macphail & Martin, 1991) but there are problems on ecological and distributional grounds in relating the two likely contenders, *Calomeria amaranthoides* and *Cassinia arcuata*, to fossil occurrences (Macphail & Martin, 1991). It might be informative to investigate extra Australian taxa producing similar pollen grains because a number, including *Ambrosia* in North America and *Stoebe* in central Africa, achieved similarly high pollen percentages during Pleistocene glacial periods.

Comparisons with extant vegetation

In very general terms environmental changes during the Neogene were responsible for the transfor-

mation of the Australian landscape from one dominated by rainforest to one composed predominantly of open vegetation communities with wetter rainforest largely restricted to present-day locally moist areas. They also resulted in the development of a strong latitudinal gradient with the segregation of rainforest elements along eastern coastal areas and development of savanna in northern Australia. Most present-day community types had developed to some degree by the end of the Pliocene, but subsequent changes in the Pleistocene had a major influence on their distributions and relative importance. Of particular significance has been the expansion of the eucalypts at the expense particularly of drier rainforests and Casuarinaceae-dominated forests. They may also recently have replaced or augmented more obviously rainforest-related trees in the savanna regions, and replaced other Myrtaceae in the canopy of wet sclerophyllous forests. Subsequent changes also include the development of temperate grasslands and the expansion of a grassy understorey to eucalypt-dominated forests and woodlands, largely at the expense of the Asteraceae. On and close to sites of sediment accumulation, sclerophyllous myrtaceous shrub vegetation is frequently recorded while swamp rainforest is now rare, even within areas that still support rainforest. Most of these changes can be attributed to an increase in burning, in which Aboriginal people may have played an important part (Hope, Chapter 15, this volume), in addition to the higher disturbance levels that would have accompanied the extreme glacial/interglacial climatic oscillations.

We thank Gary Swinton and Phil Scamp for drafting the figures and Jan Liddicut for typing the text.

REFERENCES

AMBROSE, G. J., CALLEN, R. A., FLINT, R. B. & LANGE, R. T. (1979). *Eucalyptus* fruits in stratigraphic context in Australia. *Nature*, **280**, 387–9.

BALME, B. E. & CHURCHILL, D. M. (1959). Tertiary sediments at Coolgardie, Western Australia. *Journal of the Royal Society of Western Australia*, **43**, 37–43.

BEARD, J. S. (1977). Tertiary evolution of the Australian flora in the light of latitudinal movements of the continent. *Journal of Biogeography*, **4**, 111–18.

BERGGREN, W. A., KENT, D., FLYNN, J. J. &
COUVERING, J. A. VAN (1985). Cenozoic chronology.
Geological Society of America Bulletin, **96**, 1407–8.

BINT, A. N. (1981). An early Pliocene pollen assemblage
from Lake Tay, southwestern Australia and its phytogeographic implications. *Australian Journal of Botany*, **29**,
277–91.

BOWLER, J. M. (1982). Aridity in the late Tertiary and
Quaternary of Australia. In *Evolution of the Flora and
Fauna of Arid Australia*, ed. W. R. Barker & P. J. M.
Greenslade, pp. 35–45. Frewille: Peacock Publi-
cations.

CALLEN, R. A. & TEDFORD, R. H. (1976). New late
Cainozoic rock units and depositional environments, Lake
Frome area, South Australia. *Transactions of the Royal
Society of South Australia*, **100**, 125–67.

CHUMRA, G. L. & LIU, B.-L. (1990). Pollen in the
lower Mississipi River. *Review of Palaeobotany and Palynology*, **64**, 253–61.

COLHOUN, E. A. & GEER, G. VAN DE (1986). Holo-
cene to middle last glaciation vegetation history at Tullabardine Dam, western Tasmania. *Proceedings of the Royal
Society of London B*, **229**, 177–207.

DAWSON, J. R. (1983). Late Neogene vegetation communities and climatic implications ascertained from a study
of eight cores from the Latrobe Valley depression, southeastern Australia. B.Sc. (Honours) thesis, Monash University.

DUDGEON, M. J. (1982). Stratigraphy and palaeobotany
of East and West Haldon, Main Range, southeast
Queensland. *Papers, Department of Geology, University of
Queensland*, **10**, 83–110.

FOSTER, C. B. (1982). Illustrations of early Tertiary
(Eocene) plant microfossils from the Yaamba Basin,
Queensland. *Geological Survey of Queensland, Publications*,
381, 32pp.

HAQ, B. U., HARDENBOL, J. & VAIL, P. R. (1987).
Chronology of fluctuating sea levels since the Triassic. *Science*, **235**, 1156–67.

HARRIS, W. K. (1971). Tertiary stratigraphic palynology,
Otway Basin. In *The Otway Basin of Southeastern Australia*, ed. H. Wopfner & J. G. Douglas. *Special Bulletin
of the Geological Surveys of South Australia and Victoria*,
pp. 67–87.

HEKEL, H. (1972). Pollen and spore assemblages from
Queensland Tertiary sediments. *Geological Survey of Queensland, Publications*, **355**, *Palaeontology Paper*, **30**.

HEUSSER, L. E. (1978). Spores and pollen in the marine
realm. In *Introduction to Marine Micropaleontology*, ed.
B. U. Haq & A. Boersma, pp. 327–60. New York:
Elsevier.

HILL, R. S. & MACPHAIL, M. K. (1985). A fossil flora
from rafted Plio-Pleistocene mudstones at Regatta Point,
Tasmania. *Australian Journal of Botany*, **33**, 497–517.

HILL, R. S. & MACPHAIL, M. K. (1994). Tertiary history and origins of the flora and vegetation. In *Vegetation
of Tasmania*, ed. J. B. Reid, R. S. Hill & M. J. Brown.
Hobart: Tasmanian Government Printer, in press.

HOLDGATE, G. R. & SLUITER, I. R. K. (1991). Oligocene–Miocene marine incursions in the Latrobe Valley
depression, onshore Gippsland Basin: evidence, facies
relationships and chronology. In *The Cainozoic in Australia: A Re-appraisal of the Evidence*, ed. M. A. J. Williams,
P. De Deckker & A. P. Kershaw. *Geological Society of Australia Special Publication*, **18**, 137–57.

HOLMES, W. B. K., HOLMES, F. M. & MARTIN,
H. A. (1983). Fossil *Eucalyptus* remains from the Middle
Miocene Chalk Mountain Formation, Warrumbungle
Mountains, New South Wales. *Proceedings of the Linnean
Society of New South Wales*, **106**, 299–310.

HOS, D. (1975). Preliminary investigation of the palynology
of the Upper Eocene Werillup Formation, Western Australia. *Journal of the Royal Society of Western Australia*, **58**,
1–14.

KEMP, E. M. (1978). Tertiary climatic evolution and vegetation history in the southeast Indian Ocean region. *Palaeogeography, Palaeoclimatology, Palaeoecology*, **24**, 169–208.

KERSHAW, A. P. (1976). A late Pleistocene and Holocene
pollen diagram from Lynch's Crater, north-eastern
Queensland, Australia. *New Phytologist*, **77**, 469–98.

KERSHAW, A. P. (1986). The last two glacial–interglacial
cycles from northeastern Australia: implications for climatic
change and Aboriginal burning. *Nature*, **322**, 47–9.

KERSHAW, A. P. (1988). Australasia. In *Vegetation History*,
ed. B. Huntley & T. Webb, pp. 237–306. Dordrecht:
Kluwer Academic Publishers.

KERSHAW, A. P., BOLGER, P., SLUITER, I. R. K.,
BAIRD, J. & WHITELAW, M. (1991*a*). The origin
and evolution of brown coal lithotypes in the Latrobe
Valley, Victoria, Australia. *Journal of Coal Geology*, **18**,
233–49.

KERSHAW, A. P., D'COSTA, D. M., McEWEN
MASON, J. R. & WAGSTAFF, B. E. (1991*b*). Quat-
ernary vegetation of mainland southeastern Australia.
Quaternary Science Reviews, **10**, 391–404.

KERSHAW, A. P., MCKENZIE, G. M. & MCMINN, A.
(1993). A Quaternary vegetation history of northeast
Queensland from pollen analysis of O.D.P. Site 820. *Proceedings of the Ocean Drilling Program Scientific Results*,
133, in press.

KERSHAW, A. P. & SLUITER, I. R. K. (1982). Late
Cenozoic pollen spectra from the Atherton Tableland,
north-eastern Australia. *Australian Journal of Botany*, **30**,
279–95.

KUKLA, G. (1989). Long continental records of climate
– an introduction. *Palaeogeography, Palaeoclimatology, Palaeoecology*, **72**, 1–9.

LANGE, R. T. (1978). Carpological evidence for fossil
Eucalyptus and other Leptospermae (subfamily Leptospermoideae of Myrtaceae) from a Tertiary deposit in the
South Australian arid zone. *Australian Journal of Botany*,
26, 221–33.

LULY, J. (1991). A history of Holocene changes in vegetation and climate around the Playa Lake Tyrrell, northwestern Victoria, Australia. Ph.D. thesis, Australian
National University.

326 A. P. KERSHAW *et al.*

LULY, J., SLUITER, I. R. K. & KERSHAW, A. P. (1980). Pollen studies of Tertiary brown coals: preliminary analysis of lithotypes within the Latrobe Valley, Victoria. *Monash Publications in Geography*, 23.

MACPHAIL, M. K., HILL, R. S., FORSYTH, S. M. & WELLS, P. M. (1991). A Late Oligocene–Early Miocene cool climate flora in Tasmania. *Alcheringa*, 15, 87–106.

MACPHAIL, M. K. & MARTIN, A. R. H. (1991). 'Spineless' Asteraceae. (episode 2). *Palynological and Palaeobotanical Association of Australasia Newsletter*, 23, 1–2.

MACPHAIL, M. K. & TRUSWELL, E. M. (1989). Palynostratigraphy of the central west Murray Basin. *BMR Journal of Australian Geology and Geophysics*, 11, 301–31.

MARTIN, A. R. H. (1986). Late glacial and early Holocene vegetation of the alpine zone, Kosciusko National Park. In *Flora and Fauna of Alpine Australasia: Ages and Origins*, ed. B. A. Barlow, pp. 161–70. Melbourne: CSIRO.

MARTIN, H. A. (1973). Upper Tertiary palynology in southern New South Wales. *Geological Society of Australia Special Publication*, 4, 35–54.

MARTIN, H. A. (1979). Stratigraphic palynology of the Mooki Valley, N. S. W. *Journal and Proceedings, Royal Society of New South Wales*, 112, 71–8.

MARTIN, H. A. (1980). Stratigraphic palynology from shallow bores in the Namoi River and Gwydir River Valleys, north central New South Wales. *Journal and Proceedings, Royal Society of New South Wales*, 113, 81–7.

MARTIN, H. A. (1981). Stratigraphic palynology of the Castlereagh River Valley, New South Wales. *Journal and Proceedings, Royal Society of New South Wales*, 114, 77–84.

MARTIN, H. A. (1984a). The stratigraphic palynology of the Murray Basin in New South Wales II. The Murrumbidgee area. *Journal and Proceedings, Royal Society of New South Wales*, 117, 35–44.

MARTIN, H. A. (1984b). The stratigraphic palynology on the Murray Basin in New South Wales III. The Lachlan area. *Journal and Proceedings, Royal Society of New South Wales*, 117, 45–51.

MARTIN, H. A. (1986). Tertiary stratigraphic palynology, vegetation and climate of the Murray Basin in New South Wales. *Journal and Proceedings, Royal Society of New South Wales*, 119, 43–54.

MARTIN, H. A. (1987). The Cainozoic history of the vegetation and climate of the Lachlan River Region, New South Wales. *Proceedings of the Linnean Society of New South Wales*, 117, 45–51.

MARTIN, H. A. (1988). Stratigraphic palynology of the Lake Menindee region, New South Wales. *Journal and Proceedings, Royal Society of New South Wales*, 121, 1–9.

MARTIN, H. A. (1990a). Tertiary climate and phytogeography in southeastern Australia. *Review of Palaeobotany and Palynology*, 65, 47–55.

MARTIN, H. A. (1990b). The palynology of the Namba Formation in Wooltana-1 bore, Callabona Basin (Lake Frome), South Australia and its relevance to Miocene grasslands in Central Australia. *Alcheringa*, 14, 247–55.

MARTIN, H. A. (1991a). Dinoflagellate and spore pollen biostratigraphy of the SADME MC63 bore, Western Murray Basin. *Alcheringa*, 15, 107–44.

MARTIN, H. A. (1991b). Tertiary stratigraphic palynology and palaeoclimate of the inland river systems in New South Wales. In *The Cainozoic of Australia: A Re-appraisal of the Evidence*, ed. M. A. J. Williams, P. De Deckker & A. P. Kershaw. *Special Publication of the Geological Society of Australia*, 18, 181–94.

MARTIN, H. A. & MCMINN, A. (1993). Palynology of sites 815 and 823: the Neogene vegetation history of coastal northeast Australia. *Proceedings of the Ocean Drilling Program Scientific Results*, 133, in press.

McEWEN MASON, J. R. C. (1989). The palaeomagnetics and palynology of late Cainozoic cored sediments from Lake George, New South Wales, southeastern Australia. Ph.D. thesis, Monash University.

McMINN, A. (1981). A Miocene microflora from the Home Rule Kaolin deposit. *Geological Survey of New South Wales, Quarterly Notes*, 43, 1–4.

McMINN, A. & MARTIN, H. A. (1993). Late Cainozoic pollen history from Site 765, Eastern Indian Ocean. *Scientific Reports of the Ocean Drilling Programme, Leg 123*, in press.

MILLER, K. G. & FAIRBANKS, R. G. (1985). Cainozoic ^{18}O record of climate and sea level. *South American Journal of Science*, 81, 248–9.

MILNE, L. A. (1988). Palynology of the late Eocene lignitic sequence from the western margin of the Euca Basin, Western Australia. *Memoirs of the Association of Australasian Palaeontologists*, 5, 285–310.

OWEN, J. A. (1975). Palynology of some Tertiary deposits from New South Wales. Ph.D. thesis, Australian National University.

OWEN, J. A. K. (1988). Miocene palynomorph assemblages from Kiandra, New South Wales. *Alcheringa*, 12, 269–97.

READ, J., HOPE, G. S. & HILL, R. S. (1990). Integrating historical and ecophysiological studies in *Nothofagus* to examine the factors shaping the development of cool rainforest in southeastern Australia. In *Proceedings of the Third IOP Conference*, ed. J. G. Douglas & D. C. Christophel, pp. 97–106. Melbourne: A-Z Printers.

RUSSELL-SMITH, J. (1991). Classification, species richness and environmental relations of monsoon rainforest in northern Australia. *Journal of Vegetation Science*, 2, 259–78.

SHACKLETON, N. J. & OPDYKE, N. D. (1977). Oxygen isotope and palaeomagnetic evidence for early northern hemisphere glaciation. *Nature*, 270, 216–9.

SINGH, G. & GEISSLER, E. A. (1985). Late Cainozoic history of vegetation, fire, lake levels and climate at Lake George, New South Wales, Australia. *Philosophical Transactions of the Royal Society of London*, 311, 379–447.

SINGH, G. & LULY, J. (1991). Changes in vegetation and seasonal climate since the last full glacial at Lake Frome. *Palaeogeography, Palaeoclimatology, Palaeoecology*, 84, 75–86.

SLUITER, I. R. K. (1984). Palynology of Oligo-Miocene brown coal seams, Latrobe Valley, Victoria. Ph.D. thesis, Monash University.

STOVER, L. E. & PARTRIDGE, A. D. (1973). Tertiary and Late Cretaceous spores and pollen from the Gippsland Basin, southeastern Australia. *Proceedings of the Royal Society of Victoria*, **85**, 237–86.

TRUSWELL, E. M. (1982). Palynology of a Tertiary core sample from South Alligator area, E. L. 1287. Unpublished report to Urangesellschaft Australia Pty Ltd.

TRUSWELL, E. M. & HARRIS, W. K. (1982). The Cainozoic palaeobotanical record in arid Australia: fossil evidence for the origins of an arid-adapted flora. In *Evolution of the Flora and Fauna of Arid Australia*, ed. W. R. Barker & P. J. M. Gleenslade, pp. 67–76. Frewville: Peacock Publications.

TRUSWELL, E. M., KERSHAW, A. P. & SLUITER, I. R. (1987). The Australian/southeast Asian connection: evidence from the palaeobotanical record. In *Biogeographic Evolution in the Malay Archipelago*, ed. T. C. Whitmore, pp. 32–49. Oxford: Oxford University Press.

TRUSWELL, E. M., SLUITER, I. R. & HARRIS, W. K.

(1985). Palynology of the Oligo-Miocene sequence in the Oakvale-1 corehole, western Murray Basin, South Australia. *Bureau of Mineral Resources Journal of Australian Geology and Geophysics*, **9**, 267–95.

TULIP, J. R., TAYLOR, G. & TRUSWELL, E. M. (1982). Palynology of Tertiary Lake Bunyan, Cooma, New South Wales. *Bureau of Mineral Resources Journal of Australian Geology and Geophysics*, **7**, 255–68.

VAN DER KAARS, W. A. (1991). Palynology of eastern Indonesian marine piston-cpres: a late Quaternary vegetational and climatic record for Australasia. *Palaeogeography, Palaeoclimatology, Palaeocology*, **85**, 239–302.

WILLIAMS, M. A. J. (1993). Cenozoic climatic changes in deserts: a synthesis. In *Geomorphology of Deserts*, ed. A. D. Abrahams & A. J. Parsons. London: Chapman & Hall, in press.

WOOD, G. R. (1986). Late Oligocene to Early Miocene palynomorphs from GSQ Sandy Cape 1 3R. *Geological Survey of Queensland, Publication*, **387**, 1–27.

ZAGWIN, W. H. (1985). An outline of the Quaternary stratigraphy of the Netherlands. *Geologie en Mijnbouw*, **64**, 17–24.

14 The Oligo-Miocene coal floras of southeastern Australia

D. T. BLACKBURN & I. R. K. SLUITER

This chapter presents the results of the authors' separate and conjoint studies of the palaeobotany, stratigraphy and palaeoecology of the Latrobe Valley Coal Measures, a thick sequence of coal seams and interseam clastics in the Gippsland region of southeastern Australia (Figure 14.1).

The Latrobe Valley Coal Measures are of Middle Eocene–Middle Miocene age and were deposited after the separation of Australia from Antarctica, a time of significant climatic change for the Australian continent and a period during which major global changes in sea levels and climates occurred. The Eocene–Miocene was an important time for the development of the modern Australian flora and the Latrobe Valley Coal Measures provide a window into this critical period. The study of the fossil assemblages of the coals has led to a detailed understanding of the vegetation that formed the coal measures and contributes significantly to our knowledge of the development of the modern Australian flora.

Previous workers have directed their efforts to particular aspects of the nature and relationships of the coal swamp vegetation. Early workers (Chapman, 1925a,b; Deane, 1925; Cookson, 1946, 1947, 1950, 1953, 1957, 1959; Cookson & Duigan, 1950, 1951; Cookson & Pike, 1953a,b; 1954a,b; Pike, 1953; Patton, 1958) concentrated upon the taxonomic affinities of macroscopic and microscopic plant remains. Duigan (1966) studied selected micro- and macrofossil plant taxa from

a palaeogeographical, ecological and evolutionary viewpoint. Baragwanath & Kiss (1964), Partridge (1971) and Stover & Partridge (1973) were more concerned with establishing stratigraphic relationships with the microfossils. The last of these studies, of pollen and marine microfossils, developed a long-accepted biostratigraphy for the basin.

Recently the biostratigraphy has been revised using more extensive sampling and better correlation with the geology of the basin (Haq *et al.*, 1987; Holdgate, 1985, 1992; Holdgate & Sluiter, 1991). These studies also demonstrated the significance of eustatic sea level changes in influencing sedimentation within the basin and also that significant marine transgressions can be traced in both the lithofacies and the microfossil record.

Few studies have approached the fossils from ecological or evolutionary viewpoints. Duigan (1966), in a landmark contribution, considered both macro- and microfossils in terms of their modern taxonomic relationships and the ecologies of equivalent modern vegetation types. It was concluded that *Nothofagus*, *Agathis* and Lauraceae spp. were dominant in the regional vegetation and that the coal-forming vegetation was dominated by gymnosperms (*Dacrydium*, *Araucaria*, *Agathis*) and angiosperms including Casuarinaceae, Proteaceae (*Banksia* and *Xylomelum*), Restionaceae, Centrolepidaceae and Myrtaceae.

These conclusions have been generally supported by more recent studies (Blackburn, 1980, 1981, 1985; Luly *et al.*, 1980; Sluiter & Kershaw, 1982). By combining for the first time the information from both macrofossil and microfossil studies, we now have a more complete picture of the ecology and dynamics of the coal swamp vegetation and the influences of climatic and edaphic factors. Improvements in sample collection, fossil extraction and taxonomic procedures have allowed the identification of many new taxa and the quantification of taxon–lithotype relationships.

By using short sampling intervals, continuity of sampling, accurate depth control, improved techniques for the extraction of fossils and a more refined taxonomy that combines palynomorphs and macrofossils, it has been possible to demonstrate that the variations in fossil assemblages, both vertically and laterally, were controlled by local and regional influences.

A critical and comparative study of macrofossils, microfossils, sediment lithology and the ecology of modern species equivalents was carried out in a conjoint study by the authors. The major aims of this study were:

1. To reconcile the taxonomy of palynomorphs and macrofossils by analysing the same coal samples.
2. To develop detailed descriptions of the floristics of the coal seams, in particular in relation to coal lithotypes and stratigraphy.
3. To describe the plant associations, climatic and edaphic conditions of the coal swamps and relate these to eustatic sea level changes.

REGIONAL SETTING

Southeastern Australia is an important region for the evolution of the Australian flora through the Tertiary. It was subjected to the tectonism that raised the Eastern Highlands, it has been influenced by the sea level changes which led to the opening of Bass Strait and it has experienced local tectonic movements that formed the Gippsland Basin.

The Latrobe Valley Depression lies in the western portion of the onshore Gippsland Basin, bounded to the north by the Eastern Highlands and to the south by the South Gippsland Highlands (Figure 14.1). Onshore sedimentation within the

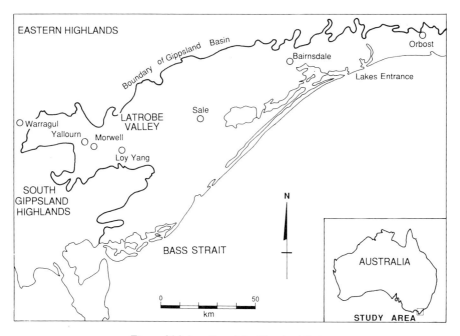

Figure 14.1 Location of the Gippsland Basin.

depression produced the Latrobe Valley Coal Measures (Thomas & Baragwanath, 1949), which are up to 700 m thick and composed of lignites, clays, sands and volcanics. These sediments overlie a Mesozoic and Palaeozoic basement. Offshore sediments of open marine environments characterised by carbonates of the Seaspray Group form lateral equivalents to the onshore coal measures. The marine units were more or less separated from the coal measures throughout the deposition of the basin sediments by a barrier of beaches and dunes (Holdgate & Sluiter, 1991), although the onshore extent of marine transgressions varied considerably.

The basal Tertiary unit of the Latrobe Valley Depression is the Middle Eocene–Early Oligocene Traralgon Formation, composed principally of clastic sediments with minor coal seams within the basin (Partridge, 1971; Hocking *et al.*, 1976). The Traralgon Formation has not been considered in this study because suitable coal samples were not available. The Traralgon Formation is overlain by clastics and coals of the Early Oligocene–Early Miocene Morwell Formation, which is in turn overlain by similar sediments of the Early–Middle Miocene Yallourn Formation (Smith, 1982). These latter geological formations contain most of the economic coal reserves of the Latrobe Valley, with individual seams reaching 100 m in thickness, with the maximum total thickness of the Morwell and Yallourn Formation coals exceeding 300 m. The coals are exposed in the Yallourn, Morwell and Loy Yang Open Cut coal mines, where they are currently being mined for electricity generation.

COAL LITHOTYPES

Coals from the Gippsland Basin are low grade lignites with low ash contents (generally less than 2%) that were deposited in thick seams of considerable continuity across the basin. The low ash and clastic content of the coals suggest that they were deposited as raised, rain-fed swamps with little fluvial influence. Stable depositional conditions led to the formation of laterally extensive layers of coals of different colours. These differently coloured layers are clearly visible in open-cut exposures and are often continuous across the full extent of the open-cut mines.

A classification of coals into groups termed lithotype classes was first described by George (1982) and used coal colour, percentage wood content (xylite) and the percentage of amorphous ground matrix (gelinite) as determinants of lithotype. It was known that the combustion properties of the coal related to its lithotype, e.g. high fusinite coals produce more ash and high xylite coals cause problems in milling because of their fibrous nature. Therefore a procedure for lithotype classification that could be used in the field or with simple laboratory tests was desirable.

Coal colour has proved to be the most useful indicator of lithotype because it can be measured easily in the laboratory and estimated reliably in the field. Coal colour as measured in the laboratory is proportional to the reflectance of a dried, powdered coal sample. The darkest coals have colour values as low as 40, while the palest coals have colour values as high as 170. Weathered coal samples are darker in colour than equivalent fresh samples. Weathering also increases the degree of gelification, tends to disrupt macrofossils and makes the coal more difficult to macerate for the recovery of fossil material. Therefore, wherever possible, fresh samples were extracted for lithotype and palaeobotanical analyses.

Secondary parameters for determining a coal sample's lithotype include xylite and gelinite. The values for these were determined from microscopic examination of coal slices using reflected light, and represent the percentage of a number of fields, covered by each parameter.

Correlation coefficients for linear regressions between colour, xylite and gelinite are given in Table 14.1. These statistics have been calculated without consideration of the distributions and variances of the data and so are not robust. They serve, however, as qualitative indicators of the relationships between coal properties.

Coal colours are negatively correlated to xylite and gelinite. This is a consequence of the colour scale assigning low values to dark coals and high values to pale coal and also xylite and gelinite

Table 14.1. *Pearson correlation coefficients for pairwise regressions of coal properties for 215 samples*

Parameters	Pearson r
Colour *vs* xylite	−0.55
Colour *vs* gelinite	−0.80
Colour *vs* Log_{10}(gelinite)	−0.84
Gelinite *vs* xylite	0.66

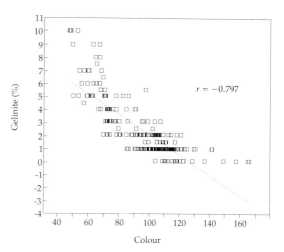

Figure 14.3 Coal colour data plotted against percentage of gelinite.

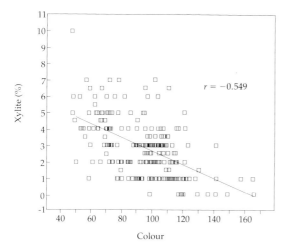

Figure 14.2 Coal colour data plotted against percentage of xylite.

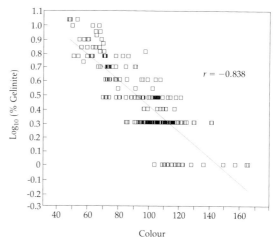

Figure 14.4 Coal colour data plotted against the logarithm (base 10) of the percentage of gelinite. This linearises the data resulting in a higher value for the regression coefficient.

being more common in darker coals. Colour, xylite and gelinite data for 215 samples are plotted as Figures 14.2, 14.3 and 14.4.

Gelinite in particular is influenced by weathering of coals after exposure. The variations in gelinite caused by weathering mean that this parameter is of little discriminatory value in classifying weathered samples. The xylite component consists principally of wood, bark and root material. The relationships between xylite and colour, and xylite and lithotype, are confused in circumstances where abundant pale-coloured root material arising from living vegetation growing on overlying horizons extends into an underlying horizon. In these situations the resulting high xylite coals may be 'lighter' in colour and yet have other characteristics that more closely ally the horizon with darker coals. Many samples are composed mostly of wood from large logs. The

colours of these samples generally bear little relationship to that of adjacent horizons, even if overall depositional conditions were the same.

Other coal components, commonly reported in coal petrographic studies, have not been used to define lithotypes. These components include sporinite, exinite, vitrinite, resinite, fusinite and alginite. Of these, fusinite and to some extent resinite have been quantified and correlated with coal depositional environments. Fusinite is most abundant in darker coals and is associated with characteristic plant assemblages. Resinite is most abundant in medium light to pale coals where it is often present as large amber bodies sometimes more than a kilogram in weight, and as dispersed resin fragments and droplets. It is assumed that the resin is derived principally from gymnosperm exudates and resin bodies in gymnosperm wood. The most likely sources for the larger amber bodies were trees of the family Araucariaceae. Sporinite and exinite consist of the resistant walls of spores and pollen grains and are the dominant component of some pale coal samples.

The need for an efficient and reliable lithotype classification system led to the studies by Blackburn (1980, 1981, 1985) of the relationships between macrofossil assemblages and lithotypes. These studies demonstrated that coal colour was correlated to the coal swamp vegetation and depositional environments. This correlation has also been addressed by Luly *et al.* (1980). Lithotype classes and typical colour ranges for the three localities forming the subject of this study are given in Table 14.2. The correlation coefficient for a regression between coal colour and lithotype assigned a rank order number (1 = Dark and 5 = Pale) is approximately 0.95 for the same three seams.

The nature and origin of lithotypes has been debated by several authors (Baragwanath, 1962; Baragwanath & Kiss, 1964; Luly *et al.*, 1980; Blackburn, 1981; Thompson, 1981; Smith, 1982; McKay *et al.*, 1985). The relationships between lithotypes and coal depositional environments were elucidated by Luly *et al.* (1980) and Blackburn (1981) and refined by Sluiter (1984) and Kershaw *et al.* (1991). Lithotypes have also been called lithofacies (Bolger, 1991), suggesting a relationship between coal colour and depositional environment.

A number of studies have suggested that Very Dark (Vdk), Dark (Dk), Medium dark (Md) and Medium light (Ml) lithotypes were formed from swamp forest vegetation, with the peat surface being generally (Dk) or seasonally (Ml) emergent. Vdk coals usually have high fusinite or charcoal contents indicative of the relatively dry conditions

Table 14.2. *Coal colours and lithotype classes for coal samples from the Yallourn, Morwell and Loy Yang localities*

Lithotype	Coal colour		
	Yallourn	Loy Yang	Morwell
Very dark (Vdk)	<50	n.a.	n.a.
Dark (Dk)	50–70	<67	50–70
Medium dark (Md)	70–85	67–95	70–85
Medium light (Ml)	85–120	95–120	85–110
Light (Lt)	120–150	120–150	110–125
Pale (Pa)	>150	>150	125–150

n.a., no coal samples of this colour class occur in the studied sections.

under which these coals formed. Coals of Light (Lt) to Pale (Pa) lithotypes were probably formed under inundated conditions, with submerged aquatic vegetation rather than the emergent woody vegetations typical of darker coals. The abundance of sporinite, exinite and resinite in pale coals and the relatively low abundance of macrofossils is indicative of the open-water depositional environment.

SAMPLE LOCATIONS

Palynological data (Sluiter, 1984) spanning the lower half of the Yallourn and most of the Morwell coal seams and interseam sediments were obtained from three core holes at Loy Yang (LY). These sections provide a punctuated history of the vegetation from the Late Oligocene to the Middle Miocene. The three cores represent a sequence of 235m, embracing the deposition of the Morwell 1B and 1A and the lower half of the Yallourn Seams. Palaeobotanical sampling of coal exposures in the Morwell and Yallourn Open Cut coal mines (MOC and YOC) has covered a significant time range, but lacks an appreciable section of Early Miocene age Morwell 1A Seam deposition due to interruption of coal sedimentation by a clastic depositional sequence. The open cut sample transects represent a discontinuous sequence of 150 m from the Morwell 1B, Morwell 1A and Yallourn Seams. Deposition of

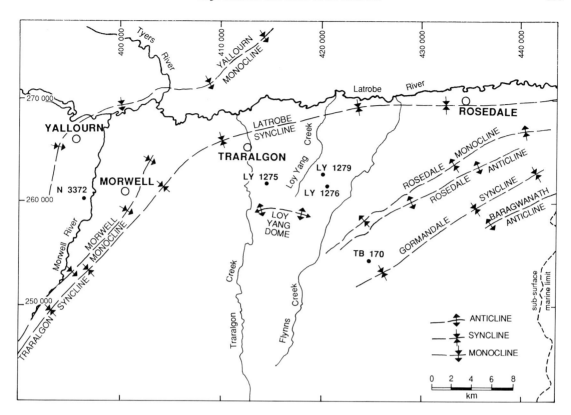

Figure 14.5 Palynological and palaeobotanical sample locations at Yallourn, Morwell and Loy Yang. Pollen samples were prepared from the three drill cores LY1275, LY1276 and LY1279. Macrofossil samples were taken from the Yallourn and Morwell Open Cut Mines.

the Yallourn Seam at Yallourn continued for approximately 25 m after it ceased at Loy Yang (G. R. Holdgate, personal communication). Palynological and palaeobotanical sample locations are shown in Figure 14.5.

STRATIGRAPHY AND SEQUENCE CORRELATIONS

Stratigraphic correlations between the three cores from Loy Yang have been made by Sluiter (1984), using five major pollen taxa/taxon groups; *Nothofagidites emarcidus*, *N. deminutus/N. vansteenisii*, Myrtaceae, Elaeocarpaceae and *Lagarostrobos* (see Figure 14.6). Large variations in relative frequencies of these taxa are taken to indicate variations in coal depositional environments and climate (Sluiter, 1984; Holdgate & Sluiter, 1991).

The quantitative pollen stratigraphy has been correlated to the independently dated stratigraphic pollen zonation for the Gippsland Basin defined by Stover & Partridge (1973) by the use of key indicator taxa. The pollen zonation for cores LY1275 and LY1276 was related to the dated stratigraphy and coal seam boundaries by Holdgate & Sluiter (1991) and Holdgate (1992). In addition, the former demonstrated from dinoflagellate assemblages that coal deposition had been interrupted a number of times by marine transgressions into the Gippsland Basin.

Within the Morwell 1B Seam of the Loy Yang section, Holdgate & Sluiter (1991) recognised eight periods of marine incursions resulting from eustatic sea level rises. These were then correlated to the Cenozoic stratigraphy of sea level changes outlined by Haq *et al.* (1987). The IIZs

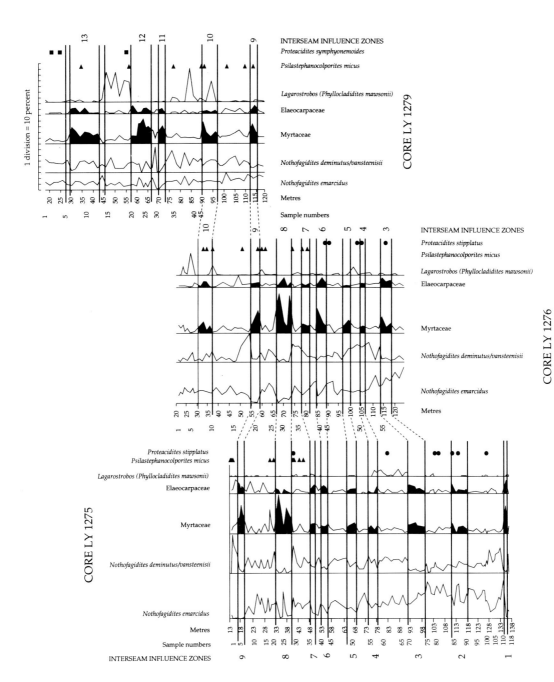

Figure 14.6 Quantitative pollen stratigraphy for Loy Yang cores LY 1275, LY 1276 and LY 1279. The location of Interseam Influence Zones is highlighted by dark shading. Interseam Influence Zones 1–8 are confidently placed, although Zones 9–13 are tentative at this stage.

(Interseam Influence Zones) were suggested to correspond more or less to the third-order cycles of marine transgression described by Haq *et al.* (1987) for the period spanning 27.5 Ma to 14 Ma. More recently work by G. R. Holdgate (unpublished data) suggests that further refinement of the Haq *et al.* (1987) sequences may be possible.

In addition to dinoflagellate evidence, the palynological, macrofossil and sedimentological records support the locations of IIZs. Sedimentological evidence is most reliable in the eastern parts of the basin where clastic sediments of marine origin are interbedded with the coal seams. These interseam sediments have been used to calibrate the IIZ stratigraphy of the eastern part of the basin. In the western part of the basin the clastic horizons corresponding to the marine incursions are generally less obvious and are concealed within more massive nonmarine clastic sedimentary accumulations or coal. For this reason the inference of the sequence positions of IIZs has been based upon studies of coal properties and fossil assemblages from the area of study.

The IIZs were originally defined within the Loy Yang section from palynological studies by Sluiter (1984). Broad correlations between Loy Yang cores and the open cut sections were postulated by Holdgate & Sluiter (1991). More detailed work by Holdgate (1992) included the expanded use of sulphur peaks from the database of the State Electricity Commission of Victoria drill cores (i.e. coal horizons that contained unusually high concentrations of sulphur, to make correlations across the basin). This demonstrated that the IIZs or marine transgressions have influenced coal deposition across the whole basin. This led Holdgate & Sluiter to develop a preliminary IIZ-based stratigraphy for the three study areas. This showed that a substantial sequence of Morwell 1A coal seam deposition recorded at Loy Yang was absent from the Morwell Open Cut section – the coal in that exposure being terminated by an erosional unconformity and overlain by a sequence of medium- to fine-grained clastics of fluvial and lacustrine origin. Conversely a sequence of the Yallourn Coal Seam was missing from the Loy Yang section, whereas deposition of Yallourn

Seam coals continued for a considerable extra period of time at Yallourn.

Figure 14.6 shows the relationships between IIZs, stratigraphy, coal seams and selected pollen taxa for cores LY1275, LY1276 and LY1279. The defined IIZs for the complete sequence including the Yallourn and Morwell Open Cut exposures are shown in Figure 14.7. The location of IIZs 9–13 are considered preliminary at this stage and await further analyses (G. R. Holdgate, personal communication).

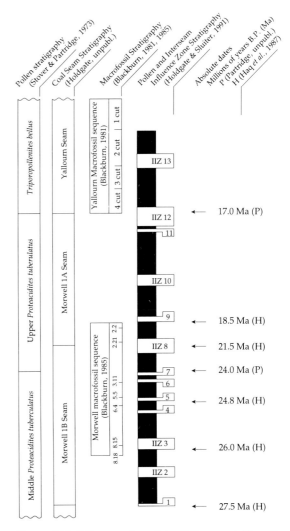

Figure 14.7 Pollen zonation of the Yallourn and Morwell seams and correlation to Interseam Influence Zones and regional pollen stratigraphy.

FLORISTIC CORRELATION AND ECOLOGICAL INTERPRETATION

The correlation of micro- and macrofossil records for the coal measures has not been previously possible because the studied coal intervals could not be placed in stratigraphic context. A major problem in relating different studies is that micro- and macrofossil data were from different coal samples. Major changes in coal floras are known to occur over vertical intervals of a metre or less. Previous workers have not recorded sample depths, so they cannot be referred to a datum, either local or Australian Height Datum (AHD, approximately equal to mean sea level). This means that only general statements can be made about the comparative floristics of different localities. In the present study, accurate depth control was maintained in all sections and so reliable comparisons could be made between the sections.

The fossils used for the study included microfossils (pollen and spores) and macrofossils (roots, wood, leaves, fruits, flowers and cones). Leaves were the most important macrofossil structures because they were abundant either as whole leaves or dispersed cuticles, could be easily separated from the coal matrix and often possessed distinctive cuticular morphologies or leaf architectures that have allowed confident assignment to modern families. Frequently, both macro- and microfossils assignable to the same taxon were associated in coal samples, e.g. Casuarinaceae fossils including pollen, anthers, foliage, fruit and wood dominated a number of samples. Because of these associations it is with a high degree of certainty that the different organs are assigned to the same biological species. Similar associations of organs have also been recorded in the Proteaceae and Gleicheniaceae. Appendix 14.1 contains a list of all taxa recorded for the Yallourn and Morwell Coal Seams as either fossil pollen or macrofossils. The coal seam and lithotype occurrences and relative abundances of all taxa are also outlined.

A significant problem in relating both macro- and microfossil floristics to the original ecology of the coal swamps arises because the ancient ecosystems cannot be observed directly and so ecological reconstructions must be based upon fossils preserved within the coal sediments. The relative abundances of fossilised and identifiable plant organs are unlikely to relate simply to the relative densities of plant taxa growing in the swamp environment. Both pollen and macrofossil assemblages are the result of:

1. Differing densities and distributions of source plants.
2. The relative contributions of these plants in terms of numbers and bulk of pollen grains, leaves etc.
3. Different modes of transport to their site of preservation.
4. Different degrees of preservation resulting not only from the nature of the pollen grains and other plant organs, but also the local depositional environment.

To allow the development of palaeoecological reconstructions it is fundamentally important to recognise the relative influences of these processes. Reconstruction requires reliable identification of fossil taxa, and an appreciation of the ecology, leaf and pollen production, and dispersal characteristics of the modern equivalents of recorded fossil taxa. It also requires studies of macro- and microfossil assemblages from the same samples to allow critical comparisons of the taxonomy and the relative contributions of the same taxa to the micro- and macrofossil assemblages. A conjoint study by the authors that addresses these requirements is described in the following section.

Resolving taxonomic relationships between the fossils and extant taxa formed a major part of the studies leading to this report. There is now abundant literature on the relationships between plant macro- and microfossil taxa from the Tertiary of southeastern Australia and extant plant species (see Chapters 10 and 11, this volume). In addition because species tend to be associative due to similar ecological requirements, the assemblages of fossils were used to determine whether particular identifications were likely. There will of course always be a group of taxa that cannot be assigned reliably to either fossil or extant groups because they may be poorly preserved, rare, nondescript or from taxa that have not been previously studied from the point of view of the palaeoecologist. Such taxa unfortunately represent lost information because potentially major components of the coal swamp floras will go unrecognised. In general, the dominant contributors to the coal floras have been identified with a high degree of certainty; the less common taxa have been identified with varying degrees of reliability. This is largely because either there was a poor propensity for preservation, e.g. due to thin cuticles or pollen grain walls, or simply their rarity means that they go unnoticed in sample preparations.

All of these factors reduce the ability of reconstructions of the coal swamp vegetation to reflect the patterns and compositions of the mosaics of plant communities that formed the coal. This is recognised and so the descriptions of coal swamp environments are based upon the dominant taxa and upon those less abundant taxa that are highly indicative of depositional conditions.

It should be pointed out that the relative frequencies of macro- and microfossils are not directly related to their contributions to the coal bulk. Microfossils represent a much greater proportion of the total coal mass in lighter coals, and particularly pale coals, than do macrofossils, because lighter coals contain little macroscopic plant material. On the other hand, dark coals are predominantly composed of macroscopic plant remains. These differences are explainable in terms of the different local environments of coal formation. Microfossil preparation techniques are aimed at separating just spores and pollen from any other plant tissues or coal groundmass. Thus, when the abundance of any given microfossil taxon is assessed, its absolute contribution to the coal mass needs to be considered in relation to the coal colour or lithotype.

CONJOINT MACRO- AND MICROFOSSIL INVESTIGATION

During 1982 a collaborative study of the palynology and palaeobotany of a selected interval of coal was conducted. The aims of this study were:

1. To record variations in coal floristics over short vertical intervals.
2. To clarify taxonomic relationships between the macro- and microfossil assemblages.
3. To characterise the floristics of individual coal strata.
4. To resolve in fine detail the floristic changes associated with lithotype boundaries.

Only through such a study could direct comparisons of the taxonomy and provenances of macro- and microfossils be made. There were several criteria that had to be fulfilled by the chosen coal section:

1. Visually obvious variations in coal properties, principally colour.
2. A major disjunction in coal deposition within the section.
3. Well-preserved fossil material.
4. Total thickness of less than 2 m to allow study of the section at approximately 4 cm resolution.

The latter requirement was a practical limit imposed by time and facilities. In addition it was desirable for the break in deposition to be of regional significance and therefore traceable across the limits of the coal exposure.

A section 1.3 m in thickness was sampled from Morwell Open Cut and contained layers of coal with abundant macrofossil remains and a distinctive layer of dark, friable, highly gelified coal of approximately 5 cm in thickness, which was clearly visible in exposures across the entire open cut mine (approximately 8 km^2). The friable layer contained medium to coarse sand grains and abundant charcoal, indicating that the layer was associated with an erosional surface. The friability and high degree of gelification indicated exposure and weathering, the abundance of clastic grains suggested erosion and the charcoal indicated dry conditions where fire was an important influence. This section lies between 4.8 m and 5.7 m below AHD. This places it within IIZ 6 (Figure 14.7). It is postulated therefore that the friable layer represents the top of an erosional surface and that the coal immediately above the erosional surface marks the beginnings of IIZ 6 and a marine transgression. The existence of a discontinuity at this location is further supported by the observation that the top few centimetres of the friable layer contained greater than 15% quartz grains in macerates. The quantity of coal eroded or alternatively the time period missing from the original peat swamp sequence is unknown. The section was subsampled by cutting it into 4 cm thick slices which were then split into three, with one third being used for lithotype determination and the remainder being macerated for micro- and macrofossil examination, respectively.

Coal colours

The section has very obvious variations in coal colour, and the sample colours are shown in Figure 14.8. Sample depths were referred to a zero datum represented by the base of the friable layer. Samples above this layer were assigned negative depths. From the base of the section there is a trend towards darker coals, although this trend is interrupted by two returns to lighter coals near depths of 40 and 10 cm. The darkest coal in the section occurs just above the zero datum between −5 and −15 cm. Above this there is a simple trend towards lighter coals, terminating at the top of the section (*ca* −70 cm) in coals that are almost at the upper colour limit (110) for the Ml lithotype.

Provenances of the fossils

Plant taxa represented by fossils in the coal seams are either local in origin, i.e. derived from plants growing on the swamp, or regional, i.e. growing marginal to, or at some distance from, the coal swamps. Taxa present as macro- but not microfossils are local and probably have poor pollen dispersal. Taxa present as micro- but not macrofossils are probably regional contributors. The best example of the latter comes from the pollen of *Nothofagus* subgenus *Brassospora*. This is abundant in most coal samples contributing from 7% to 50% of all pollen in samples from this section; but it has not been identified as a macrofossil and is therefore unlikely to have grown in the coal

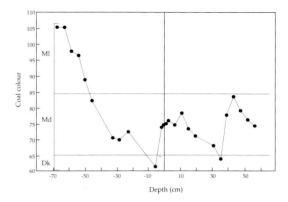

Figure 14.8 Coal colours for conjoint sample, Morwell Open Cut. Ml, Medium light; Md, Medium dark; Dk, Dark.

swamp. Casuarinaceae, however, are abundant both as micro- and macrofossils and are certain to have been growing on the swamp, though this does not preclude the family from being distributed regionally as well. Other local contributors include Gleicheniaceae, Araucariaceae (*Araucaria* and *Agathis*), Cupressaceae, Podocarpaceae (*Dacrycarpus* and *Phyllocladus*), Cunoniaceae, Elaeocarpaceae, Saxifragaceae (*Quintinia*), Myrtaceae, Proteaceae, Restionaceae and Sparganiaceae/Typhaceae. The specific distributions of major taxa are considered in the following section.

Plant assemblages

The fossil assemblages of the studied coal swamp section exhibit variations in a number of dominant taxa that are highly correlated to coal colours and presumably coal depositional environments. Major changes are observable in the contributions of dominant macro- and microfossil plant groups throughout the section and, most obviously, in the vicinity of the friable layer (zero datum). Figures 14.9, 14.10 and 14.11 are plots of the relative frequencies of individual macro- and microfossil taxa. Figure 14.9 includes those taxa that tend to be most abundant below the zero datum, Figure 14.11 is of taxa that tend to be most abundant above the datum and Figure 14.10 is of taxa which have either generalised distributions or are restricted to the region around the datum.

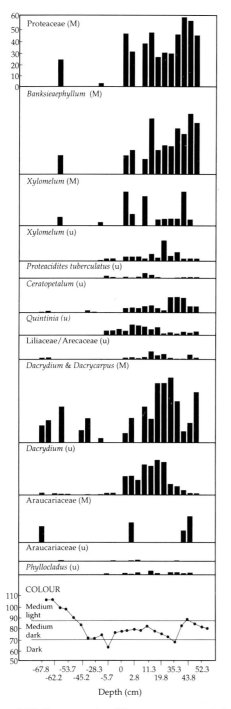

Figure 14.9 Frequency plots (% occurrence on vertical axis) for macro- and microfossil taxa (indicated by M and m, respectively), which occur mostly below the zero datum, and approximate location of the peak of IIZ 6.

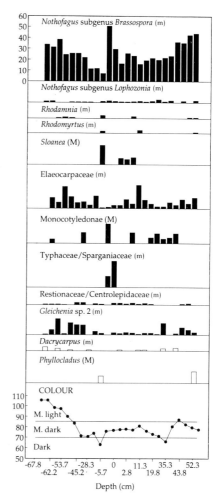

Figure 14.10 Frequency plots of taxa with generalised or restricted distributions in the section.

Plant taxa dominating both micro- and macrofossil assemblages of this section include ferns, conifers and angiosperms, and in particular Podocarpaceae, Casuarinaceae, Cunoniaceae, Elaeocarpaceae, Myrtaceae and Proteaceae, and Monocotyledonae. Other taxa including Gleicheniaceae, Fagaceae and Saxifragaceae (*Quintinia*) are common as pollen but uncommon or rare as macrofossils, whereas for the Blechnaceae, Zamiaceae and Araucariaceae the converse is true.

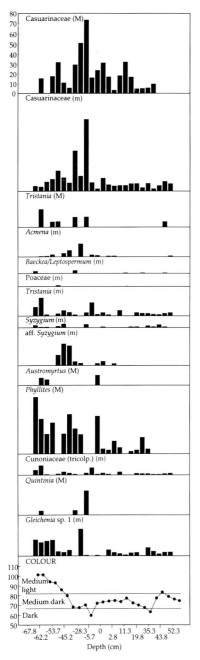

Figure 14.11 Frequency plots of taxa that occur mostly above the zero datum (peak of IIZ 6).

The following descriptions of the distributions of individual dominant taxa are arranged in descending order of their overall numerical contributions to the fossil assemblages.

Nothofagus

Pollen grains of several taxa of the genus *Nothofagidites* are abundant throughout the section, with best representation just above the zero datum of the section. This is within the friable and highly fusinitic horizon that characterises the middle region of the section.

No macrofossils assignable to *Nothofagus* have been recorded. Given the known occurrence of this genus in other deposits, particular attention was given to any cuticular remains with superficial similarity to those known for the genus.

Macrofossils certainly ascribable to *Nothofagus* are well known from Eocene–Oligocene and Pliocene–Quaternary deposits in Tasmania and New Zealand (Hill, 1983*a,b*, 1984, 1987, 1991; Kovar *et al.*, 1987), with most occurrences in clastic sediments – typically carbonaceous clays. Despite the lack of *Nothofagus* macrofossils, pollen evidence from the Loy Yang cores indicates that a member of *Nothofagus* subgenus *Lophozonia* was a component of the coal swamp flora. The taxon is typically present at levels of up to 5% of the pollen sum in the Morwell seam Dk and Md lithotypes, which contain charcoal. The capacity to withstand fire and resprout from epicormic shoots is present within modern species of the subgenus (e.g. *N. moorei* and *N. cunninghamii*) according to Howard (1981).

It is considered unlikely that any other *Nothofagus* spp. were present on the coal swamp, despite the abundance of pollen of subgenus *Brassospora*. A similar situation occurs today in New Guinea, where modern pollen rain studies (Walker & Flenley, 1979) have shown *Nothofagus* subgenus *Brassospora* pollen to be transported by wind over very long distances. This pollen can be overrepresented with respect to its relative presence in hinterland vegetation. The consistent occurrence of very high *Nothofagus* subgenus *Brassospora* pollen values in Morwell 1B coal especially is suggested by Sluiter (1984) to have been aided by the contemporaneous lack of other families with high pollen dispersal capabilities, such as Euphorbiaceae, Urticaceae/Moraceae, Ulmaceae and Poaceae.

Subsidiary support for a nonswamp, wind-transported origin for this *Nothofagus* pollen is forthcoming from the relationship the taxon group exhibits with coal colour. It was noted above (p. 337) that the absolute contributions of microfossils to individual coal horizons is related to coal colour because the percentages of the total coal bulk represented by macroscopic plant remains is inversely proportional to coal colour. Paler coals have a relatively low macroscopic tissue component as a result of their deposition in more open water environments where leaf-litter and wood deposition rates are likely to be scarcely greater than decomposition rates. It is likely therefore that, over the extent of the section of interest, inputs of *Nothofagus* pollen from regional sources occurred at a nearly constant rate. This is of particular interest because, at least over limited intervals, this constant background 'rain' of *Nothofagus* subgenus *Brassospora* pollen could potentially be used to infer the relativities between local and regional contributions of other taxa. Taxa whose relative pollen frequency curves do not follow that of *Nothofagus* are likely to be predominantly local in origin.

The peak abundance of *Nothofagus* subgenus *Brassospora* is at the top of the friable, zero datum horizon. It is possible that the peat surface may

have been dry at this time with a reduced diversity of plants. Under such conditions, it is conceivable that regional pollen contributors would have been of greater significance in the preserved assemblages.

Casuarinaceae

Macrofossils assignable to *Gymnostoma* included leafy shoots, dispersed cuticles, complete anthers and rare female cones. Pollen extracted from the anthers was identified as *Haloragacidites harrisii*. Note that apparent differences in macro- and microfossil distributions (Figure 14.12) are a result of Podocarpaceae pollen (*Dacrydium* and *Dacrycarpus*) being over-represented. This has depressed the relative frequencies for Casuarinaceae pollen in the lower half of the section. The distribution of Casuarinaceae is almost the inverse of that of *Nothofagus*. It is certain that Casuarinaceae were local to the coal swamps and important in the vegetation.

Gymnostoma fossils are abundant and widely distributed through the section, but the peak abundances occur just above the top of the friable horizon; that is, immediately following the marine transgression of IIZ 6. It is likely that *Gymnostoma* would have been a dominant component of the

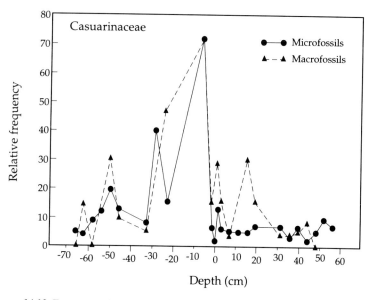

Figure 14.12 Frequency plot of Casuarinaceae macro- and microfossils around the zero datum (peak of IIZ 6).

vegetation at this time. It is absent from the bottom of the section where assemblages were dominated by Proteaceae and Podocarpaceae. The fossil taxon of *Gymnostoma* was probably an opportunistic plant coloniser during periods of major change in coal-depositional environments and this is supported by its modern distribution. Extant taxa occur on the margins of tropical – subtropical rainforests in Australia, Indonesia, New Guinea, New Hebrides and New Caledonia, and Fiji. Some species occur along and within stream channels, on lake margins, on back-beach sands and in shallow freshwater swamps. There is some evidence that Casuarinaceae in general occur in coal sections where colours fluctuate markedly with depth or where there is evidence for an influence of fire in the vegetation.

Casuarinaceae fossils assignable to both *Allocasuarina* and *Gymnostoma* are common throughout the coal seams. The former is most common in the Yallourn Seam, but it never reaches the abundances of *Gymnostoma*, the cuticles of which can represent about 80% of the macrofossil flora in some samples from the Morwell Seam, including the study section.

The Casuarinaceae is one of five dominant macrofossil plant families within the fossil assemblages of the Morwell Coal Seams. The others are Araucariaceae, Cunoniaceae, Podocarpaceae and Proteaceae.

Podocarpaceae

The dominant podocarp macrofossils in the section are *Dacrydium rhomboideum* and *Dacrycarpus latrobensis* (Figure 14.13), with *Phyllocladus morwellensis* a secondary component. Fossils consist of wood, roots, leafy shoots, dispersed cuticles, fruits and pollen. Locally *D. rhomboideum* formed thick accumulations of leafy shoots with three-dimensional preservation and its wood is locally abundant. The taxon is very easily identified in the field and is an important indicator of Md to Dk lithotypes. It is often found associated with leaf-litters of *Banksieaephyllum laeve*, *B. pinnatum* and, less commonly, *B. obovatum* and *Phyllocladus morwellensis*.

Phyllocladus is a minor component of the pollen flora. Macrofossils were recorded from only two samples that lay immediately above and below the limits for *Phyllocladus* pollen. In both samples, foliage of the taxon was easily identifiable on

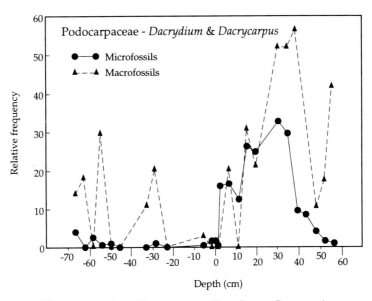

Figure 14.13 Frequency plot of Podocarpaceae (*Dacrydium* + *Dacrycarpus*) macro- and microfossils around the zero datum (peak of IIZ 6).

broken coal surfaces and associated with shoots of *D. rhomboideum*. The lack of correspondence between the macro- and microfossil records for this genus has not been explained. *Phyllocladus* as a macrofossil is uncommon but widespread in the Yallourn and Morwell coal seams, predominantly in Mdk and Ml lithotypes and has been found in abundance in the interseam Yallourn Clay. Locally, in the coal seams, it forms leaf-litters in association with *Banksieaephyllum laeve*, *Gymnostoma* and *Dacrydium rhomboideum*. It is closely related in its phyllode and cuticular morphology to the extant species *P. hypophyllus* (Hill, 1989), which grows in moist, high altitude forests from New Guinea to Borneo, often in association with species of *Nothofagus* subgenus *Brassospora*. Other related species grow in cool–cold temperate forests, often on organic substrates, in Tasmania and New Zealand.

Dacrydium and *Dacrycarpus* are much more abundant than *Phyllocladus*, have similar distributions as both macro- and microfossils and have distributions that are almost the inverse of that of the Casuarinaceae. These podocarps are most abundant below the zero datum, with relative frequencies similar to those of pollen and cuticles. In the region of the erosional zone, they are absent as macrofossils and occur in low numbers as pollen. The absence of macrofossils certainly means these genera were absent from the local vegetation near the sample site. This does not imply that they were entirely absent from the coal swamp basin during the peak of the transgression. As has been mentioned, the friable horizon is observable across very long distances in the Morwell Open Cut. Those podocarps are clearly indicators of a significant depositional change, but this does not imply that coal deposition and the growth of major species of the coal vegetation did not stop entirely, either across the whole basin, or simultaneously. It is reasonable to suppose that pockets of coal swamp taxa may have survived at different places and different times.

Above the top of the friable horizon, pollen from these taxa represents typically less than 3% of the pollen flora, while dispersed cuticles represent up to 30% of the macrofossil material.

These differences probably result from the abundance of *Nothofagus* and Casuarinaceae pollen in this part of the section, so that Podocarpaceae pollen frequencies appear to be low.

Extant *Dacrydium* and *Dacrycarpus* spp. may be shrubs or more commonly canopy trees in, or marginal to, rainforest or swamp forest. *Dacrycarpus* now occurs from southernmost China to Fiji and New Zealand, including Vanuatu and New Caledonia, the highest diversity being in New Guinea, often in cool, montane forests (Hill & Carpenter, 1991). *Dacrydium* species occur today throughout subtropical Southeast Asia to New Caledonia, Fiji and New Zealand, frequently ascending to subalpine forests (Page, 1990). Neither genus is extant in Australia, although both were common during the Cenozoic (Wells & Hill, 1989; Hill & Carpenter, 1991).

Proteaceae

The Proteaceae are represented as macrofossils by wood, leaves and dispersed cuticles of two genera, *Banksieaephyllum* and *Xylomelum*. Three species of the former genus have been recorded in the section, *B. laeve*, *B. pinnatum* and *B. obovatum*, and their distributions are shown in Figure 14.14. *Banksieaephyllum* and *Xylomelum* almost always occur together in the section and they share dominance of the macrofossil assemblages below the zero datum with *Dacrydium* and *Dacrycarpus* and are commonly associated with *Ceratopetalum* (Cunoniaceae). These Proteaceae are restricted mostly to Md coals within the Morwell Coal Seam and it appears that they are best developed in sequences that have stable depositional environments.

Banksieaephyllum as a whole includes taxa with affinities to modern *Banksia* and *Dryandra* spp. The fossil taxa from the Morwell Seam are almost certainly congeneric with *Banksia*, although their species relationships are unknown. Modern *Banksia* spp. are major components of dry sclerophyllous forests and heathlands on soils with low nutrient status, frequently occurring as dominants in coastal dune heaths. They also occur in swampy environments in southeastern Australia.

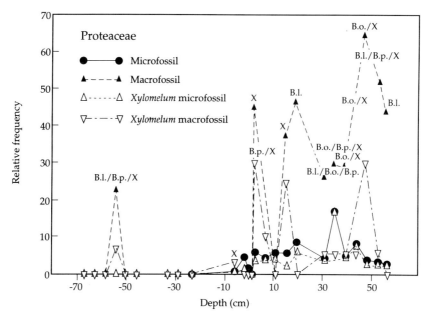

Figure 14.14 Frequency plot of Proteaceae macro- and microfossils around the zero datum (peak of IIZ 6). B.l., *Banksieaephyllum laeve*; B.o., *B. obovatum*; B.p., *B. pinnatum*; X, *Xylomelum*.

Cunoniaceae

This family is represented by macrofossils of *Phyllites yallournensis* and pollen of *Ceratopetalum* and tricolporate pollen (Figure 14.15). *Phyllites* is very common throughout the Yallourn and Morwell Coal Seams where it is an important indicator of Md to Ml lithotypes. It has been ascribed to the Cunoniaceae on the basis of its leaf architecture and cuticular morphology. The pinnately lobed leaves are very distinctive and the rounded pinnae are easily identifiable in the field.

Its association throughout the section with the tricolporate pollen type suggests that the leaves and pollen may be derived from the same plant species. Within the section, *Phyllites* and the tricolporate pollen taxon are dominant above the friable horizon. They also occur at the zero datum, a distribution that is similar to those of *Gleichenia* and several Myrtaceae taxa.

Ceratopetalum pollen is common below the zero datum and is locally abundant. Its distribution closely matches that of *Dacrydium* pollen and it was probably a major component of the coal swamp vegetation in the lower part of the section.

No macrofossil material ascribable to the extant genus *Ceratopetalum* has been recognised, but this is mostly a consequence of a lack of comparative studies between the fossil material and extant taxa.

Myrtaceae

The family is represented by a large number of macro- and microfossil taxa, with the dominants being related to *Tristania*, *Syzygium* and *Acmena* (Figure 14.16).

Dispersed cuticles ascribable to *Tristania* and pollen of *Myrtaceidites parvus* occur mainly above the friable horizon. *Tristania* cuticles are moderately common in Md and Ml lithotypes throughout the Morwell Seam. Extant members of the genus generally occur as canopy dominants of rainforest and rainforest alliances, with a number of species including *Tristania laurina* and *T. neriifolia* known to grow in wet, organic, low nutrient status substrates.

Pollen with affinities to *Syzygium* (*Myrtaceidites mesonesus*) is locally abundant above the friable horizon, where it is codominant with *Phyllites*. It has not been possible to refine the comparative

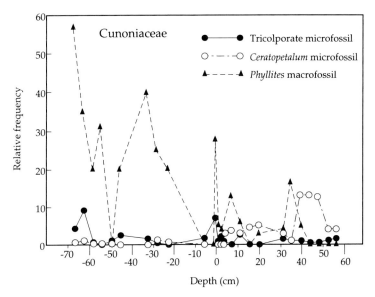

Figure 14.15 Frequency plot of Cunoniaceae macro- and microfossils around the zero datum (peak of IIZ 6).

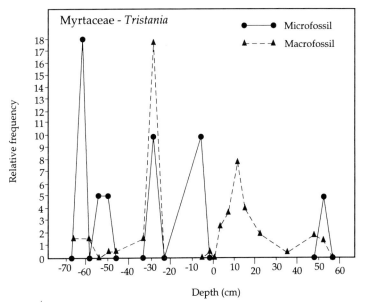

Figure 14.16 Frequency plot of *Tristania* (Myrtaceae) macro- and microfossils around the zero datum (peak of IIZ 6).

macro- and microfossil taxonomy beyond genus level. Extant species are distributed widely in warm temperate–subtropical rainforests and wet sclerophyllous forests, in well-drained–swampy sites throughout the Indo-Pacific region. In

Australia, *Syzygium* is found mostly as canopy components of rainforests.

Fossil pollen of *Rhodamnia* and *Rhodomyrtus* are rare in the section, but are most abundant on either side of the zero datum. Extant Australian species of these genera grow as canopy dominants of subtropical–warm temperate rainforests of the east coast.

Dispersed cuticles and pollen with affinities to *Acmena* are common above the friable horizon and are associated with the other myrtaceous taxa. Extant Australian *Acmena* spp. are widely distributed in rainforests of warm temperate–tropical origin.

Elaeocarpaceae

Pollen of this taxon is abundant in the section where Casuarinaceae and *Phyllites* are relatively uncommon. It is also rare in samples within, and at the top of, the friable horizon. Macrofossils of the family assignable to *Sloanea* occur in a limited range of samples around the zero datum. Pollen of the *Sloanea* type has not been recorded, although other pollen types that are comparable with the genus *Elaeocarpus* are common throughout the Yallourn and Morwell Coal Seams. Elaeocarpaceae pollen is also particularly abundant at major seam boundaries. Macrofossils including fruits of *Elaeocarpus* are abundant near the base of the Yallourn Seam, in Dk coals with abundant charcoal and associated with *Gleichenia* aff. *dicarpa*, but are not known from the study section.

Cuticles very similar to those of the extant species *Sloanea woolsii*, a warm temperate rainforest plant, are widespread, but uncommon in Md to Ml lithotypes in the lower half of the Morwell 1B Seam.

Comprehensive data on the comparative morphologies of extant Elaeocarpaceae leaves and cuticles are not available. It is possible therefore that representatives of the family have not been identified in the study.

Gleicheniaceae

Spores of two taxa were recorded from the section (Figure 14.17). The first, *Gleicheniidites circinidites*, is found in most samples, and is abundant above the top of the friable horizon. It is rare in the region of the zero datum. The second, *Dictyophyllidites arcuatus*, has a similar distribution, but is relatively more abundant near the base of the section. The former taxon has affinities to the extant genus *Gleichenia*, the latter has affinities to extant *Dicranopteris*.

The family is widespread throughout the coal seams and is known from the two pollen taxa and three macrofossil taxa, although no macrofossils were found in the study section. One macrofossil taxon, restricted to the Yallourn Seam, and found

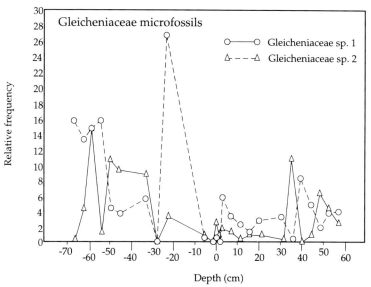

Figure 14.17 Frequency plot of *Gleichenia* microfossils around the zero datum (peak of IIZ 6).

locally in great abundance as rhizomes, pinnae, pinnules and spores, has close affinities to the extant *Gleichenia dicarpa*, which occurs in permanently wet areas of open, sclerophyllous forests and is highly tolerant and possibly dependent upon natural fires. It has been found most abundantly in Dk coals with high charcoal contents. Pinnules are often preserved by being partly charcoalified and the scaly, woody rhizomes form laterally extensive, almost monospecific, horizons.

The other macrofossils have affinities with extant species more commonly found in shaded, wetter, warm–cool temperate forests. *Gleichenia dicarpa* forms a dense, scrambling understorey, but taxa from more humid forests occur as scattered clumps and rarely contribute significantly to the leaf-litter. The absence of evidence for local fire and the thin cuticles of these taxa means that they are unlikely to be preserved as macrofossils. The occurrence of spores should be taken as a reliable indicator of the local occurrence of the family, whereas the absence of macrofossils may be due to the fact that conditions were not suitable for the preservation of the delicate tissues.

Araucariaceae

Pollen of *Araucaria* and *Agathis* occurs at low abundances (less than 2%) in scattered samples in the section, typically in Md lithotypes. They are most abundant below the zero datum. Macrofossils of *Araucaria ligniticii*, as leafy shoots and dispersed cuticles were locally abundant (up to 20%) and most common below the zero datum, forming substantial leaf-litter in places. Although not recovered as dispersed cuticles, well-preserved leafy shoots of *Araucaria* were seen within the friable horizon, where they were associated with high charcoal concentrations and often exhibited evidence of charring. Within the section there is a poor correspondence between the distributions of araucarian macro- and microfossils, and this has not been explained. However the fossils are certainly local in origin.

Extant species of *Araucaria* and *Agathis* occur in a wide range of rainforest environments, including montane rainforest, wet coastal forests, swamps and drier rainforest thickets. They are typically large, long-lived, emergent trees and usually dominate the canopy.

Araucaria ligniticii is small leaved and is placed in the section *Eutacta* of the genus, which includes the extant species *A. cunninghamii*, from which the fossil is clearly distinct. Some related extant species occur in relatively dry environments where the soil is wet and poorly drained. The fossil taxon is common–locally abundant in Md and Dk lithotypes throughout the Yallourn and Morwell Coal Seams. Its fossilised wood is very abundant in these lithotypes in the Morwell Seam and large logs, up to a metre in diameter, are common and the source of large resin bodies. *Araucaria* was undoubtedly a major component of the coal swamp vegetation over a long period of time.

Araucariaceae is well known in the Australian Tertiary record. Cookson & Duigan (1951) described the Latrobe Valley taxa, while Hill & Bigwood (1987) described six taxa from Tasmania and related these to other fossil and extant taxa, including those from the Latrobe Valley. The latter authors hypothesised that the modern distribution of Araucariaceae, with most species occurring in tropical and subtropical climates, is relict and that the past distributions may have included more temperate climates. The lithotype occurrences of fossil *Araucaria* and its association with other taxa suggests that it grew in an environment with at least seasonally water-logged, low nutrient, organic substrates. The abundance of charcoal in some samples where *Araucaria* was abundant suggests that the environment may have been seasonally dry.

Saxifragaceae

Quintinia has been recorded as dispersed cuticles and pollen. It is known from cuticles only above the friable zone, where it can represent as much as 20% of the macrofossil material. It is most abundant just above the top of the friable zone.

Pollen of *Quintinia psilatispora* is common below the top of the friable zone. The almost complete lack of correspondence between the macro- and microfossil records of this family has not been explained. Dispersed cuticles are assignable to the genus with certainty because of their highly characteristic trichomes that are most closely related to the extant species *Q. sieberi* and *Q. verdoni*. The pollen of extant species is indistinguishable from the fossil pollen.

Extant *Quintinia* spp. are normally secondary canopy trees in simple notophyll rainforests in New South Wales, Queensland and New Guinea. *Quintinia* macrofossils are rare in Ml and Md lithotypes in the Morwell Open Cut. The pollen is widely distributed throughout the coal seams and is locally common in the Ml and Md lithotypes in the Morwell 1A and 1B Seams.

Monocotyledonae

Typhaceae/Sparganiaceae macrofossils and pollen are common near the zero datum. Below this level, Liliaceae/Arecaceae pollen is common, while Restionaceae/Centrolepidaceae occur throughout as macro- and microfossils, with their greatest abundances above the friable horizon (Figure 14.18).

The Typhaceae/Sparganiaceae pollen is assigned to the fossil taxon *Sparganiaceaepollenites barungensis*, while the macrofossils that include

leaves, dispersed cuticles and fruits have been assigned to *Typha*. Extant species of these taxa typically form dense reedy swamps in shallow water or fringe open-water bodies. They are tolerant of seasonal drying and burning. Within the Morwell Seam, *Typha* macrofossils occur mostly in dark lithotypes and are frequently associated with abundant charcoal. Three-dimensional preservation of leaf tissues in carbonised fossils has been reported from both the Yallourn and Morwell Seams.

Ecological interpretations

It is clear from the distributions of taxa demonstrated in Figures 14.9, 14.10 and 14.11 that taxa fall into three main groups that reflect their distributions within the coal section. These groups are:

1. Those taxa that occur mostly below the zero datum.
2. Those that occur mostly above the friable horizon.
3. Those which have generalised or restricted distributions.

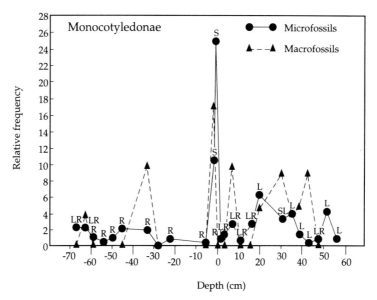

Figure 14.18 Frequency plot of monocotyledon macro- and microfossils around the zero datum (peak of IIZ 6). L, Liliaceae/Arecaceae; R, Restionaceae/Centrolepidaceae; S, Sparganiaceae/Typhaceae.

These three groups suggest very different vegetation types at different stages in the deposition of the short sequence of coal.

Below the zero datum, the vegetation was dominated by Proteaceae and Podocarpaceae, with *Banksieaephyllum laeve* and *Dacrydium rhomboideum* forming a shrub or small-tree storey. Locally there were probably individual emergents of *Ceratopetalum*, *Quintinia* and *Araucaria*, with an understorey dominated by Liliaceae in association with Restionaceae and/or Centrolepidaceae. *Gleichenia* was an accessory taxon in places.

Above the zero datum, the vegetation was dominated by Cunoniaceae (*Phyllites yallournensis*) and Casuarinaceae (*Gymnostoma*), with an abundance and diversity of Myrtaceae. The plant form of *Phyllites* is not known, although extant members of the family that produce tricolporate pollen types such as *Ackama* and *Pullea* are canopy components of a variety of rainforest types in New Zealand, New Caledonia, New Guinea and Australia. It is possible that *Gymnostoma* may have formed almost monospecific stands of medium trees, scattered in mosaics across the coal swamp. Gleicheniaceae sp. 1 was locally abundant and probably formed a scrambling understorey. Myrtaceae was diverse and medium–tall shrubs and trees of *Tristania*, *Syzygium* and *Acmena* would have been prominent. *Baeckea/Leptospermum* probably occurred in scattered heaths, with wiry Restionaceae as an understorey in more open areas of vegetation.

Taxa that grew generally across the coal swamp included Elaeocarpaceae, *Gleichenia* and Restionaceae/Centrolepidaceae. The pollen record also reveals that *Nothofagus* subgenus *Brassospora* was an important regional contributor. Taxa that grew around the friable zone are probably indicative of radically different conditions of coal deposition. The flora is depauperate and fire is considered to have been a significant influence. If we ignore the regional *Nothofagus* component, the vegetation at this time was probably a reed and rush swamp, with some Cunoniaceae shrubs and stands of Casuarinaceae.

Modern equivalents to a mixed Proteaceae/Podocarpaceae forest are not known, but the taxa and coal properties suggest that conditions may

have been warm, wet and climatically stable. Fire does not seem to have been a major influence. Associations of *Gymnostoma* and Cunoniaceae are known from scrubby forests in Australia, Fiji, New Guinea and New Caledonia, where locally they form distinct riparian and swamp margin vegetations. These extant vegetation types usually have a monocotyledonous understorey, with Typhaceae, Restionaceae and Cyperaceae being important components. Where Casuarinaceae are less abundant, Myrtaceae and Cunoniaceae combine to dominate a closed canopy vegetation with emergent medium–tall trees and a fern understorey. This vegetation type is also well developed on the margins of rainforest in subtropical eastern Australia, and in coastal environments behind mangroves and on low hill slopes in New Caledonia.

It is clear that the coal swamp vegetation included species associations that are unknown in the modern Australasian vegetation and that many fossil taxa cannot be confidently related to extant species. *Banksia* and *Dacrydium* spp. occur as the dominant components of individual horizons in the coal seams. There is no overlap between geographical distributions of these genera in any modern flora. This underlines the emphasis in the present study of making reasonable comparisons between fossil floras and modern vegetations and highlighting where these comparisons fail.

There is considerable scope for detailed studies of the coal vegetations and their plant associations as explicit environmental indicators. Far more complete studies of the pollen and foliar morphology of modern species are required before the vegetation of the Latrobe Valley coal swamps can be described in full.

PALAEOBOTANY OF THE YALLOURN AND MORWELL COAL SEAMS

The results of the conjoint study presented above have been derived from detailed studies of just 1.3 m of the coal exposure. This is a very short sequence when compared to some 300 m of coal analysed from open cut and drill core material.

The significance of the detailed study of the short sequence lies in its contribution to calibration of the macrofossil record against the microfossil record, and the recognition of characteristic plant assemblages. This knowledge has been applied to a study of the fossil assemblages for the whole of the Yallourn and Morwell seams.

The coal seams contain a large number of plant taxa. There are 148 recorded macro- and microfossil taxa, representing at least 100 biological taxa. The general distributions of these taxa in terms of coal seams and lithotypes are given in Appendix 14.1. From the conjoint study it is apparent that a subset of taxa can be used to define characteristic vegetation assemblages and environments of deposition. The dominant, characteristic or otherwise significant taxa are discussed in greater detail below.

Macro- and microfossil taxonomy

Of the list of 148 identified macro- and microfossil taxa (see Appendix 14.1), 90 are known as microfossils and 58 as macrofossils, with 19 micro- and macrofossil taxa identified to the same species. It is likely that the actual number of taxa that grew at various times during the development of the coal seams is much greater than this. The limitations of sampling from drill cores and open cut transects, the spatial variability in plant communities, vagaries of preservation and chemical maceration procedures means that many uncommon or non-woody taxa are likely to go unrecognised. Although incomplete, the assemblage of identified taxa greatly increases our knowledge of the coal swamp vegetation.

Interseam Influence Zones

The likely stratigraphic positions of IIZs in the Yallourn and Morwell Coal Seams are shown in Figures 14.6 and 14.7, above. The single zone (IIZ 6) studied in the conjointly sampled section has given us an understanding of vegetation and depositional environment dynamics prior to, and following the onset peak of, a single marine transgression. The actual conditions experienced at the

height of the marine transgression are difficult to interpret. At locations closer to the palaeoshoreline, such as at Loy Yang and Rosedale, the presence of dinoflagellates confirms the likelihood of marginal marine conditions at the time of deposition of IIZ 6. The lack of marine indicators precludes this possibility at the more continental Morwell Open Cut, although the presence of quartz grains in the coal matrix suggests that some fluvial influence may have occurred.

Ecology of the Yallourn and Morwell Seam coal swamps

The ecology of the coal swamps forming the Yallourn and Morwell Seams has been most completely described by Sluiter (1984) and Blackburn (1985). They demonstrated general relationships between fossil assemblages, depositional conditions and lithotypes. The conjoint study described earlier was able to demonstrate some clear relationships between plant communities and depositional conditions. The relationships between lithotypes and plant communities are less clear.

It must be remembered that our studies have been of a few narrow, vertical coal sections. It has not been possible systematically to study lateral variations in fossil assemblages. It is therefore difficult to develop a picture of the mosaic of plant communities.

Yallourn Clay Seam and basal coals

An understanding of the patterns of some plant communities has been gained from a 1 m coal sequence at the base of the Yallourn Open Cut and exposures of the Yallourn Clay in the Yallourn and Morwell Open Cuts.

The basal Yallourn Seam section contained lithotypes ranging from Vdk to Pa. It had been exposed for about four years, so the coal was well weathered. This weathering made lithotype banding clearer and the weathering allowed individual bedding planes to be discerned. The individual lithotype bands could be traced horizontally for about 500 m. Macrofossils were exceptionally well preserved and easily identifiable in the field. Basal carbonaceous clays of the interseam Yallourn Clay were also exposed on the working bench of the coal mine. The Yallourn Clay sequence has been described as indicative of a depositional hiatus between the Morwell 1A Coal Seam and the Yallourn Coal Seam (Hocking *et al.*, 1976). It is more accurately described as a basal clay to the Yallourn Coal Seam because it becomes more carbonaceous upwards and grades into the coal seam proper. The complete depositional sequence of the Yallourn Clay is exposed as overburden in the Morwell Open Cut and is illustrated in Figure 14.19.

The palynoflora of the Yallourn Clay includes a regional *Nothofagus* element, with local and regional components of *Quintinia*, *Dacrycarpus latrobensis*, *Podocarpus*, *Phyllocladus*, *Agathis*, Casuarinaceae, Cunoniaceae (*Ackama/Pullea* and *Ceratopetalum*) and Myrtaceae (*Syzygium*) abundant (Greenwood, 1981). Pollen data from a 160 cm

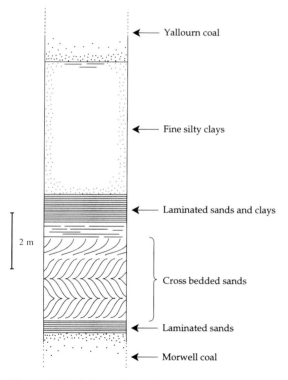

Figure 14.19 Sedimentology of the Yallourn clay (after Greenwood, 1981).

Table 14.3. *Frequencies as a percentage of the pollen sum of the dominant pollen taxa in four samples from the Yallourn Clay in the Yallourn open cut exposure. Depths are from the bottom of the sampled interval, with the uppermost sample corresponding roughly to the clay/coal boundary*

Taxon	Sample depths			
	160 cm	120 cm	80 cm	40 cm
Nothofagus subgen. *Brassospora*	10.3	11.8	6.3	20.8
Nothofagus subgen. *Lophozonia*	0.5	2.0	10.0	12.5
Syzygium complex	16.5	17.0	14.5	4.5
Elaeocarpus	9.5	27.5	13.0	1.8
Quintinia	4.0	1.8	2.3	1.5
Sapindaceae	0.3	2.5	5.0	3.0
Helicia complex	—	0.3	2.3	2.8
Casuarinaceae	13.0	4.5	0.8	3.5
Ackama/Pullea complex	1.8	3.5	4.0	2.0
Podocarpus	26.0	4.5	8.0	21.0
Phyllocladus	2.8	1.5	1.8	1.3
Agathis/Araucaria	3.5	2.5	2.0	1.3
Total Fagaceae	10.8	13.8	16.3	33.3
Total Myrtaceae	23.5	22.5	18.0	7.0
Total Proteaceae	—	0.8	3.5	—
Total Coniferae	32.5	10.0	25.0	—

interval near the top of the Yallourn Clay (I. R. K. Sluiter, unpublished data) are summarised for the dominant taxa in Table 14.3. The uppermost sample (160 cm) was a clayey coal, while the lowermost was a finely laminated, slightly carbonaceous, silt.

The macrofossil assemblages of the Yallourn Clay are dominated by Podocarpaceae (*Dacrycarpus*, *Phyllocladus*), Myrtacaeae, Casuarinaceae (*Gymnostoma*), Lauraceae and Elaeocarpaceae with Zamiaceae (*Pterostoma*, aff. *Macrozamia*; Hill, 1980) as an uncommon but significant element (Greenwood, 1981). The presence of *Pterostoma* is of interest in that it has been previously described only from Eocene deposits at Anglesea in Victoria and Nerriga in New South Wales (Hill, 1980). Araucariaceae is represented by rare leaves and cuticles of *Agathis* sp. aff. *yallournensis*. Proteaceae, which are a dominant element of the Yallourn Seam coals, are rare.

In the Yallourn exposure the uppermost metre of the Yallourn Clay contained woody tree stumps and roots associated with leaves of *Dacrycarpus latrobensis*. The stumps, typically 10 to 15 cm in diameter were about 2 to 5 m apart. This suggests that the *Dacrycarpus* trees were of the order of 5 m in height. Casuarinaceae occurred as foliar remains, female cones and pollen; related modern species are typically of similar stature. The Myrtaceae was dominated by *Syzygium* pollen with *Baeckea/Leptospermum*, *Acmena*, *Melaleuca* and *Rhodamnia* present at between 0.5% and 2.0% of the assemblages. This presents a picture of a vegetation with an emergent storey of *Dacrycarpus*, Casuarinaceae and Myrtaceae and an understorey dominated by myrtaceous shrubs. Equivalent modern vegetation types are found within, and marginal to, closed canopy rainforests in eastern Australia, the North Island of New Zealand and New Caledonia in warm temperate–subtropical climatic regions.

The clays grade upwards into the coal proper, becoming more fine grained and carbonaceous. The pollen and macrofossil contents and diversity also increase from clay to coal.

YALLOURN SEAM COALS

The basal coals have high fusinite contents (up to 56% of the coal maceral) and consist of distinct layers dominated by Gleicheniaceae foliage and roots, monocotyledonous rhizomes and leaves including Typhaceae and Cyperaceae, Proteaceae leaves and wood, *Phyllocladus* foliage, Casuarinaceae and *Gymnostoma* foliage, cones and anthers, *Elaeocarpus* fruits and abundant macroscopic fungal structures and spores. Fern sporangia assignable to *Gleichenia* and *Blechnum* are also common.

These layers, and in particular layers dominated by *Gleichenia*, Casuarinaceae and *Elaeocarpus*, can be traced horizontally for up to 500 m. The homogeneity of the vegetation on this scale lends some confidence to the interpretation of plant associations within the coal seams.

These coal layers contain a number of taxa that, in modern vegetation, commonly occur in sclerophyllous woodlands and shrublands, swamp forests and open-water swamp environments. The coal laminations suggest that depositional conditions were undergoing significant changes with time and that the coal floras were developing in response to those changes.

The base of the Yallourn Seam corresponds closely to IIZ 12. The change in sedimentation demonstrates a radical change from depositional environments dominated by clastics derived from riparian environments and sediments from the nearby highlands to increasingly stable coal swamp conditions. The complexity of the changes in depositional environments and floristics of the newly forming coal swamp has defied detailed analysis using the available data. The interseam clastics at this location were apparently deposited during a period of marine transgression. The high fusinite concentrations in the basal coals and their association with Gleicheniaceae, Typhaceae and Restionaceae indicate that the early Yallourn Seam vegetation consisted of reedy swamps with abundant xeric ferns, which were seasonally drying and largely controlled by a natural fire regime. Another interval where fusinite concentrations were very high occurred at IIZ 13 (fusinite = 35–60%). This interval is also characterised by Dk and Md coals and abundant monocotyledons (Typhaceae, Sparganiaceae, Restionaceae) and Proteaceae (*Banksieaephyllum*, *Xylomelum*). Myrtaceae in these fusinitic coals is typified by genera that are components of modern swamp heaths (*Baeckea*, *Leptospermum*, *Melalueca*). Hepaticae, in particular Jungermanniales, are also common as carbonised thalli.

The remainder of the Yallourn Seam is composed principally of Md to Lt coals, and is dominated by Podocarpaceae (*Dacrydium* and *Dacrycarpus*, with locally common *Phyllocladus*), Araucariaceae (*Agathis* and locally *Araucaria*), and Oleaceae (*Oleinites*). Myrtaceae are also widely distributed and include a range of taxa assignable to modern genera including *Syzygium*, *Tristania* and *Acmena* (Figure 14.20).

The darker coals with their moderate to high

fusinite contents suggest seasonally drying conditions, with fire as a major determinant of the vegetation types. These assemblages have many similarities to modern Australian, New Zealand and New Caledonian heaths and maquis vegetation.

Md and paler coals have low fusinite contents and are dominated by conifers. These have many taxonomic and structural similarities to some extant warm–cool temperate floras of southeastern Australia and New Caledonia.

The relationship between the succession of vegetation types and water levels and the influence of fire has been described by Blackburn (1981) and Luly *et al.* (1980). Their interpretations of typical pyric and hydroseral successions are illustrated in Figures 14.21 and 14.22.

MORWELL SEAM COALS

The base of the Morwell Coal Seam exposed in Loy Yang drill core LY1275 is dominated by pollen of Myrtaceae and Elaeocarpaceae (Sluiter, 1984). This association is reflected in the pollen assemblages of the basal coals of the Morwell Open Cut (stratigraphically some 10 m higher) and the Yallourn Open Cut (stratigraphically 100 m higher). It seems that the Myrtaceae/Elaeocarpaceae association is typical of the transition between clastic and coal deposition throughout the deposition of the Yallourn and Morwell Coal and Clay sequences. Modern equivalents to this flora occur in, and marginal to, closed canopy rainforests in eastern Australia, New Zealand, New Guinea, Indonesia and Malaysia.

The basal metre of the Morwell 1A Coal Seam in the Morwell Open Cut is dominated by macrofossils of Proteaceae (*Banksieaephyllum laeve* and *B. pinnatum*), Myrtaceae (*Baeckea* and *Acmena*) with taxa related to *Orites/Darlingia* (Proteaceae), *Melaleuca*, *Neomyrtus*, *Tristania* (Myrtaceae) and Elaeocarpaceae as uncommon elements. The pollen assemblages are dominated by Myrtaceae and Elaeocarpaceae.

This mix of species does not occur in any modern Australian vegetation type; however, the association of *Banksieaephyllum*, *Baeckea* and *Melaleuca* is comparable to modern swamp heath vegetation (Groves, 1981).

The Morwell Coal Seam macrofossil flora is dominated by Podocarpaceae (*Dacrycarpus* and *Dacrydium*) and Proteaceae (*Banksieaephyllum*), with Araucariaceae (*Araucaria*, *Agathis*), Cunoniaceae (*Phyllites*), Oleaceae (*Oleinites*), Saxifragaceae (*Quintinia*) and Myrtaceae (*Tristania*, *Eugenia* and *Acmena*) as common representatives. *Azolla* spores and *Typha* pollen, explicit indicators of open-water environments, occur in Ml to Pa coals.

The floras of the Morwell Coal Seam were probably dominated by closed canopy forests, with emergent elements from the families Araucariaceae and Myrtaceae, a shrubby storey of Podocarpaceae, Proteaceae, Oleaceae and Saxifragaceae and an understorey of *Typha* and other aquatics, including *Azolla* in lighter-coloured coals. Nonxerophytic Gleicheniaceae and Blechnaceae occurred locally and are indicative of seasonally inundated conditions. The generally low fusinite contents of Morwell Seam coals suggest that fire was of low frequency and intensity.

The fossil assemblages of the Morwell Seam suggest much less seasonal environments than for the Yallourn Seam. Fire, which was a major influence on the Yallourn Seam floras, is far less significant in the deposition of the Morwell B Seam, although of increasingly greater importance in the deposition of the Morwell 1A Seam. The Morwell Seam floras contained fewer xeric elements than Yallourn, with *Dacrycarpus* and Proteaceae the dominants throughout. A range of myrtaceous genera occurring in modern warm temperate–subtropical closed forests also characterise the Morwell Seam. There is much less variation in coal colours and, in particular, fewer abrupt changes in lithotypes than are found in the Yallourn Seam. Together, this information suggests more stable depositional conditions.

The current depositional environments in which modern equivalents of the fossil taxa from the Yallourn and Morwell Seams are found have been summarised by Luly *et al.* (1980) and are given in Appendix 14.1.

(a)

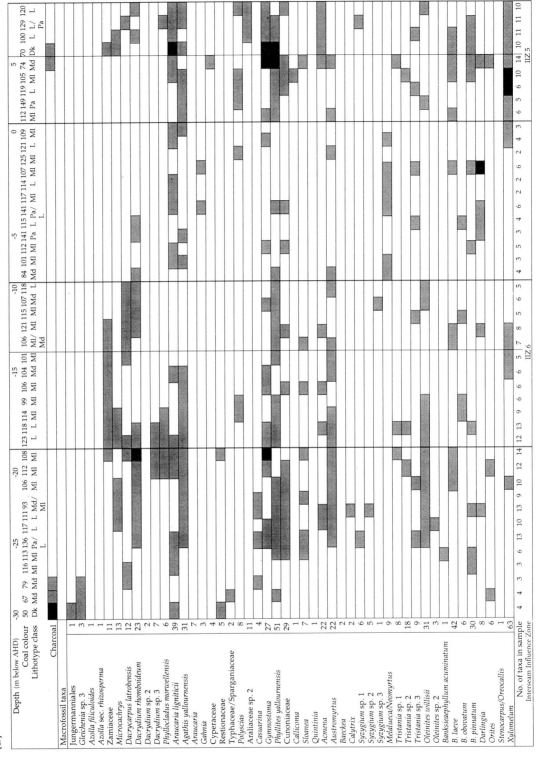

(b)

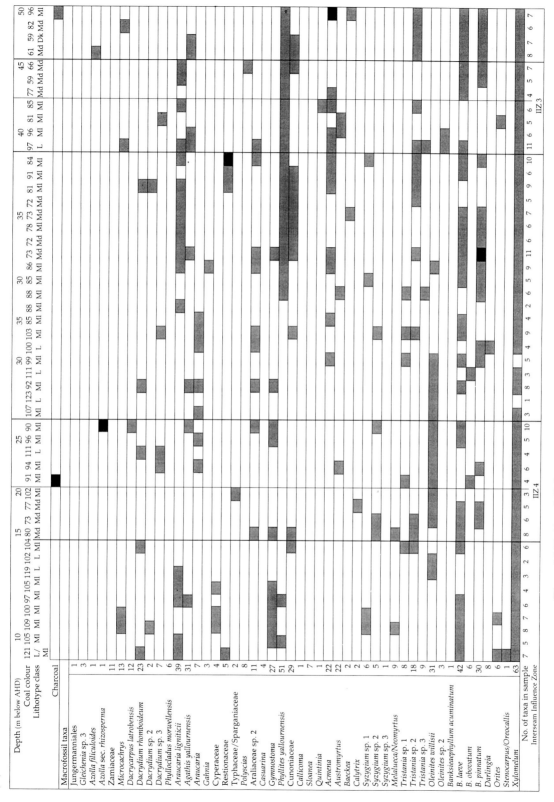

Figure 14.20 (a) and (b) Stratigraphic distributions of major macrofossil taxa in the Yallourn Seam. For lithotype class abbreviations, see Table 14.1.

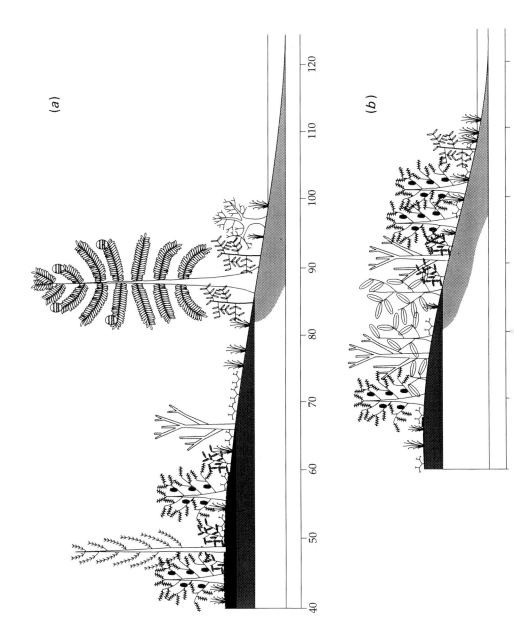

(a)

(b)

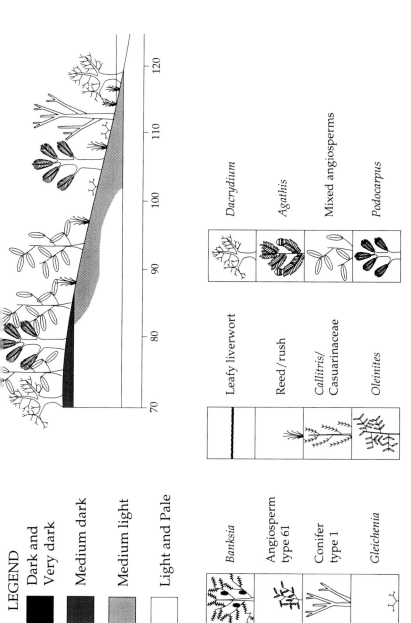

Figure 14.21 Hydroseral and pyric succession for three intervals from Yallourn Open Cut: (a) to (c) are arranged in order of decreasing fire influence. Water levels indicate probable permanent inundation. (*a*) Succession under the influence of frequent fires. The margins of open water are essentially a *Dacrydium*–*Oleinites*–*Agathis*–reed/rush swamp. On drier areas above this is a reed/rush–*Gleichenia* moor. This grades into a *Banksia*–*Gleichenia*–conifer type 1–reed/rush scrub. Rare *Callitris* and Casuarinaceae occur on the driest parts with leafy liverworts. (*b*) Succession under the influence of infrequent fires. The margins of open water are occupied by an *Oleinites*–*Gleichenia*–reed/rush swamp. On drier areas above this is a *Banksia*–*Gleichenia*–angiosperm type 61–conifer type 1 scrub. Above this again is a region having a mixed angiosperm scrub. On the driest parts *Banksia*–*Gleichenia*–reed/rush scrub dominates. (*c*) Succession under the influence of very infrequent fires. The margins of the open water are colonised by a *Dacrydium*–*Gleichenia*–conifer type 1–reed/rush swamp. On the drier areas there is a mixed agiosperm–*Podocarpus*–*Dacrydium* scrub.

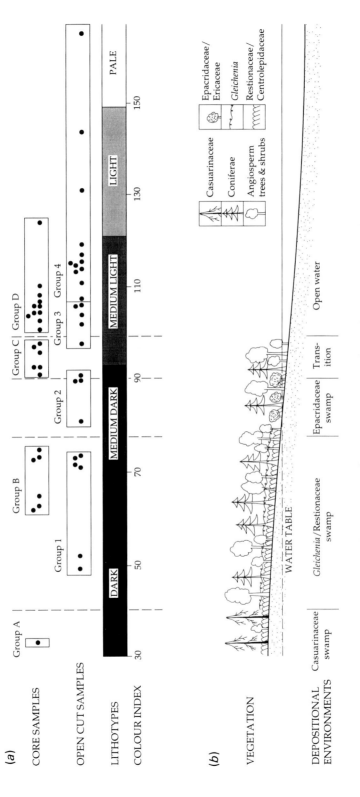

Figure 14.22 (*a*) Position of core and open cut samples in relation to brown coal lithotypes and colour index scale. (*b*) Presumed vegetation and depositional environments derived from pollen sample groups in relation to brown coal lithotypes and colour index scale.

CONCLUSION

This combined macro- and microfossil study of the Latrobe Valley coal floras demonstrates conclusively the value of such an approach when appropriate sediments are available. The correlation between the two approaches, which has often been considered notoriously unreliable, is shown here to have real value in interpreting several aspects of past environmental conditions. The complexity of the floral associations through time and space are now being unravelled and offer an unparalleled opportunity to examine the evolution of the Australian flora during a critical phase in climatic and edaphic transition. Future studies will sharpen the current picture, and should be encouraged as a major opportunity to examine the effect of long-term environmental change on species and community evolution.

Appendix 14.1. *Macro- and microfossil taxa recorded from the Yallourn and Morwell Coal Seams*

	Fossil taxa	Occurrence	Comments
	Bryophyta		
m	*Sphagnum* sp. (= *Stereisporites* sp.)	Rare Dk and Md, M1B, M1A; locally common Dk, M1A	Bryophytes are rarely recorded as macrofossils because of their fragility
M	aff. Jungermanniales	Rare Dk and Md, Y	Found only as carbonised tissues in high fusinite coals
M	aff. *Ricciocarpus*	Very rare Dk, Y	Found only as carbonised thalli
	Lycophyta		
	Lycopodiaceae		
m	*Lycopodium* sp. aff. *laterale* (= *Latrobosporites crassus*)	Rare M1B and Y; uncommon M1A	Lycopod macrofossils are not found because of their fragility
m	Lycopodiaceae sp. 1 (= *Foveotriletes balteus*)	Locally common Dk and Md, M1A, M1B, Y	Spores are likely to be overrepresented
m	Lycopodiaceae sp. 2 (= *Herkosporites elliottii*)	Locally common Dk and Md, M1A, M1B, Y	
	Pteridophyta		
	Gleicheniaceae		
m	Gleicheniaceae sp. 1 (= *Gleicheniidites circinidites*)	Spores locally abundant Dk and Md, M1A, M1B, Y	Spores are often overrepresented
M	aff. *Gleichenia dicarpa*	Dk and Md, M1A, M1B, Y; macrofossils locally dominant Dk and Md, Y	Generally associated with high fusinite content
m	Gleicheniaceae sp. 2 aff. *Dicranopteris* (= *Dictyophyllidites arcuatus*)	Spores and macrofossils locally abundant in Dk and Md, M1A, M1B	
M	*Gleichenia* sp. 2	Locally abundant Dk and Md, Y	
M	*Gleichenia* sp. 3	Locally common Md and M1, M1A	
	Cyatheaceae		
m	*Cyathea* sp. aff. *australis* (= *Cyathidites subtilis*)	Rare at base of M1B and M1A	
	Lophosoriaceae		
m	*Lophosoria* sp. (= *Cyatheacidites annulatus*)	Rare at base of M1A	
	Dicksoniaceae		
m	*Dicksonia* sp. (= *Trilites tuberculiformis*)	Rare throughout M1A and Y	
	Blechnaceae		

Appendix 14.1. (*cont.*)

	Fossil taxa	Occurrence	Comments
m	Blechnaceae sp. (= *Laevigatosporites ovatus*)	Rare, locally common M1B, M1A and Y	
M	aff. *Blechnum*	Locally abundant in thin horizons Md, Dk, M1A	Never found in cuticle preparations
	Polypodiaceae		
m	Polypodiaceae (= *Polypodiisporites* sp.)	Rare M1B, M1A and Y	
	Ophioglossaceae		
m	aff. Ophioglossaceae	Rare Y	
	Salviniaceae		
M	*Azolla* sp. 1 aff. *filiculoides* (*Azolla* section *Azolla*)	Rare Md, M1B	*Azolla* is indicative of quiet, open-water environments
M	*Azolla* sp. 2 (*Azolla* section *Rhizosperma*)	Rare Md, M1A	
	Cycadophyta		
M	aff. Zamiaceae (similar to *Macrozamia* and *Bowenia*)	Rare Md, M1A	
	Coniferophyta		
	Cupressaceae		
m	Cupressaceae/Taxodiaceae	Rare M1B, M1A; uncommon to locally common M1A and Y	Cupressaceae pollen is generally overrepresented, but macrofossils confirm local presence
M	*Callitris* sp.	Uncommon Y	
	Podocarpaceae		
m	*Dacrycarpus* sp. (= *Dacrycarpites australiensis*)	Rare M1B, uncommon to locally common M1A and Y	
M	*Dacrycarpus latrobensis*	Common to abundant Md and M1, M1A and Y	
m	*Dacrydium* sp. aff. *cupressinus* (= *Dacrydiumites florinii*)	Rare M1B, uncommon to common M1A and Y	Equivalent to macrofossil type *Dacrydium rhomboideum*, although possibly representative of several biological species
m	*Podocarpus/Dacrydium* spp. (= *Podocarpidites*)	Uncommon to common M1B, M1A and Y	Comparable to some species of *Podocarpus* and *Dacrydium*
m	*Lagarostrobos* sp. aff. *franklinii* (*Phyllocladidites mawsonii*)	Locally common M1 and Lt, M1B; common to abundant M1 and Lt, M1A and Y	
M	*Dacrydium rhomboideum*	Common M1A and Y; locally common top of M1B	Equivalent to palynomorph type *Dacrydiumites florinii*
M	aff. *Dacrydium* sp. 2	Rare throughout M1B and M1A	
M	aff. *Dacrydium* sp. 3	Rare throughout M1B and M1A	
m	*Phyllocladus* sp. (= *Phyllocladidites paleogenicus*)	Moderately common M1B, M1A and Y	
M	*Phyllocladus morwellensis*	Locally common to abundant M1A and Y	
m	*Microcachrys* sp. aff. *tetragona* (= *Microcachrydites antarcticus*)	Rare Md and Dk, M1B, M1A and Y	
m	Podocarpaceae sp. (= *Podosporites microsaccatus*)	Rare M1B, M1A and Y	

Appendix 14.1. (*cont.*)

	Fossil taxa	Occurrence	Comments
m	*Podocarpus* sp.	Rare to moderately common Md and Dk, M1A and Y	Usually associated with high charcoal, Md and Dk lithotypes
	Araucariaceae		
m	*Araucaria* sp.	Rare M1B, M1A and Y	
M	*Araucaria lignitici*	Common to locally abundant Md and Dk, M1B, M1A and Y	
M	aff. *Araucaria* sp.	Rare at top of M1B and M1A	
m	*Agathis* sp.	Rare M1B, M1A and Y	
M	*Agathis yallournensis*	Rare M1B, M1A and Y	
	Anthophyta–Monocotyledonae		
	Cyperaceae		
M	aff. *Gahnia* sp.	Rare M1A and M1B	
M	Cyperaceae	Rare at base of M1A	
	Restionaceae		
m	Restionaceae/Centrolepidaceae	Restricted to Dk and Md, uncommon M1B, locally abundant M1A and Y	
M	aff. Restionaceae	Rare Dk and Md, M1B and M1A	
	Poaceae		
m	Poaceae (= *Graminidites media*)	Very rare, M1B, M1A and Y	
	Sparganiaceae or Typhaceae		
m	Sparganiaceae/Typhaceae (= *Sparganiaceaepollenites barungensis*)	Rare Md and Dk, M1B, M1A and Y; locally abundant Dk, M1A	
M	aff. Typhaceae or Sparganiaceae	Mostly rare M1A and Y; locally abundant Dk, M1A	
	Liliaceae or Arecaceae		
m	Liliaceae/Arecaceae (= *Liliacidites* spp.)	Rare M1B, M1A and Y	
	Araceae		
m	aff. Araceae	Very rare, base M1A	This taxon is possibly an algal cyst
	Anthophyta–Dicotyledonae		
m	Anacardiaceae/Simaroubaceae	Very rare M1A and Y	
	Apocynaceae		
m	aff. *Alyxia*	Very rare top of M1B	
	Aquifoliaceae		
m	*Ilex* (= *Ilexpollenites anguloclavatus*)	Rare M1B, M1A and Y	
	Araliaceae		
m	aff. *Mackinlaya*	Rare M1B, M1A and Y	It is likely that several genera of this family are also present, including *Polyscias* and *Pseudopanax*
M	aff. *Polyscias*	Rare M1 and Lt, M1A	
M	aff. Araliaceae sp. 2	Rare in mostly Md, M1B and M1A	
	Araliaceae or Goodeniaceae		
m	aff. Araliaceae/Goodeniaceae (= *Rhoipites sphaerica*)	Rare M1B, M1A and Y	Compared to *Olea* by Cookson (1947) but more likely to have affinities with these families based on pollen morphological similarities

Appendix 14.1. (*cont.*)

	Fossil taxa	Occurrence	Comments
	Caryophyllaceae		
m	aff. Caryophyllaceae (= *Periporopollenites demarcatus*)	Rare M1B and M1A, very rare Y	
	Casuarinaceae		
m	Casuarinaceae (= *Haloragacidites harrisii*)	Common and locally abundant, M1B, M1A and Y	Casuarinaceae genera cannot be separated palynologically
M	*Allocasuarina* sp.	Common and locally abundant, M1B, M1A and Y	
M	*Gymnostoma* sp.	Moderately common M1A and Y	
	Cunoniaceae		
m	Cunoniaceae (tricoplorate)	Common M1B, M1A and Y	
M	*Phyllites yallournensis*	Common to abundant Md and M1, M1B, M1A and Y	
m	aff. *Ceratopetalum*	Common, Dk and Md, M1B, M1A and Y	
m	aff. *Geissois*	Moderately common M1B, M1A and Y; locally common M1, Lt and Pa, M1B, M1A and Y	
	Droseraceae		
m	Droseraceae	Very rare Md and Dk, base M1A	
	Ebenaceae		
m	aff. *Diospyros*	Very rare M1B and M1A	
m	aff. *Diospyros*	Rare Y	
	Elaeocarpaceae		
m	Elaeocarpaceae	Common M1, M1B, M1A and Y; abundant at major seam boundaries	Pollen is of the *Elaeocarpus* type and not *Sloanea*
M	*Elaeocarpus* sp. 1	Locally common Dk coals with charcoal, Y; mostly at major seam boundaries	
M	*Elaeocarpus* sp. 2	As for sp. 1	
M	aff. *Sloanea*	Uncommon M1, M1A	
	Epacridaceae		
m	Epacridaceae/Ericaceae (= *Ericipites crassiexinus*)	Rare Md and Dk, M1B; moderately common Md and Dk, M1A and Y	
M	aff. *Brachyloma*	Rare M1B, M1A and Y	
	Euphorbiaceae		
m	*Austrobuxus* sp. aff. *swainii* (= *Malvacipollis subtilis*)	Rare M1B, M1A and Y	
m	*Macaranga/Mallotus* (= *Tricolporopollenites endobalteus*)	Rare M1B, M1A and Y	
	Fagaceae		
m	*Nothofagus* subgenus *Brassospora* (= *Nothofagidites*)	Abundant M1B, M1A and Y; contains 6 form taxa; common between seam boundaries	The pollen is considered to be regionally derived
m	*Nothofagus* subgenera *Nothofagus* and *Fuscospora*	Rare M1B, M1A and Y; two form taxa	Pollen is regionally derived
m	*Nothofagus* subgenus *Lophozonia*	Rare M1B, M1A and Y; locally common in Md and Dk with charcoal	One taxon indistinguishable from modern taxa. Locally represented

Appendix 14.1. (*cont.*)

	Fossil taxa	Occurrence	Comments
	Gyrostemonaceae		
m	Gyrostemonaceae	Rare M1B, M1A and Y	Possibly regionally derived
	Haloragaceae		
m	*Haloragis*	Very rare M1A and Y	
	Loranthaceae		
m	Loranthaceae (= *Gothanipollis bassensis*)	Rare M1B, M1A and Y	
	Meliaceae		
m	aff. *Dysoxylum*	Rare M1B, M1A and Y	
	Myrsinaceae		
m	aff. *Rapanea*	Rare M1B, M1A and Y	
	Myrtaceae		
m	aff. *Austromyrtus* (= *Myrtaceidites verrucosus*)	Rare M1B, M1A and Y	
M	aff. *Austromyrtus*	Rare M1, M1A	
m	aff. *Acmena*	Moderately common to locally common, M1B, M1A and Y	
M	aff. *Acmena*	Common Md, M1A	
m	aff. *Baeckea/Leptospermum*	Moderately common M1B, M1A and Y	
M	aff. *Baeckea*	Rare Md and M1, M1B	
M	aff. *Calytrix*	Rare Md, M1A, M1B	
m	aff. *Eucalyptus* (= *Myrtaceidites eucalyptoides*)	Rare M1B, M1A and Y	
M	aff. *Metrosideros*	Rare Y	
m	aff. *Syzygium* (= *Myrtaceidites mesonesus*)	Moderately common to common, M1B, M1A and Y	
M	aff. *Syzygium* sp. 1	Moderately common M1, M1B and M1A	
M	aff. *Syzygium* sp. 2	Moderately common Md, M1B and M1A	
M	aff. *Syzgium* sp. 3	Locally common Md, M1A	
m	aff. *Tristania/Metrosideros* (= *Myrtaceidites parvus* forma *anesus*)	Moderately common to common, M1B, M1A and Y	
M	aff. *Tristania* sp. 1	Moderately common top of M1B and M1A	
M	aff. *Tristania* sp. 2	Moderately common M1B and M1A	
M	Myrtaceae taxon 45	Rare M1, Y	
M	Myrtaceae taxon 46	Rare M1A	
	Oleaceae		
M	*Oleinites willisii* (aff. *Notelaea*)	Common Md and M1, locally abundant at top of M1B, M1A and Y	
	Polygalaceae		
m	aff. Polygalaceae (= *Polycolporopollenites esobalteus*)	Rare M1B, M1A and Y	
	Potiliaceae		
m	*Fagraea*	Rare Y	
	Proteaceae		

Appendix 14.1. (*cont.*)

	Fossil taxa	Occurrence	Comments
m	*Banksia/Dryandra* sp. (= *Banksieaeidites elongatus*)	Locally common in Dk, M1A and Y; very rare Dk, M1B	
M	*Banksieaephyllum acuminatum*	Abundant Md, Y	
M	*Banksieaephyllum angustum*	Abundant Dk and Md, Y with charcoal	
M	*Banksieaephyllum laeve*	Common, locally abundant M1, M1B, M1A and Y	
M	*Banksieaphyllum obovatum* (Cookson & Duigan, 1950)	Rare M1, M1A and Y; top M1B	
M	*Banksieaephyllum pinnatum*	Rare Md and M1, M1B, M1A and Y	
m	aff. *Musgravea* (= *Banksieaeidites arcuatus*)	Very rare M1B	
M	aff. *Conospermum*	Rare Y	
m	aff. *Carnarvonia* (= *Proteacidites stratosus*)	Very rare M1A	
M	aff. *Darlingia*	Moderately common, top of M1B and M1A	
m	aff. *Embothrium* (= ?*Triporopollenites ambiguus*)	Very rare M1B	
m	aff. *Helicia/Orites*	Rare M1B, M1A and Y	
m	*Symphyonema* (= *Proteacidites symphyonemoides*)	Rare Y	
M	aff. *Orites*	Moderately common Md, M1B and M1A	
M	aff. *Stenocarpus/Oreocallis*	Rare M1 and Lt, M1A	
M	aff. *Xylomelum*	Moderately common to locally common, M1 and Md, M1B, M1A and Y	
m	Proteaceae taxon (= *Proteacidites* cf. *callosus*)	Rare M1B, M1A and Y	
m	Proteaceae taxon (= *Proteacidites obscurus*)	Very rare, lower M1B and M1A; common upper M1A and Y	
m	Proteaceae taxon aff. *Petrophile* (*Proteacidites rectomarginis*)	Very rare M1B, M1A and Y	
m	Proteaceae taxon (= *Proteacidites stipplatus*)	Very rare M1B	
m	Proteaceae taxon (= *Proteacidites tuberculatus*)	Rare M1B; very rare M1A and Y	
m	*Beauprea* (= *Beaupreadites elegansiformis*)	Rare M1B, M1A and Y	
M	Proteaceae taxon 58	Rare Md, M1A and Y	
M	Proteaceae taxon 59	Rare M1, M1A and Y	
M	Proteaceae taxon 60	Rare Md, M1A and Y	
	Rhamnaceae		
m	aff. *Emmenosperma*	Rare M1B, M1A and Y	
	Rubiaceae		
m	aff. *Coprosma*	Rare Y	
	Rutaceae		
m	aff. *Acronychia*	Rare M1B, M1A and Y	
m	aff. *Flindersia*	Rare M1B, M1A and Y	
m	aff. *Euodia*	Rare M1B and M1A	

Appendix 14.1. (*cont.*)

	Fossil taxa	Occurrence	Comments
m	*Melicope* sp. aff. *broadbentiana*	Rare Lt, M1B and M1A; moderately common M1 and Lt, Y	
	Santalaceae		
m	aff. *Exocarpus*	Very rare M1B and M1A	
	Sapindaceae		
m	aff. *Dodonaea* (= *Dodonaea sphaerica*)	Very rare M1B	
m	aff. Tribe Cupanieae (= *Cupanieidites orthoteichus*)	Rare Md and Dk, M1A and Y	
	Sapotaceae		
m	Sapotaceae (= *Sapotaceoidaepollenites rotundus*)	Moderately common M1B, M1A and Y; locally common M1 and Lt	
	Saxifragaceae		
m	*Quintinia* (= *Quintinia psilatispora*)	Moderately to locally common, M1B, M1A and Y	
M	*Quintinia* sp.	Very rare M1 and Md, M1B and M1A	
	Sphenostemonaceae		
m	*Sphenostemon*	Rare M1B, M1A and Y	
	Sterculiaceae		
m	aff. *Brachychiton*	Very rare M1B/M1A interseam	
	Ulmaceae		
m	aff. *Celtis*	Very rare M1B, M1A and Y	
	Urticaceae/Moraceae		
m	aff. Urticaceae/Moraceae	Rare M1B, M1A and Y; moderately common at base of M1A	
	Vitaceae		
m	Vitaceae	Rare M1B, M1A and Y	
	Winteraceae		
m	aff. *Tasmannia* (= *Gephyrapollenites calathas*)	Very rare M1B, M1A and Y	

Note that microfossils are indicated by 'm', macrofossils by 'M'. Abbreviations for coal lithotypes and seams are as used throughout the text. Form taxon numbers are from Blackburn (1980, 1981, 1985).

REFERENCES

BARAGWANATH, G. E. (1962). Some aspects of the formation and nature of brown coal, and the behaviour of brown coal ash in water tube boilers with special reference to Victorian deposits. *Proceedings, Australian Institute of Mining and Mineralogy*, **202**, 131–249.

BARAGWANATH, G. E. & KISS, L. (1964). *Report on Palynological Investigations of Victorian Brown Coals*. Melbourne: State Electricity Commission of Victoria, Scientific Division.

BLACKBURN, D. T. (1980). *Megafossil/lithotype Corre-lations in a Transect of the Yallourn Open cut Coal Mine. Palaeobotanical Project – Report 1*. Melbourne: State Electricity Commission of Victoria.

BLACKBURN, D. T. (1981). *Floristic Control on Lithotype Banding within the Yallourn Coal Seam, Victoria: Evidence from Megafossil Assemblages. Palaeobotanical Project – Report 2*. Melbourne: State Electricity Commission of Victoria.

BLACKBURN, D. T. (1985). *Palaeobotany of the Yallourn and Morwell Coal Seams. Palaeobotanical Project – Report 3*. Melbourne: State Electricity Commission of Victoria.

BOLGER, P. F. (1991). Lithofacies variations as a consequence of Late Cainozoic tectonic and palaeoclimatic events in the onshore Gippsland Basin. In *The Cainozoic in Australia: A Re-appraisal of the Evidence*, ed. M. A. J. Williams, P. De Deckker & A. P. Kershaw. *Geological Society of Australia Special Publication*, 18, 158–80.

CHAPMAN, F. (1925a). Notes on the brown coal from Morwell, South Gippsland. *Geological Survey of Victoria Bulletin*, 45, 485–87.

CHAPMAN, F. (1925b). On some seed-like bodies in the Morwell brown coal. *Geological Survey of Victoria Bulletin*, 45, 487–89.

COOKSON, I. C. (1946). Pollens of *Nothofagus* Blume from Tertiary deposits in Australia. *Proceedings of the Linnean Society of New South Wales*, 71, 49–63.

COOKSON, I. C. (1947). On fossil leaves (Oleaceae) and a new type of fossil pollen grain from Australian brown coal deposits. *Proceedings of the Linnean Society of New South Wales*, 72, 185–97.

COOKSON, I. C. (1950). Fossil pollen grains of protea-ceous type from Tertiary deposits in Australia. *Australian Journal of Scientific Research*, Series B, 3, 166–77.

COOKSON, I. C. (1953). The identification of the sporo-morph *Phyllocladidites* with *Dacrydium* and its distri-bution in southern Tertiary deposits. *Australian Journal of Botany*, 1, 64–70.

COOKSON, I. C. (1957). On some Australian Tertiary spores and pollen grains that extend the geological and geographical distribution of living genera. *Proceedings of the Royal Society of Victoria*, 69, 41–53.

COOKSON, I. C. (1959). Fossil pollen grains of *Notho-fagus* from Australia. *Proceedings of the Royal Society of Victoria*, 71, 25–30.

COOKSON, I. C. & DUIGAN, S. L. (1950). Fossil Banksieae from Yallourn, Victoria, with notes on the mor-phology and anatomy of living species. *Australian Journal of Scientific Research*, Series B, 3, 133–65.

COOKSON, I. C. & DUIGAN, S. L. (1951). Tertiary Araucariaceae from south-eastern Australia, with notes on living species. *Australian Journal of Scientific Research*, Series B, 4, 415–49.

COOKSON, I. C. & PIKE, K. M. (1953a). A contribution to the Tertiary occurrence of the genus *Dacrydium* in the Australian region. *Australian Journal of Botany*, 1, 474–84.

COOKSON, I. C. & PIKE, K. M. (1953b). The Tertiary occurrence and distribution of *Podocarpus* (section *Dacry-carpus*) in Australia and Tasmania. *Australian Journal of Botany*, 1, 71–82.

COOKSON, I. C. & PIKE, K. M. (1954a). The fossil occurrence of *Phyllocladus* and two other podocarpaceous types in Australia. *Australian Journal of Botany*, 2, 60–7.

COOKSON, I. C. & PIKE, K. M. (1954b). Some dicoty-ledonous pollen types from Cainozoic deposits in the Australian region. *Australian Journal of Botany*, 2, 197–219.

DEANE, H. (1925). Fossil leaves from the open cut, state brown coal mine, Morwell. *Records of the Geological Sur-vey of Victoria*, 4, 492–8.

DUIGAN, S. L. (1966). The nature and relationships of the Tertiary brown coal flora of the Yallourn area in Vic-toria, Australia. *Palaeobotanist*, 14, 191–201.

GEORGE, A. M. (1982). Latrobe Valley Brown Coal – lithotypes: macerals: coal properties. *Australian Coal Geol-ogy*, 4, 111–30.

GREENWOOD, D. R. (1981). The Miocene Fossil Flora of the Yallourn Clay and its relationships to the associated Morwell and Yalbourn Coal Floras. B.Sc. Hons. thesis, Adelaide University.

GROVES, R. H. (ed.) (1981). *Australian Vegetation*. Cam-bridge: Cambridge University Press.

HAQ, B. U., HARDENBOL, J. & VAIL, P. R. (1987). Chronology of fluctuating sea levels since the Triassic. *Sci-ence*, 235, 1156–67.

HILL, R. S. (1980). Three new Eocene cycads from east-ern Australia. *Australian Journal of Botany*, 28, 105–22.

HILL, R. S. (1983a). *Nothofagus* macrofossils from the Ter-tiary of Tasmania. *Alcheringa*, 7, 169–83.

HILL, R. S. (1983b). Evolution of *Nothofagus cunninghamii* and its relationship to *N. moorei* as inferred from Tas-manian macrofossils. *Australian Journal of Botany*, 31, 453–65.

HILL, R. S. (1984). Tertiary *Nothofagus* macrofossils from Cethana, Tasmania. *Alcheringa*, 8, 81–6.

HILL, R. S. (1987). Discovery of *Nothofagus* fruits corre-sponding to an important Tertiary pollen type. *Nature*, 327, 56–8.

HILL, R. S. (1989). New species of *Phyllocladus* (Podocar-paceae) macrofossils from southeastern Australia. *Alcher-inga*, 13, 193–208.

HILL, R. S. (1991). Tertiary *Nothofagus* (Fagaceae) macro-fossils from Tasmania and Antarctica and their bearing on the evolution of the genus. *Botanical Joural of the Linnean Society*, 105, 73–112.

HILL, R. S. & BIGWOOD, A. J. (1987). Tertiary gymno-sperms from Tasmania: Araucariaceae. *Alcheringa*, 11, 325–35.

HILL, R. S. & CARPENTER, R. J. (1991). Evolution of *Acmopyle* and *Dacrycarpus* (Podocarpaceae) foliage as inferred from macrofossils in south-eastern Australia. *Australian Systematic Botany*, 4, 449–79.

HOCKING, J. B., GLOE, C. S. & THRELFALL, W. F. (1976). Gippsland Basin. In *Geology of Victoria*, ed. J. G. Douglas & J. A. Ferguson. *Geological Society of Australia Special Publication*, 5, 248–73.

HOLDGATE, G. R. (1985). Latrobe Valley brown coals, their geometry and facies equivalents as a guide to depo-sitional environment. *Australian Coal Geology*, 5, 53–68.

HOLDGATE, G. R. (1992). Effects of relative sea level changes on brown coals in the Latrobe Valley, Victoria. *Proceedings, Fifth Australian Coal Science Conference* pp. 271–9. University of Melbourne, Australian Institute of Energy.

HOLDGATE, G. R. & SLUITER, I. R. K. (1991). Oligo-cene–Miocene marine incursions in the Latrobe Valley depression, onshore Gippsland Basin: evidence, facies relationships and chronology. In *The Cainozoic in Aus-*

tralia: *A Re-appraisal of the Evidence*, ed. M. A. J. Williams, P. De Deckker & A. P. Kershaw. *Geological Society of Australia Special Publication*, **18**, 137–57.

HOWARD, T. M. (1981). Southern closed-forests. In *Australian Vegetation*, ed. R. H. Groves, pp. 102–20. Cambridge: Cambridge University Press.

KERSHAW, A. P., BOLGER, P., SLUITER, I. R. K., BAIRD, J. & WHITELAW, M. (1991). The origin and evolution of brown coal lithotypes in the Latrobe Valley, Victoria, Australia. *Journal of Coal Geology*, **18**, 233–49.

KOVAR, J. B., CAMPBELL, J. D. & HILL, R. S. (1987). *Nothofagus ninnisiana* (Unger) Oliver from Waikato Coal Measures (Eocene–Oligocene) at Drury, Auckland, New Zealand. *New Zealand Journal of Botany*, **25**, 79–85.

LULY, J., SLUITER, I. R. K. & KERSHAW, A. P. (1980). Pollen studies of Tertiary brown coals: preliminary analysis of lithotypes within the Latrobe Valley, Victoria. *Monash Publications in Geography*, **23**.

McKAY, G. H., ATTWOOD, D. H., GAULTON, R. J. & GEORGE, A. M. (1985). The cyclic occurrence of brown coal lithotypes. *State Electricity Commission of Victoria, Department of Resources and Development, Report* 50/85/93.

PAGE, C. N. (1990). Podocarpaceae. In *The Families and Genera of Vascular Plants*, vol. I *Pteridophytes and Gymnosperms*, ed. K. Kubitzki, pp. 332–46. Berlin: Springer-Verlag.

PARTRIDGE, A. D. (1971). Stratigraphic palynology of the onshore Tertiary sediments of the Gippsland, Basin, Victoria. M.Sc. thesis, University of New South Wales.

PATTON, R. T. (1958). Fossil wood from Victorian brown coal. *Proceedings of the Royal Society of Victoria*, **70**, 129–43.

PIKE, K. M. (1953). Fossil fruiting cones of *Casuarina* and *Banksia* from Tertiary deposits in Victoria. *Proceedings of the Royal Society of Victoria*, **65**, 1–8.

SLUITER, I. R. (1984). Palynology of Oligo-Miocene brown coal seams, Latrobe Valley, Victoria. Ph.D. thesis, Monash University.

SLUITER, I. R. & KERSHAW, A. P. (1982). The nature of Late Tertiary vegetation in Australia. *Alcheringa*, **6**, 211–22.

SMITH, G. C. (1982). A review of the Tertiary–Cretaceous tectonic history of the Gippsland basin and its control on coal measure sedimentation. *Australian Coal Geology*, **4**, 1–38.

STOVER, L. E. & PARTRIDGE, A. D. (1973). Tertiary and Late Cretaceous spores and pollen from the Gippsland Basin, southeastern Australia. *Proceedings of the Royal Society of Victoria*, **85**, 237–86.

THOMAS, D. E. & BARAGWANATH, W. (1949). Geology of the brown coals of Victoria. *Mining and Geological Journal of Victoria*, **13**, 28–55.

THOMPSON, B. R. (1981). The Gippsland Sedimentary Basin. Ph.D. thesis, University of Melbourne.

WALKER, D. & FLENLEY, J. R. (1979). Late Quaternary vegetation of the Enga Province of upland Papua New Guinea. *Philosophical Transactions of the Royal Society of London B*, **286**, 265–344.

WELLS, P. M. & HILL, R. S. (1989). Fossil imbricate-leaved Podocarpaceae from Tertiary sediments in Tasmania. *Australian Systematic Botany*, **2**, 387–423.

15   Quaternary vegetation

G. S. HOPE

The Quaternary is the period of modern life in which all the kinds of plants and animals still living have evolved, or have continued from the Tertiary unaffected by new environments. Technically, the beginning of the Quaternary has been defined as the last phase of the Matuyama reversed magnetism epoch after the Olduvai event, which finished at 1.62 million years (Ma). Others relate it to the first appearance of arctic marine faunas about 2.2 Ma in the mid-latitude North Atlantic (Bowen, 1978). The period therefore breaks with the definitions of older periods based on widespread evolutionary changes and instead uses climatic changes, such as the growth of icesheets and spread of cold surface water, to establish the chronology. Cores from the sea-floor show that parts of the tropical Pacific were little affected, but the lock up of snow in icecaps left the oceans enriched in the heavy isotope of oxygen (^{18}O), so that a record of ice on the earth is well preserved in marine sediments. This record shows a series of cyclic changes on time scales of about 100 ka, with phases of maximum and minimum ice extent (stadial and interglacial periods, respectively), each occurring for 10% of each cycle, the remainder being cool interstadials (Chappell & Shackleton, 1986). Over 20 of these alternations have been identified, showing that for the last 2.3 Ma the earth has been very sensitive to minor changes in thermal equilibria, forced, in part at least, by variations in the season and amount of solar radiation reaching the earth (Chappell & Grindrod, 1983).

There are only two divisions in the Quaternary: the Pleistocene from 2.2 (or 1.62) Ma to only 10 ka ago, and the Holocene, our present interglacial, which is the last 10 000 years. The Pleistocene represents the period of establishment of our present landscapes, climatic patterns and types of variability, and the adaptation of the Tertiary biota to these new environments. The term Holocene means life as at present and this is generally true in relation to evolution – the time has been far too short for major new species to appear. However, in terms of changes to populations and particularly to plant and animal communities, the Holocene has been far from stable, and presents us with some novel communities and others not seen since the last interglacial, 126 ka ago.

If we want to understand the modern vegetation of Australia, with its puzzles of distribution, floristic patterns and connections, we can look backwards from the present to ask about its continuity through time and sometimes the length of its occupation of the same place. Quaternary vegetation history differs from that of previous epochs in being concerned with comparing species still living, or which lived recently, with fossil flora. This provides a powerful tool, in that we can flesh out meagre fossil records by comparison with the richness of extant communities. However, it is also now realised that our Holocene is an interglacial that represents unusually warm conditions compared to those prevailing for 85% of the past million years or so. Interglacials take up only about 10–15% of the time of the Quaternary, so that communities that form in them may be atypical, composed of species that have to survive much longer times of cool, dry conditions. Such species may have contributed to quite different communities during glacials and hence a fossil occurrence of a modern species need not imply the environmental niche presently occupied. For example, the koala fossils from semiarid Lake Mungo do not prove the existence of tall eucalypt forests near at hand 40 ka ago.

Biota of the Quaternary have had to adapt to some of the most radical changes since the evolution of the advanced groups in the Mesozoic. In general, it is assumed that migration of pre-adapted elements was minor, although Macphail *et al.* (Chapter 10, this volume) show that at least one alpine species (*Astelia alpina*) arrived here from New Zealand. By the start of the period, high levels of Poaceae, Asteraceae and Casuarinaceae pollen indicate that seasonally dry conditions with widespread droughts were well established inland, and eucalypt-dominated communities were to be found closer to the mountains. Semi-deciduous rainforests and noneucalypt savannas occupied the tropics and stretched across the Arafura Plain. Closed forests still persisted in New Guinea and on the Eastern Highlands down to Tasmania. At Lake George, New South Wales, in the most reliably dated site at the start of the Pleistocene, McEwen Mason (1991) found remarkably high levels of Asteraceae and Poaceae pollen replacing *Nothofagus* subgenus *Brassospora* forests in an environment that seems to have its closest analogy today in New Caledonia – subtropical, probably with moderate seasonal but reliable rainfall (Read *et al.*, 1990). Martin (1989) has already noted this change in central New South Wales, where eucalypts were already prominent in the Pliocene.

In Tasmania, Hill & Macphail (1985, 1994) commented on Early or mid-Pleistocene fossil plants from Macquarie Harbour. The pollen includes traces of *Eucalyptus*, Poaceae, Asteraceae and Chenopodiaceae but is dominated by *Allocasuarina*, which seems to have occupied the site with cool temperate *Lagarostrobos franklinii*–*Nothofagus* forest nearby. This is the earliest record of a modern type of Tasmanian rainforest

and it includes *Quintinia*, a genus now confined to more temperate montane rainforest in New South Wales, Queensland and New Guinea. Some sclerophyllous groups, such as two species of *Banksia* (now extinct; Jordan & Hill, 1991), and epacrids are present as well. The Pleistocene therefore continues a process of loss of diversity in the wet communities and reassortment in the sclerophyll that had commenced much earlier.

Faunas also demonstrate the continuing expansion into the Pleistocene of open conditions: the kangaroos, diprotodons, short-faced kangaroos and open country flightless birds all actively radiated during the Pliocene and early Quaternary (Archer & Clayton, 1984). The newly arrived rodents have achieved their greatest diversity of new species in the arid zone, presumably because they had the opportunity to occupy the new niches of sandhills and gibber (rocks) that appeared in the Quaternary. Genera such as *Pseudomys* and *Rattus* have also adapted to the new (or certainly greatly expanded) cold mountain grasslands and shrublands.

The cycles of glacials and interglacials, with corresponding alternations in sea levels, rainfall, evaporation, temperature and possibly the frequencies of severe frost, drought, winds and storm events, imposed a new tempo of climatic variability on the Tertiary biota. The time scales, although involving tens of thousands of years, seem to have been short enough to discourage specialisation to stable niches. The Quaternary rewarded generalists such as *Eucalyptus* and species with the ability to migrate or disperse readily. The climatic changes caused major shifts in the boundaries of the major Australian biomes. For plants this meant widespread repeated annihilation of populations and complete ecosystems. This process caused extinction and range fragmentation, but it also may have provided opportunities for speciation by creating small, isolated populations often in areas with slightly different climates, geology and species associates. A response to such pressures is the rapid appearance of new groups able to evolve in small, isolated populations. The classic Australian example is *Acacia*, with over 640 species. These plastic groups are extraordinarily well represented in Australia, where the climatic stresses seem not to have wiped out entire forest types, as occurred in Arctic regions. The migration of plants from outside Australia has been very limited, so it has been local genera that have diversified into many new niches.

The final major event for the Australian vegetation was the migration of humans, the ultimate Quaternary generalist species, around 50 ka ago. To what degree preindustrial humans in Australia and New Guinea had an impact on species and vegetation is an ongoing controversy, but for some regions at least it is clear that human-controlled burning created managed landscapes (Dodson, 1992). The extinction of all large browsers and most grazers and predators has occurred since the arrival of people and this must also have influenced vegetation structure (Flannery, 1990). Imposed on top of the extreme climatic change of the last glacial, human activities accelerated the rate of environmental change for biota.

The result of these processes is the present-day vegetation, which forms six distinctive areas or floristic provinces: Tumbuna, Papua, Torresia, Eremaea, southwest Australia and Bassiana (Figure 15.1). These correspond to the major modern climatic regions of Australia–New Guinea. If we analyse these provinces in terms of their vegetation histories, it is apparent that some have been much more affected by Quaternary change than others, and they differ in the age of their development as well. As might be expected, the less changed, more stable regions contain the greatest number of species that relate directly to Tertiary times (Melville, 1975), and may also contain newly evolved species (Hill & Read, 1987).

Australia is so large and variable that we need the present variety of pattern and the persistence of small areas of vegetation out of balance with most of a region to remind us that past histories based on chance finds of fossils will conceal similar variability. The methods of studying vegetation history in the Quaternary use the same materials as the more distant past, that is micro- and macrofossil remains, but the existence of present-day stands of vegetation and of abundant sites of deposition allow the calibration of the microfossil

records so that vegetation structure can be inferred even though pollen is not usually identifiable to the species level. This means that Quaternary pollen analysis can be used to comment on the history and degree of stability of plant communities, as long as the link to modern ecology is retained. The location of sites is strongly biassed to cool wet areas in the mountains of New Guinea, down the eastern coast and ranges, and in Tasmania, with a scatter of sites elsewhere (Figure 15.2). The voluminous literature on the Quaternary is available from a major database on Quaternary climates maintained by the Bureau of Mineral Resources, Canberra (Bleys et al., 1991).

MAJOR CHANGES OF THE QUATERNARY

Cooler times: the mountains and the coast

The volume of ice in the earth's icecaps through time can be inferred from oxygen isotope analysis of deep-sea cores. Although cooling is apparent almost throughout the Tertiary, major variations in ice volumes are restricted to the Quaternary. So the major fluctuations in sea level are also Quaternary phenomena, reflecting the amount of land ice at any one time. The oceans were lowered by 120–140 m at times of maximum ice lock up and were 5–8 m higher than at present in the warmest interglacials. This has created a restricted but widespread new niche, the coastal zone, in which continuous change and seral responses are dominant. The other niche to appear and expand as a result of cold was the alpine (above treeline) niche, which fluctuated even more widely than the coastal. For the last glacial, some altitudinally limited vegetation boundaries descended more than 800 m in altitude, which was about the same as the lowering of the snowline (Galloway, 1986).

The present coastal zone has developed since the rise of the sea to its current level. The previous interglacial, at 126 ka ago, was the last time that the sea had reached this level, and surpassed it by a few metres. By 7 ka ago, the sea had flooded

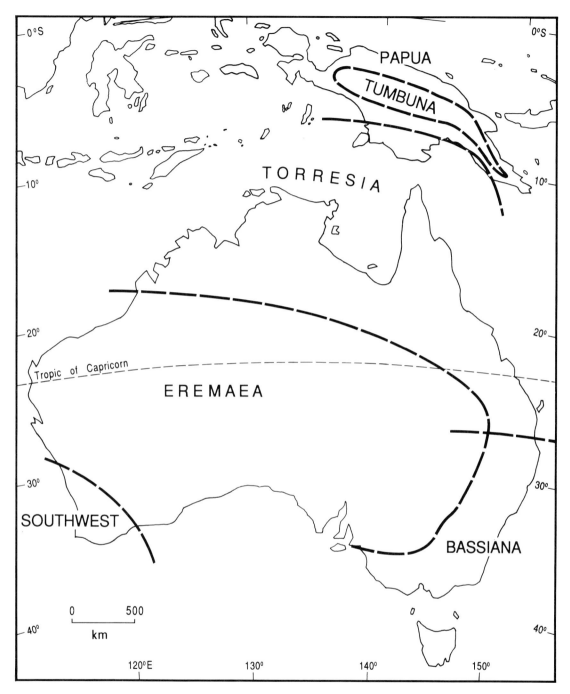

Figure 15.1 The biological provinces of the Australian region.

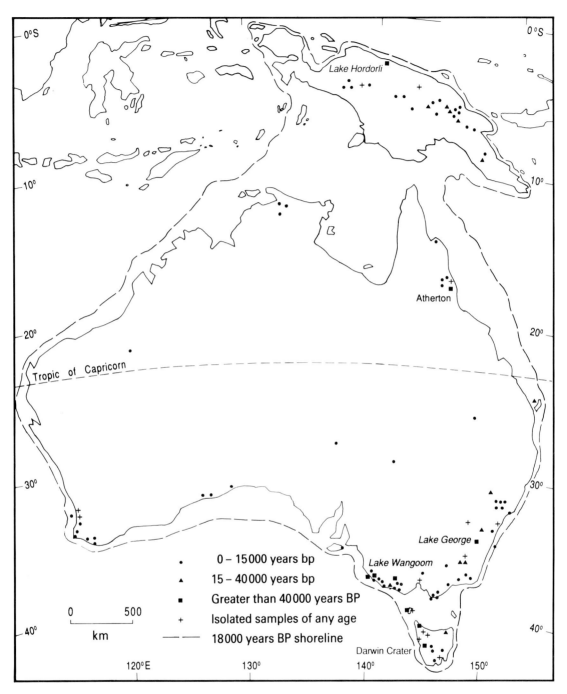

Figure 15.2 Sites of Quaternary palaeoecological investigation in Australia and New Guinea. (After Dodson, 1989.)

into every valley cut below its present level, and in most areas it is still infilling the valleys with swamps, lagoons and saltmarshes. Beach sand, brought onshore by the rising sea level, formed large dunes that were invaded and stabilised by plants once the sand supply diminished. Previous high sea levels have also left dunes and sandplains in many areas, and the long exposure to leaching has provided poor siliceous sands that support only specialised heaths or woodlands. The resultant mosaic of communities developing on cliffs, in marshes or on sand combines with a strong zonation in substrate stability and salt to provide very complex vegetation, with many locally endemic communities. The coastal flora is derived from pre-adapted genera of swamps, sand heaths and river edges. As is true of other new habitats, immigrant groups are well represented, because new environments provide vacant niches and unstable habitats reward migratory ability that will be a characteristic of immigrants. Migratory ability means that, in general, the least stable communities, such as strandline or saltmarsh, have the most widespread species. This contrasts with the more stable but isolated siliceous sand communities, which have many localised species.

Vegetation histories from coastal lagoons have been obtained in many localities, but are generally limited to the last 7 ka. Offshore cores from Narooma, on the south coast of New South Wales, go back to 14 ka, but are not very different from present-day samples (P. Roy personal communication). The lowest sections show mainly saltmarsh pollen with inputs from eucalypt forest. However Roy's analysis of foraminifera in this core suggests that sea temperatures could have been 2 deg.C warmer than at present, a most surprising result, given that the glacial maximum had only just passed, and land conditions were generally a lot cooler than at present. The answer may well lie in land-bridges that blocked off the Torres and Bass Straits at times of lower sea level. These barriers kept the eastern current flowing southwards close to the continent, resulting in a coastal strip of mild climate along the southern coast of present New South Wales, and the eastern coast of Victoria and Tasmania.

Similar results from coastal swamps and caves in Western Australia (Martin, 1973; Balme et al., 1978) and the Northern Territory (Torgersen et al., 1988; Shulmeister, 1991) show very little evidence for temperature change in the last glacial. In those areas there are clear indications of greater aridity at times of lower sea levels but these can be explained by the greater distance to the contemporaneous coast, which gave more inland characteristics.

Evidently the climates and arrangement of vegetation during glacial times in the Quaternary are not simply a shift downhill of those of the present day. If we needed any further reminders, then a consideration of the mountain biomes provides more examples. When it was realised that New Guinea, Tasmania and a tiny area of the Australian alps had experienced glaciation it was easy to calculate the mean lowering of the snowline required (Galloway, 1965). Varying estimates were obtained from different areas, because the snowline is a complex integration of temperature, snowfall, snow drift and exposure to melting. However the maximum figures indicate that the snowline was about 1000 m lower than at present, corresponding to temperatures approximately 6−7 deg.C cooler than present. In places such as Mt Kosciusko, smaller changes are apparent, suggesting that past conditions were drier as well as having depressed mean annual temperatures.

Many vegetation histories are now available from the mountains of Tasmania, Victoria and New South Wales (e.g. Kershaw et al., 1986). These are mostly restricted to the last 12−15 ka or so and show that, at the peak of the last glacial 18 ka ago, there was aridity as well as cold. Treeless vegetation dominated by Asteraceae and Poaceae extended across most of southeastern Australia, down to sea level on the southern coast of Victoria (e.g. Hope, 1978; Dodson, 1989; Figure 15.3). Eucalypts were present but apparently scattered or restricted to sheltered positions. Faunal records from lowland areas contain more ecological groups of animals at a given locality than occur at any equivalent place today. For example, 16 ka ago on Kangaroo Island, South Australia, pygmy possums now restricted to

Figure 15.3 Composite shrubs in grasslands near Great Lake, Tasmania, provide an analogue of Late Pleistocene upland and Bassian shrub steppe.

Figure 15.4 Windswept coastal shrublands on the west coast of Tasmania provide a complex mosaic that may be structurally similar to larger areas of Southern Australia during the last glacial.

Tasmania occurred together with rodents found in red seif dune habitats today (Hope *et al.*, 1978). This mixed faunal ecology suggests a mosaic of vegetation types from dense scrubs to open shrublands (Figure 15.4). Such a mosaic may have been a response to lowered precipitation and evaporation. The latter raised groundwater levels; perhaps ridges and dune crest habitats were desert like, while swales supported lagoons and dense scrubs.

In southeastern Australia the long cores from Lake George, New South Wales (Singh & Geissler, 1985), Lake Wangoom, Victoria (Kershaw et al., 1990), and Darwin Crater, Tasmania (Colhoun, 1988), provide records through more than one glacial–interglacial cycle. Well-wooded periods alternate with more open steppe, or grassland with increased alpine components. The alpine shrubs and herbs are also prominent in part of the wooded periods, often with cool temperate rainforest components. *Eucalyptus* dominated the vegetation on several occasions at all three sites. At Lake George and Lake Wangoom the past phases of *Eucalyptus* never match the dominance in the present interglacial, but at Darwin Crater parallels to the present vegetation arose several times.

In New Guinea two major changes occurred: a depression of the treeline to below 2200 m (compared to 3900 m today) and an expansion of lowland grasslands and seasonally dry communities at the expense of lowland tropical rainforest. No very long cores have been collected to date from New Guinea, but the last 30 ka are well documented (Hope, 1986). The 1700 m depression in the treeline suggests that glacial-time conditions differed from the present in not allowing trees to form a closed forest as they do in the subalpine today. Periodic disturbance by drought or frost may have been a cause (Hope, 1989; Figure 15.5).

The records show that cold-adapted vegetation is restricted today, and in southeastern Australia generally occupies relatively wet mountain tops that favour dense sclerophyllous shrublands. This may explain the low endemism of this alpine flora, because many species may have been recruited from woodlands and forests. The formerly widespread alpine communities of glacial times are retained only on drier mountains and consist of mires and tussock grasslands with numerous composite herbs and shrubs and some seasonal species. Major elements of this cold steppe vegetation persist in lower altitude frost hollows such as the Central Plateau of Tasmania and the Monaro of New South Wales, and this leads to the impression of poor obligate alpine floras and

Figure 15.5 *Cyathea* tree fern-dominated grasslands above 3000 m in New Guinea are created by burning but seem to have recruited species that formerly made up natural above-treeline communities.

faunas. Occurrences of treeline species down to sea level are common in southeastern Australia, e.g. *Gentianella diemenensis*, *Eucalyptus pauciflora* or *Euphrasia* spp. while others display remarkably broad altitudinal ranges and ecological plasticity, e.g. *Allocasuarina verticillata*, *Acaena novaezelandiae* or *Banksia marginata*. This evidence for widespread treelessness suggests that during the height of the glacial the treeline had virtually disappeared, allowing cold dry steppe to expand across a very large area, accompanied by unstable slopes and alluvial fan activity down to about

500 m above sea level (Barlow, 1986; Hope, 1989). Only in western Tasmania and the crest of the Great Eastern Scarp was there an alpine zonation like that of the present, with a largely woody vegetation changing in stature with altitude. Western Tasmania remained wet enough for the specialised subalpine gymnosperms and deciduous beech to survive to the present, although they were forced down to 200 m above sea level during both the last ice advance and the end of the previous interglacial (Colhoun, 1985).

The survival of a wide range of rainforests and tall eucalypt forests in southeastern Australia, differentiated north to south and altitudinally, shows the influence of the eastern fall of the coastal escarpment in providing reliable orographic rainfall and sheltered sites. At full glacial times there would have been an abrupt boundary from these forests to the treeless cold dry tablelands, exposed to the westerly winds. The period from 26 to about 12.5 ka was typical of these conditions in southern New South Wales.

However, not all of the last glacial penalised forests. *Nothofagus*, presumably *N. cunninghamii*, was living at an altitude of 1850 m on Mt Kosciusko 35 ka ago and survived through several early glacial phases near Lake George (Kershaw *et al.*, 1986). It had an expanded range in Victoria prior to the peak of the last glacial (M. Mackenzie, personal communication). Although beech has a discontinuous distribution in southeastern Australia today, there may even have been recent connections across the Bass Strait, for large *Nothofagus* logs have been found on King Island dated to 38 ka ago, and an expansion of wet forests compared to today is indicated at this time at Pulbeena Swamp in northwest Tasmania (Colhoun *et al.*, 1982). This suggests that the montane rainforests were well developed over a wide altitudinal range in southeastern Australia for lengthy periods between stadials (Hope & Kirkpatrick, 1989). A similar conclusion was reached for the mountains of northeast Queensland, where gymnosperm-rich forests containing *Araucaria, Agathis* and *Dacrydium* occur from the end of the last interglacial until about 38 ka ago (Kershaw, 1986). For this area Kershaw hypothesised that rainfall was less

than at present but that, none the less, expansion of rainforest occurred as a result of cooler conditions with less evaporation. Similarly the past wider extent of *Nothofagus* in southern Australia does not imply heightened rainfall but rather a more even rainfall, no substantial drought, and a reduced fire regime (Kershaw *et al.*, 1990).

There are no data about the warm rainforests of *Eugenia* and *Elaeocarpus* that today extend to the Bass Strait islands down the eastern coast of Australia in discontinuous patches (P. Cullen, personal communication). It is clear that these have not had a continuous distribution during the present interglacial, although they may have expanded when conditions seem to have been warmer and slightly wetter along the coast. In the Holocene the period from about 7000 to 4500 BP is notable for high levels of pollen of the understorey shrub *Pomaderris*, which is even noticeable in New Zealand cores (Macphail, 1981). The last time when extensive connections may have existed may date back to the early Quaternary or even before that, but some expansion at least occurred during the last glacial, perhaps as a result of the strengthened warm current noted above. Evidence for this comes from Tasmania, where isolated Late Pleistocene fossils of macropods such as *Aepyprymnus rufescens*, now restricted to subtropical forests of New South Wales, have been found. Colhoun *et al.* (1982) noted that there were expanded wet forests along the northwest coast of Tasmania around 70–40 ka ago, suggesting more available moisture prior to the last glacial advance.

The Quaternary history of the eucalypt forests is also rather speculative. Many of the alpine communities and the montane tall wet forests of southeastern Australia are shared with Tasmania, for example alpine ash and mountain ash forests (Figure 15.6). Endemic eucalypts are concentrated in the drier southeast of Tasmania and in the alpine and escarpment crest of the mainland. Pollen diagrams are not helpful in defining the kinds of eucalypt present, although the associated species show that dense eucalypt forests have probably always been retained on the eastern coast of Wilson's Promontory (Ladd, 1979). It is currently assumed that the Bassian disjunctions of

Figure 15.6 The *Eucalyptus regnans* forests of central Tasmania are disjunct from mainland examples in Victoria. They possibly represent an early stage in the development of eucalypt communities as they are killed by frequent fires but flourish when rare fires open up a rainforest canopy.

the ash forests occurred by Early to mid-Pleistocene times, since they support some locally endemic species, including arboreal mammals such as Leadbeater's possum (*Gymnobelideus leadbeateri*). However, the forests would have been favoured by the same conditions that permitted the spread of rainforests in the earlier part of the last glacial. At Darwin Crater in Western Tasmania, there are at least four phases of tall eucalypt forest dominance within the Late Pleistocene (Colhoun, 1988). These forests have presumably survived in the region and expanded out periodically whenever conditions have been warm enough and dry enough to encourage fires at infrequent intervals. It is possible that rainforest retreated and was replaced by ash forests about 4 ka ago in central Tasmania because of an increase in fire frequency (Macphail, 1979).

The pollen diagrams from Victoria and Lake George, combined with modern eucalypt distributions, show that open eucalypt woodlands were very widespread across the Bassian Plain and along the Western Slopes as far as Queensland. Direct evidence is scanty, but some of these woodlands supported cold-tolerant species such as the *Pimelea pauciflora* found in Late Pleistocene clays at Spring Creek in western Victoria (Flannery & Gott, 1984) and possibly *Astelia* and *Plantago*. In the Coorong of southeastern South Australia, snow gum (*E. pauciflora*) was apparently widespread before 10 ka ago (Dodson, 1975).

We should be cautious in assuming that some novel kind of subalpine woodland had displaced all other eucalypt communities from the area. Many probably persisted in sheltered sites. Today the most widespread eucalypt communities are the savanna box woodlands and their associated grasslands. These communities have presumably advanced to higher altitudes in the Holocene, probably by a process of tree invasion and range retreat by shade-intolerant steppe species (Hope & Kirkpatrick, 1989). The ranges of a number of herbs and low shrubs became disjunct and some extinction may have occurred in seasonal geophytes. Examples include species of *Swainsona*, *Leptorhynchos*, and other Asteraceae, *Bulbine*, *Discaria*, *Scleranthus* and *Hymenanthera*. Some species of Chenopodiaceae and Asteraceae seem to have become extinct in this process, and many of the herb species are now rare and endangered.

We can therefore reconstruct full glacial conditions of southeastern Australia as consisting of very open steppe grasslands with annuals and perennial geophytes and drought-deciduous shrubs. Some of these plants are now restricted to the alpine zone, but most are preserved in frost and drought-prone montane grasslands of rainshadows, such as the Monaro of New South Wales and the Midlands and Central Plateau of Tasmania. These communities were presumably also extensive at times of rainforest expansion, but excluded by forests or shrublands from the wetter mountains (Macphail, 1986). The dense *Allocasuarina nana* heaths of the eastern scarp must have been more extensive under harsh and windy conditions. Many of the present open forests, for example gum-stringy bark (e.g. *Eucalyptus mannifeora*, *E. macrorhyncha*), ironbark (*E. sideroxylon*) and mountain gum (*E. dalrympleana*) forests, were presumably favoured by wet glacial conditions, but reduced in area during the cold dry maxima.

It is clear that at least one major woodland became restricted in southern Australia during the last glacial maximum. This is the riparian forests of river red gum along the Murray and its tributaries. *Eucalyptus camaldulensis* is the most widespread eucalypt in Australia but occurs in several forms. The occurrence of widespread source-bordering dunes and lunettes on streams and lakes after 30 ka ago and a changed river regime prior to the Holocene suggests that the bank stability provided by riparian trees was absent. *Eucalyptus camaldulensis* is derived from the tropical red gum group and its southern form is less tolerant of very high temperatures than the northern form,

but also is frost tender. It may be an example of a species that has adapted to Pleistocene change but which has been unable to exploit a cold riparian niche. However, the role of raised water tables and consequent high soil salinity in the Murray Basin may be contributory. It is curious that late glacial wood of river red gum was recovered from the channel sediments of the Yarra River in Melbourne, and that it was associated with fossils of *Sphagnum* (Duigan & Cookson, 1956).

Nix (1982) has pointed out that a cold-limited growing season changes the favoured strategies for exploiting moisture. In water-limited vegetation, tall grasses (usually C_4) will be favoured in tropical regions because they can respond quickly to favourable moisture conditions. In cold areas with winter rains, however, the dormant vegetation cannot intercept the water, which infiltrates to greater depth and favours perennial shrubs and trees with well-established root systems. Colder conditions of glacial times might therefore be anticipated to cause a substantial advance of southern shrub steppe at the expense of tropical grassland formations (Singh & Luly, 1991).

DRIER TIMES AND AUSTRALIAN DESERT BIOMES

The evidence for substantial changes due to cold are restricted to upland southeastern Australia, but the modern biota, landforms and fossil evidence show that change was widespread across the whole continent in the Pleistocene. This change towards aridity was well advanced by the start of the Quaternary, as shown by the evolution of central Australian large browsers and grazers, but further faunal development took place in the Pleistocene, resulting in specialised genera such as *Diprotodon*, with some species adapted to the shrub steppe and others to desert.

The biogeography of Eremaea presents several puzzles. A few genera and families are very large, with local endemism evident, e.g. *Eremophila*, *Cassia*, *Acacia*, some Chenopodiaceae and Amaranthaceae. Specialised grasses such as the hummock grasses (*Triodia*) and cane grass (*Zyg-*

ochloa paradoxa) fill the small shrub niche across the continent. Several communities are dominant over large areas but absent from apparently suitable habitats. An example of disjunction is *Acacia peuce*, which occurs only in three widely separated stands around the northern Simpson Desert. These are the symptoms of a very young habitat that has been stressed or partitioned (Barker & Greenslade, 1982). Few physical barriers exist, other than the vast size of the area. However, soils are quite variable, including calcareous, gypsiferous and sandy clay soils, and the shield area ironstone laterites. The pattern of seif dunes is generally relict at present, with well-developed soils and orientation often not in balance with modern sand-moving winds. The pattern of dunes became finally established within the last 500 ka. This is indicated by dunefields overunning lake beds that contain the palaeomagnetic reversal dated at approximately 700 ka, thermoluminescence dates of 290 ka for sand burial and an evident increase in aeolian dust accession on the northwest shelf after 500 ka (Wasson, 1986). The dunes are partially reactivated at times during the glacial, as shown by seif dunes that extend below present sea level in northwest Australia, and by records of aeolian dust dated to 18 ka in New South Wales, Victoria and the Tasman Sea.

It therefore seems that the Pleistocene continues a trend towards aridity in Australia, with enlargement of the dunefield on several occasions within the Pleistocene and the loss of the large Tertiary lakes. Some species are older than these events and are now relictual. The best-known examples are the oasis species such as the palm *Livistona mariae*. The distribution of *Acacia peuce* correlates with noncalcareous soils and this species presumably pre-dates the drying and deflation of the lakes of the Lake Eyre Basin. Many plants in these categories have growth related to summer rainfall regimes. We can thus speculate that the flora of arid Australia is derived partly from immigrants, such as the daisies, salt bush and grasses, and partly by adaptation of seasonal tropical forest (dry scrub) taxa to a regime of long droughts. Families contributing to the arid shrub and tree floras are often also promi-

nent in rainforests, e.g. Pittosporaceae, Rutaceae, Sapindaceae, Sterculiaceae, Bombacaceae, Meliaceae, Cupressaceae, Casuarinaceae, Malvaceae, Papilionaceae and Mimosaceae (Figure 15.7). Many of these species are not cold tolerant, and the southern arid lands instead are dominated by specialised shrub eucalypts with lignotubers, the mallees. This adaptation is also common in cold moist mountain summits, reflecting a common strategy to enhance regeneration and survival in growth-limited areas.

Quaternary plant fossils are uncommon in the semiarid regions of Australia, and pollen is of limited use because the desert plants respond to dry conditions by ceasing to flower and reducing density rather than replacement by other species. In addition ephemerals or drought evaders are important there, and so it is difficult to identify climatic change in the core of the desert. Singh has argued for change in the balances between summer and winter rainfall reaching Lake Frome, with a general retreat of summer flowering grasses during dry or cold times. This is held to reflect a failure of the monsoon and favours perennial shrublands of Asteraceae and Chenopodiaceae that are adapted to winter rainfall and cold conditions (Singh & Luly, 1991). Martin (1973) suggested that on the Nullarbor the coastal zones migrated southwards as sea level fell, causing localities to become more open at 20 ka ago.

More dramatic changes have occurred around the margins of the desert lands. The widespread treelessness of southeastern Australia, which reflects drought and raised saline watertables as well as cold, has already been described. Modern disjunctions, such as isolated stands of arid hummock grasses along the Glenelg River (near Portland, Victoria), white box (*Eucalyptus alba*) and *Callitris glauca* in the Snowy and upper Hunter river valleys, and a patch of mallee in well-watered southern Victoria, suggest that at times the semiarid zone has been very extensive. From the Glenelg River, fossil desert animals have been recovered from cave sediments dating from 30 to 12 ka (Hope, 1983), and seif dunes cross Kangaroo Island in South Australia (Bowler & Wasson, 1984). A retreat of moist forests is also marked

Figure 15.7 Arid zone woodlands are dominated by species derived from rainforest families. *Atalaya hemiglauca* (Sapindaceae)–*Pittosporum philliraeoides* (Pittosporaceae) woodland, Strezlecki Desert, eastern South Australia.

at the Atherton Tablelands after 35 ka ago, and Kershaw (1986) estimated a drop in precipitation to 30% of present at that time. This requires a shift of vegetation boundaries eastwards of only 30 km at this locality, but much more extensive adjustments to vegetation zones seem to have occurred across the plains of Northern Australia at this time. Cores from Lake Carpentaria suggest only open grassland at 20 ka ago (Torgersen *et al.*, 1988). Marine cores taken by the Snellius expedition from the Timor Sea show an increase in Chenopodiaceae, Asteraceae and a decline in *Eucalyptus* after 40 ka ago until the Holocene (van der Kaars, 1989). This supports the model of vegetation change suggested by Nix & Kalma (1972) for an advance of non-eucalypt arid communities almost to the Arafura shelf, representing migration of about 400 km northwards. At this

time monsoon forests would have been quite restricted, although Arnhem Land obviously acted as a major refuge. We do not yet know what the vegetation response to early glacial conditions may have been, although higher sea levels and lower temperatures may have contributed to rainfall and available moisture.

Periods of heightened aridity and possibly stronger winds have thus been important in stressing Eremaea and may be one of the causes of the patchiness of community growth across the area. Fossil soils and some dating of dunes show repeated episodes of sand dune activity in the past (Wasson, 1986). Development towards withstanding both short- and long-term change is characteristic of evolution in the arid areas.

ARRIVAL OF HUMANS AND VEGETATION CHANGE

It is not excessively anthropocentric to claim that the most significant Cenozoic migration event for the Australian vegetation was the arrival of people more than 40 ka ago. We were preceded here in the Miocene by well-dispersed placental mammals such as whales and bats, in the Pliocene by the murid rodents, and we ran a close race with stegodonts in the late Quaternary, which we probably won by cheating (we ate them). Nevertheless, human activity has caused massive changes to floristics and structure through fire, faunal change and plant introductions. Of course it can be argued that the arrival of the advanced plant families and genera such as grass, daisies, saltbush and possibly *Acacia* were events of greater magnitude, but the significant species are autochthonous species that have evolved following the establishment of aliens. Humans had occupied all available niches by 35 ka ago, including New Guinea, montane Tasmania and the arid zone (Dodson, 1992). The limitations of radiocarbon dating make an assessment of earlier occupation rather hazardous. It seems clear, however, from thermoluminescence and uranium series dates at selected localities, that coastal New Guinea and Australia had scattered occupation by 50 ka ago and some claims for earlier dates have been made.

The effect of occupation on vegetation structure and plant community formation is currently being argued. With the exception of closed non-seasonal rainforests such as those of New Guinea, the Queensland mountains and Tasmania, all Australian vegetation tolerates fire, and some communities are adapted to particular fire regimes. Perhaps early sparse settlement had little effect on the environment, since deliberately lit fires would not have made much difference. However, the change to vegetational structure when Aboriginal firing was changed after European contact is well documented in many localities, such as northwestern Tasmania, the Brigalow scrub and Kakadu. In terms of fossil evidence, some of the best comes from large offshore islands

that were not occupied at the time of European arrival. Clark (1983) examined a 7000 year lagoon record on the eastern end of Kangaroo Island, only 14 km from mainland South Australia. She showed that a significant change in fire regime occurred after 2.6 ka ago when occasional fires yielding large amounts of charcoal and soil erosion started to occur. Prior to this, the fires seemed to have been frequent, but yielded only minor amounts of charcoal and no significant erosion products. Clark proposed that the island was abandoned 2500 years ago for unknown reasons. The result of the 'natural' fire regime was to allow a dense eucalypt forest to occupy all sites. Meanwhile, on the mainland, open savanna woodlands of fire-tolerant box species above grassland had developed, and this difference seems to have been the result of a continuing firing on the mainland. Traditional Aboriginal groups of central and northern Australia are quite clear that management of land involves careful burning, and the staggering area of hummock grass in central areas may be the result of mosaic burning. In Kakadu, the difference between hot wildfires, which tend to occur at the end of the dry season in November, and human-lit patch fires, which burn the tall moist grass at the start of the dry season in May, is acknowledged (e.g. Gill *et al.*, 1981). In subtropical climates in Fiji and New Caledonia equivalent to coastal Queensland, people arrived later, at about 3 ka ago, and caused a dramatic shift from closed forest to open savanna (G. S. Hope, unpublished data). The seasonally dry subtropics have enough moisture to allow trees to flourish, but the dry season also provides a period of vulnerability to fire. Clearly the balance throughout most of Australia is in favour of open vegetation, reflecting a high frequency fire regime; how far this is a human artefact is debatable.

At the time of human arrival, Australia supported large browsing and grazing animals in the now extinct families Diprotodontidae, Palorchestidae and the subfamily Sthenuridae. Kangaroos larger than any presently known in the genera *Macropus* and *Protemnodon* were also present. The niches occupied by these animals were in general the open communities from central Australia to the alpine steppes of Tasmania, although some species occupied forests in New Guinea. The causes of the almost complete loss of the large browsing niche are also debated, but the timing of the extinctions appears to be drawn out from perhaps 50 to 30 ka ago for the largest species, and continuing to the end of the Pleistocene for the smaller. Although the process by which extinction took place is unknown, the effects on the vegetation would have been dramatic (Flannery, 1990). Large open areas in the semiarid and subalpine regions may have been controlled by grazing, as the African tropical plains are supposed to be today. The chief effect is to reduce tree and shrub growth to a few lucky survivors, widely separated by grasslands. Although Flannery has argued that Australia has always been depauperate in vertebrate faunas due to intrinsic habitat poverty, this is contradicted by the dramatic success of introduced grazers and browsers, such as goats, horses and camels, that have invaded the Australian deserts and subalpine woodlands. The rapid evolution of mammals and birds from the Pliocene onwards also argues for significant opportunities for specialisation in these niches. The loss of the animals may have encouraged more perennial shrubs and trees to survive and compete for water, with the reduction in ephemerals and shallow-rooted plants. This trend might have encouraged fire and the growth of fire-tolerant or evading groups such as the mallee eucalypts at the expense of taxa such as *Callitris*. Our impressions of the semiarid shrublands of *Acacia*-rich formations as natural may therefore have to be amended because other potential dominants have been forced out by insufficient competitive ability or fire tolerance. The spread of woody weeds in western New South Wales after overgrazing of palatable shrubs by sheep provides another possibility in that our precontact semiarid vegetation may be relatively palatable. The abrupt shift in saltbush dominance from palatable to noxious species in the semiarid and from species-rich herbfields towards coarse tussock in subalpine areas occurred very quickly following stock (and rabbit) introduction. These changes indicate how sensitive the vegetation is to increases in grazing pressures.

THE ADAPTATIONS OF THE QUATERNARY

Unravelling the skein

The problem of assessing the responses of vegetation to climatic change, humans, fire and faunal loss is that they are all compounded. The vegetation histories that cover the last 40 ka demonstrate the frequent presence of fire. At Lake George this increases markedly in the Holocene to levels not apparent in the previous records. At Lynch's Crater on the Atherton Tableland of North Queensland, on the other hand, burning was apparent after 38 ka ago, but diminished in the Holocene as wet vegetation reinvaded the area. Bass Strait shows abundant evidence for fire in a grassy steppe 25 ka ago (Hope, 1978), but a similar vegetation on the Central Plateau of Tasmania in the Holocene does not seem to undergo burning, although a grassy cover is maintained (Thomas & Hope, 1994). Some fires are evident over the several glacial/interglacial cycles recorded at Darwin Crater, and there is no marked increase in the last 50 ka (Colhoun, 1988). This matches a preliminary analysis of a deep-sea core from the Tasman Sea that shows only a minor increase in carbonised particles over the past 150 ka (G. S. Hope, unpublished; see p. 389).

The obvious solution to discriminating the effects of human arrival from climatic change is to compare the Holocene with the last interglacial, 129 ka ago, and the penultimate glacial maximum, at around 135 ka ago, with the well-understood changes of 25 to 12 ka ago. Sea level records, and the occurrence of corals as far south as Newcastle, hint that the peak of the last interglacial was slightly warmer than at any time in the Holocene (Roy & Thom, 1981). However, there are few useful last interglacial records. On the Atherton Tablelands, and at Darwin Crater, vegetation cover seems fairly equivalent to that of the Holocene. Some gymnosperm species that subsequently became extinct were present at Atherton (Dodson, 1992). Both these sites are wet forests that seem to have been unaffected by humans in the Holocene. The grassland–woodland sites of Lake George and Lake Wangoom show a similar pattern, but there are some distinctions between the interglacials (Kershaw *et al.*, 1990). At Lake George the supposed last interglacial is marked by declines in Casuarinaceae and an increase in *Eucalyptus* and grass, as well as the appearance of carbonised particles (fine charcoal from fires). By the present interglacial, one type of Casuarinaceae appears to be extinct and carbonised particles reach historical high levels. Singh & Geissler (1985) have suggested that these changes were caused by an increase in fire consistent with human occupation. However, there are significant peaks of carbonised particles further back in the Lake George record, but the sequence is not well dated and seems to contain significant gaps. Wright (1986) has pointed out that the pollen and charcoal do not yet provide good evidence for very early occupation and no direct occupation traces have been located. Lake Wangoom does not change greatly in vegetation structure between interglacials, but Edney *et al.* (1990) have pointed out that Asteraceae become less important and Poaceae more so, which may be consistent with an increase in fire frequencies.

Clearly many more interglacial records must be found and interrogated before we can assess our own period with any confidence. Although the data are still contradictory, there is the possibility that sclerophyllous forests are now more fire adapted and dominated by eucalypts than was previously the case. This is a possible explanation for the manifest rapid differentiation occurring in *Eucalyptus* at the present time. But it is still impossible to say that this commenced with the advent of humans, and for some communities it has probably been developing over the last million years, as climates have dried, soil mulches been destroyed and nutrients flushed away.

Similarly it is difficult to assess whether plant extinction has been more marked during the last glacial cycle than occurred previously. Kershaw (1984) has reviewed plant extinctions and range

retreats and noted some clear losses such as *Dacrydium*, large-grain Casuarinaceae, a small spine alpine Asteraceae, and possible shrub steppe Chenopodiaceae in Tasmania. Range retreats, such as *Phyllocladus* being lost from Victoria (Churchill & Dodson, 1980), *Cordyline* from the Atherton Tableland (Kershaw, 1985) and *Nothofagus* from southern New South Wales (Kershaw *et al.*, 1986), also occur with the onset of the last glacial peak about 30 ka ago, or in the rapid warming approaching the start of the Holocene. These are palynologically obvious taxa, and the vast majority of extinctions probably will never be known. Extinctions do appear to be marked in the last glacial compared to earlier times, but this conclusion may be shown to be incorrect when we learn more about the past. It is certain that many rainforest communities had all but disappeared by the time the late Quaternary record commenced. Lake George, one site that traverses the entire Quaternary, is silent for over one million years due to the fact that the lake sediments have been oxidised and reddened by long periods of soil formation. If seasonal, relatively subtropical, conditions are assumed for periods in the early and mid-Quaternary, this must have been the time when drought first became characteristic (McEwen Mason, 1991; Martin, 1989). Some of these early droughts may have resulted in widescale devastation to plant ranges, with consequent extinctions. This possibility is speculative so far.

The last glacial maximum

In Australia, the Holocene/Pleistocene boundary is generally very clearly marked. Most upland sites do not extend back beyond 11 ka, due to gravels or soils underlying swamp and lake sediments. In fact the last glacial, 26 to *ca* 12.5 ka, stands out as the most extreme time visible in the palynological and limnological record (De Deckker, 1986). This is surprising, given that the episode is not particularly distinctive in the deep-sea cores, compared to previous periods of ice lock-up. Colhoun (1991) believed that ice extent in Tasmania around 85 ka ago (and also in earlier glaciations) probably exceeded that of the last glacial maximum. Contrasting to this is the fact that the fossil lakes of western New South Wales are mostly less than 120 ka old and appear to be a new landscape response to river and groundwater changes (Bowler, 1980). Taken together, the vegetational response seems to reflect colder but drier times than ever before experienced. This may indicate a relatively recent expansion of the range of the Poaceae and Asteraceae flora that today characterises the dry uplands of southeastern Australia. Similarly, the peak of the last glacial may represent the first time that montane rainforest has become so disjunct and also may mark the retreat of monsoon jungle and the deciduous scrubs of Northern Australia to restricted refuges (Russell-Smith et al., 1992).

Directions in Quaternary evolution

It is apparent that the question of how the Quaternary vegetation developed has a multiplicity of answers, depending on the history of the site and its access to preadapted biota. Judged by the floras of the present day, the six biogeographical provinces all experienced some degree of increased change. Tumbuna, Papua and southwestern Australia were probably the least stressed provinces and their overall diversity probably continued to increase throughout the Quaternary. Not only would extinction rates have been low in these provinces, but, as Walker (1982) has pointed out for the north and Hopper (1979) for the southwest, mild climatic changes caused opportunities for both migration and allopatric speciation due to isolation. A powerful impetus for local speciation is the gentle change to local habitat, which provides sites for natural experiments in migration and variation in populations. Thus, Hopper explained the very high diversity of the central zone of southwestern Australian heaths by considering that this zone has fluctuated sufficiently to encourage speciation. The driest and wettest areas would also vary, but not sufficiently to isolate populations and encourage novelty. In New Guinea, continuing dramatic landscape change and alternations in temperature seem to have led

to rapid local speciation in at least some genera and families (Walker & Hope, 1982).

Some vegetation types in Torresia have been greatly stressed, while others have been fairly stable. The evidence for open woodland and retreat of wet forests on the Atherton Tableland demonstrates that massive migrations have occurred. The disjunctions in the range of depauperate evergreen rainforests, including the dry types, indicates that little speciation has occurred under the late Quaternary regime. But some dry forests and coastal scrubs display some local endemism. These communities have managed to recolonise burnt and dry niches very successfully, although they are now impeded by the arrival of placental grazers. Some communities, e.g. Brigalow and Callitris intratropica, are disjunct at present, but able to dominate large areas, and presumably are well adapted to the stresses of aridity. The absence of many physical or thermal barriers in this province means that the main stress has been changing aridity. This stress is usually modified by local conditions so that refugia persist on sites with permanent groundwater.

Eremaea and Bassiana have been influenced by cold as well as aridity, and some areas so altered that complete changes of plant formation have been effected. These regions generally experienced levels of stress that exceeded the modest change that can be a cause of increased diversity. Presumably extinction rates have exceeded speciation rates in these cases. The diverse pattern of microendemism along the eastern scarp of the continent is an exception to this, and may be evidence for a gradient in stability from the moist slopes to the open cold dry steppes of the tablelands. In the southeastern mountains above 500 m altitude, the complex alternations of cold moist, cold dry and warm moist (but drought prone) conditions would have prevented almost any vegetation community from enjoying a long tenure on a given site. Plants in this zone show considerable plasticity and it is possible that in many cases species have been capable of contributing to several vegetation formations that thus varied in structure and dominance even though retaining a range of species in common. Hope & Kirkpatrick (1989) have suggested that snowgum (*E. pauciflora*) woodlands and subalpine grasslands are examples of this.

Australia does not seem to have been stressed to the point of massive extinction in the Quaternary in the way that the Arctic or Boreal regions have, with consequent simple communities of generalists. Australian ecosystems perhaps retain or exceed the levels of community diversity developed in the Tertiary, despite the new climates and instability.

The Holocene

The last 10 ka seems curiously undramatic and familiar. Alpine studies in New Guinea, Mt Kosciusko and Tasmania (Macphail, 1979; Hope, 1986; Kershaw *et al.*, 1986; Markgraf *et al.*, 1986; Kershaw & Strickland, 1990) indicate warming conditions and invasion of the alpine tundra by herbfields after 13–14 ka ago, and temperatures equivalent to those of the present occur by 9.5 ka ago. The delay in the arrival of sea levels to about their present position for another 2500 years influences coastal rainfall in some sites, such as the Nullarbor. Adjustment by vegetation to interglacial climates takes varying amounts of time. As argued for Kangaroo Island earlier, in some cases the adjustment has been hindered by fire. However, in western Tasmania the paradox exists that the amelioration of the climate led to the exclusion of humans from large areas, as closed forests and scrubs invaded former open alpine communities (Cosgrove *et al.*, 1990). The same process may have occurred in New Guinea as well (Hope, 1986). For burnable areas, the expansion of forest has been hindered or even halted, and it is in a state of dynamic equilibrium or even gradual adjustment. For example, an area such as the Monaro is capable of supporting eucalypt open forest, and in fact does so on favourable sites. But thin soils and the effects of cold, drought and fires has resulted in widespread grasslands with miserable tortured specimens of *Eucalyptus rubida* and *E. pauciflora* scattered about, perhaps the van-

guard of an advancing treeline for which ten millennia have still not provided enough time. Many pollen diagrams from the Holocene establish the continuity of Pleistocene vegetation to the present, such as some of the western Victorian lakes studied by, for example, Dodson (1975, 1979), D'Costa *et al.* (1989), and Edney *et al.* (1990). Singh & Luly (1991) studied Lake Frome, a desert playa lake, and found relatively undramatic changes other than supposed increased densities of desert species, and possible shifts in the balances between grassland and shrubland. These occur through the Holocene, and remind us that the adjustment by the arid vegetation to warmer conditions was probably quite quick, since outlier populations probably survived in the southern deserts throughout the glacial.

The final event of the Quaternary was the arrival of agriculture and direct disturbance of the ground by ploughing and overgrazing. Agriculture (and even commercial forestry) is an attempt to manipulate ecosystems to favour short-lived species for their fruits, leaves or wood. The species selected are usually aggressive colonisers and they came complete with a suite of their competitors, defined as weeds because of their success in the same environment as wanted plants. With the exception of the poorest habitats, weed invasion of Australia has been dramatic, though less so in New Guinea. But the experience of the Quaternary, with its gradually increasing tempo of change, has produced native species that are well adapted to stress, and which survive well in competition with introductions. The grazed rangelands of Australia may in some ways resemble the Pleistocene shrub steppe communities. The current remobilisation of the desert dunefields by feral grazers and browsers seems to demonstrate that pre-extinction Pleistocene steppe could also have been influenced by mammal utilisation. Clearing, firing and grazing of *Eucalyptus* woodlands may tend to recreate types of savanna or shrublands that existed in the past. In that sense the last part of the Holocene has continued to be a major laboratory of evolution of adaptation to disturbance.

CONCLUSIONS

The boundary of the Quaternary is probably not especially dramatic in Australia, where the long-term evolution of the climate towards more arid environments had been proceeding for the previous ten million years, and there was no sudden onset of glaciation. The continent is large enough to cover a wide range of climates, so that habitats were preserved for species capable of migration. The eastern ranges and Tasmanian west coast remained wet and preserved rainforests reasonably intact. The most dramatic event was the expansion of the longitudinal dunefields, indicating a vegetational cover of less than 10% and strong winds, probably about 700 ka ago. Subsequent development of the arid-adapted vegetation perhaps means that cover will not drop as low again. The biota have become adapted to relatively slow but complex cycles of climatic change and to the appearance of short growing seasons in the south.

At least some species have acquired ecological (and probable genetic) plasticity and have the ability to contribute to different kinds of vegetation at various levels of dominance. These strategies have been encouraged by the intrinsic variability of many Australian environments, leading to such characteristic innovations as mallee (lignotuberous) shrubs and trees and hummock grasses. Many of the semiarid and woodland communities have relatively simple floristics, and this may reflect widespread, highly stressed environments in the past. Other communities with a high level of local species are more likely to have survived in intact stands but have probably been isolated for long periods. Where endemic species from isolated taxonomic groups (palaeoendemics) are important, the vegetation will in many cases be relictual and occupy habitats that have had stable cores throughout the Quaternary.

The effect of humans on the environment has been to favour species that have become well adapted to Pleistocene variability by developing the mechanisms of fire tolerance or avoidance and migratory ability. Human disturbance may have favoured *Eucalyptus* over some other taxa in wood-lands and open forest. The removal of many large grazers and browsers and all large predators may have swung some areas towards shrubland growth and stabilised some landscapes. This interference has now been reversed by the introduction of a complete range of placental species (excepting large carnivores). A fundamental shift in vegetation cover towards ephemeral species and lower cover is currently taking place.

I am grateful to Jon Luly, Bob Hill, Greg Jordan, Bren Weatherstone and Peter Kershaw for their cruel but justified trashing of early drafts of this chapter. I also owe a debt to Jim Bowler, Eric Colhoun, John Dodson, Dave Gillieson, Jamie Kirkpatrick, Brian Lees, Mick MacPhail, Gus van de Geer and Peter White for their various works in the Quaternary. I defy anyone to singlehandedly write the story of the Australian Quaternary and really appreciate all the insights derived from geomorphologists, biologists, ecologists, prehistorians and oceanographers too numerous to mention.

REFERENCES

ARCHER, M. & CLAYTON G. (1984). *Vertebrate Zoogeography and Evolution in Australasia*. Perth: Hesperian Press.

BALME, J., MERRILEES, D. & PORTER, J. K. (1978). Late Quaternary mammal remains, spanning about 30 000 years, from excavations in Devil's Lair, Western Australia. *Journal of the Royal Society of Western Australia*, **16**, 33–65.

BARKER, W. R. & GREENSLADE, P. J. (eds.) (1982). *Evolution of the Flora and Fauna of Arid Australia*. Frewville: Peacock Publications.

BARLOW, B. A. (ed.) (1986). *Flora and Fauna of Alpine Australasia*. Australia: CSIRO.

BLEYS, E., HUNT, G. R. & TRUSCOTT, M. (1991). *Quaternary climate in Australia – A Bibliography*. Bureau of Mineral Resources Record.

BOWEN, D. Q. (1978). *Quaternary Geology*. Oxford: Pergamon.

BOWLER, J. M. (1980). Quaternary chronology and palaeohydrology in the evolution of the Mallee landscapes. In *Aeolian Landscapes in the Semi-arid Zones of Southeastern Australia*, ed. R. R. Storrier & M. E. Stannard, pp. 69–71. Wagga Wagga: Australian Society of Soil Sciences.

BOWLER, J. M. & WASSON, R. (1984). Glacial age environments of inland Australia. In *Late Cainozoic Palaeoclimates of the Southern Hemisphere*, ed. J. C. Vogel, pp. 183–208. Rotterdam: A. A. Balkema.

CHAPPELL, J. M. A. & GRINDROD, A. (eds.) (1983). *CLIMANZ 1: Late Quaternary Climatic History of Aus-*

tralia, New Zealand and Surrounding Seas. Canberra: Department of Biogeography and Geomorphology, Australian National University.

CHAPPELL, J. M. A. & SHACKLETON, N. J. (1986). Oxygen isotopes and sea level. *Nature*, **334**, 137–40.

CHURCHILL, D. M. & DODSON, J. R. (1980). The occurrence of *Phyllocladus aspleniifolius* (Labill.) Hook. f. in Victoria prior to 11 000 BP. *Muellaria*, **4**, 277–84.

CLARK, R. L. (1983). Pollen and charcoal evidence for the effects of Aboriginal burning on the vegetation of Australia. *Archaeology in Oceania*, **18**, 32–7.

COLHOUN, E. A. (1985). Pre-last glaciation maximum vegetation history at Henty Bridge, Western Tasmania. *New Phytologist*, **100**, 681–90.

COLHOUN, E. A. (ed.) (1988). *Cainozoic Vegetation of Tasmania*. Newcastle: Department of Geography, University of Newcastle.

COLHOUN, E. A. (1991). *Climate During the Last Glacial Maximum in Australia and New Guinea. Australian and New Zealand Geomorphology Group Special Publication* 2.

COLHOUN, E. A., GEER, G. VAN DE & MOOK, W. G. (1982). Stratigraphy, pollen analysis, and palaeoclimatic interpretation of Pulbeena Swamp, northwestern Tasmania. *Quaternary Research*, **18**, 108–26.

COSGROVE, R., ALLEN, F. J. & MARSHALL, B. (1990). Palaeo-ecology and Pleistocene human occupation in south central Tasmania. *Antiquity*, **64**, 59–78.

D'COSTA, D. M., EDNEY, P. A., KERSHAW, A. P. & DE DECKKER, P. (1989). Late Quaternary palaeo-ecology of Tower Hill, western Victoria, Australia. *Journal of Biogeography*, **16**, 461–82.

DE DECKKER, P. (1986). What happened to the Australian aquatic biota 18 000 years ago? In *Limnology in Australia*, ed. P. De Deckker & W. D. Williams, pp. 487–96. Melbourne: CSIRO.

DODSON, J. R. (1975). Vegetation history and water fluctuations at Lake Leake, south-eastern South Australia. II. 50 000 BP to 10 000 BP. *Australian Journal of Botany*, **23**, 815–31.

DODSON, J. R. (1979). Late Pleistocene vegetation and environments near Lake Bullenmerri, Western Victoria. *Australian Journal of Ecology*, **4**, 419–27.

DODSON, J. R. (1989). Late Pleistocene vegetation and environmental shifts in Australia and their bearing on faunal extinctions. *Journal of Archaeological Science*, **16**, 207–17.

DODSON, J. R. (1992). *The Native Lands – Prehistory and Environmental Change in the South West Pacific*. Melbourne: Longman Cheshire.

DUIGAN S. L. & COOKSON I. C. (1956). The occurrence of *Azolla filiculoides* and associated vascular plants in a Quaternary deposit in Melbourne, Victoria. *Proceedings of the Royal Society of Victoria*, **69**, 5–13.

EDNEY, P. A., KERSHAW, A. P. & DE DECKKER, P. (1990). A late Pleistocene and Holocene vegetation and environment record from Lake Wangoom, western plains

of Victoria, Australia. *Palaeogeography, Palaeoclimatology, Palaeoecology*, **80**, 325–43.

FLANNERY, T. M. (1990). Pleistocene faunal loss: implications of the aftershock for Australia's past and future. *Archaeology in Oceania*, **25**, 45–67.

FLANNERY, T. M. & GOTT, B. (1984). The Spring Creek locality, south-western Victoria: a late surviving megafaunal assemblage. *Australian Zoologist*, **21**, 385–422.

GALLOWAY, R. W. (1965). Late Quaternary climates in Australia. *Journal of Geology*, **73**, 603–18.

GALLOWAY, R. W. (1986). Australian snowfields past and present. In *Flora and Fauna of Alpine Australia*, ed. B. A. Barlow, pp. 27–36. Melbourne: CSIRO.

GILL, A. M., GROVES, R. H. & NOBLE, I. R. (eds.) (1981). *Fire and Australian Biota*. Canberra: Australian Academy of Science.

HILL, R. S. & MACPHAIL, M. K. (1985). A fossil flora from rafted Plio-Pleistocene mudstones at Regatta Point, Tasmania. *Australian Journal of Botany*, **33**, 497–517.

HILL, R. S. & MACPHAIL, M. K. (1994). Tertiary history and origins of the flora and vegetation. In *Vegetation of Tasmania*, ed. J. B. Reid, R. S. Hill & M. J. Brown. Tasmania: Government Printer, in press.

HILL, R. S. & READ, J. (1987). Endemism in Tasmanian cool temperate rainforest: alternative hypotheses. *Botanical Journal of the Linnean Society*, **95**, 113–24.

HOPE, G. S. (1978). The late Pleistocene and Holocene vegetational history of Hunter Island, north-western Tasmania. *Australian Journal of Botany*, **26**, 493–514.

HOPE, G. S. (1986). Development of present day biotic distributions in the New Guinea mountains. In *Flora and Fauna of Alpine Australia*, ed. B. A. Barlow, pp. 129–45. Melbourne: CSIRO.

HOPE, G. S. (1989). Climatic implications of timberline changes in Australasia from 30 000 BP to present. In *CLIMANZ 3*, ed. T. Donnelly & R. Wasson, pp. 91–9. Canberra: CSIRO, Division of Water Research.

HOPE, G. S. & KIRKPATRICK, J. B. (1989). The ecological history of Australian forests. In *Australia's Ever Changing Forests*, ed. K. Frawley & N. Semple, pp. 3–22. Canberra: Publication 1, Department of Geography and Oceanography, Australian Defence Forces Academy.

HOPE, J. H. (1983). The vertebrate record, 15–10 ka, Kangaroo Island – western Victoria. In *CLIMANZ 1: Late Quaternary Climatic History of Australia*, ed. J. M. A. Chappell & A. Grindrod, p. 77. Canberra: Department of Biogeography and Geomorphology, Australian National University.

HOPE, J. H., LAMPERT, R. J., EDMONSON, L., SMITH, M. J. & VAN TETS, G. F. (1978). Late Pleistocene remains from Seton rock shelter, Kangaroo Island, South Australia. *Journal of Biogeography*, **4**, 363–85.

HOPPER, S. D. (1979). Biogeographical aspects of speciation in the Southwest Australian flora. *Annual Review of Ecology and Systematics*, **10**, 399–422.

JORDAN, G. J. & HILL, R. S. (1991). Two new *Banksia* species from Pleistocene sediments in western Tasmania. *Australian Systematic Botany*, 4, 499–511.

KERSHAW, A. P. (1984). Late Cenozoic plant extinctions in Australia. In *Quaternary Extinctions. A Prehistoric Revolution*, ed. P. S. Martin & R. G. Klein, pp. 691–707. Tucson: University of Arizona Press.

KERSHAW, A. P. (1985). An extended late Quaternary vegetation record from northeastern Queensland and its implications for the seasonal tropics of Australia. *Proceedings of the Ecological Society of Australia*, 13, 179–89.

KERSHAW, A. P. (1986). The last two glacial–interglacial cycles from northeastern Australia: implications for climatic change and Aboriginal burning. *Nature*, 322, 47–9.

KERSHAW, A. P., BAIRD, J., D'COSTA, D. M., EDNEY, P. A., PETERSON, J. A. & STRICKLAND, K. M. (1990). A comparison of long Quaternary pollen records from the Atherton and Western Plains volcanic provinces, Australia. In *The Cainozoic of the Australian Region: A Reappraisal*, ed. M. A. J. Williams, P. De Deckker & A. P. Kershaw, pp. 288–301. Melbourne: Geological Society of Australia.

KERSHAW, A. P., McEWEN MASON, J. R., McKENZIE, G. M., STRICKLAND, K. M. & WAGSTAFF, B. E. (1986). Aspects of the development of cold-adapted flora and vegetation in the Cenozoic of southeastern mainland Australia. In *Flora and Fauna of Alpine Australasia*, ed. B. Barlow, pp. 147–60. Melbourne: CSIRO.

KERSHAW, A. P. & STRICKLAND, K. M. (1990). The development of alpine vegetation on the Australian mainland. In *Proceedings of the First Fenner Conference: The Scientific Significance of the Australian Alps*, ed. R. Good, pp. 113–26, Canberra: Australian Academy of Science.

LADD, P. G. (1979). A Holocene vegetation record from the eastern side of Wilson's Promontory, Victoria. *New Phytologist*, 82, 265–76.

MACPHAIL, M. K. (1979). Vegetation and climates in southern Tasmania since the last glaciation. *Quaternary Research*, 11, 306–41.

MACPHAIL, M. K. (1981). Fossil *Pomaderris apetala*-type pollen in north-west Nelson: reflecting extension of wet sclerophyll forests in south-eastern Australia? *New Zealand Journal of Botany*, 19, 17–22.

MACPHAIL, M. K. (1986). 'Over the top': pollen based reconstructions of past alpine floras and vegetation in Tasmania. In *Flora and Fauna of Alpine Australasia*, ed. B. Barlow, pp. 173–204. Melbourne: CSIRO.

MARKGRAF, V., BRADBURY, J. P. & BUSBY, J. R. (1986). Palaeoclimates in southwestern Tasmania during the last 13,000 years. *Palaeos*, 1, 368–80.

MARTIN, H. A. (1973). Palynology and historical ecology of some cave excavations in the Australian Nullarbor. *Australian Journal of Botany*, 21, 221–33.

MARTIN, H. A. (1989). Evolution of Mallee and its environment. In *Mediterranean Landscapes in Australia*, ed. J. C. Noble & R. A. Bradstock, pp. 83–92. Melbourne: CSIRO.

McEWEN MASON, J. R. C. (1991). The late Cainozoic magnetostratigraphy and preliminary palynology of Lake George, New South Wales. In *The Cainozoic in Australia: A Re-appraisal of the Evidence*, ed. M. A. J. Williams, P. De Deckker & A. P. Kershaw, pp. 195–209. Melbourne: Geological Society of Australia.

MELVILLE, R. (1975). The distribution of Australian relict plants and its bearing on angiosperm evolution. *Botanical Journal of the Linnean Society*, 71, 67–88.

NIX, H. A. (1982). Environmental determinants of biogeography and evolution in Terra Australis. In *Evolution of the Flora and Fauna of Arid Australia*, ed. W. R. Barker & P. J. M. Greenslade, pp. 47–66. Frewville: Peacock Publications.

NIX, H. A. & KALMA, J. D. (1972). Climate as a dominant control in the biogeography of northern Australia and New Guinea. In *Bridge and Barrier: The Natural and Cultural History of Torres Strait*. ed. D. Walker, pp. 61–92. Canberra: Department of Biogeography and Geomorphology, Australian National University.

READ, J., HOPE, G. S. & HILL, R. (1990). The dynamics of some *Nothofagus* dominated rain forests in Papua New Guinea. *Journal of Biogeography*, 17, 185–204.

ROY, P. S. & THOM, B. G. (1981). Late Quaternary marine deposits in New South Wales and southern Queensland – an evolutionary model. *Journal of the Geological Society of Australia*, 28, 471–89.

RUSSELL-SMITH J., LUCAS D. E., BROCK J. & BOWMAN D. M. J. S. (1992). *Allosyncarpia*-dominated rainforest in monsoonal northern Australia. *Journal of Vegetation Science*, 4, 67–82.

SHULMEISTER J. (1991). Late Quaternary environmental history of Groote Eylandte, Northern Australia. Ph.D. thesis, Australian National University.

SINGH, G. & GEISSLER, E. (1985). Late Cainozoic history of vegetation, fire, lake levels and climate at Lake George, N. S. W., Australia. *Proceedings of the Royal Society of London B*, 311, 379–447.

SINGH, G. & LULY, J. (1991). Changes in vegetation and seasonal climate since the last full glacial at Lake Frome, South Australia. *Palaeogeography, Palaeoclimatology, Palaeoecology*, 84, 75–86.

THOMAS, I. & HOPE, G. S. (1994). An example of Holocene stability from Camerons Lagoon, a new treeline site on the Central Plateau, Tasmania. *Australian Journal of Ecology*, 19, in press.

TORGERSEN, T., LULY, J., DE DECKKER, P., JONES, M. R., SEARLE, D. E., CHIVAS, A. & ULLMAN, W. J. (1988). Late Quaternary environments of the Carpentaria Basin, Australia. *Palaeogeography, Palaeoclimatology, Palaeoecology*, 67, 245–61.

VAN DER KAARS, W. A. (1989). Aspects of Late Quaternary palynology of eastern Indonesian deep sea cores. *Netherlands Journal of Sea Research*, **24**, 495–500.

WALKER, D. (1982). Speculations on the origin and evolution of Sunda and Sahul rain forests. In *Biological Diversification in the Tropics*, ed. G. T. Prance, pp. 554–75. New York: Columbia University Press.

WALKER, D. & HOPE, G. S. (1982). Late Quaternary vegetation history. In *Biogeography and Ecology of New Guinea*, ed. J. L. Gressitt, pp. 263–85. The Hague: Junk.

WASSON, R. J. (1986). Geomorphology and Quaternary history of the Australian continental dunefields. *Geographical Review of Japan*, **59**, 55–67.

WRIGHT R. (1986). How old is Zone F at Lake George? *Archaeology in Oceania*, **21**, 138–9.

Note added in proof

Kershaw *et al.* (1993) note a dramatic increase in black material after 160 ka in a core from the continental slope. They suggest that a fundamental shift in burning regime occurred at that time.

KERSHAW, A. P., MCKENZIE, G. M. & MCMINN, A. (1993). A Quaternary vegetation history of northeastern Queensland from pollen analysis of ODP site 820. *Proceedings of the Ocean Drilling Program. Scientific Results*, **133**, in press.)

16 The history of selected Australian taxa

R. S. HILL

In examining the fossil record of the Australian flora since the arrival of angiosperms, several taxonomic groups loom large, either because they have an extensive and informative fossil record, or because they are prominent in the living vegetation and are selectively sought in the fossil record. The aim here is to consider some taxa that cover each of these areas in order to complement the vegetation reconstructions discussed in earlier chapters. There are several candidates that fall into the category of having an extensive fossil record, but the outstanding one is *Nothofagus*, which dominates many palynofloras, and is also well represented in the macrofossil record. The Podocarpaceae, Araucariaceae, Proteaceae and Casuarinaceae are also considered here because they have a mixture of good pollen and macrofossil records, and important evolutionary arguments can be based on these records. The prime example of a taxon that is prominent in the living vegetation and is actively sought in the fossil record is *Eucalyptus*, and its pollen and macrofossil record will also be considered, although it is much smaller than those of the other taxa. In choosing these taxa, several other notable groups have been excluded. In most cases this is because the record is biassed to either pollen (e.g. *Acacia*, Chenopodiaceae) or macrofossils (e.g. Lauraceae, cycads) and a combined data set cannot be supplied. There is no doubt that the greatest weakness lies with the macrofossil record, especially for those

taxa that produce entire-margined, medium-sized leaves. Taxonomic research on these groups should be a priority for the future.

NOTHOFAGUS

Nothofagus has been described as the key genus in the study of southern hemisphere plant evolution and biogeography (van Steenis, 1971, 1972). There are a number of reasons for this.

1. It has a completely southern hemispheric distribution, whereas its closest relatives are, and probably always have been, northern hemispheric.
2. It occurs in all the major Gondwanic land masses except Antarctica, where it has an extensive fossil record, and Africa and India, where it has never been recorded as an autochthonous fossil.
3. Fruits are extremely poorly dispersed (Rodway, 1914) and cannot survive long periods in sea water (Preest, 1963). It is generally agreed that long-distance fruit dispersal is unlikely to have been an important agent for *Nothofagus* dispersal, although Macphail *et al*. (Chapter 10) provide important data suggesting that long-distance dispersal has operated in *Nothofagus*.
4. Its fossil record is extensive and well studied.

It is only recently that the fossil record has been properly exploited in reconstructing the past his-

Table 16.1. Nothofagus *subgenera as defined by Hill & Read (1991) and pollen types as defined by Dettmann* et al. *(1990). Subgeneric names are used in the text in place of pollen-type names to avoid confusion*

Nothofagus			
Subgenus *Nothofagus*	Subgenus *Fuscospora*	Subgenus *Lophozonia*	Subgenus *Brassospora*
Five species in South America; *N. fusca* type b pollen	Five species in South America, New Zealand and Australia; *N. fusca* type a pollen	Six species in South America, New Zealand and Australia; *N. menziesii*-type pollen	Nineteen species in New Guinea and New Caledonia; *N. brassii*-type pollen

tory of *Nothofagus*. Earlier attempts at determining the centre of origin and migration routes of *Nothofagus* often relied heavily on the distribution and interrelationships of extant species, and the literature contains an array of possible options, with almost every conceivable centre of origin being invoked at one time or another (e.g. Cranwell, 1963; Darlington, 1965; van Steenis, 1971, 1972; Raven & Axelrod, 1972; Hanks & Fairbrothers, 1976; Schuster, 1976). A recent revision of the vast fossil pollen record (Dettmann *et al.*, 1990) and descriptions and reviews of macrofossils (e.g. Romero & Dibbern, 1985; Tanai, 1986; Hill, 1991) have allowed a far more detailed interpretation of the past distribution and evolution of *Nothofagus*, although some problems remain. In order fully to integrate the pollen and macrofossil record, a brief description of the current infrageneric classification of *Nothofagus* is required.

Infrageneric classification

Nothofagus is usually assigned to the Fagaceae, although some recent evidence supports the erection of the monogeneric family Nothofagaceae (e.g. Nixon, 1989). Furthermore, Nixon concluded that *Nothofagus* is more closely related to Betulaceae than to Fagaceae, although there is other conflicting evidence, summarised by Hill, (1993).

There have been three different infrageneric classifications proposed for *Nothofagus* (van Steenis, 1953; Philipson & Philipson, 1988; Hill & Read, 1991). A major difficulty with the earlier classifications was that palynologists recognised three distinct pollen morphologies amongst the

extant and fossil species, only one of which correlated with formal infrageneric groupings. In van Steenis' classification, the primary infrageneric division was based on the dichotomy created by the deciduous or evergreen habit. Recognition of four distinct vernation patterns in *Nothofagus* (Philipson & Philipson, 1979, 1988), in conjunction with several other characteristics of leaf and cupule morphology, led Hill & Read (1991) to hypothesise that the evergreen habit is polyphyletic. This may explain the lack of correlation between classifications based primarily on the deciduous–evergreen dichotomy and the pollen groupings. The most recent infrageneric classification (Hill & Read, 1991) correlates well with the established but recently revised pollen groupings (Dettmann *et al.*, 1990; Table 16.1). With the infrageneric taxonomy and pollen groupings now in harmony, more order is apparent in the fossil record of *Nothofagus*.

Origin, dispersal and diversification

The centre of origin and migration routes of *Nothofagus* must be compatible with that of related taxa in Fagaceae and Betulaceae, which are regarded as the most closely related families. Burger (1981, 1990) and Truswell *et al.* (1987) recently focussed attention on the possible interchange of plants between Southeast Asia and Australia during the Late Cretaceous–early Tertiary. They suggest that a complex of continental fragments probably provided a kind of stepping stone path since the Early Cretaceous, and especially in the Late Cretaceous and early Tertiary. Truswell *et al.* (1987) reported several floral examples from the fossil pollen record as probable exploiters of this route in both directions, but they explicitly excluded *Nothofagus*, since its highly distinctive pollen has not been recorded in the appropriate sediments. However, recent macrofossil records from the southern hemisphere provide another option. As well as an abundance of *Nothofagus* macrofossils, there are leaves and wood in Late Cretaceous–early Tertiary sediments that are regarded as fagaceous, but not *Nothofagus* (Romero, 1986; Romero & Dibbern, 1985; Birkenmajer & Zastawniak, 1989*a*; M. S. Pole, personal communication). Furthermore, a leaf from the Late Cretaceous Winton Formation in Queensland (R. S. Hill, unpublished data) appears to be fagaceous, and Pole (1992) has recorded Late Cretaceous leaves from New Zealand that are morphologically similar to Betulaceae. This suggests that a Cretaceous fagalean complex may have existed in Gondwana during the early history of *Nothofagus*, and possibly preceded the origin of *Nothofagus*.

This fagalean complex, with a generalised pollen type, may have existed in the Southeast Asian–Australian region during the Cretaceous, giving rise to the Betulaceae and Fagaceae in the northern part of this region, where the fossil record of the Betulaceae at least is both early and diverse (Crane, 1989). Part of the complex may have migrated into high southern latitudes, via the route suggested by Burger (1981) and Truswell *et al.* (1987; Figure 16.1), there to evolve into *Nothofagus*. Such an hypothesis is compatible with the presence of Late Cretaceous–early Tertiary 'fagaceous–betulaceous-like' macrofossils in the southern hemisphere, explains the lack of previous geographical links between *Nothofagus* pollen and that of closely related taxa, and is also compatible with the hypothesis of Dettmann *et al.* (1990) that *Nothofagus* arose at high southern latitudes. Although it seems clear from the extensive fossil pollen record that *Nothofagus* had a high latitude southern origin during the Late Cretaceous (early Campanian) within the Weddellian province (Dettmann *et al.*, 1990; see below), the sparse Cretaceous pollen record from the northern Australian–Southeast Asian region may mask a possible centre of origin for *Nothofagus* there. Clearly more fossil evidence is required to test this hypothesis rigorously.

Case (1988, 1989) extended the Weddellian Zoogeographic Province (Figure 16.2) to a 'biogeographic province' and hypothesised that it was a centre of diversification during the Late Cre-

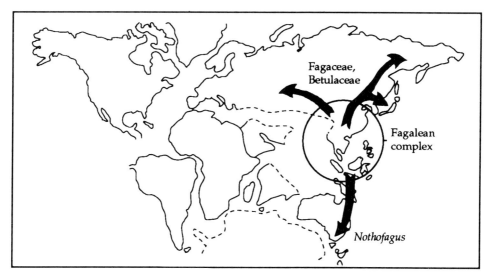

Figure 16.1 Map of possible Early Cretaceous route of Fagalean lineages from a centre of origin in Southeast Asia. (Adapted from Truswell *et al.*, 1987.)

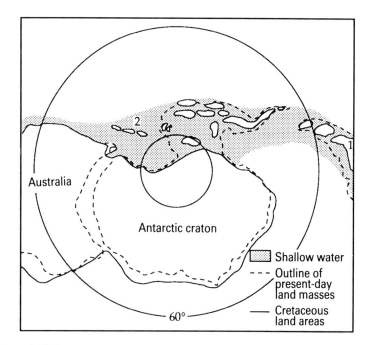

Figure 16.2 Reconstruction of the Weddellian Zoogeographic Province (stippled area), during the Late Cretaceous and early Tertiary. The numbered land masses are: 1, South America; 2, New Zealand. (Adapted from Case (1988), who considers that the stippled area should represent a continuous, coastal environment bordered by shallow seas.)

taceous and Tertiary, and that part of the evolving biota included *Nothofagus*. The region extends from southern South America, along the Antarctic Peninsula and West Antarctica to Tasmania and southeastern Australia and New Zealand. Case hypothesised that many novel species arose and survived in this region, primarily due to the low extinction rates that characterise high latitude environments (Hickey *et al.*, 1983; Jablonski *et al.*, 1983). Such an hypothesis is consistent with the *Nothofagus* pollen record for the region.

The most complete fossil record of *Nothofagus* is supplied by pollen. Of the eight pollen types now recognised in fossil and living *Nothofagus*, two are nominated as 'ancestral' and appear first in the fossil record (Dettmann *et al.*, 1990). Ancestral type a, represented by the single fossil species *Nothofagidites senectus*, occurs over wide areas of southern Gondwana in sediments as old as early Campanian. Two other fossil species, *Nothofagidites endurus* (from southeastern Australia) and *N.*

kaitangensis (from New Zealand), are assigned to ancestral type b and first occur in the mid-Campanian. Ancestral pollen types a and b are considered to be most closely related to pollen of *Nothofagus* subgenera *Fuscospora* and *Nothofagus*, respectively (Dettmann *et al.*, 1990).

Following the appearance of *Nothofagus* in the fossil record, there were spatially and temporally distinctive events of rapid diversification. The oldest known occurrences of the pollen types belonging to the extant subgenera are from the late Campanian of the Antarctic Peninsula and from the Maastrichtian of southern South America, a region that was a centre of Late Cretaceous diversity for *Nothofagus* (Dettmann *et al.*, 1990). The significantly younger occurrences in southern Australia and New Zealand of *Nothofagus* subgenus *Nothofagus* (Paleocene), *Nothofagus* subgenus *Lophozonia* (Eocene) and *Nothofagus* subgenus *Brassospora* (Eocene in New Zealand, late Maastrichtian in Australia) imply a

migrational lag and routes involving western Antarctica (Dettmann *et al.*, 1990). Since pollen of
Nothofagus subgenus *Brassospora* spp. first appears
in the west Antarctica–southern South America
region, the subgenus probably originated there.
However, there is a restricted diversity of this pollen type in that region and it is unknown post-
Oligocene (Dettmann *et al.*, 1990). Rapid
diversification of *Nothofagus* subgenus *Brassospora*
occurred during the Eocene–Oligocene in the
Australian–New Zealand region (Dettmann
et al., 1990) and its pollen type dominates pollen
assemblages, particularly in southeastern Australia during this time (Martin, 1978; Macphail
et al., Chapter 10, this volume; Kershaw *et al.*,
Chapter 13, this volume). Most species producing
Nothofagus subgenus *Brassospora*-type pollen
became extinct in southern Australia and New
Zealand during the late Tertiary, presumably in
response to climatic change, although some
species remained in Tasmania until the Pleistocene (Hill & Macphail, 1994).

The most common *Nothofagus* macrofossils are
leaves, which represent both deciduous and evergreen forms (Romero & Dibbern, 1985; Tanai,
1986; Birkenmajer & Zastawniak, 1989*a,b*; Hill,
1991). Outside of Tasmania, and to some extent
southeastern mainland Australia, good organic
preservation of *Nothofagus* leaves is rare. This is
unfortunate because the infrageneric affinities of
Nothofagus leaves are difficult to determine on
gross leaf morphology and venation pattern alone
(Hill, 1991). Reproductive structures are generally uncommon macrofossils, and the woody
cupules, sometimes containing fruits, have been
recovered only from Tasmania (Hill, 1987, 1991)
and one site on mainland southeastern Australia
(Christophel, 1985).

By far the most informative macrofossil record
of *Nothofagus* comes from Tasmania. Two of the
four extant subgenera (*Lophozonia* and *Fuscospora*)
now occur in Tasmania, with the others restricted
to New Guinea–New Caledonia (*Brassospora*)
and southern South America (*Nothofagus*),
respectively. However, all subgenera occur in
Tasmanian Oligocene sediments, and apparently
coexisted, since they are found in small sedimen-

tary units with spatially restricted catchments
(Hill, 1991). Biogeographically, the most significant Tasmanian macrofossils are:

1. Leaves of *N. cethanica* (Hill, 1984).
2. Leaves, cupules and fruits of subgenus *Nothofagus* (Hill, 1987, 1991).
3. Leaves and cupules of more than one species
of subgenus *Brassospora* (Hill, 1987, 1991).

Nothofagus cethanica is very similar to *N. fusca*, currently a New Zealand endemic, and unlike *Nothofagus* spp. on any other land mass. Prior to the
discovery of the Tasmanian fossils of subgenus
Nothofagus there was only limited pollen evidence
for the past existence of this subgenus far outside
its current range. The macrofossils of subgenus
Brassospora are virtually unique and provide the
only substantial evidence of the gross morphology
of the plants that produce the most abundant pollen. Tasmanian Oligocene *Nothofagus* macrofossils expose several formidable palaeoecological
problems, including the reasons for:

1. The extinction of *N. cethanica* in Tasmania,
coupled with the survival of *N. fusca* in similar
latitudes in New Zealand.
2. The extinction of *Nothofagus* subgenus *Nothofagus*, coupled with its survival in similar (?or
more southerly) latitudes in South America.
3. The co-occurrence of species in all four
subgenera within very close proximity, and
possibly within the same vegetation. This
includes such apparently bizarre mixtures as
Nothofagus subgenus *Brassospora* (now in tropical montane vegetation) with *N. gunnii* (an
extant Tasmanian subalpine or alpine deciduous shrub or small tree), which has been
recorded at Cethana (Hill, 1984; Carpenter,
1991).
4. The apparent ability of evergreen members of
subgenus *Lophozonia* to evolve substantially (in
leaf form at least) and remain in the region to
the present, coupled with the extinction of
most other species.
5. The survival to the present, with unchanged
gross leaf morphology, of the winter deciduous
N. gunnii, while numerous evergreen species

in the same and two other subgenera have become extinct, and members of the fourth subgenus have survived only after substantial change in leaf morphology.

The great enigma of the *Nothofagus* record is the mid-Tertiary abundance of *Nothofagus* subgenus *Brassospora*-type pollen in southern Australia and New Zealand, coupled with the paucity of associated macrofossils. Two distinct types of cupule and leaf of subgenus *Brassospora* have been described from a small sedimentary unit in northwestern Tasmania (Hill, 1987, 1991), and probable leaves and a single cupule from a more diverse suite of fossils in north-central Tasmania (Carpenter, 1991), but they are the only conclusive nonpollen record of the subgenus. Their co-occurrence with all other *Nothofagus* subgenera has led to analyses of prevailing palaeoclimates and experiments on the physiological tolerances of living species (Read & Hill, 1989; Read & Hope, 1989; Read, 1990; Read *et al.*, 1990), resulting in the hypothesis that the Oligocene climate in Tasmania had several features with no modern analogue (Hill, 1990*a*; Read *et al.*, 1990). This climate included mild temperatures with severely compressed extremes, which led to a relatively frost-free environment, and year-round rainfall with very high humidity. The high humidity is evidenced by the diverse and dense epiphytic fungal populations on many of the leaves (Hill, 1990*a*). This climate probably allowed compression of ecological zonation, in a manner similar to that described by Hyndman & Menzies (1990) for the extant rainforests of the Ok Tedi headwaters in New Guinea. Within the Oligocene Tasmanian rainforests, *Nothofagus* subgenus *Brassospora* probably occurred in the canopy, so that juvenile individuals of this type matured with protection against any climatic extremes provided by adult trees. *Nothofagus* spp. in other subgenera occurred in more exposed areas such as ridgetops and lake and river margins, where both juvenile and adult plants were exposed to any prevailing climatic extremes and had to be tolerant of them to survive. With climatic deterioration later in the Tertiary, and in particular a widening of tempera-

ture extremes and a decrease in rainfall, which became markedly seasonal, *Nothofagus* subgenus *Brassospora* spp. were gradually eliminated from the region and were replaced in the canopy by more tolerant *Nothofagus* spp. from other subgenera. Continuing climatic change forced the extinction of some of these species, whereas others, notably in subgenus *Lophozonia*, evolved in response to the changing conditions (Hill, 1991). An alternative hypothesis is that extinction of *Nothofagus* subgenus *Brassospora* from southeastern Australia resulted from its failure to keep pace with the climatic fluctuations that became more pronounced in the region from the mid-Miocene (Read *et al.*, 1990), which was first proposed to account for the extinction of the subgenus from New Zealand (Wardle, 1968). These hypotheses will undoubtedly be refined as future research provides more palaeobotanical and ecophysiological data, but they are consistent with results of past physiological research on extant species from across the latitudinal range. Another possibility is that the frequency of wide-scale disturbance was relatively high during the Oligocene, favouring *Nothofagus* spp. (Hill, 1987), many of which have a disturbance-based ecology (e.g. Veblen & Ashton, 1978; Veblen *et al.*, 1981; Read *et al.*, 1990). The later decrease in disturbance frequency may have forced the demise of these species. Finally, changes in atmospheric CO_2 levels may have altered the competitive abilities of the component taxa in Tasmanian Oligocene rainforests, but data are absent.

There is little macrofossil evidence for evolution in *Nothofagus*, probably because the bulk of the record occurs in the Cenozoic. Pollen evidence (e.g. Dettmann *et al.*, 1990) demonstrates that most major evolutionary events within the genus were complete by the end of the Cretaceous and the common early Tertiary macrofossils support this, in that they are often morphologically similar to extant species. The earliest known species of subgenus *Brassospora* (Oligocene) had serrate-margined leaves and cupule valves that enclosed the fruits (Hill, 1991), although a recently discovered cupule has much reduced valves (R. S. Hill, unpublished data). Serrate-margined leaves and large cupule valves may represent the condition ancestral to the extant species in the subgenus that have finely serrate or entire-margined leaves and cupules that range from robust valves fully enclosing the fruits to those where cupule valves are absent. However, these fossils substantially postdate the origin of the subgenus and may represent derived forms peculiar to the Tasmanian region. More widespread and conclusive fossil evidence suggests that in south-eastern Australia (including Tasmania) and probably also in New Zealand, evergreen species in subgenus *Lophozonia* responded to Cenozoic climate change by significantly reducing their leaf area, forming the characteristic microphyllous species that dominate microthermal rainforest in the regions (Hill, 1991; Figure 16.3). *Nothofagus moorei*, which has a leaf form very similar to the widespread early Tertiary broad-leaved ancestral species in Australia and possibly New Zealand (it may even have dispersed to New Zealand from Australia over the Tasman Sea in the early Tertiary), is extant in isolated high altitude microthermal rainforest at mid-latitudes on the eastern Australian coast. In contrast, the only extant deciduous species outside South America, the high altitude Tasmanian endemic *N. gunnii*, has not altered its general leaf morphology since the Early Oligocene at least (Hill, 1991). This may be because leaves of this species were and are absent during the rigours of the increasingly severe Tasmanian winters.

Thus, the fossil record of *Nothofagus* has proved to be extremely useful for reconstructing past distributions of the genus and, to a lesser extent, major evolutionary events. Perhaps the major contribution that research on this genus will make in the future is in the reconstruction of past climates and vegetation types because it appears to be extremely conservative, especially in the Cenozoic, and modern physiological data can be applied to the past with reasonable confidence.

PODOCARPACEAE

The family Podocarpaceae consists of 18 extant genera and more than 150 species. By far the largest genus is *Podocarpus*, with more than half the total number of species, with the next largest being *Dacrydium* with 16. Only six of these genera are now present in Australia, with most species restricted to Tasmania and northeast Queensland. This is only a small remnant of a once highly diverse podocarpaceous flora, which has left an extraordinary fossil record.

The origins and relationships of the Podocarpaceae are obscure, although Miller (1988) notes that Podocarpaceae and Cephalotaxaceae consistently occur close to one another in his cladograms. The genus *Rissikia*, from the Triassic of Madagascar, South Africa, Australia and Antarctica (Townrow, 1967a), indicates that the family had evolved by the onset of the Mesozoic (Miller, 1988). Other early far-southern conifers, such as the Jurassic *Nothodacrium* (Townrow, 1967b) and *Mataia* (Townrow, 1967a) are probably podocarps, but their relationship to extant genera requires clarification (Stockey, 1990). Podocarps were still prominent in the Cretaceous, although in Australia they are assigned to extinct genera, e.g. *Bellarinea barklyi* from the Early Cretaceous of Victoria (Drinnan & Chambers, 1986). The pollen record at this time is more useful. For example, Dettmann & Jarzen (1990) documented the Cretaceous pollen records of the extant genera *Dacrycarpus*, *Dacrydium* and *Podocarpus* from the Antarctic/Australian rift valley. They note that *Dacrycarpus* pollen occurs in Late Cretaceous sediments in southern Australia, where it ranges into the Pliocene. This is the earliest known occurrence of this pollen type and probably the earliest fossil record, as *Dacrycarpus* macrofossils reported by Florin (1940) from the Jurassic of India must be considered extremely unlikely. '*Dacrydium*' pollen, as defined by Dettmann & Jarzen (1990), is produced by more than one extant genus. One morphotype, with fossils designated as *Lygistepollenites*, is consistent with extant *Dacrydium* and is widely dispersed in Late Cretaceous sediments of southern Gondwana, with the oldest occurrences in the Coniacian–Santonian of southern Australia and the Antarctic Peninsula. The other '*Dacrydium*' pollen type, which is consistent with extant genera such

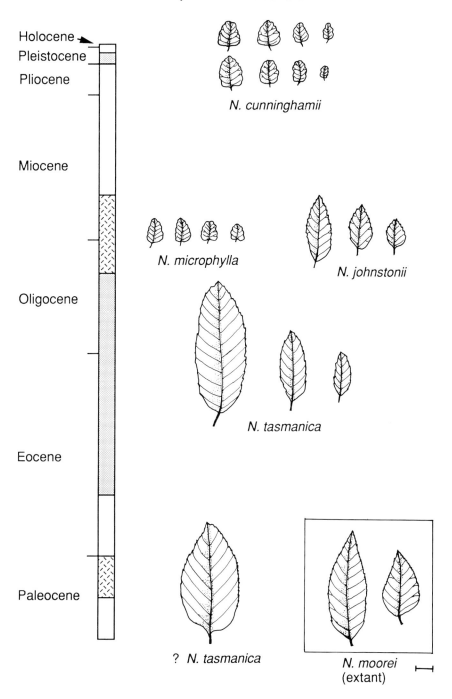

Figure 16.3 Fossil and extant species of *Nothofagus* subgenus *Lophozonia* in southeastern Australia. The shaded areas next to the time scale show the approximate stratigraphic ranges of the fossils. The two rows of *N. cunninghamii* leaves at the top represent the extant species (top row) and Early to mid-Pleistocene leaves (bottom row). Scale bar represents 1 cm.

as *Halocarpus* in New Zealand, is similar to *Podocarpus* pollen and has fossil counterparts in Jurassic sediments. *Podocarpidites ellipticus* 'is similar in general form to, but not diagnostic of, that shed by *Podocarpus* and some species of *Dacrydium*' (Dettmann & Jarzen, 1990). Thus, Dettmann & Jarzen note that the Early Jurassic to Recent range of this species in Australia cannot be taken as indicating *Podocarpus* and/or '*Dacrydium*'.

The Australian Cenozoic pollen record of the Podocarpaceae is abundant and diverse and some taxa warrant special mention. Playford & Dettmann (1979) noted that the fossil species *Phyllocladidites mawsonii* was produced by plants identical with, or very closely related to, *Lagarostrobos franklinii*, which is currently restricted to riverine communities in southwest Tasmania. This pollen type has been prominent in many vegetation and climatic reconstructions (e.g. Kemp, 1978). It is therefore of particular interest that the macrofossil record of *Lagarostrobos* is so sparse. Wells & Hill (1989) assigned one specimen from the Oligocene of northwest Tasmania to a new species of *Lagarostrobos*, but it is not particularly similar to *L. franklinii*, and similarly Carpenter (1991) has suggested an affinity to *Lagarostrobos* for a specimen from Oligocene sediments in northern Tasmania. In other instances the lack of *Lagarostrobos* macrofossils is difficult to explain. For instance, Macphail *et al.* (1991) recorded 45% *P. mawsonii* pollen at the Late Oligocene–Early Miocene Monpeelyata flora in central Tasmania and noted that many grains were apparently immature or aberrant, suggesting a very local source. However, despite the presence of hundreds of podocarpaceous macrofossils, none has so far been related to *Lagarostrobos*, and all have affinities with genera with well-known and distinctive pollen types. Given that *L. franklinii* today is a riverine species, it must be assumed that the ecology of at least one of the species that produced this pollen type in the past was quite different, since *L. franklinii* twigs are common in extant water bodies and are found frequently in Tasmanian Quaternary sediments (e.g. Wells & Hill, 1989; Jordan *et al.*, 1991).

A similar case is presented by the pollen species *Microcachrydites antarcticus*, first described by

Cookson (1947). This pollen type, which ranges in Australia from the Jurassic to Miocene, is considered to have affinities with extant *Microcachrys tetragona*, a Tasmanian endemic alpine shrub. There can be no doubt that the taxon (?taxa) that produced *M. antarcticus* pollen had a much wider ecological range than *M. tetragona*, but that does not mean it was morphologically distinct. Blackburn (1985) described a leafy twig from the Latrobe Valley coal as having affinities to *Microcachrys*, and it certainly has more than a superficial resemblance to that genus. If this fossil is *Microcachrys*, then it provides evidence of a plant very similar in leaf arrangement and morphology to the extant alpine shrub, but which must have had a quite different ecological niche. A similar situation occurs in the lowland Early to mid-Pleistocene Regatta Point sediments in western Tasmania, where *M. tetragona* twigs are relatively common (G. J. Jordan, personal communication) in what is interpreted as a predominantly lowland closed-forest vegetation. Perhaps the extant species is the result of compression towards one end of a formerly much broader ecological niche as the result of increased competition, perhaps in response to climatic change or changes in atmospheric CO_2 (maybe as recently as the Quaternary glaciations) or evolution of more specific niches in the expanding angiosperm flora. A similar ecological compression may have occurred in the genus *Microstrobos*. The two extant species are morphologically distinct from one another, and one, *M. niphophilus*, is an endemic alpine shrub or small tree in Tasmania. Macrofossils virtually identical with this species have been recovered from the Monpeelyata sediments in Tasmania (Wells & Hill, 1989), in a region that is considered to have supported a cool climate flora (Hill & Gibson, 1986; Macphail *et al.*, 1991). However, another specimen of the same species has been recovered from lowland Oligocene sediments in northwestern Tasmania in a complex rainforest association (Wells & Hill, 1989; Carpenter *et al.*, Chapter 12 this volume). This suggests a much broader ecological niche during the Oligocene for a species with foliage very similar to extant *M. niphophilus* and a subsequent compression of this niche towards the cooler end.

Many other extant podocarpaceous genera are recorded as macrofossils in Australia during the early Tertiary. Most of them, e.g. *Podocarpus, Falcatifolium, Acmopyle, Dacrycarpus, Dacrydium, Lepidothamnus* and *Phyllocladus*, occur in southeastern Australia, but *Retrophyllum* has been recovered from the southwest. There are also some extinct genera, e.g. *Coronelia* (Townrow, 1965), *Smithtonia* and *Willungia* (Hill & Pole, 1992). Two of these genera, *Acmopyle* and *Dacrycarpus*, contain fossil species that exhibit convergent evolution in response to climate change during the course of the Tertiary.

Acmopyle is very restricted today, with one species in New Caledonia and the other in Fiji. However, it is known as a macrofossil from several early Tertiary sites in southeastern Australia and demonstrates an elegant pattern of evolution, presumably in response to declining water availability (Figure 16.4). The oldest *Acmopyle* macrofossils (Late Paleocene, southern New South Wales) are fully amphistomatic, with stomata equally distributed over both leaf surfaces. This suggests that water was plentiful throughout the year for this species and that water loss was never a serious problem (Hill & Carpenter, 1991). Therefore

the plant has maximised its stomatal distribution in an attempt to maximise photosynthesis and growth. By the Early Eocene *Acmopyle* had begun to lose stomata on one leaf surface (Regatta Point and Buckland in Tasmania), suggesting that control of water loss was developing as a serious problem; the species had also adapted morphologically by reducing their stomatal distribution, presumably with the consequence that net photosynthesis was also reduced (Hill & Carpenter, 1991). The plant was clearly adapting to a harsher environment but at a cost in terms of potential growth. This trend continued throughout the Eocene, and by the Late Eocene one species of *Acmopyle* had evolved to the point where stomata were reduced to one leaf surface (Loch Aber, Tasmania). The last record of *Acmopyle* in southeastern Australia is from the Early Oligocene, and it can be assumed that at about this time the climate changed to such a degree that *Acmopyle* was no longer able to evolve a competitive response to it, and thus became extinct in the region. The two living species of *Acmopyle* both have a stomatal distribution similar to that of the Early Eocene species in Tasmania – there are stomata on both leaf surfaces, but they are relatively restricted on

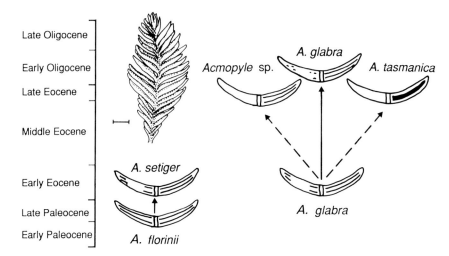

Figure 16.4 Stomatal distribution on leaves of fossil species of *Acmopyle* from southeastern Australia. The leaf 'pair' in each drawing represents the functional 'upper' surface (left-hand side) and 'lower' surface (right-hand side) of a single leaf. The stomatal distribution is shown in black on the leaf surfaces. Broken lines represent uncertain phylogenetic relationships. The entire short shoot, which illustrates the general leaf arrangement, belongs to *A. florinii*. Scale bar represents 5 mm.

one of them. This suggests that the prevailing climate over the range of living *Acmopyle* is suboptimal, but that it is not as severe as the genus can withstand.

Dacrycarpus illustrates an interesting combination of morphological changes that are probably responses to both temperature and rainfall changes. *Dacrycarpus* has two foliage types. In one type, the leaves are bilaterally flattened, so that each functional leaf surface is composed of both adaxial and abaxial leaf surfaces. Furthermore, the whole short shoot (a single season's growth) is flattened into two dimensions, so that the short shoot probably functions like a single broad angiosperm leaf. As well as this, *Dacrycarpus* produces small, scale-like, bifacially flattened leaves that are arranged spirally around the axis in an imbricate arrangement. In most living species these two foliage forms appear apparently at random on the plant, although the bilaterally flattened foliage is often considered to represent a juvenile phase.

Dacrycarpus macrofossils are particularly abundant in early Tertiary sediments in southeastern Australia (Cookson & Pike, 1953; Wells & Hill, 1989; Hill & Carpenter, 1991), and demonstrate a clear trend in foliage evolution (Figure 16.5). In Early Eocene sediments, both foliage types are very common, and in each stomata occur all over both leaf surfaces. However, in progressively more recent sediments there is a trend in two aspects of leaf morphology. Firstly, the bilaterally flattened foliage becomes less common, and by the end of the Oligocene is no longer present as a macrofossil. Secondly, in the bifacially flattened foliage, stomates become progressively more restricted to the inner (adaxial) leaf surface, and by the end of the Oligocene are only found in that position. No *Dacrycarpus* macrofossils have been found in Tasmania after the earliest Miocene, but pollen is recorded up until the Early to mid-

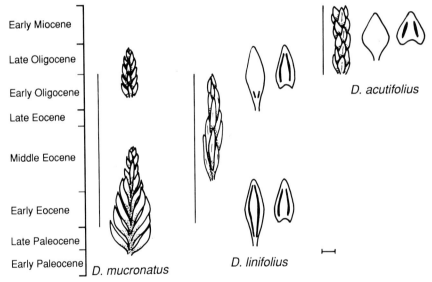

Figure 16.5 Stratigraphic distribution of selected fossil *Dacrycarpus* spp. in southeastern Australia. The line to the left of each species shows the approximate stratigraphic range. *Dacrycarpus mucronatus* shows the change from a combination of bilaterally and bifacially flattened foliage (Early Eocene) to predominantly bifacially flattened foliage (Early Oligocene). *Dacrycarpus linifolius* has been recovered only as bifacially flattened foliage, but the stomatal distribution varies through time. To the right of the short shoot are stylised drawings of leaves, showing the abaxial surface (left-hand side) and adaxial surface (right-hand side). The stomatal distribution is shown in black on the leaf surface. Note the reduction in stomata on the abaxial surface from the Early Eocene to the Early Oligocene. *Dacrycarpus acutifolius* has a more restricted stratigraphic range, but is typical of several Late Oligocene–Early Miocene species in having only bifacially flattened leaves and stomates restricted to the adaxial surface. Scale bar represents 1 mm.

Pleistocene (Hill & Macphail, 1985, 1994), suggesting that *Dacrycarpus* may have been a victim of the Quaternary glacial cycles in Australia. Thus, in this genus we see reduction in leaf area similar to that noted earlier for *Nothofagus*, presumably in response to declining mean annual temperature, as well as a restriction in the distribution of stomata, which is interpreted as a response to the onset of seasonal rainfall and possibly a decrease in total annual rainfall as well. The other interesting feature of *Dacrycarpus* is that these trends are seen in more than one phylogenetic line, suggesting that these were general convergent trends in foliage in response to common external factors (Hill & Carpenter, 1991).

The fossil record of the Podocarpaceae is very productive. There is no doubt that further work will improve our understanding of phylogenetic relationships among fossil and extant species, and this will allow refinement of evolutionary studies on the response of these taxa to climatic change. The Podocarpaceae appears to be a particularly important family for study of the effect of very long-term climate change on leaf morphology and, ultimately, on vegetation form and taxonomic make-up during the Cenozoic.

ARAUCARIACEAE

The two extant genera in the Araucariaceae, *Agathis* and *Araucaria*, are confined to the southern hemisphere, where they occur as trees sometimes reaching 60 m or more (Florin, 1963; Stockey, 1982). There are 13 extant species of *Agathis*, three of which occur in northeastern Australia (Whitmore, 1980). *Araucaria* is divided into four sections containing 22 species, 2 of which occur in northeastern Australia and 1 on Norfolk Island (Hill & Bigwood, 1987). The current centre of diversity for both genera is New Caledonia, with 5 *Agathis* and 14 *Araucaria* spp.

The family was more diverse and widespread during the Mesozoic, and extended into the northern hemisphere during the Jurassic (Stockey, 1982). A large number of fossils have been assigned to the Araucariaceae, many of which must be regarded as extremely doubtful. The pol-

len record is extensive and widespread, but, because of its uniformity, is not helpful in tracing particular phylogenetic lines within the family (Martin, 1978; Stockey, 1982). In this brief review only some of the more certain and relevant records are considered.

Araucaria has a much more extensive fossil record than *Agathis*. However, this record is difficult to interpret because of the number of conifers with superficially similar foliage and the proliferation of generic names used to encompass them. The pre-Jurassic record is particularly uncertain and is not considered reliable (Stockey, 1982). Stockey (1982) considered that *Araucarites phillipsii*, the cone and cone scale associated with *Brachyphyllum mamillare* from the Yorkshire Jurassic, is part of the same species, which has strong affinities to *Araucaria* section *Eutacta*. Stockey described several other *Araucarites* spp. from widespread Jurassic and Cretaceous localities, which have less certain affinities to this section of *Araucaria*.

Drinnan & Chambers (1986) described an *Araucaria* sp. from the Early Cretaceous Koonwarra fossil bed, which they compared with *A. heterophylla*, and assigned to section *Eutacta*. Species that can be assigned to section *Eutacta* are widespread across South America–Antarctica–Australia–New Zealand during the Tertiary (Florin, 1940) and are particularly prominent in southeastern Australia (summarised by Hill & Bigwood, 1987; Hill, 1990b). Organically preserved specimens have been described from the Early Eocene of Tasmania (Hill & Bigwood, 1987), Oligocene of Victoria (Cookson & Duigan, 1951) and Oligocene – earliest Miocene of Tasmania (Hill, 1990b). There are also several undescribed species in Tasmania from Eocene and Oligocene sediments (R. S. Hill, unpublished data). One of the interesting features of the Tertiary *Araucaria* section *Eutacta* record in Tasmania is that in several deposits up to three species apparently co-occurred. The overall species diversity in Tasmania during the Oligocene is comparable with New Caledonia today, and suggests that the current concentration of species there may be due to the extant climate–vegetation associations rather than to any historical centre of origin.

Tasmania in the early Tertiary may have been similarly conducive to a high species diversity in this section of *Araucaria*.

Section *Intermedia* of *Araucaria*, which is currently restricted to New Guinea, is represented in the fossil record by the ? Late Cretaceous *A. haastii* from New Zealand (Ettingshausen, 1891; Stockey, 1982; Hill & Bigwood, 1987). Section *Columbea*, with two extant species in South America, is represented in the fossil record by *A. balcombensis* from the Tertiary of Victoria, and *A. hastiensis* from the Middle–Late Eocene of Tasmania (Hill, 1990*b*). *Araucaria nathorstii*, from the Tertiary of Argentina (Menendez & Caccavari, 1966) may also belong to section *Columbea*, although this has not been determined with certainty (Hill & Bigwood, 1987).

Section *Bunya* of *Araucaria* is restricted to a single extant species, *A. bidwillii*, in Queensland. This section was more widespread during the Mesozoic, even into the northern hemisphere, and more diverse, and remains indicate that the cone morphology in this section was similar to *A. bidwillii* by the Jurassic (Stockey, 1982). Stockey noted that *A. bidwillii* represents the last species of a very distinctive group of araucarians that is of major significance in determining evolutionary relationships within the genus.

The fossil record of *Agathis* is restricted to the southern hemisphere, and more particularly to Australia and New Zealand (Florin, 1963). The oldest record of *Agathis* in this region is *A. jurassica* from the Jurassic Talbragar Fish Bed Flora in New South Wales (White, 1981). This fossil species consists of leaf impressions and associated cone scales. However, Stockey (1982) noted that the cone scales also show close similarities with *Araucaria* section *Eutacta*, and there must be considerable doubt about the taxonomic connection between the leaves and these reproductive structures; later, Stockey (1990) suggested that the foliage of *A. jurassica* may in fact be podocarpaceous. Florin (1940) noted the possible presence of *Agathis* in the Late Cretaceous of Australia, but his record requires confirmation. Christophel (1981; Chapter 11, this volume) recorded the presence of *Agathis* in the Middle Eocene Maslin Bay sediments, but this taxon has not yet been described. The Late Eocene Vegetable Creek sediments in northern New South Wales contain leaf and cone scale impressions that are certainly *Agathis*, although there is no confirmatory cuticular preservation. Well-described and organically preserved *Agathis* leaves have been described from probable Oligocene sediments in Victoria (Cookson & Duigan, 1951) and Tasmania (Hill & Bigwood, 1987; Carpenter, 1991), and Middle Eocene–Oligocene sediments in southwestern Australia (Hill & Merrifield, 1993).

Bigwood & Hill (1985) described leaves from Early Eocene sediments in Tasmania and placed them in the genus *Araucarioides*. These leaves were considered to have features of both *Agathis* and *Araucaria*. Stockey (1990) hypothesised that this genus, coupled with the large number of 'araucarian' cone scales in northern hemisphere localities, 'suggests that the Araucariaceae may have included an additional genus in the past'. Cantrill (1991) described *Podozamites taenioides* from mid–late Albian sediments in Victoria. This species, based on foliage, is very similar to the Tertiary species assigned to *Araucarioides*, and may represent an older record of the same genus. Cantrill (1991) noted that some Australian *Podozamites* spp. were similar in gross morphology to *Agathis*. It may be that this group of taxa, which extended into the Early Eocene in Tasmania, included the taxon ancestral to *Agathis*, which probably evolved in the Australia–New Zealand region sometime in the Jurassic–Cretaceous.

Clearly, the present diversity of the Araucariaceae, with the exception of New Caledonia, is only a small fraction of the past, but it represents a family that figured largely during the formative stages of the extant Australian vegetation. Although many fossils in the Araucariaceae are yet to be described or fully understood, there is no doubt of the importance of Australia in unravelling the phylogenetic history of this important family of currently southern hemisphere conifers.

PROTEACEAE

The Proteaceae is one of the major Australian families. Worldwide there are about 75 genera of evergreen trees and shrubs, with a strong representation in tropical and subtropical closed forests, as well as in sclerophyllous heaths and woodlands on oligotrophic soils (Johnson & Briggs, 1981). The family has been divided into five subfamilies (Johnson & Briggs 1975), all of which occur in Australia. The extant species of the Proteaceae have been extremely well studied and, primarily due to the efforts of Johnson & Briggs (1963, 1975, 1981), the phylogeny of the family is well understood. Given its mostly southern hemispheric distribution, and its strong centres of diversity in Australia (east coast and southwest corner) and the southern tip of Africa, the Proteaceae is an extremely interesting family in terms of both its evolution and past distribution. The fossil record is invaluable in reconstructing this history.

Johnson & Briggs (1975) undertook an extremely detailed analysis of the living species and proposed several hypotheses relating to the evolution and past distribution of parts of the family. Although the fossil record at the time was not particularly useful for testing these hypotheses, more recent fossil data have proved to be strongly supportive of some of these hypotheses. The intention here is to concentrate on the Australian fossil history of the family, but some examples of the value of fossil data in testing well-established hypotheses are worth mentioning. For example, Johnson & Briggs (1975) noted that 'The ancestors of Cenarrheninae, Dilobeiinae, Proteeae, Hicksbeachiinae, Gevuininae, Macadamiinae and Roupalinae . . . must have evolved by the early Upper Cretaceous.' Furthermore, they considered that *Dilobeia*, Hicksbeachiinae, Gevuininae and Macadamiinae were probably distributed by land in the mid-Cretaceous rather than by distance dispersal over marine barriers at a later date. More recently, Dettmann (1989) noted that during the Campanian–Maastrichtian an array of Proteaceae were introduced into Antarctica, including possible *Macadamia* (Macadamiinae), *Gevuina/Hicksbeachia* (Gevuininae/Hicksbeachiinae), *Knightia*, *Xylomelum* and *Beauprea* (Cenarrheninae). She considered that southeastern Australia may have been the site of evolution of the New Caledonian and New Zealand *Knightia* and the *Gevuina/Hicksbeachia* alliance, which today has a disjunct distribution between northeastern Australia and Chile. Furthermore, she suggested an introduction of *Macadamia* into Antarctica and southern Australasia by the Campanian. Dettmann & Jarzen (1990, 1991) proposed this region as a source area for *Beauprea* as well. Thus, Johnson & Brigg's hypothesis has, in large part, been substantiated by fossil evidence.

Cretaceous pollen data provide vital evidence for the time and place of appearance of proteaceous groupings (Dettmann, 1989; Dettmann & Jarzen, 1990, 1991), and led to the hypothesis that the family originated in northern Gondwana (Dettmann, 1989). Furthermore, Dettmann & Jarzen (1991) have reported pollen of both rainforest and sclerophyllous types of Proteaceae in the Late Cretaceous of the Otway Basin in southeastern Australia, strengthening the argument for a diversification of the family at that time. However, in the early Tertiary the pollen and macrofossil record of the family expanded enormously.

Martin (1978, 1982) noted the major increase in diversity and abundance of Proteaceae during the Late Paleocene–Eocene. Although the familial affinity of some of these fossil pollen types is in doubt (Martin, 1973; Martin & Harris, 1974), many 'are undoubtedly true Proteaceae' (Martin & Harris, 1974). The abrupt disappearance of many of these pollen species at about the end of the Eocene is apparent from stratigraphic range charts (e.g. Stover & Partridge, 1973). This high diversity and abundance of proteaceous pollen has been treated with some suspicion for two reasons:

1. Relatively low percentages of proteaceous pollen are retrieved from surface samples in extant dry Australian forests where the family is abundant and diverse (Martin, 1978). Martin (1982) considers that it is possible that the early Tertiary species were trees that would produce more pollen than a comparable extant sclerophyllous shrub layer, but notes that Kershaw (1970) has recorded low pollen counts from extant proteaceous trees in Queensland rainforests. Martin further suggested that many of the taxa may

have dominated the vegetation, but see (2), below.

2. Christophel & Blackburn (1978) noted that the Middle Eocene Maslin Bay flora contained 35–40% *Proteacidites* spp., but relatively few proteaceous macrofossils. They and Christophel (1981), who noted the same trend at Nerriga, considered that one factor that may be contributing to this discrepancy is the difficulty experienced in identifying the pollen, but this is unlikely to be of major significance.

There is little doubt that the majority of the great diversity and abundance of pollen assigned to 'Proteaceae' in the early Tertiary came from proteaceous plants, but there is at present no reasonable explanation for their demise or the paucity of proteaceous macrofossils in the deposits that yield the pollen.

One possibility that has not been fully investigated is a shift in pollination strategies. Extant Proteaceae rely on biotic pollination, by insects (which is considered to be the ancestral state), birds or animals (Johnson & Briggs, 1975). However, it is possible that wind pollination may have been significant in an early Tertiary group of proteaceous species that are now extinct. The evidence for this is circumstantial. About half the proteaceous species that became extinct in southeastern Australia by the end of the Eocene had pollen grains with diameters in the range 20–40 µm, which Muller (1979) considered optimal for wind pollination. The other species all had much larger grains (some over 100 µm in diameter), which may reflect early animal pollination strategies. The smaller proteaceous species are often quite abundant, comparable with other, well-known wind-pollinated taxa in the same sediment, e.g. species of *Nothofagus* or Casuarinaceae. Faegri & van der Pijl (1979) give some examples of species, in families that are basically entomophilous, that have apparently become secondarily anemophilous; the same may have occurred in the Proteaceae. This goes some way to explaining the abundance of pollen but paucity of associated macrofossils. An even more convincing demonstration of the phenomenon of pollen

dominance and paucity of associated macrofossils is supplied by *Nothofagus* subgenus *Brassospora* (see earlier discussion). The reason for the extinction of a large number of Proteaceae by the end of the Eocene is uncertain, but may be tied to climatic change, either directly, due to the climatic tolerances of the species involved, or indirectly. Indirect effects of climatic change include either a change in pollination strategies due to an increase in animal and/or bird pollinators, or a change in vegetation type, which made wind pollination a less competitive strategy. For example, if the Proteaceae were not canopy trees, wind pollination would be a viable method only if the vegetation was relatively open. The development of closed vegetation would mean that canopy or forest edge taxa (e.g. *Nothofagus*, Casuarinaceae and the conifers) could still utilise wind pollination, but subcanopy taxa (the now extinct Proteaceae) would become extinct. Faegri & van der Pijl (1979, p. 48) noted that in dense forest vegetation 'wind is so slight and infrequent that anemophily is contra-indicated'.

Another hypothesis to explain the reduction in wind pollination in understorey plants is that during the very early Tertiary the southern Australian vegetation was at very high latitudes (Wilford & Brown, Chapter 2, this volume), and the vegetation existed under polar photoperiods. With the sun at a low angle in the sky during summer, and tracking an almost circular path around the horizon during the day, it is likely that the structure of forests was quite different from that observed in wet regions of Australia today. It is probable that relatively widely spaced conical trees dominated, and thus wind pollination was a viable strategy for understorey plants below the open canopy. As Australia moved into lower latitudes, the sun angle increased and a closed-forest structure would have developed as the most efficient way to utilise the incoming solar radiation. This may have been critical for the loss of wind pollination as a viable strategy in understorey plants. This hypothesis is tentative at present, but is the first that explains the observed data.

It should be noted that there were probable bird-pollinated Proteaceae present in Australia by the Eocene. McNamara & Scott (1983) described a *Banksia* infructescence from Early Eocene silcretes in the Stirling Ranges of northwestern Australia and noted that one follicle had apparently been removed from the infructescence in a fashion similar to that employed by parrots today. This suggests a plant–animal interaction that may have extended to pollination; given the current pollination strategies of *Banksia* (Hopper, 1980; Hopper & Burbidge, 1982; Turner, 1982), either bird or small mammal pollination is indicated. Christophel (1984) described an inflorescence closely related to the extant genera *Musgravea* and *Austromuellera* from the Middle Eocene Anglesea sediments that again suggests animal (?probably bird) pollination.

Some Oligocene macrofloras differ from the preceding Eocene macrofloras in containing diverse and numerous proteaceous remains. The best examples are the Latrobe Valley coal in Victoria (Cookson & Duigan, 1950; Blackburn, 1985; Hill & Christophel, 1988; Hill, 1990*c*; Blackburn & Sluiter, Chapter 14, this volume), Cethana in Tasmania (Hill & Christophel, 1988; Carpenter & Hill, 1988; Carpenter, 1991; Carpenter *et al.*, Chapter 12, this volume), and West Dale in southwestern Australia (Hill & Merrifield, 1993). Many of the proteaceous taxa in these deposits can be assigned to extant genera, and in particular the *Banksia/Dryandra* group (fossils are assigned to *Banksieaephyllum* or *Banksieaeformis* depending on the preservation or otherwise of cuticle). Leaves of this type are first recorded in the Late Paleocene of southern New South Wales (R. S. Hill & J. G. Jordan, unpublished data) and are present in several Eocene sites in southeastern Australia (Hill & Christophel, 1988). However, the number of species, and the diversity in single deposit peaks in the Oligocene, may also be related to a change in pollination mechanisms (see Carpenter *et al.*, Chapter 12, this volume).

Hill (1990*a*) has hypothesised that the *Banksieaephyllum* fossils demonstrate the probable pathway for the development of the sclerophyllous heath flora that is prominent in Australia today in areas with oligotrophic soils and a mediterranean climate (hot, dry summers and mild, wet winters).

The first *Banksieaephyllum* leaves in the fossil record have several typically sclerophyllous characters (thick, highly vascularised leaves etc.), suggesting that they occurred on oligotrophic soils. However, they occur in typical rainforest associations and have relatively few morphological adaptations for stomatal protection. This suggests that water was always plentiful, which is in keeping with the climatic reconstruction for that time (Quilty, Chapter 3, this volume). However, Hill (1990*a*) suggested that such leaves may have been pre-adapted to xeromorphic conditions that increased in intensity and spread geographically later in the Tertiary. For example, the *Banksieaephyllum* spp. in the Latrobe Valley coal show a mixture of leaf morphologies. According to Blackburn (1985) those associated with rainforest species still show little or no stomatal protection, whereas those associated with more open (xerophytic) vegetation show some adaptations for stomatal protection (e.g. dense trichomes, stomata in pits between the veins, revolute leaf margins). Such adaptations must be associated with reduced water availability, since they must substantially reduce the CO_2 uptake (and hence photosynthesis) and must be advantageous to the plant if protection against excessive transpirational loss was of particular importance. Among modern *Banksia* and *Dryandra* spp., several extreme forms of stomatal protection occur, including those mentioned above, although often developed to a much greater degree (Figure 16.6). While it is tempting to suggest that this represents an evolutionary progression from the early Tertiary forms, which were pre-adapted to some extent to xeromorphic conditions, through to the extant forms, with their often extreme adaptations, there has been no phylogenetic analysis of fossil *Banksieaephyllum* leaves to test this hypothesis. As with the Casuarinaceae (see later), we cannot discount an alternative hypothesis, that extremely xeromorphic *Banksia/Dryandra* spp. evolved at a relatively early stage in restricted dry microsites, or in remote areas in central Australia and later expanded into southern and eastern Australia, replacing the less xeromorphic forms. In either case, this group offers an excellent opportunity to

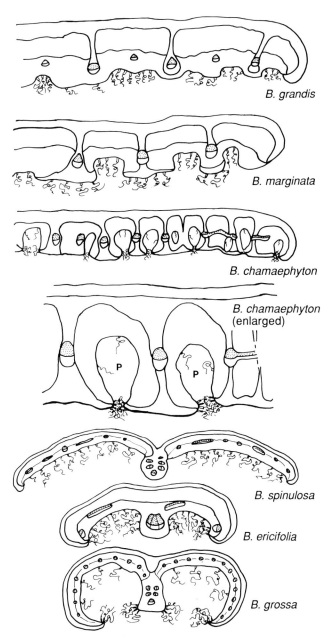

Figure 16.6 Drawings of transverse sections of leaves of several extant *Banksia* spp. In all cases the stomatal surface (abaxial) is the lower surface. The top three species show a range among broad-leaved species from those with minor depressions containing stomata and trichomes (*B. grandis*) to those with more deeply developed pits (*B. marginata*) to the extreme case of stomatal pit development (*B. chamaephyton*). In *B. chamaephyton* the stomata occur in large balloon-shaped pits (P), which extend more than half way through the leaf. The opening of the pit is protected by tufts of trichomes. The lower three species are examples of relatively narrow leaves that protect the stomata by means of revolute margins. These species range from only slightly revolute (*B. spinulosa*) to an intermediate phase (*B. ericifolia*) to the extreme case of highly revolute leaves (*B. grossa*) with dense bands of trichomes along the mid-vein and leaf margins protecting the opening of the grooves in the leaf containing the stomata. This is a similar strategy to the tufts of trichomes in *B. chamaephyton*.

study the response of a restricted taxon to major climatic change over a long period.

The fossil record for most other proteaceous genera is more restricted, but overall this record is very impressive. The pollen record is well known, but still in need of revision in order to determine the true affinities of some species (Dettmann & Jarzen, 1990). The leaf record is also impressive, has grown considerably over the last few years and promises further expansion in the near future. Reproductive structures are fewer but offer a particularly important source of phylogenetic information. There are still undescribed taxa of proteaceous reproductive structures (e.g. Hill & Macphail, 1983; Christophel, 1984) and we can expect further progress in this direction over the next few years. A detailed analysis of this record will undoubtedly lead to a much greater understanding of the past history of this major Australian family.

CASUARINACEAE

The family Casuarinaceae comprises 96 extant species in four genera that are confined to the Malesian–Australian–Melanesian region, except for the essentially littoral species *Casuarina equisetifolia* (Johnson & Wilson, 1989). The family is widespread in Australia, and ranges through most vegetation types, although it is restricted to one small population in rainforest. The affinities of the family are obscure, but it has an impressive fossil record. Pollen of the Casuarinaceae is readily identifiable, but unfortunately the generic affinities cannot be determined. According to Johnson & Wilson (1989) fossil pollen has been reliably reported from South Africa (Coetzee & Muller, 1984, Coetzee & Praglowski, 1984), Argentina (Archangelsky, 1973) and on the suboceanic Ninetyeast Ridge in the Indian Ocean (Kemp & Harris, 1977). It is also common in Australia and New Zealand from the Paleocene onwards (Martin, 1978; Mildenhall, 1980). The latest occurrence of Casuarinaceae pollen in New Zealand is Early Pleistocene (Mildenhall, 1980).

Macrofossils of Casuarinaceae are relatively abundant, and convey much more taxonomic information than does pollen. In Australia the macrofossil record extends from the Early Eocene onwards (although a specimen described by Douglas (1978) from Victoria may extend the record into the Paleocene, according to Blackburn (1985)) and demonstrates some interesting trends. The earliest records (Eocene–Oligocene) are all *Gymnostoma*. Three species have been described in detail (Scriven & Christophel, 1990), although none of the fossil *Gymnostoma* species has yet been named. *Gymnostoma* extends into the Miocene in the Latrobe Valley coal (Blackburn, 1985) and the Etadunna Formation in central Australia (Greenwood *et al.*, 1990), but it has not been recorded as a macrofossil after that time. *Gymnostoma australianum* still occurs in Australia, in a small patch of rainforest on the Atherton Tableland (Fig. 16.7).

Macrofossil records of other casuarinacean genera in Australia are sparse. Blackburn (1985) described cladodes and female cones from the Yallourn and Morwell coal seams as *Casuarina*, and suggested that they were similar to extant species such as *C. equisetifolia*. Patton (1936) noted the presence of cladodes and females cones of Casuarinaceae from the possibly Miocene siliceous sandstones at Limestone Reserve in Victoria. These specimens appear most likely to belong to *Allocasuarina*, although a reinvestigation is nececessary to confirm this. Blackburn (1985) also noted the presence of cladodes and cones that he considered to be almost certainly *Allocasuarina* in the Pliocene of Victoria (Deane, 1904) and the ?Miocene of South Australia. Greenwood *et al.* (1990) noted the presence of what are probably *Allocasuarina* cones in Miocene sediments in central Australia, where they co-occur with *Gymnostoma*.

Thus the macrofossil record from central, coastal southern and eastern Australia suggests that *Gymnostoma* was dominant from the Early Eocene (or ?Paleocene) until the Oligocene and then there was a relatively short transition during the Miocene to dominance by *Casuarina* and *Allocasuarina*. Blackburn (1985) noted this transition in the Latrobe Valley coal, where *Gymnostoma*

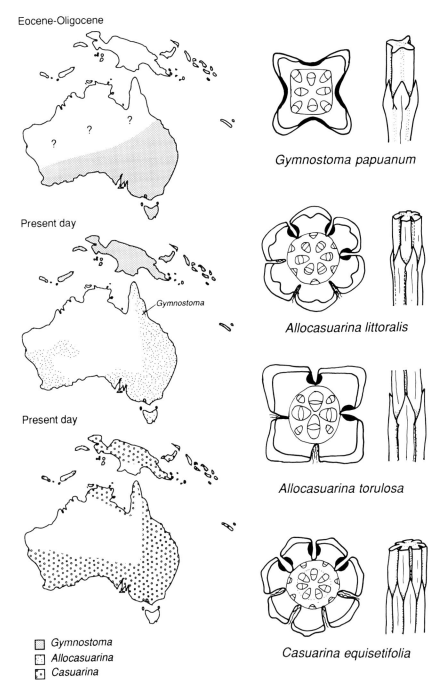

Eocene-Oligocene

Present day

Present day

▨ Gymnostoma
▩ Allocasuarina
▨ Casuarina

Gymnostoma papuanum

Allocasuarina littoralis

Allocasuarina torulosa

Casuarina equisetifolia

Figure 16.7 Distribution of the three major Casuarinaceous genera during the Eocene–Oligocene and present day. The population of *Gymnostoma* shown in northeastern Australia in the present day is *G. australianum*. Examples of transverse sections through cladodes of species from each genus are shown on the right-hand side. Stomatal distribution is shown by thick black lines on the surface. Trichomes are shown in some stomatal pits. Note the relatively superficial stomata in *Gymnostoma* compared with the other genera.

dominates over *Casuarina* at Morwell, but the reverse is true at Yallourn.

Hill (1990*a*) considered that the Casuarinaceae family was an excellent illustration of the effect of developing aridity in southern Australia during the Tertiary. In the early Tertiary, when the climate was uniformly wet, *Gymnostoma* was the only Casuarinaceae identified, and it occurs in many macrofossil deposits. Although *Gymnostoma* is quite scleromorphic in its reduced and virtually nonfunctional leaves and woody photosynthetic stem, it is not particularly adapted to prevent water loss. The very name of the genus describes its unprotected stomata (Figure 16.7), which are present in rows on the usually glabrous stem surface. Thus *Gymnostoma* can be considered to be a genus that may be adapted to oligotrophic soils, and in particular the characteristically phosphorus-poor Australian soils, but is not particularly xeromorphic. In contrast, both *Casuarina* and *Allocasuarina* have well-protected stomata, in deep grooves along the stem, which are often filled with trichomes (Figure 16.7). These genera are not only scleromorphic, but xeromorphic as well, i.e. they have morphological adaptations to prevent excessive water loss. These adaptations must surely have evolved in response to continually dry conditions, since they must severely disrupt the entry of CO_2 into the stems. Such an adaptation must therefore confer a great benefit to the plant in order to overcome the disadvantage in terms of lost productivity.

The apparent changeover in dominance during the Miocene is therefore readily explainable in terms of climatic change. As the climate dried in southern Australia, *Gymnostoma* was no longer competitive and its niche was taken over by *Casuarina* and *Allocasuarina*. It is tempting to consider that *Gymnostoma* may have been ancestral to *Casuarina* and/or *Allocasuarina*, but there is no evidence for this. When *Casuarina* and *Allocasuarina* first appear in the Australian macrofossil record they are recognisable as the extant genera, and, although Christophel (1980) has suggested that some of the fossils from the Latrobe Valley coal may be intermediate, there has been no further report of this nature; Blackburn (1985)

confidently assigned all specimens from that locality to extant genera.

There are two options to explain this rapid change in dominant casuarinacean genera in southern Australia during the Miocene. It is possible that *Casuarina* and *Allocasuarina* evolved relatively late in the Tertiary, especially in comparison with *Gymnostoma*, and rapidly exploited new niches as they became available during the Miocene spread of aridity. However, I consider it more likely that *Casuarina* and *Allocasuarina* are also ancient genera, perhaps as old as *Gymnostoma*, but without the associated macrofossil record. During the early Tertiary, or perhaps even earlier, *Gymnostoma* was common in the vast tracts of rainforest that occurred around coastal southern and eastern Australia. However, even during this period of high water availability there must have been dry microsites in this coastal belt (e.g. ridgetops, sand dunes), and in these xeric microsites it is possible that the more xeromorphic *Casuarina* and *Allocasuarina* became established. Even if this occurred, it is unlikely that plants in such microsites would leave a macrofossil record, being remote from sedimentary environments. However, such plants would have been very well placed to expand rapidly when the climate dried later in the Tertiary. There may also have been more extensive areas suitable for these genera in inland Australia during the early Tertiary, but the only reliable macrofossil record of the family from central Australia is a report of two species of *Gymnostoma* from the Middle Eocene Nelly Creek flora (Christophel *et al.*, 1992).

Extra-Australian records

There is a small but very significant extra-Australian macrofossil record of the Casuarinaceae. Frenguelli (1943) described *Casuarina patagonica* from Miocene sediments in southern South America (since revised to 59 million years (Early Eocene), E. J. Romero, personal communication). The illustrated specimen consists of a short branch bearing four female cones. Christophel (1980) considered that it was possible that this specimen belonged to a non-*Gymnostoma* form of Casuarinaceae, but it was 'equally if not more likely' to have affinities with *Gymnostoma*. Johnson & Wilson (1989) unequivocally assign this species to *Gymnostoma*. E. J. Romero (personal communication) has collected two more specimens of Casuarinaceae female cones, from Early Eocene sediments in South America, that are about twice the size of *C. patagonica* and have several other distinguishing features. Romero concludes that these specimens represent a new species of *Gymnostoma* that is distinct from *C. patagonica*.

Campbell & Holden (1984) have described two casuarinacean species from Miocene sediments in the South Island of New Zealand. They considered that the characteristics of the female cone of *Casuarina stellata* are consistent with *Gymnostoma*, a conclusion supported by Johnson & Wilson (1989). However, the fruiting cones of *C. avenacea* clearly do not belong to *Gymnostoma*. Johnson & Wilson (1989) considered that this specimen should be assigned to *Allocasuarina*, but Johnson (1991) has revised this opinion and the generic affinities of this species are less certain. Some specimens of *C. avenacea* clearly have more than four pairs of bracts in a whorl of the female fructification, but one specimen appears to have only four, and superficially resembles *Gymnostoma*. This has implications when we consider the affinities of *C. patagonica*. Although the parts of that specimen all appear to be in whorls of four, there is at least the possibility that this is only part of the range exhibited by the species and that it was, in fact, not *Gymnostoma*.

Thus the extra-Australian macrofossil record of Casuarinaceae demonstrates that *Gymnostoma* was present in South America during the Early Eocene and New Zealand during the Miocene. Non-*Gymnostoma* Casuarinaceae was also present in New Zealand in the Miocene, and possibly also in South America in the Early Eocene. The presence of Casuarinaceae pollen in New Zealand since the Paleocene suggests a long history in that land mass. Thus we can be sure that *Gymnostoma* at least was widespread across the southern hemisphere during the early Tertiary and it is probable that its origin precedes the separation of Australia

and New Zealand from Antarctica. There is less convincing evidence to suggest that at least one other genus (possibly *Casuarina*) has a similarly long history, although more recent long-distance dispersal from Australia may play a role here. This fossil record makes a particularly interesting comparison with *Eucalyptus* (discussed below), especially its apparently contemporaneous extinction in New Zealand during the Early Pleistocene.

EUCALYPTUS

Eucalyptus has justifiably been called the 'universal Australian' (Pryor & Johnson, 1981), with its more than 500 species, which are nearly all endemic to Australia. In form, the species range from low shrubs to the tallest flowering plants in the world. Lange (1980) noted that eucalypts have been estimated to contribute 75% of Australian vegetation, and at the wetter margins of the continent they dominate nearly all vegetation except rainforest and allied mesic types. Only in the arid Australian interior are eucalypts generally lacking in dominance.

The macrofossil record

The genus is highly diversified (Johnson, 1972; Pryor & Johnson, 1981) and, in conjunction with its current dominance, it would be expected to yield a complex and challenging fossil record. That this is not the case has been well documented, most notably by Lange (1978, 1980, 1982), who made a detailed but futile search for organically preserved eucalypt remains among coastal Tertiary dispersed cuticles. However, *Eucalyptus* has a macrofossil record, and it is more detailed than is usually acknowledged. The widespread ignorance of this record is as much a commentary on the mystique surrounding this genus in Australia as a sign of scientific rigour. When Maiden (1922) produced his *Critical Revision of the Genus* Eucalyptus, he included descriptions of 19 macrofossil species, as well as four records of sub fossils that were assigned to extant species. In stark contrast, Lange (1980) believed there was 'no Tertiary record of *Eucalyptus* that could withstand close taxonomic scrutiny' and Christophel (1989) considered that the only macrofossil records of *Eucalyptus* worth mentioning were 'definitely eucalyptoid' Miocene leaf and fruit impressions from Chalk Mountain in New South Wales. Christophel also noted that, by the Pliocene, *Eucalyptus* 'is well represented', although to my knowledge there is only one macrofossil record of this age (*E. pluti*; McCoy, 1876).

Clearly, after the promising start reported by Maiden (1922), something went seriously wrong with the Australian *Eucalyptus* macrofossil record. Part of the problem extends from the general view of the quality of taxonomic work on Australian Tertiary plant macrofossils undertaken during the late nineteenth and early twentieth centuries. In Australia, Deane (1896, 1900) launched a vigorous attack on the taxonomic work of von Ettingshausen (1888), noting for instance in 1896 that von Ettingshausen 'has expressed views as to the origin of the vegetation of Australia, and of the rest of the world, which appear to be entirely erroneous'. E. W. Berry (in Maiden, 1922), in an apparent defence of von Ettingshausen, noted that, apart from the work of that taxonomist and of Mueller, 'There have been more worthless articles written about the Cretaceous and Tertiary floras of Australia than any other equal area of the earth's surface'. These two examples represent only a small fraction of the criticism aimed at Australian macrofossil palaeobotany at that time. This ongoing criticism, from both within and without, did enormous damage to Australian palaeobotany. Although there was undoubtedly some poor taxonomic work done during these pioneer days, there was much good work carried out as well. The strength of criticism seems to have brought research almost to a halt from the 1920s until the 1950s, when Cookson resurrected macrofossil research. It is not surprising that few people wished to become involved in a field associated with such a stigma. Furthermore, living plant taxonomists lost confidence in palaeobotany.

That this stigma endures in the minds of some palaeobotanists is evidenced by the comments of Lange (1980) and Christophel (1989) mentioned earlier. *Eucalyptus* seems to have suffered particu-

larly heavily. Some of the fossil species described by early researchers are clearly not *Eucalyptus*. The oft-cited example is *E. praecoriacea*, which was described by Deane (1902a). Deane considered this species to be similar to the extant *E. coriacea* (now *E. pauciflora*). Hill (1980) considered that the type specimen of *E. praecoriacea* is almost certainly part of a frond of a zamioid cycad. However, Christophel's (1981) suggestion that Deane 'had but to glance at the well preserved cuticular structure to realize that he was dealing with a cycad-like plant and not a eucalypt' was unfair, since the organic remains on the fossil are very fragmentary and cell detail has never been successfully observed from attempted preparations. It is instructive to note that Patton (1919), in a paper describing Tertiary *Eucalyptus* leaves, commented that *E. praecoriacea* 'is a very doubtful eucalypt'. This one piece of poor taxonomy should not evince universal contempt. In fact many of the early reports of *Eucalyptus* macrofossils appear to be very convincing, and in the case of the Berwick Quarry fossils described by Deane (1902b) a reinvestigation (Pole *et al.*, 1993) has confirmed the generic identity. Thus, the specimens described during the late nineteenth and early twentieth centuries are badly in need of re-examination, and such work may help to resurrect the reputations of some of Australia's pioneer palaeobotanists.

This review is restricted to those reports that are considered to be reliable by modern standards, and consequently many of the previously defined species are disregarded. This should not be taken as a lack of faith in the identity of these fossils, but a lack of sufficient information. In studying the macrofossil record of *Eucalyptus*, there is almost as much to be gained from noting where it does not occur as where it does, and this is also considered.

At present, the oldest reliably dated and described macrofossil associated with *Eucalyptus* is a tree stump, which is probably *in situ* and is enclosed in a 21 million years old basalt in the upper Lachlan Valley of New South Wales. The wood is reasonably well preserved and has been identified by Bishop & Bamber (1985) as 'Myrtaceae with affinities to *Eucalyptus* B' (of Dadswell,

1972). Unfortunately, confirmation of this by identification of diffuse parenchyma cells was not possible. Other wood nearby was identified as *Nothofagus* and *Acacia*, but they were not necessarily closely associated in the source vegetation. Holmes *et al.* (1982) described *Eucalyptus bugaldiensis*, from the Middle Miocene Chalk Mountain Formation in the Warrumbungle Mountains of New South Wales, based on impressions of umbelliferous fruits. They also described two *Eucalyptus* leaf types from impressions and noted that at least two of the phylogenetic lines proposed by Johnson (1972) for *Eucalyptus* were present at the site. They also concluded that the fossils exhibited some features of advanced states in both fruits and leaf venation. The fossils occur in association with predominantly mesophyllous leaves that Holmes *et al.* (1982) interpreted as rainforest, although they cautioned that this should not be used as evidence that eucalypts in the Middle Miocene were adapted to rainforest conditions.

Deane (1902*b*) described seven species of *Eucalyptus* from Middle Miocene sediments at Berwick in Victoria, and Holmes *et al.* (1982) noted the presence of a leaf with *Eucalyptus*-like cuticle and an impression of a eucalypt-like fruit in the flora. This site has recently undergone extensive re-collection, and several leaves with reasonably well-preserved cuticle have been recovered which have clear affinities with *Eucalyptus* (Pole *et al.*, 1993). The remainder of the flora has the character of rainforest. The only other *Eucalyptus* macrofossils from the coastal belt of eastern Australia that are universally accepted are the Pliocene leaves named *E. pluti* from Daylesford in Victoria (McCoy, 1876). McCoy noted that these leaves are almost identical in size and shape with extant *E. globulus*, but he noted some minor differences in the venation pattern. Despite the lack of organic preservation, the generic identity of this species has never been challenged, although that may well be due to its relatively recent age, since similarly convincing but older fossil leaves have been rejected.

The macrofossil record from central Australia casts further light on the history of *Eucalyptus*, but the evidence is frustratingly uncertain. Lange (1978) described several beautifully preserved casts of eucalypt-like fruits, which are only part of a more diverse collection from Island Lagoon, near Woomera. Unfortunately, there is poor age control on this diverse assemblage of eucalypts, with the suggestion that they have either an Eocene–Oligocene age, or a Miocene age (Ambrose *et al.*, 1979). Greenwood (1991) reported 'the earliest record' of *Eucalyptus* macrofossils from the Middle Eocene silcretes in the Eyre Formation in central Australia. However, he neither described nor illustrated these macrofossils, and his earlier work on the flora (Greenwood *et al.*, 1990) mentioned only leaves assigned to '*Eucalyptophyllum*', about which he concluded 'it is not possible to determine the affinities of this leaf type'. If unequivocal *Eucalyptus* fossils have been found in Middle Eocene sediments in central Australia, their formal description should be a high priority.

It is constructive to consider where *Eucalyptus* macrofossils have not been found in the Australian Tertiary. Lange (1980) systematically searched dispersed cuticle from several near-coastal Eocene or Oligocene floras and found only one cuticle that approached that of a eucalypt on general appearance (among 600 examined from Maslin Bay). Since that time extensive research on coastal Paleocene to Oligocene macrofloras in coastal southeastern Australia (including Tasmania) has failed to uncover a single eucalypt specimen. It is also notable that *Eucalyptus* has not been reported from the Latrobe Valley coal macroflora (Blackburn & Sluiter, Chapter 14, this volume). Allied to this is the argument presented by Archer *et al.* (1991) that the relative rarity of early koalas in Oligo-Miocene rainforests may have been due to the rarity of eucalypts in that vegetation.

Therefore the evidence, sparse as it is, seems to favour the hypothesis proposed by Lange (1980) that in the mid-Tertiary the continental margins of Australia supported only mesic non-eucalypt vegetation, while eucalypts contributed to the more xeric vegetation in the interior. With progressive development of central Australian aridity, eucalypts were displaced to the continental margins, while most mesic vegetation was eliminated altogether. It is informative that the Miocene coastal sites where *Eucalyptus* has been recorded do seem to contain rainforest vegetation that is less mesomorphic in character and may represent a transition to drier climate forest.

The pollen record

The early pollen record of *Eucalyptus* is even more enigmatic than the macrofossil record. Part of the reason for this is that fossil pollen that has sometimes been referred to *Eucalyptus* is also similar to *Angophora*, *Syncarpia* and possibly *Metrosideros* (Martin, 1978). Martin (1981) noted that 'It is unfortunate that pollen cannot be used as evidence of the development of the modern *Eucalyptus* flora.' However, since that report there has been an increase in the confidence of assessment of certain types of pollen with *Eucalyptus*.

Cookson & Pike (1954) described the species *Myrtaceidites eucalyptoides* for pollen grains with probable affinities with *Eucalyptus* and they noted its range as Pliocene to Pleistocene. However, this pollen type, which is referred to as the *Angophora*/bloodwood eucalypt type of pollen (Martin, 1989; Chapter 7, this volume) is now recorded from the Late Paleocene of the Lake Eyre Basin in inland Australia (I. R. K. Sluiter & N. F. Alley, unpublished data). Harris (1965) described *M. tenuis* and noted that this species is a very rare form that is similar to forms related to *Eucalyptus* by Cookson & Pike (1954). *Myrtaceidites tenuis* does not extend beyond the Eocene, and its possible affinity with *Eucalyptus* is particularly interesting because the eucalypt-type does not appear again in southeastern Australia until well into the Oligocene.

Although the general type of *Eucalyptus* pollen is difficult to identify at the generic level in the fossil record, there are specific eucalypt pollen types, with distinctive morphology, that offer more hope. An excellent example is *E. spathulata*, which is unique in having distinctive, large verrucae/rugulae (Martin & Gadek, 1988). When similar forms occur in the fossil record their affinity with

Eucalyptus can be assumed with more confidence. Martin & Gadek (1988) have identified *E. spathulata*-like pollen from three Pliocene sites in Australia, one in the southwest and two in the southeast. This species, which is now restricted to southwest Australia, is considered by Martin & Gadek to be the living remnant of a larger group of species that may have filled niches quite different from that of the extant species. Martin (1989) noted that this fossil pollen type is used as a marker of the Miocene/Pliocene boundary in the Gippsland Basin (Victoria), where it extends into the Pleistocene.

In more recent sediments, there is excellent evidence for a dramatic increase in abundance of *Eucalyptus* associated with increased charcoal levels. Although this occurs at different times in different parts of Australia, this is compelling evidence for the influence of Aborigines in artificially increasing the fire frequency (Singh *et al.*, 1980; Kershaw, 1986). The current superdominance of *Eucalyptus* in Australia may therefore be a relatively recent phenomenon, the result of the extraordinary adaptation of the genus to fire (Jackson, 1968) coupled with an artificial increase in ignition associated with the the arrival of humans. A drying climate probably assisted this event, since the vegetation was then much more prone to continue to burn once it was lit.

Extra-Australian records

An intriguing aspect of the fossil record of *Eucalyptus* is its apparent presence in both South America and New Zealand. Frenguelli (1953) described a group of three fruits from Miocene sediments in Patagonia as *Eucalyptus patagonica*. The age of these sediments is currently being revised and is considered to be probably Eocene (E. J. Romero, personal communication). Johnson & Briggs (1984) discussed this fossil and others which had been collected subsequently. They noted that Frenguelli's specimen 'could conceivably belong among the more generalized members of *Symphyomyrtus*', but they did not consider that the other specimens they saw belonged to the *Eucalyptus* alliance. E. J. Romero (personal

communication) has collected further leaf specimens that he believes are *Eucalyptus*, although he
considers it possible that they belong to an
ancestor or sister group. Future research on
these fossils should be of particular importance
to those with an interest in the origin and evolution of *Eucalyptus*.

Pole (1989) described leaves from Early
Miocene sediments in the South Island of New
Zealand that are 'identical to many *Eucalyptus*
species', and noted their occurrence with eucalypt-like fructifications. This record is far more
reliable at present than that from South America,
especially when it is considered in conjunction
with the pollen record. Mildenhall (1980) lists the
stratigraphic range of *Eucalyptus* pollen in New
Zealand as Miocene–Early Pleistocene, suggesting a relatively recent extinction.

These extra-Australian records are of particular significance in tracing the past history of the
genus. The New Zealand records can be considered in one of two ways. Either they represent
long-distance dispersal from Australia, in the
Miocene or earlier, or they represent part of an
ancient lineage that was present across at least
part of Gondwana prior to separation of New Zealand and Australia but did not assume prominence
(and thus appear in the fossil record) until climatic
changes favoured its expansion. It is impossible at
present to determine between these options, but
the South American record will prove crucial. If
further work determines a link with *Eucalyptus*,
then an ancient (?Cretaceous) Gondwanic distribution for this now characteristically Australian
genus would have to be considered the most likely
hypothesis. The recent extinction of *Eucalyptus*
from New Zealand is probably tied to the effects
of the Quaternary glaciations.

REFERENCES

AMBROSE, G. J., CALLEN, R. A., FLINT, R. B. &
  LANGE, R. T. (1979). *Eucalyptus* fruits in stratigraphic
  context in Australia. *Nature*, 280, 387–9.
ARCHANGELSKY, S. (1973). Palinologia del Paleoceno
  de Chubut. I. Descripciones sistematicas. *Ameghiniana*,
  10, 339–99.
ARCHER, M., HAND, S. J. & GODTHELP, H. (1991).
  Riversleigh. Sydney: Reed.
BIGWOOD, A. J. & HILL, R. S. (1985). Tertiary arauc-
  arian macrofossils from Tasmania. *Australian Journal of
  Botany*, 33, 645–56.
BIRKENMAJER, K. & ZASTAWNIAK, E. (1989a). Late
  Cretaceous–Early Tertiary floras of King George Island,
  West Antarctica: their stratigraphic distribution and
  palaeoclimatic significance. In *Origins and Evolution of
  the Antarctic Biota*, ed. J. A. Crame. *Geological Society
  of London Special Publication*, 47, 227–40.
BIRKENMAJER, K. & ZASTAWNIAK, E. (1989b). Late
  Cretaceous and Early Neogene vegetation history of the
  Antarctic Peninsula sector. Gondwana break-up and Ter-
  tiary glaciations. *Bulletin of the Polish Academy of Earth
  Sciences*, 37, 63–88.
BISHOP, P. & BAMBER, R. K. (1985). Silicified wood
  of Early Miocene *Nothofagus*, *Acacia* and Myrtaceae (aff.
  Eucalyptus B) from the Upper Lachlan Valley, New South
  Wales. *Alcheringa*, 9, 221–8.
BLACKBURN, D. T. (1985). *Palaeobotany of the Yallourn
  and Morwell Coal Seams. Palaeobotanical Project – Report 3.*
  Melbourne: State Electricity Commission of Victoria.
BURGER, D. (1981). Observations on the earliest angio-
  sperm development with special reference to Australia. *Pro-
  ceedings, Fourth International Palynological Conference,
  Lucknow (1976–77)*, 3, 418–28.
BURGER, D. (1990). Early Cretaceous angiosperms from
  Queensland, Australia. *Review of Palaeobotany and Paly-
  nology*, 65, 153–63.
CAMPBELL, J. D. & HOLDEN, A. M. (1984). Miocene
  casuarinacean fossils from Southland and Central Otago,
  New Zealand. *New Zealand Journal of Botany*, 22,
  159–67.
CANTRILL, D. J. (1991). Broad leafed coniferous foliage
  from the Lower Cretaceous Otway Group, southeastern
  Australia. *Alcheringa*, 15, 177–90.
CARPENTER, R. J. (1991). *Palaeovegetation and environ-
  ment at Cethana, Tasmania.* Ph.D. thesis, University of
  Tasmania.
CARPENTER, R. J. & HILL, R. S. (1988). Early Tertiary
  Lomatia (Proteaceae) macrofossils from Tasmania, Aus-
  tralia. *Review of Palaeobotany and Palynology*, 56, 141–50.
CASE, J. A. (1988). Paleogene floras from Seymour Island,
  Antarctic Peninsula. *Geological Society of America, Memoir*,
  169, 523–30.
CASE, J. A. (1989). Antarctica: the effects of high latitude
  heterochroneity on the origin of the Australian mar-
  supials. In *Origins and Evolution of the Antarctic Biota*, ed.
  J. A. Crame, *Geological Society of London Special Publication*,
  47, 217–26.
CHRISTOPHEL, D. C. (1980). Occurrence of *Casuarina*
  megafossils in the Tertiary of south-eastern Australia. *Australian Journal of Botany*, 28, 249–59.
CHRISTOPHEL, D. C. (1981). Tertiary megafossil floras

of Australia as indicators of floristic associations and palaeo-climate. In *Ecological Biogeography of Australia*, ed. A. Keast, pp. 379–90. The Hague: W. Junk.

CHRISTOPHEL, D. C. (1984). Early Tertiary Protea-ceae: the first floral evidence for the Musgraveinae. *Australian Journal of Botany*, 32, 177–86.

CHRISTOPHEL, D. C. (1985). First record of well-preserved megafossils of *Nothofagus* from mainland Aus-tralia. *Proceedings of the Royal Society of Victoria*, 97, 175–8.

CHRISTOPHEL, D. C. (1989). Evolution of the Austra-lian flora through the Tertiary. *Plant Systematics and Evolution*, 162, 63–78.

CHRISTOPHEL, D. C. & BLACKBURN, D. T. (1978). Tertiary megafossil flora of Maslin Bay, South Australia: a preliminary report. *Alcheringa*, 2, 311–19.

CHRISTOPHEL, D. C., SCRIVEN, L. J. & GREEN-WOOD, D. R. (1992). An Eocene Megafossil flora from Nelly Creek, South Australia. *Transactions of the Royal Society of South Australia*, 116, 65–76.

COETZEE, J. A. & MULLER, J. (1984). The phytogeo-graphic significance of some extinct Gondwana pollen types from the Tertiary of the Southwestern Cape (South Africa). *Annals of the Missouri Botanical Garden*, 71, 1088–99.

COETZEE, J. A. & PRAGLOWSKI, J. (1984). Pollen evidence for the occurrence of *Casuarina* and *Myrica* in the Tertiary of South Africa. *Grana*, 23, 23–41.

COOKSON, I. C. (1947). Plant microfossils from the lig-nites of Kerguelen Archipelago. *BANZ Antarctic Research Expedition 1929–31 Reports*, Series A, 2, 127–42.

COOKSON, I. C. & DUIGAN, S. L. (1950). Fossil Banksieae from Yallourn, Victoria, with notes on the mor-phology and anatomy of living species. *Australian Journal of Scientific Research*, Series B, 3, 133–65.

COOKSON, I. C. & DUIGAN, S. L. (1951). Tertiary Araucariaceae from south-eastern Australia, with notes on living species. *Australian Journal of Scientific Research*, Series B, 4, 415–49.

COOKSON, I. C. & PIKE, K. M. (1953). The Tertiary occurrence and distribution of *Podocarpus* (section *Dacry-carpus*) in Australia and Tasmania. *Australian Journal of Botany*, 1, 71–82.

COOKSON, I. C. & PIKE, K. M. (1954). The fossil occurrence of *Phyllocladus* and two other podocarpaceous types in Australia. *Australia Journal of Botany*, 2, 60–7.

CRANE, P. R. (1989). Early fossil history and evolution of the Betulaceae. In *Evolution, Systematics, and Fossil His-tory of the Hamamelidae*, vol. 2 'Higher' Hamamelidae, ed. P. R. Crane & S. Blackmore. *Systematics Association Special Volume*, 40B, 87–116. Oxford: Clarendon Press.

CRANWELL, L. M. (1963). *Nothofagus*: living and fossil. In *Pacific Basin Biogeography*, ed. J. L. Gressitt, pp. 387–400. Hawaii: Bishop Museum Press.

DADSWELL, H. E. (1972). *The Anatomy of Eucalypt Woods*.

Forest Products Laboratory, Division of Applied Chemis-try, Technical Paper No. 66. Australia: CSIRO.

DARLINGTON, P. J. (1965). *Biogeography of the Southern End of the World*. Cambridge, MA: Harvard University Press.

DEANE, H. (1896). President's address. *Proceedings of the Linnean Society of New South Wales*, 10, 619–67.

DEANE, H. (1900). Observations on the Tertiary flora of Australia, with special reference to Ettingshausen's theory of the Tertiary cosmopolitan flora. *Proceedings of the Lin-nean Society of New South Wales*, pp. 463–75.

DEANE, H. (1902a). Notes on the fossil flora of Pitfield and Mornington. *Records of the Geological Survey of Victoria*, 1, 15–20.

DEANE, H. (1902b). Notes on the fossil flora of Berwick. *Records of the Geological Survey of Victoria*, 1, 21–32.

DEANE, H. (1904). Further notes on the Cainozoic flora of Sentinel Rock, Otway coast. *Records of the Geological Survey of Victoria*, 1, 212–6.

DETTMANN, M. E. (1989). Antarctica: Cretaceous cradle of austral temperate rainforests? In *Origins and Evolution of the Antarctic Biota*, ed. J. A. Crame, *Geological Society of London Special Publication*, 47, 89–105.

DETTMANN, M. E. & JARZEN, D. M. (1990). The Antarctic/Australian rift valley: Late Cretaceous cradle of northeastern Australasian relicts? *Review of Palaeobotany and Palynology*, 65, 131–44.

DETTMANN, M. E. & JARZEN, D. M. (1991). Pollen evidence for Late Cretaceous differentiation of Protea-ceae in southern polar forests. *Canadian Journal of Botany*, 69, 901–6.

DETTMANN, M. E., POCKNALL, D. T., ROMERO, E. J. & ZAMALOA, M. d. C. (1990). *Nothofagidites* Erdt-man ex Potonié, 1960; a catalogue of species with notes on the paleogeographic distribution of *Nothofagus* Bl. (southern beech). *New Zealand Geological Survey Paleontol-ogical Bulletin*, 60, 1–79.

DOUGLAS, J. G. (1978). Victoria's oldest flowers. *Vic-torian Naturalist*, 95, 137–40.

DRINNAN, A. N. & CHAMBERS, T. C. (1986). Flora of the Lower Cretaceous Koonwarra Fossil Bed (Korumburra Group) south Gippsland. In *Plants and Invertebrates from the Koonwarra Fossil Bed, South Gippsland, Victoria*, ed. P. A. Jell & J. Roberts. *Memoirs of the Associ-ation of Australasian Palaeontologists*, 3, 1–77.

ETTINGSHAUSEN, C. VON (1888). Contributions to the Tertiary flora of Australia. *Memoirs of the Geological Survey of New South Wales*, 2, 1–189.

ETTINGSHAUSEN, C. VON (1891). Contributions to the knowledge of the fossil flora of New Zealand. *Trans-actions of the New Zealand Institute*, 23, 237–310.

FAEGRI, K. & PIJL, L. VAN DER (1979). *The Principles of Pollination Ecology*. Oxford: Pergamon Press.

FLORIN, R. (1940). The Tertiary fossil conifers of south Chile and their phytogeographical significance. *Kung-*

liga Svenska Vetenskapakademiens Handlingar, **19**, 1–107.

FLORIN, R. (1963). The distribution of conifer and taxad genera in time and space. *Acti Horti Bergiani*, **20**, 121–312.

FRENGUELLI, J. (1943). Restos de *Casuarina* en el Mioceno de el Mirador Patagonia central. *Notas del Museo de la Plata*, **8**, 349–54.

FRENGUELLI, J. (1953). Restos del género *Eucalyptus* en el Mioceno del Neuquén. *Notas del Museo Universidad Nacional de Eva Perón*, **16**, 209–13.

GREENWOOD, D. R. (1991). Middle Eocene megafloras from central Australia: earliest evidence for Australian sclerophyllous vegetation. *American Journal of Botany Supplement*, **78**, 114–15.

GREENWOOD, D. R., CALLEN, R. A. & ALLEY, N. F. (1990). *The Correlation and Depositional Environment of Tertiary Strata Based on Macrofloras in the Southern Lake Eyre Basin, South Australia. Department of Mines and Energy South Australia, Report Book* 90/15.

HANKS, S. L. & FAIRBROTHERS, D. E. (1976). Palynotaxonomic investigation of *Fagus* L. and *Nothofagus* Bl.: light microscopy, scanning electron microscopy, and computer analyses. *Botanical Systematics*, **1**, 1–141.

HARRIS, W. K. (1965). Basal Tertiary microfloras from the Princeton area, Victoria, Australia. *Palaeontographica Abt. B*, **115**, 75–106.

HICKEY, L. J., WEST, R. M., DAWSON, M. R. & CHOI, D. K. (1983). Arctic terrestrial biota: paleomagnetic evidence of age disparity with mid-northern latitudes during the Late Cretaceous and Early Tertiary. *Science*, **221**, 1153–6.

HILL, R. S. (1980). Three new Eocene cycads from eastern Australia. *Australian Journal of Botany*, **28**, 105–22.

HILL, R. S. (1984). Tertiary *Nothofagus* macrofossils from Cethana, Tasmania. *Alcheringa*, **8**, 81–6.

HILL, R. S. (1987). Discovery of *Nothofagus* fruits corresponding to an important Tertiary pollen type. *Nature*, **327**, 56–8.

HILL, R. S. (1990a). Evolution of the modern high latitude southern hemisphere flora: evidence from the Australian macrofossil record. In *Proceedings, Third IOP Conference*, ed. J. G. Douglas & D. C. Christophel, pp. 31–42. Melbourne: A-Z Printers.

HILL, R. S. (1990b). *Araucaria* (Araucariaceae) species from Australian Tertiary sediments – a micromorphological study. *Australian Systematic Botany*, **3**, 203–20.

HILL, R. S. (1990c). Tertiary Proteaceae in Australia: a re-investigation of *Banksia adunca* and *Dryandra urniformis*. *Proceedings of the Royal Society of Victoria*, **102**, 23–8.

HILL, R. S. (1991). Tertiary *Nothofagus* (Fagaceae) macrofossils from Tasmania and Antarctica and their bearing on the evolution of the genus. *Botanical Journal of the Linnean Society*, **105**, 73–112.

HILL, R. S. (1993). *Nothofagus*: evolution from a southern perspective. *Trends in Ecology and Evolution*, **7**, 190–4.

HILL, R. S. & BIGWOOD, A. J. (1987). Tertiary gymnosperms from Tasmania: Araucariaceae. *Alcheringa*, **11**, 325–35.

HILL, R. S. & CARPENTER, R. J. (1991). Evolution of *Acmopyle* and *Dacrycarpus* (Podocarpaceae) foliage as inferred from macrofossils in south-eastern Australia. *Australian Systematic Botany*, **4**, 449–79.

HILL, R. S. & CHRISTOPHEL, D. C. (1988). Tertiary leaves of the tribe Banksieae (Proteaceae) from south-eastern Australia. *Botanical Journal of the Linnean Society*, **97**, 205–27.

HILL, R. S. & GIBSON, N. (1986). Macrofossil evidence for the evolution of the alpine and subalpine vegetation of Tasmania. In *Flora and Fauna of Alpine Australasia: Ages and Origins*, ed. B. A. Barlow, pp. 205–17. Melbourne: CSIRO.

HILL, R. S. & MACPHAIL, M. K. (1983). Reconstruction of the Oligocene vegetation at Pioneer, north-east Tasmania. *Alcheringa*, **7**, 281–99.

HILL, R. S. & MACPHAIL, M. K. (1985). A fossil flora from rafted Plio-Pleistocene mudstones at Regatta Point, western Tasmania. *Australian Journal of Botany*, **33**, 497–517.

HILL, R. S. & MACPHAIL, M. K. (1994). Origin of the Flora. In *Vegetation of Tasmania*, ed. J. B. Reid, R. S. Hill & M. J. Brown. Hobart: Government Printer, in press.

HILL, R. S. & MERRIFIELD, H. (1993). An Early Tertiary macroflora from West Dale, south-western Australia. *Alcheringa*, **17**, 285–326.

HILL, R. S. & POLE, M. S. (1992). Leaf and shoot morphology of extant *Afrocarpus*, *Nageia* and *Retrophyllum* (Podocarpaceae) species, and species with similar leaf arrangement from Tertiary sediments in Australasia. *Australian Systematic Botany*, **5**, 337–58.

HILL, R. S. & READ, J. (1991). A revised infrageneric classification of *Nothofagus* (Fagaceae). *Botanical Journal of the Linnean Society*, **105**, 37–72.

HOLMES, W. B. K., HOLMES, F. M. & MARTIN, H. A. (1982). Fossil *Eucalyptus* remains from the Middle Miocene Chalk Mountain Formation, Warrumbungle Mountains, New South Wales. *Proceedings of the Linnean Society of New South Wales*, **106**, 299–310.

HOPPER, S. D. (1980). Bird and mammal pollen vectors in *Banksia* communities at Cheyne Beach, Western Australia. *Australian Journal of Botany*, **28**, 61–75.

HOPPER, S. D. & BURBIDGE, A. A. (1982). Feeding behaviour of birds and mammals on flowers of *Banksia grandis* and *Eucalyptus angulosa*. In *Pollination and Evolution*, ed. J. A. Armstrong, J. M. Powell & A. J. Richards, pp. 67–75. Sydney: Royal Botanic Gardens.

HYNDMAN, D. C. & MENZIES, J. I. (1990). Rain forests of the Ok Tedi headwaters, New Guinea: an ecological analysis. *Journal of Biogeography*, **17**, 241–73.

JABLONSKI, D., SEPKOSKI, J. J., BOTTJER, D. J. & SHEEHAN, P. M. (1983). Onshore–offshore patterns in the evolution of Phanerozoic shelf communities. *Science*, **222**, 1123–5.

JACKSON, W. D. (1968). Fire, air, water and earth – an elemental ecology of Tasmania. *Proceedings of the Ecological Society of Australia*, **3**, 9–16.

JOHNSON, L. A. S. (1972). Evolution and classification in *Eucalyptus*. *Proceedings of the Linnean Society of New South Wales*, **97**, 11–29.

JOHNSON, L. A. S. (1991). Casuarinaceae – some clarifications. *Australian Systematic Botany Society Newsletter*, **67**, 25–6.

JOHNSON, L. A. S. & BRIGGS, B. G. (1963). Evolution in the Proteaceae. *Australian Journal of Botany*, **11**, 21–61.

JOHNSON, L. A. S. & BRIGGS, B. G. (1975). On the Proteaceae – the evolution and classification of a southern family. *Botanical Journal of the Linnean Society*, **77**, 83–182.

JOHNSON, L. A. S. & BRIGGS, B. G. (1981). Three old southern families – Myrtaceae, Proteaceae and Restionaceae. In *Ecological Biogeography of Australia*, ed. A. Keast, pp. 429–69. The Hague: W. Junk.

JOHNSON, L. A. S. & BRIGGS, B. G. (1984). Myrtales and Myrtaceae – a phylogenetic analysis. *Annals of the Missouri Botanical Garden*, **71**, 700–56.

JOHNSON, L. A. S. & WILSON, K. L. (1989). Casuarinaceae: a synopsis. In *Evolution, Systematics, and Fossil History of the Hamamelidae*, Vol. 2 'Higher' Hamamelidae, ed. P. R. Crane & S. Blackmore, pp. 167–88. Oxford: Clarendon Press.

JORDAN, J. G., CARPENTER, R. J. & HILL, R. S. (1991). Late Pleistocene vegetation and climate near Melaleuca Inlet, south-west Tasmania, as inferred from fossil evidence. *Australian Journal of Botany*, **39**, 315–33.

KEMP, E. M. (1978). Tertiary climatic evolution and vegetation history in the southeast Indian Ocean region. *Palaeogeography, Palaeoclimatology, Palaeoecology*, **24**, 169–208.

KEMP, E. M. & HARRIS, W. K. (1977). The palynology of Early Tertiary sediments, Ninetyeast Ridge, Indian Ocean. *Palaeontological Association of London, Special Paper in Palaeontology*, **19**, 1–69.

KERSHAW, A. P. (1970). Pollen morphological variation within the Casuarinaceae. *Pollen et Spores*, **12**, 145–61.

KERSHAW, A. P. (1986). Climatic change and aboriginal burning in north-east Australia during the last two glacial/interglacial cycles. *Nature*, **322**, 47–9.

LANGE, R. T. (1978). Carpological evidence for fossil *Eucalyptus* and other Leptospermeae (Subfamily Leptospermoideae of Myrtaceae) from a Tertiary deposit in the South Australian arid zone. *Australian Journal of Botany*, **26**, 221–33.

LANGE, R. T. (1980). Evidence for lid-cells and host-specific microfungi in the search for Tertiary *Eucalyptus*. *Review of Palaeobotany and Palynology*, **29**, 29–33.

LANGE, R. T. (1982). Australian Tertiary vegetation. In *A history of Australasian Vegetation*, ed. J. M. B. Smith, pp. 44–89. Sydney: McGraw-Hill.

MACPHAIL, M. K., HILL, R. S., FORSYTH, S. M. & WELLS, P. M. (1991). A Late Oligocene–Early Miocene cool climate flora in Tasmania. *Alcheringa*, **15**, 87–106.

McCOY, F. (1876). Prodromus of the palaeontology of Victoria. *Geological Survey of Victoria, Special Publications, Decade IV*.

McNAMARA, K. J. & SCOTT, J. K. (1983). A new species of *Banksia* (Proteaceae) from the Eocene Merlinleigh Sandstone of the Kennedy Range, Western Australia. *Alcheringa*, **7**, 185–93.

MAIDEN, J. H. (1922). *A Critical Revision of the Genus Eucalyptus*, vol. IV. Sydney: Government Printer.

MARTIN, A. R. H. (1973). Reappraisal of some palynomorphs of supposed proteaceous affinity. I. The genus *Beaupreadites* Cookson ex Couper and the species *Proteacidites hakeoides* Couper. *Geological Society of Australia Special Publications*, **4**, 73–8.

MARTIN, A. R. H. & HARRIS, W. K. (1974). Reappraisal of some palynomorphs of supposed proteaceous affinity. *Grana*, **14**, 108–13.

MARTIN, H. A. (1978). Evolution of the Australian flora and vegetation through the Tertiary: evidence from pollen. *Alcheringa*, **2**, 181–202.

MARTIN, H. A. (1981). The Tertiary flora. In *Ecological Biogeography of Australia*, ed. A. Keast, pp. 393–406. The Hague: W. Junk.

MARTIN, H. A. (1982). Changing Cenozoic barriers and the Australian paleobotanical record. *Annals of the Missouri Botanical Garden*, **69**, 625–67.

MARTIN, H. A. (1989). Evolution of mallee and its environment. In *Mediterranean Landscapes in Australia*, ed. J. C. Noble & R. A. Bradstock, pp. 83–91. Melbourne: CSIRO.

MARTIN, H. A. & GADEK, P. A. (1988). Identification of *Eucalyptus spathulata* pollen and its presence in the fossil record. *Memoirs of the Association of Australasian Palaeontologists*, **5**, 311–27.

MENENDEZ, C. A. & CACCAVARI, M. A. (1966). Estructura epidermica de *Araucaria nathorsti* Dus. del Terciaro de Pico Quernado, Rio Negro. *Ameghiniana*, **4**, 195–9.

MILDENHALL, D. C. (1980). New Zealand late Cretaceous and Cenozoic plant biogeography: a contribution. *Palaeogeography, Palaeoclimatology, Palaeoecology*, **31**, 197–233.

MILLER, C. N. (1988). The origin of modern conifer families. In *Origin and Evolution of Gymnosperms*, ed. C. B. Beck, pp. 448–86. New York: Columbia University Press.

MULLER, J. (1979). Form and function in angiosperm pollen. *Annals of the Missouri Botanical Garden*, **66**, 593–632.

NIXON, K. C. (1989). Origins of Fagaceae. In *Evolution, Systematics, and Fossil History of the Hamamelidae*, vol. 2 'Higher' Hamamelidae, ed. P. R. Crane & S. Blackmore, pp. 23–43. Oxford: Clarendon Press.

PATTON, R. T. (1919). Notes on eucalypt leaves occurring in the Tertiary beds at Bulla. *Proceedings of the Royal Society of Victoria*, **31**, 362–3.

PATTON, R. T. (1936). A fossil *Casuarina*. *Proceedings of the Royal Society of Victoria*, **49**, 36–9.

PHILIPSON, W. R. & PHILIPSON, M. N. (1979). Leaf vernation in *Nothofagus*. *New Zealand Journal of Botany*, **17**, 417–21.

PHILIPSON, W. R. & PHILIPSON, M. N. (1988). A classification of the genus *Nothofagus* (Fagaceae). *Botanical Journal of the Linnean Society*, **98**, 27–36.

PLAYFORD, G. & DETTMANN, M. E. (1979). Pollen of *Dacrydium franklinii* Hook. f. and comparable early Tertiary microfossils. *Pollen et Spores*, **20**, 513–34.

POLE, M. (1989). Early Miocene floras from central Otago, New Zealand. *Journal of the Royal Society of New Zealand*, **19**, 121–5.

POLE, M. (1992). Cretaceous macrofloras of Eastern Otago, New Zealand: angiosperms. *Australian Journal of Botany*, **40**, 169–206.

POLE, M. S., HILL, R. S., GREEN, N. & MACPHAIL, M. K. (1993). The Miocene Berwick Quarry flora – rainforest in a drying environment. *Australian Systematic Botany*, in press.

PREEST, D. S. (1963). A note on the dispersal characteristics of the seed of the New Zealand podocarps and beeches and their biogeographical significance. In *Pacific Basin Biogeography*, ed. J. L. Gressitt, pp. 415–24. Hawaii: Bishop Museum Press.

PRYOR, L. D. & JOHNSON, L. A. S. (1981). *Eucalyptus*, the universal Australian. In *Ecological Biogeography of Australia*, ed. A. Keast, pp. 501–36. The Hague: W. Junk.

RAVEN, P. H. & AXELROD, D. I. (1972). Plate tectonics and Australasian paleobiogeography. *Science*, **176**, 1379–86.

READ, J. (1990). Some effects of acclimation temperature on net photosynthesis in some tropical and extratropical Australasian *Nothofagus* species. *Journal of Ecology*, **78**, 100–12.

READ, J. & HILL, R. S. (1989). The response of some Australian temperate rain forest tree species to freezing temperatures and its biogeographical significance. *Journal of Biogeography*, **16**, 21–7.

READ, J. & HOPE, G. S. (1989). Foliar frost resistance of some evergreen tropical and extratropical Australasian *Nothofagus* species. *Australian Journal of Botany*, **37**, 361–73.

READ, J., HOPE, G. S. & HILL, R. S. (1990). Integrating historical and ecophysiological studies in *Nothofagus* to examine the factors shaping the development of cool rainforest in southeastern Australia. In *Proceedings, Third IOP Conference*, ed. J. G. Douglas & D. C. Christophel, pp. 97–106. Melbourne: A-Z Printers.

RODWAY, L. (1914). Botanic evidence in favour of land connection between Fuegia and Tasmania during the present floristic epoch. *Papers and Proceedings of the Royal Society of Tasmania*, pp. 32–4.

ROMERO, E. J. (1986). Fossil evidence regarding the evolution of *Nothofagus* Blume. *Annals of the Missouri Botanical Garden*, **73**, 276–83.

ROMERO, E. J. & DIBBERN, M. (1985). A review of the species described as *Fagus* and *Nothofagus* by Dusen. *Palaeontographica Abt B*, **197**, 123–37.

SCHUSTER, R. M. (1976). Plate tectonics and its bearing on the geographical origin and dispersal of angiosperms. In *Origin and Early Evolution of Angiosperms*, ed. C. B. Beck, pp. 48–138. New York: Columbia University Press.

SCRIVEN, L. & CHRISTOPHEL, D. C. (1990). A numerical taxonomic study of extant and fossil *Gymnostoma*. In *Proceedings, Third IOP Conference*, ed. J. G. Douglas & D. C. Christophel, pp. 137–47. Melbourne: A-Z Printers.

SINGH, G., KERSHAW, A. P. & CLARK, R. (1980). Quaternary vegetation and fire history in Australia. In *Fire and the Australian Biota*, ed. A. M. Gill, R. H. Groves & I. R. Noble, pp. 23–54. Canberra: Australian Academy of Science.

STEENIS, C. G. G. J. VAN (1953). Results of the Archbold expeditions. Papuan *Nothofagus*. *Journal of the Arnold Arboretum*, **34**, 301–74.

STEENIS, C. G. G. J. VAN (1971). *Nothofagus*, key genus of plant geography, in time and space, living and fossil, ecology and phylogeny. *Blumea*, **19**, 65–98.

STEENIS, C. G. G. J. VAN (1972). *Nothofagus*, key genus to plant geography. In *Taxonomy, Phytogeography and Evolution*, ed. D. H. Valentine, pp. 275–88. London: Academic Press.

STOCKEY, R. A. (1982). The Araucariaceae: an evolutionary perspective. *Review of Palaeobotany and Palynology*, **37**, 133–54.

STOCKEY, R. A. (1990). Antarctic and Gondwana conifers. In *Antarctic Paleobiology*, ed. T. N. Taylor & E. L. Taylor, pp. 179–91. New York: Springer-Verlag.

STOVER, L. E. & PARTRIDGE, A. D. (1973). Tertiary and late Cretaceous spores and pollen from the Gippsland Basin, southeastern Australia. *Proceedings of the Royal Society of Victoria*, **85**, 237–86.

TANAI, T. (1986). Phytogeographic and phylogenetic history of the genus *Nothofagus* Bl. (Fagaceae) in the southern hemisphere. *Journal of the Faculty of Science, Hokkaido University, Series IV*, **21**, 505–82.

TOWNROW, J. A. (1965). Notes on Tasmanian pines. I. Some Lower Tertiary podocarps. *Papers and Proceedings of the Royal Society of Tasmania*, **99**, 87–107.

TOWNROW J. A. (1967a). On *Rissikia* and *Mataia* podocarpaceous conifers from the Lower Mesozoic of southern lands. *Papers and Proceedings of the Royal Society of Tasmania*, **101**, 103–36.

TOWNROW, J. A. (1967b). The *Brachyphyllum crassum*

complex of fossil conifers. *Papers and Proceedings of the Royal Society of Tasmania*, **101**, 149–72.

TRUSWELL, E. M., KERSHAW, A. P. & SLUITER, I. R. (1987). The Australian–south-east Asian connection: evidence from the palaeobotanical record. In *Biogeographical Evolution of the Malay Archipelago*, ed. T. C. Whitmore, pp. 32–49. Oxford: Oxford University Press.

TURNER, V. (1982). Marsupials as pollinators in Australia. In *Pollination and Evolution*, ed. J. A. Armstrong, J. M. Powell & A. J. Richards, pp. 55–66. Sydney: Royal Botanic Gardens.

VEBLEN, T. T. & ASHTON, D. H. (1978). Catastrophic influences in the vegetation of the Valdivian Andes, Chile. *Vegetatio*, **36**, 147–67.

VEBLEN, T. T., DONOSO, Z. C., SCHLEGEL, F. M. & ESCOBAR, R. B. (1981). Forest dynamics in south-central Chile. *Journal of Biogeography*, **8**, 211–47.

WARDLE, P. (1968). Evidence for an indigenous pre-Quaternary element in the mountain flora of New Zealand. *New Zealand Journal of Botany*, **6**, 120–5.

WELLS, P. M. & HILL, R. S. (1989). Fossil imbricate-leaved Podocarpaceae from Tertiary sediments in Tasmania. *Australian Systematic Botany*, **2**, 387–423.

WHITE, M. E. (1981). Revision of the Talbragar Fish Bed Flora (Jurassic) of New South Wales. *Records of the Australian Museum*, **33**, 695–721.

WHITMORE, T. C. (1980). A monograph of *Agathis*. *Plant Systematics and Evolution*, **135**, 41–69.

Taxonomic index

General index